D1083717

ADAPTIVE BEHAVIOR AND LEARNING

ADAPTIVE BEHAVIOR AND LEARNING

J. E. R. STADDON
Duke University

CAMBRIDGE UNIVERSITY PRESS
Cambridge
London   New York   New Rochelle
Melbourne   Sydney

Published by the Press Syndicate of the University of Cambridge
The Pitt Building, Trumpington Street, Cambridge CB2 1RP
32 East 57th Street, New York, NY 10022, USA
296 Beaconsfield Parade, Middle Park, Melbourne 3206, Australia

© Cambridge University Press 1983

First published 1983

Printed in the United States of America

Library of Congress Cataloging in Publication Data
Staddon, J. E. R.
Adaptive behavior and learning.
Includes bibliographical references and index.
1. Learning in animals. 2. Adaptation (Biology)
I. Title. [DNLM: 1. Behavior, Animal. 2. Learning.
QL 785 S776a]
QL785.S8   1983   591.51   83-5206
ISBN 0 521 25699 2   hardcovers
ISBN 0 521 27658 6   paperback

To
I. F. R.
D. N. S.
and
DB

CONTENTS

PREFACE

Those facts are to be regarded as important which lead to the establishment of general laws ... Facts are important, in the inductive sciences, solely in relation to theories; and new theories give importance to new facts.

Bertrand Russell

There are at least four ways to write a textbook. One can emphasize the history of a field, its techniques, the facts it has turned up, or the theoretical principles that tie them together. I have tried to emphasize principles in this book. Where no principles are available, I have described the awkward facts and suggested the kinds of theory that might explain them. I have described experimental methods only as much as necessary to appreciate important experimental results. Those methods are explained in some detail, because the results of experiments often depend heavily on details of procedure. Much of the theory in the book is careful analysis of procedures, rather than grand speculations about animal nature.

Many facts are intrinsically interesting: The luxuriance and diversity of the Brazilian rainforest astonished European explorers such as Von Humboldt and Darwin when they first saw it. Exotic animals and plants, unusual habits of life, and the very large and the very small command interest apart from their meaning in some larger scheme of things. Some of the facts of animal psychology are interesting in themselves: The behavior of apes, monkeys, and wolves – animals like us or like our pets; the singing of birds and much of their courtship; the exotic senses of bats and sidewinders; and the behavior of bees all engage general interest. Natural history, the cataloging of facts like these, has an honorable and important past and was a necessary precursor to more abstract scientific endeavors.

The facts of learning are not like this, unfortunately. Because learning is a relation between past and present it cannot be directly observed. Nor can much useful information be gotten just by watching animals as they learn. Controlled experiments are necessary, with all that this implies by way of procedural detail and abstract argument. Experiments on learning rarely excite spontaneous interest; their justification is the hypotheses they test and the theories to which they give rise. For these reasons I felt to be under no obligation to describe experimental results for their own sake. I describe experimental results only when they illuminate some general principle, or provide information necessary for later work.

I have also been sparing with historical material, for the following reasons. The history of animal learning is a tangled skein, with strands from both zoology and psychology. The psychological side is marked by divergent and well-defined schools identified with particular experimental methods and even species, united only in their neglect of biology. Many of the terms and methods developed by these schools are significant largely in relation to outmoded theories. Even the term *reinforcement*, still widely used and now passed into common speech, reflects a view of learning with which many would quarrel. The zoological study of learning has been less burdened with controversy, but is also much weaker in its experimental analysis. Konrad Lorenz thought that he could see directly in the acts of his jackdaws the learned and the instinctive elements. He felt no need to do the tedious and difficult experiments necessary to prove an apparently obvious ontogeny.

Sometimes the best way to understand a field is to study how it got where it is: Many concepts are most easily communicated by studying their history. I don't believe that this is generally true of learning and adaptive behavior. In this field, at least, detailed discussion of historical origins is as likely to confuse as instruct.

This said, there are nevertheless cases where some historical explanation is helpful. It would be unthinkable to write about animal learning without mentioning Darwin and Sherrington, Pavlov and Jennings, Hull, Tolman, and Skinner. Many questions derive some of their interest from the setting in which they arose. I discuss nothing whose interest is solely historical; but where a topic has an important historical aspect it is discussed, either in the text or the extensive Notes that follow each chapter. I have included most historical material in headed sections in the Notes.

The main difference between this book and others on animal learning is its emphasis on functional as well as mechanistic explanations for behavior. By *functional* I mean explanations in terms of outcomes, either evolutionary outcomes (Darwinian fitness) or outcomes in the life of the individual (goals or motives, reinforcers or "preference structures" – take your pick). Functional explanations, in this sense, have been rather slighted by psychologists, for three reasons: First, mechanistic (causal) explanations seem much more scientifically respectable, perhaps because they promise a link to underlying neurophysiology. Second, psychologists have always had too much respect for philosophy, and the positivists scared them by pointing out the evils of teleology. Functional explanation looked like teleology to a lot of people and, respectability being everything in science, they wanted none of it. Third, functional explanation brings in evolutionary biology, hence many psychologists decided that this wasn't their business. I argue in the text that reinforcement, the central behaviorist notion, is in fact a functional explanation (and often a good one), but this was disguised by assuming mechanisms for reinforcement that were not adequately defined or tested.

Francis Bacon wrote, "The truth shall sooner follow error than confusion." I have tried hard to follow this prescription for clarity over comprehensiveness. The history of this field is now long enough and muddy enough that one must often make a choice between covering everything, and confusing the reader, and presenting only a biased sample, and being clear. The book is not a survey of all that has been said or done in animal learning; I have not attempted to update

Mackintosh's monumental work, or the earlier texts of Kimble and Hilgard and Marquis. This book does not cover everything or everybody; although I believe that every major topic receives some attention. I hope instead to have provided a relatively simple theoretical framework into which the major facts of learning can fit. I entrust future research with eliminating what is false and showing what needs amendment.

The book is written at two levels, represented by the main text and the Notes at the end of each chapter. The body of the text contains the major arguments and should be accessible to advanced undergraduates with some training in natural science. I assume no prior knowledge of psychology, but an acquaintance with graphs and the algebra of simple functions – the kind of sophistication acquired after a course or two of college math, chemistry, physics, or engineering – will make things much easier. There is a little elementary differential calculus in the text, but it can be skipped over and the arguments should be clear without it.

The main text contains few names and references – these are supplied in the Notes after each chapter and References at the end of the book. Bacon also wrote, "Few follow the things themselves, more the names of the things, and most the names of their masters." He knew the weaknesses of scholarship: How much easier it is to learn a few names and definitions than to grasp the ideas behind the names! The apparatus of scholarly attribution is important to the progress of science and to those going on in research. It is merely distracting to most students, whose purpose should be to understand the principles and major facts of a field. Students often ask which names they should remember – inviting the equally rash replies of "all" or "none." They can safely be required to learn the modest number of names in the main text, and I hope that the paucity of scholarly detail will make the facts and conceptual matter stand out more clearly.

The extensive Notes sections serve three purposes: They supply the detailed references lacking in the text; they supplement the meager historical discussion in the text; and they take up topics at a more advanced or detailed level – giving proofs, qualifications, or references for what in the text is merely asserted. The Notes, many of which bear subject headings, also permit the instructor to assign additional material in areas of specialized interest. Without the Notes, the book can serve as a text for an upperclass first course in animal learning and behavior; with the Notes, and some reading of original sources, it can serve as a graduate text.

The book is probably too long for a one-semester undergraduate course. A one-semester course emphasizing operant behavior and behavioral ecology is adequately covered by assigning Chapters 1–10 and selected sections in Chapter 11 (skipping the more mathematical bits). A course emphasizing learning and classical conditioning can be based on Chapters 1, 2, 4, 5, 8, 10, 12, 13, and 14.

Topics standard in most learning courses – historical introduction, definitions of operant and Pavlovian conditioning, basic procedures, stimulus control, and memory acquisition and extinction – are covered in Chapters 1, 4, 5, 10, 12, 13, and 14.

Chapters 7, 9, and 11 contain some quantitative arguments and may present difficulties for students with no natural-science background. These chapters require more support from the lecturer than the others.

The book contains few allusions to neuroscience, although much exciting work with behavioral implications is now going on in that area. The omission is intentional. We still need to know much more at a "black-box" level about what animals are doing in learning experiments. What is now known still makes little real contact with physiological work, despite several decades of earnest attempts. More important, the volume of work in neuroscience is now so great, and the information required for its comprehension so specialized, that I could not do justice to it within a reasonable space. It seemed better to omit references to neuroscience where it sheds no special light on the organization of behavior, rather than to add token paragraphs that distract rather than inform.

Finally, the book is about adaptive behavior and learning in *animals* not, or not especially, about learning in people. Animals have always struck me as especially worth studying because they are complicated enough to be interesting, but simple enough to offer hope of finding out something in a reasonable time. People are interesting all right, but the prospect of finding out something of enduring scientific value by studying only people seems to me remote. "The proper study of mankind is man," is a popular quote, but it was written by a poet, not a scientist. The history of science offers opposing testimony: Human relevance has usually been a poor guide to fruitful research. Pope was right in this sense, that if you want to find out something about people that will be of immediate, practical use, then looking at people is likely to pay off sooner than looking elsewhere. But if you want to build a scientific understanding of the evolution and meaning of intelligence then you must study animals – for themselves, not for what they can tell you about depression or your personal problems. The eventual payoff will indeed be an understanding of people as well as beasts, but this may not be arrived at soon – and it may be impeded by too urgent a concern with human problems and their solution.

No man is an island and no book springs fully formed from its author's brain. Many have contributed, directly or indirectly, to this book, and I thank here all I can remember. I apologize alike to those I have failed to recall, and those whose advice I have failed to take.

I thank first my teachers at Harvard and M.I.T.: Richard Herrnstein, Peter van Sommers, Fred Skinner, Larry Stark, Donald Griffin, and others in Memorial Hall and the Electronic Systems Lab.

Many colleagues have been kind enough to read particular chapters: I thank Irving Diamond, Carol Eckerman, Richard Ettinger, James Gould, Norman Guttman, John Hinson, John Horner, Alasdair Houston, Nancy Innis and her students, Gregory Kimble, Gregory Lockhead, Tony Nevin, Alliston Reid, Sam Revusky, David Rubin, Philip Teitelbaum, John Vaughn, Meredith West, and Cliff Wing. John Hinson has also sustained the computers without which I am helpless. I am grateful to several anonymous reviewers for their comments, well- and ill-judged. I thank Eric Wanner for early encouragement, and Edna Bissette for unfailing and good-humored aid over many years. Susan Milmoe and Rhona Johnson at Cambridge University Press have been a pleasure to work with during the production of the book. Finally I thank several cohorts of Psychology/ Zoology 101 students, denied a "real" textbook, who have struggled uncomplainingly through xeroxed versions of this work.

I am grateful to Duke University for a gracious environment conducive to scholarship, to the John Simon Guggenheim Memorial Foundation for a fellowship that allowed me to finish the book, and to the National Science Foundation for research support over many years.

JOHN STADDON
Durham, North Carolina
June 1983

1

THE EVOLUTION, DEVELOPMENT, AND MODIFICATION OF BEHAVIOR

NICHES: SIMILARITIES AND DIFFERENCES

Organisms are machines designed by their evolution to play a certain role. This role, together with the environment within which it is played, is called the organism's *niche*.[1] For example, most cats – tigers, leopards, mountain lions – play the role of solitary hunters; wolves and wild dogs are social hunters; antelope are social grazers; and so on. The niche defines the patterns of adaptive behavior essential to an animal's survival and reproduction.

For simple niches, such as those filled by most nonsocial invertebrates, direct responses to particular kinds of stimulation are all that is required. The animal need keep no record of its *past history* in order to succeed; it is sufficient that it avoid bad things and approach good ones. A modest memory for the immediate past allows the creature to respond to changes in stimulation. Direct stimulus-response mechanisms, plus some sensitivity to rates of change, are sufficient for a wide range of surprisingly intelligent behavior. Adaptive mechanisms that require little or no dependence on history are discussed in Chapters 2 and 3.

As the niche grows more complex, adaptive behavior depends more and more on the animal's past. The greater flexibility that this allows carries with it two kinds of cost: First, the animal must *have* a past if its behavior is to be guided by it. This implies a lengthening of infancy and adolescence, which necessarily delays reproductive maturity, and puts the individual at a reproductive-fitness disadvantage compared to others quicker on the draw – it is sometimes better to be dumb and fast than intelligent and slow. Second, there is a growing bookkeeping cost. The behaviors acquired through past experience, and some representation of the environments in which they are appropriate, must be "stored," with minimal duplication, in such a way that the animal has ready access to the most appropriate action. Representing data in the most flexible and economical way is a problem that also confronts human filing systems. Much work in computer science is concerned with "data-base management," as this is termed. The difficulties encountered in designing efficient and flexible data-base-management systems show that early learning theories greatly underestimated the information processing task implied by the behavior of mammals and birds.

Situations rarely recur in precisely the same form; and only some of the differences between situations are important for action. Hence, the animal's representation of past environments must also allow it to behave appropriately in environments *similar* to those it has already encountered. Just what *similar* means, and how it is determined both by the animal's evolutionary history and its own experience, is one of the most intriguing unsolved questions in animal behavior. These issues are taken up in Chapters 10, 13, and 14.

When niches grow more complex, the need for simple mechanisms does not diminish – even human beings need reflexes, for example – but, in addition, more complex, history-dependent processes are required.

An animal's past experience can affect its future in a variety of ways. The simplest way to make sense of these is the supposed dichotomy between *learned* and *innate* behavior. Innate behavior is completely independent of experience, and learned behavior is, well, learned. Of course, nothing is truly innate, in the sense of being independent of any experience, but many things are almost independent of any *particular kind* of experience. For example, many small invertebrates avoid light; they need no special training, no nasty shock in a lighted place, to show this pattern. Most mammalian reflexes are of this sort: As soon as an infant can move at all, it will automatically withdraw its hand from the fire. The knee jerk to a tap, pupillary contraction to a bright light, and many other reflexes are all concomitants of development in a variety of environments, common to all normal members of the human species. I discuss reflexes in Chapter 2.

But there are many effects of experience that do not fit into the innate–learned dichotomy. For example, age slows responses and hardens joints, fatigue reduces muscular strength, hunger (food deprivation) and thirst change preferences in systematic, reversible ways, and so on; a number of other, developmental effects will be discussed shortly. None of these corresponds to the usual meaning of the term *learning*, which refers to a more *specific* and only partly reversible change, often related to a positive or negative outcome: The animal learns where food is to be found or to avoid the predator. This book is primarily concerned with learning in this sense, but the category is not exact – simply because we do not really know what learning is. Indeed, there is probably no single process that underlies it. Experience can change behavior in many ways that manifestly do not involve learning, as well as in ways where we are not sure. In other words, there is no hard-and-fast line separating learning from other kinds of behavioral change. There is no neat dichotomy between "learned" versus "innate" behavior; rather, there is a spectrum of ways in which past experience affects future behavior, and learning is perhaps the most interesting, and certainly the least understood, of these.

The innate–learned dichotomy, nevertheless, refers to a useful distinction better expressed by the term *canalization*. A structure or behavior is said to be canalized if its development is almost independent of a particular experience or environment. Some things develop in almost any environment: Characteristics such as the four-chambered heart of mammals, or bilateral symmetry, are strongly canalized, in the sense that just about any environment that allows the organism to develop at all will also be sufficient to permit their development. A trait such as competence in the English language, or the ability to do algebra, is not canalized at all, because it is critically dependent on a particular environment. Competence in *some* language is an intermediate case: Evidently just about any linguistic environment is sufficient to produce language learning in a normal infant, even in the absence of explicit instruction. In a similar way, male chaffinches and white-crowned sparrows will develop some adult song if they can listen to a model at the critical time in their first year of life, but the kind of song they develop depends on the model, as well as the species. Language and song development are canalized, but the particular song or language to be learned is not.

What an animal learns, and the way that it learns it, is much affected by its niche. Because niches differ in many respects, so, too, do learning mechanisms. Since niches do not differ in every respect, there are also similarities among learning mechanisms.

Space and time are common to all niches. In consequence, a wide range of animal species adapt to the temporal and spatial properties of the environment in similar ways. There are some general rules that apply across niches: Old information is generally less useful than new information; consequently animals forget, and they forget less about things they have learned recently. Conversely, the environment of an animal around the time of birth usually has a special significance, and things learned at that time may be especially resistant to change. Food, water, sex, and habitat are vitally important to all species. Hence these things are better remembered than "neutral" events and have special properties as guides of behavior.

This book is mainly concerned with the way that animals adapt to these things that are common to all niches.[2] The major emphasis is on adaptation that depends on learning about rewards and punishments.

PHILOSOPHICAL BACKGROUND

Methodological behaviorism

Animals and people seem to have purposes, beliefs, attitudes, and desires; they seem to know some things and not others, to want some things and disdain others, and so on. These are what philosophers call *intentional systems*. Intentionality may seem to set psychology apart from the physical and biological sciences. After all, the chemist does not worry about the beliefs of his compounds nor is the physicist concerned about the purposes of protons or the quirks of quarks. Does this mean that psychology is not scientific? Does it mean that it is different in kind from the physical sciences? Not at all; the difference is in the richness of behavior of the things studied, their sensitivity to their environment, and the dependence of present behavior on past experience. The language of intentionality is simply the everyday way that we deal with complex historical systems. I am typing this book with the aid of a microcomputer that has an operating system called "CP/M." Look how the instruction manual refers to CP/M and its associated programs: "CP/M could not find a disk...," "PIP assumes that a...character...," "CP/M does not know that...," "Seven commands are recognized by CP/M...." No one assumes that there is a little man or woman, complete with "real" knowledge, beliefs, desires, and understanding, inhabiting the microchips. Anything that responds to varied stimuli in varied ways, especially if its behavior depends upon past history, is understood at a commonsense level in intentional terms.

The most striking examples are provided by chess-playing programs. A good one elicits precisely the same kinds of comment we would use for a human player: "It is attacking the queen," "It wants to gain control of the center of the board," and so on. Yet no one doubts that the underlying program provides a perfectly precise and mechanical account of the machine's behavior.[3]

But do machines *really* have beliefs, attitudes, and so on? Aren't we just begging the question by talking about smart machines? There are two questions here: First, are there such "things" as beliefs, desires, and so forth? And second, if so, do machines possess them? There are two schools of thought on these questions: the first answers "yes" to the first question and "no," or at least "probably not," to the second; the second views the questions as irrelevant. Human beings have real desires, attitudes, and so on, the first view holds, and it is the business of real psychologists to study them. Sometimes attitudes and beliefs are deemed to be worthy of study in their own right. More commonly, perhaps, they are studied as *causes of action;* people do what they do because they believe what they believe. This approach leaves little room for work with nonhuman animals, tends to keep things at a verbal level, and attends first to the "meaning" of people's actions, verbal and otherwise, rather than focusing on the details of the actions themselves. This is the psychology of paper-and-pencil test, of interview, and of verbal report.

The practical utility of this view cannot be denied. The power of advertising rests, in some measure, on the correct assessment of people's attitudes to, for example, bodily functions and products that promise to diminish, enhance, or in some other way modify them. Nevertheless, it has both experimental and theoretical limitations. The experimental problem derives from the difficulty of separating *correlation* from *causation.* This is an old question: Do we run because we are afraid, or are we afraid because we run? "Fear" is a property of the subject's internal state, not something external that the experimenter can manipulate directly. Consequently, one can never be certain that the running and the fear are not both caused simultaneously by the same external conditions. The problem is not insuperable. There are ways that intentional terms like "fear," "hope," and so on can be made methodologically respectable and tied to observables. The theoretical question is whether the labor involved is worth it. The whole enterprise rests on the presupposition that familiar intentional terms such as "fear," "belief," "attitude," and the like form the very best basis for theoretical psychology.

Yet our discussion of clever computer programs showed that such terms represent a primitive kind of explanation at best. They enable someone ignorant of the details of the program to make some sort of sense of what the machine is doing. But full understanding rarely reveals anything in the program that corresponds directly to intentional terms, useful though they may be in the absence of anything better. It is rash, therefore, to base a program of psychological research on the assumption that intentional terms represent the ultimate form of explanation.

Thus, the second answer to the question "Are there *really* such things as beliefs, desires and so on?" is "Maybe...but who cares?" Obviously these terms are useful ways of coping with some complex systems. But the "really" question is metaphysical, and we have no reason to suppose that these terms will prove especially useful in unraveling the mechanisms of behavior – which are what we are really interested in. Hence, we will be concerned with the *behaviors* of people and animals, measured pretty much in physical terms – that is, with a minimum of interpretation. This is termed *methodological behaviorism*, and is the dominant stance among psychologists and biologists interested in animal behavior.[4]

Methodological behaviorism is not untheoretical; it simply takes no advance position on the nature of appropriate theory. In particular, it does not presume that psychological theory should be based on intentional language.

Two kinds of explanation

The ultimate explanation of the chess-playing program is in terms of the program itself, the individual instructions that determine each move as a function of prior moves by both players. But it is usually convenient, when designing such a program as well as when trying to understand it, to divide it into two parts: a part that generates potential moves, and a part that evaluates each move in terms of a set of criteria. The dichotomy between *variation* and *selection* was discovered by Darwin and Wallace as their theory of evolution by natural selection, but the distinction is more general: All adaptive, purposive behavior can be analyzed in this way. The dichotomy leads to two kinds of explanation for adaptive behavior: *causal* or *mechanistic* explanations, which define both the rules by which behaviors are generated (rules of variation) and the rules by which adaptive variants are selected (selection rules); and *functional* explanations, which just specify (perhaps in simplified form) the selection rules. Mechanistic accounts deal only in *antecedent* causes; functional accounts in terms of *final* outcomes. Thus the form of the shark is explained functionally by its hydrodynamic efficiency, the taking by a chess program of its opponent's queen in terms of the improved position that results.

As we will see, the selection rules for learning cannot be stated as explicitly as the rule of natural selection. Indeed, even that rule is now much less clear than it was in days before we were aware of the problem of the unit of selection (individual organisms succeed or fail to reproduce, but it is individual genes that are passed on – what, then, is selected?). Consequently, functional explanations for adaptive behavior are often stated in terms of goals, purposes, or reinforcers (rewards and punishments), which act as guides of behavior. These notions can be formalized in terms of some kind of *optimality theory* that makes goals explicit and shows how conflicting goals are to be reconciled. The general idea is that animals act so as to maximize something that relates to inclusive fitness, such as net rate of food acquisition, number of offspring, or territory size. I return to the relation between optimality accounts and selection rules in a moment.

Functional explanations can, in principle, be reduced to mechanistic ones: Given perfect understanding of the principles of genetics and development, and complete information about evolutionary history, we can, in principle, reconstruct the process by which the shark achieved its efficient form. For this reason the biologist Pittendrigh (1958) suggested the label *teleonomic* (as opposed to *teleological*) for such accounts. Teleological explanations are not acceptable because they imply final causation – the shark's streamlining is teleologically explained by Mother Nature's hydrodynamic foresight. Teleonomic accounts relate form and hydrodynamics through the mechanisms of variation and natural selection. Teleonomic functional accounts are philosophically respectable; teleological ones are not. In practice, of course, the necessary detailed information about mechanisms is often lacking so that we must settle for functional accounts and hope that they are teleonomic ones.

Functional explanations have sometimes been criticized as being "Just-so" stories, because they are so flexible – adaptive significance, or an unsuspected reward or punishment, can be conjured up to explain almost anything. There are two answers to this criticism: Functional explanations often lead to mechanistic explanations; and functional explanations can explain relationships that cannot be explained in any other way.

Functional accounts are often way stations to mechanistic explanations. In studies of learning they help identify important variables and draw attention to the *constraints* that limit animals' ability to attain functional goals. These constraints, in turn, provide clues to underlying mechanisms. For example, mammals and birds can easily learn to use stimuli as guides to the availability of food; a hungry pigeon has no difficulty learning that a peck on a red disk yields food whereas a peck on a blue disk does not. But they are much less capable of using *past* stimuli as guides. In the *delayed-match-to-sample* task, one of two stimuli is briefly presented, then after some delay both are presented, and a response to the one that matches the first is rewarded. Delays of more than a few seconds between sample and choice presentations impair gravely most animals' ability to choose correctly. This is a memory constraint. Other psychological constraints have to do with animals' ability to process information, and with their perceptual and motor abilities. Identification of limitations of this sort is the first step toward understanding behavioral mechanisms.

In addition to internal (psychological) constraints, there are also constraints imposed by the environment. For example, the animal cannot do more than one thing at a time, so that total amount of activity is limited; spatial arrangements limit the order in which food sites can be visited and the time between visits. *Reinforcement schedules*, either natural (as in picking up grain, one peck per grain, or in natural replenishment processes) or artificial (ratio and interval schedules, for example), further constrain the distribution of activities. Functional explanations, precisely expressed in the form of optimality theory, allow, indeed force, one to take account of these external constraints.

Functional explanations do one thing that no mechanistic explanation can: They can explain similar outcomes produced by different means. For example, the eyes of vertebrates and octopi are very similar in many ways: Both have lenses, a retina, and some means of limiting the amount of light that can enter. This convergence cannot be explained by a common ancestry or any similarity of developmental mechanisms. The *only* explanation we can offer for this astonishing similarity is the common function of these organs as optical image-formers. Because convergence is such a common phenomenon in evolutionary biology, it is no wonder that functional explanations are so common and so powerful there.

It is traditional in psychology to look down somewhat on functional accounts (although they often come in by the back door, in the form of vaguely expressed reinforcement theories). Indeed, one of our most influential figures boasts in his memoirs that in planning his major work he deliberately avoided any discussion of adaptiveness. Few maintain that position today. Looking at behavior in terms both of its adaptive (evolutionary) function and in relation to current goals (reinforcers) is useful in identifying important variables and in distinguishing environmental from psychological constraints. Functional and mechanistic theories are on an equal footing in this book.

The idea that organisms attain goals, either through natural selection for the best form of wing or individual reinforcement of the most effective foraging strategy, derives naturally from the selection/variation idea: A wide range of variants occurs, the best (in terms of flight efficiency or eating frequency) are preferentially selected, the next round of variants contains a few that do even better, and so on. This process will, indeed, lead to better adaptation only if two things are true: We have the selection rule right – that better fliers really have more offspring; and that the *right variants occur*. In other words, an animal may fail to behave in what seems to us the optimal fashion either if we have misread what it is trying to achieve (the selection rule), or because it never generates the necessary behavioral variant: The most efficient foraging strategy cannot be selected (reinforced) if it never occurs. Memory constrains behavior in ways that prevent animals from developing certain kinds of foraging patterns – patterns that require memorization of complicated sequences, for example. These patterns will not occur, even in situations where they would be optimal. Thus, failures to optimize are, if anything, even more informative than successes, because they offer clues to the underlying behavioral mechanisms. Optimality theories to explain how animals adapt to reward and punishment are discussed in Chapters 6–10.

EVOLUTION AND DEVELOPMENT

The processes of individual development, *ontogeny*, are the product of past evolution and they also limit future evolutionary possibilities. Unlike human machines, natural machines – animals and plants – manufacture themselves. This process limits their potential and often incorporates the effects of experience in ways that contrast with, and thus help define, learning.

Organisms change throughout their lifetimes, and the processes by which they change are the outcome of past evolution. As Darwin pointed out, organisms bear their evolutionary history both in their structure and in the manner of its development. Rudimentary organs provide some of the most striking examples. The human vermiform appendix, the rudimentary breasts of male mammals, the vestigial lung of snakes (that have only one functional lung), the teeth of fetal whales that vanish in the adult, the uncut teeth of unborn calves – none has any function in the adult, yet they remain: "[They] may be compared with the letters in a word, still retained in the spelling, but become useless in the pronunciation, but which serve as a clue for its derivation."[5] There are behavioral parallels in the inappropriate "grass-flattening" of domestic dogs, and exaggerated fears (of the dark, or of strangers, for example) in human children. In many cases these vestigial behaviors disappear with age, as in some of Darwin's examples.

These examples illustrate the half-truth that "ontogeny recapitulates phylogeny," that is, the idea that the stages through which an organism passes, from embryo to adult, represent a history of the race in an abbreviated form. Gill-slits in the human fetus were once taken to mean that the fetus at that stage resembles the ancient fish from which mammals are descended. The actual relations between ontogeny and phylogeny are more complicated and derive from the fact that evolution acts via the mechanisms of development.

Development can be compared to a railroad switchyard in which incoming cars on a single track are sorted by a branching arrangement of switchoffs so that each car arrives at a different destination. At conception the organism is essentially undifferentiated and "pluripotent," that is, many things are possible. With progressive cell divisions, there is increasing differentiation and the options for further development are reduced – the railroad car has passed through several switchoffs and is closer to its final destination. This process of progressively finer differentiation, and the concomitant reduction in future options, takes place throughout life. Eventually the car enters the final stretch of track that terminates in death – which is not a wearing out, but the largely predetermined end of a course charted by prior evolution. Typical life span, like other characteristics, is determined by its costs and benefits, weighed in the delicate balance of natural selection.

Genes determine the direction of the successive switches that occur throughout ontogeny. We don't yet know exactly how this works. A recent theoretical account begins: "Despite our relatively detailed understanding of molecular biology, the processes which control the development of a multicellular organism from a single cell, the fertilized egg, are almost completely unknown" (Caplan & Ordahl, p. 120). Nevertheless, one thing is clear: The genetic changes that provide the raw material for evolution act not directly on morphology or behavior, but on the course of development – a stage may be added or missed entirely, stages may be accelerated or retarded. These changes in the path of development are the raw material for the formation of new species. For example, if the genital system matures relatively faster than the rest of the body, the result may be a sexually mature "larval" animal, as in the case of the Mexican axolotl (*Ambystoma tigrinum*), a salamander that can become sexually mature while still a tadpole. Continued selection might well fix a change of this sort, so that the terrestrial stage is completely abolished and a new species of entirely aquatic amphibian is the result.

It is easy to see that this process will leave traces of a species' past evolutionary history in the path of development of an individual organism. For example, the immature form of a fish such as the angelfish (*Pterophyllum scalare*), which is strongly laterally compressed in the adult, or of flatfish such as the flounder (*Bothus lunatus*), which is vertically compressed and has lost bilateral symmetry by having both eyes on the same side of the head – is quite normal looking, with the "typical" elongated, bilaterally symmetrical fish shape. Presumably the abnormal body form arose via genetic changes that acted to modify growth gradients in the ancestral species at a relatively late stage of development. Thus the immature forms of these "abnormal" species provide a partial record of the immature forms of the ancestral species from which they derive.

This view of evolutionary action implies that related species with different bodily forms should often be transformable one into the other by stretching or compressing along the three bodily axes. This was first pointed out by the British biologist D'Arcy Wentworth Thompson; Figure 1.1 shows a couple of fish examples.

The action of a gene depends on its environment, which includes the rest of the genotype, the cell of which it is a part, the constitution of neighboring cells, circulating metabolites such as hormones, and the neurotransmitters released by nerve impulses. For example, during the development of the fruit fly *Drosophila* a specific section of cytoplasm (the polar cytoplasm) influences the nuclei that migrate through it to differentiate into the reproductive cells. If the polar cytoplasm is removed from the egg so that migrating nuclei do not encounter it, the reproductive cells do not develop and a sterile animal is the result. Polar cytoplasm affects the expression during development of genes responsible for the reproductive system.

Because the organism's internal environment is affected intimately by its external environment, the course of development is a joint product of genotype

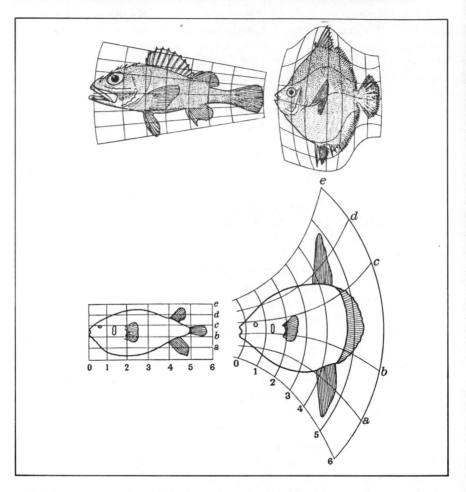

Figure 1.1. Two pairs of fish whose shapes can be related by coordinate transformation. Top left: *Scorpaena* sp.; top right: *Antigonia capros*; bottom left: *Diodon* sp.; bottom right: *Orthagoriscus mola*. (Taken from D'Arcy Thompson, 1961, pp. 300–301, original edition, 1917.)

and the environment in which the organism grows up. In effect, therefore, the action of genes and the action of the environment are symmetrical: Each depends on the other. The successive switches that constitute development are joint effects of environment and genotype. The main difference between environmental and genetic effects is that, since the genotype is fixed but the environment can vary, environmental effects on behavior may be *reversible*. Because the sensorimotor systems that bring the animal into contact with its environment develop with age, the effects of environment on development are likely to become richer and more subtle as the organism grows older. The effects of the environment on an embryo or a newborn may be great, but they are unlikely to involve the

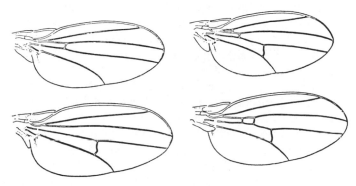

Figure 1.2. Four typical *Drosophila* wing types induced by heat stress. (Venation phenocopies; from Waddington, 1960, p. 394.)

transmission of as much information as interactions later in life. On the other hand, environmental effects are likely to be self-limiting owing to the accumulation of irreversible changes, so that mid-life may often be the time of maximal sensitivity to external influences.

Gross morphological changes are not readily reversible. This is why traces of an organism's evolution are retained in its development. But behavior, and presumably the changes in brain state that correspond to behavior, are, almost by definition, easily altered. Consequently it is not at all clear what we should expect of the relation between behavioral and morphological development. Is it reasonable to assume, for example, that behaviors that appear early in ontogeny only to disappear later tell us the same sort of thing as human fetal gill-slits? Were these the behaviors of our immature ape ancestors? In the case of some primitive reflexes this may be a reasonable guess. For things like fear of strangers or of the dark, we cannot be sure.[6] This question will not be settled until we understand how brain structure and physiology relate to behavior. How is past experience represented neurophysiologically? How does the current environment interact with this representation, and the animal's motivational state, to produce action? Unfortunately, we are a very long way from answering these questions.

Epigenesis and genetic assimilation

The subtleties of gene–environment interaction are illustrated nicely by some ingenious experiments of the geneticist Waddington. Using fruit flies (*Drosophila melanogaster*), Waddington showed that environmental changes can act as a sort of probe to uncover latent characteristics of the genotype. In one experiment, fruit-fly pupae were subjected to heat shock at an age when this treatment was known to produce adults with various kinds of altered wing-vein pattern (these variants are called *venation phenocopies*, see Figure 1.2). Individuals showing a particular kind of phenocopy were then selectively bred together. Soon, strains were produced that responded to this kind of stress with high frequencies of the phenocopy. Continued intense selection made it possible to produce strains in which the selected-for abnormality appeared even in the *absence* of heat stress. Thus, phenotypic variants produced by an environmental probe, when bred together, eventually yield genotypes that show the variants even without any probe. Waddington called this effect *genetic assimilation.*

There are many other examples. Consider the familiar phenomenon of the formation of skin calluses in response to (and as a protection against) abrasion. Calluses form much more readily on some parts of the body, such as the hands and feet, than on others – reflecting differential selection pressure. Indeed, calluses will form on the feet of a bedridden person (i.e., in the absence of the usual environmental stimulus). The phenomenon even shows an evolutionary vestige, in the form of knuckle calluses, which form spontaneously in many individuals. They are now of no use, but presumably served an adaptive function in our knuckle walking ancestors. Evidently past selection for callus development in response to abrasion leads, in some people, to the spontaneous development of calluses.[7]

In higher animals, training procedures may be regarded as genetic probes whose effects can be interpreted in the same way as Waddington's heat shock. For example, the ease with which an individual learns some skill, such as music or mathematics, may be an indication of how close his genotype is to one that would produce this behavior with minimal or no training. One might speculate that a program of selective breeding for precocious musical ability would eventually lead to Mozarts capable of writing sonatas at six, and finally, perhaps, to infants capable of spontaneous musical expression of a high order in the absence of any explicit training.[8]

This may seem improbable, either because it seems unreasonable that something as complex as musical proficiency should be entirely innate, or because we are accustomed to think of *learned* and *innate* as opposites. There is no basis for either objection. For example, sheep dogs are selected for their ability to learn to herd sheep; and training a professional sheep dog takes years. Nevertheless, components of herding, such as circling and attempting to group people or animals, appear spontaneously in pets that have never been specifically trained. Speech is learned, yet adult intonation patterns, and an enormous number of phonemes (speech-sound units), occur spontaneously in infants. Precocial birds learn to identify their own species early in life, via the process termed *imprinting;* nevertheless, the type of object accepted for imprinting, especially later sexual imprinting, cannot deviate too much from the natural stimulus if it is to be effective. Young male swamp sparrows (*Melospiza georgiana*) learn their song from adult males, but are quite selective about what they will accept as a model. In all these cases, the learned ability is put together with ingredients provided innately.

Very complex behavior may be innately programmed, and complexity by itself is an unreliable guide as to whether something is learned or innate. Almost all the intricate behavior of insects develops independently of experience, for example. Some songbirds, such as the song sparrow (*Melospiza melodia*), develop their elaborate song with almost no specific experience, whereas others, such as the chaffinch (*Fringilla coelebs*), require early exposure to their much simpler species' song if they are later to sing normally.

Most dramatic of all, perhaps, is the inheritance of navigational ability by migratory birds. The golden plover (*Pluvialis dominicus*) breeds in northern Alaska and migrates during the fall to Argentina by way of Labrador, then returns in the spring across land, over Central America heading north and west along the Mississippi River. The bobolink (*Dolichnyx oryzivorus*) travels from Maine to Brazil; Wilson's petrel (*Oceanites oceanicus*) from the Falkland Islands (near the Antarctic) to Newfoundland. Some marine animals, for example, whales and salmon, perform comparable feats of migration. Migratory birds use such navigational-aid cues as sun direction, assessed both directly and through sky polarization on partly cloudy days; time (via an internal clock keyed to the light–dark cycle); the direction of the earth's magnetic field; very low-frequency sounds (infrasound), such as the sound of surf on a distant beach (which may be audible hundreds of miles away); visual-terrain cues; and perhaps other features not yet identified. These cues are combined to guide flight in ways that are far from being understood, even in the short flights of homing pigeons. The capacity to carry out these long migrations is but little dependent on experience in many solitary species. The complexity of a behavior pattern is an unreliable clue to its developmental origins.

Environmental effects on development can occur at any time, from conception to old age. For example, experiments by the pioneering developmental psycho-biologist Zing-Yang Kuo showed that the passive movement of the head and beak in embryonic chicks caused by their heartbeat in the egg plays a role in the later development of pecking. Gottlieb has demonstrated that ducklings' ability to follow the repetitive call of their mother depends for its selectivity on *prehatching* experience with sound frequencies above 1500 Hz. Normal ducklings, given a choice between a recording of the normal call and the call with the higher frequencies filtered out, reliably chose the normal call. However, birds deprived of hearing the mother while in the egg choose both alternatives equally. Under normal circumstances all ducklings receive the required experience, both by hearing their own calls while in the egg, and by hearing the calls of their siblings in adjacent eggs.[9]

These dependencies of later behavior on apparently unrelated earlier experience may seem odd and even capricious, but they fit in well with the *epigenetic* view of development: the now well-accepted idea that development is the outcome of continuous interactions between a genetically encoded program and the environment of the developing organism, rather than the unfolding of a pre-formed and predetermined entity. Gene action is a strictly conditional business, dependent on the gene environment and thus, in many cases, on the environment of the organism. Natural selection favors any genetic change that reliably has a beneficial effect on the phenotype. If the expression of that genetic change in the phenotype depends on the presence of a particular environmental feature, then as long as that feature is a reliable accompaniment of normal development, the gene will be favored and an environmental dependence will become established.

Imprinting is the best-known example of this kind of dependency. Precocial birds, such as chicks and ducklings, will generally form a permanent attachment to individuals that they see and can follow during the first day or two of life. This is part of the process by which these species learn to identify their own kind. Species identification in these animals might have developed in several ways. For example, the ducklings might be provided at hatching with an essentially built-in "template" enabling them immediately to recognize conspecifics. Many species are provided with such a template – almost all insects, and brood parasites such as the cuckoo (*Cuculus canorus*) and North American cowbird (*Molothrus ater*), which never see their own parents and could not function without the innate ability to recognize their own species. But even with such a template, the ducklings would also require a propensity to follow (following the mother when she calls is essential if the duckling is not to end up inside a predator). Following the mother implies some learning so that the animal doesn't follow *any* female duck. But given the existence of some learning mechanism, it is obviously parsimonious to arrange that the young animals learn not only the particular individual that they must follow, but also the characteristics of that individual's species. Under normal circumstances, of course, the first individual that the chick or duckling sees is a member of its own species. If not, its future is likely to be dim; only individuals whose first experience is of their own parents are likely to contribute to future generations. An efficient solution to the species-identification problem in precocial species, therefore, is the existence of a critical period during the first days of life when the individual learns about the characteristics of its parents by following them. Most genes act only during specific stages of development, so that small changes in genotype may have been necessary to change from the built-in template kind of development to imprinting. As imprinting evolved, less and less of the template would be necessary and mutations tending to degrade it would not be selected against.[10]

Vestiges of the template mechanisms from which imprinting may have evolved can still be detected. Ducklings imprint most rapidly to stimuli resembling members of their own species, and weak imprinting to a severely abnormal stimulus, such as a moving box, can be overcome later by exposure to the natural stimulus. I return to imprinting in Chapter 13.

SUMMARY

The introductory chapter of any book must always anticipate the presuppositions of its readers. I have assumed that many readers will know somewhat more about psychology than biology, but will still hold (one hopes weakly) commonsense mentalist views about the causes of human and animal behavior. Hence my emphasis on animals as machines, and the discussion of intentional terms and their uncertain status as explanations of behavior. Some other readers are likely to have encountered the extreme ethological dogma that insists on the uniqueness of each species and the consequent impossibility of a general psychology of learning. Hence, it seemed important to note the similarities among niches and the constraints that limit all information-processing systems, because these provide the basis for such a general psychology.

The distinction between learned versus innate behavior is another commonplace that requires modification. Some years ago, a perceptive psychologist (Verplanck, 1955) wrote a paper entitled "Since learned behavior is innate, and vice versa, what now?" that defined the problem: Learning depends upon inherited mechanisms and is constrained by them. Moreover, past experience affects later behavior in many ways, only a few of which we call "learning." The best way to get a feeling for the range of possibilities is to look at ontogeny, how it reflects past evolution, and how it incorporates the effects of environment, sometimes in an apparently capricious and idiosyncratic way. All this leads to the epigenetic view of development: Changes in morphology and behavior during ontogeny reflect a process of differentiation in which some options are chosen and others given up, guided at every instant by the joint effects of genotype and environment.

Beyond that, the chapter sets the stage for those that follow by distinguishing functional from mechanistic explanations and showing why each is useful. Optimality theory allows functional explanations to be formulated precisely and forces us to specify the things that are important to an animal, as well as the internal and external constraints that limit the range of behavioral variation. These constraints, in turn, provide clues to the underlying mechanisms that allow the animal to behave in a goal-directed way.

The next chapter begins the story of how animals can adapt to a variable world by describing the very simplest adaptive mechanisms: the processes that plants and single-celled animals use to find a congenial habitat, and the automatic, protective reflexes of higher animals.

NOTES

1. *Niche* is one of those essentially undefinable terms that are, nevertheless, essential to what might be termed the "sciences of organized complexity," such as ecology, psychology, and behavioral biology. Like most such terms, it is best defined by example. It is pretty obvious that the talents required of a good leopard are quite different from those needed by an effective antelope. Among the former are powerful means of attack, a digestive system attuned to meat, and a visual system adapted to attend to one thing at a time. Among the latter are a good means of evading attack, a lengthy gut able to cope with the poor diet provided by grazing, and a visual system able to detect threat from any quarter. Thus, the claws and teeth of the leopard, its forward-facing eyes and short digestive tract, as well as the rapid and maneuverable running of the antelope, its lengthy digestive tract and sideways-facing eyes, all find a functional explanation.

The behavioral adaptations required by different niches are usually less apparent than morphological differences, especially if they involve differences in the way that past experience affects present potential. The match between adaptation and niche is no less close because it is hard to see, however.

The basis for the modern idea of niche is Darwin's discussion of an organism's "place in the economy of nature." References to later work, including mathematical definitions of the concept, can be found in any standard ecology text.

2. The division of interest between those features of adaptive behavior that differ among niches and those that are common to all niches corresponds roughly to the division between ethologists and animal psychologists. In days when psychologists were less aware of biology than they are now, learning theorists rallied around the search for "general laws of learning." The discovery of types of learning specific to particular situations or species gradually made this position untenable. There are general laws, but they seem to reflect commonalities among niches or general features of all information-processing systems, rather than a common plan of construction – as the earlier view implied. In biological terms, the resemblances are a mixture of *convergence* and *homology* (the composition of the mixture being largely unknown in most cases) rather than the pure homology implied by the general-law idea (see also note 6).

3. Excellent, readable accounts of intentional systems, chess-playing machines, and the like, appear in the philosopher Daniel Dennett's book *Brainstorms* (1978), a collection of essays on psychology and artificial intelligence.

4. Of course, human psychology must eventually come up with an explanation for *why* intentional terms are so useful and ubiquitous as makeshift explanations. Perhaps the answer is that intentional accounts are just fuzzy functional explanations, and thus the best that one can do without detailed knowledge of behavioral mechanisms.

Methodological behaviorism is to be the contrasted with *radical* behaviorism, the position advocated most forcefully by B. F. Skinner. Radical behaviorism asserts that it is unnecessary to go significantly beyond the level of behavioral description to account for all behavior. The position made some sense in reaction against rampant mentalism, but makes none now. To pursue the computer analogy, it is like asserting that the chess-playing program can be explained entirely in terms of its inputs and outputs and direct (stimulus-response) links between them. For critiques (sympathetic and otherwise) of various aspects of behaviorism see Chomsky (1959), Dennett (1978), and Staddon (1967, 1973).

The philosophy of behaviorism was crystallized by a book by John Broadus Watson, *Psychology from the Standpoint of a Behaviorist* (1919), a polemical attack on the then prevailing phenomenological view that took subjective experience (with all its problems of intersubjective reliability) more or less at face value. This book is the source of a dominant movement in recent American psychology. It has given rise to a number of subsidiary streams, from Skinner's radical behaviorism, at one extreme, to various eclectic movements that are willing to explain human and animal behavior in terms of expectancies, attitudes, "means-end-readinesses," and the like. Good accounts of these historical trends are available in a number of books, most notably Boring (1957) and Herrnstein and Boring (1965). For a witty, clear, and controversial account of the antecedents of these movements see Bertrand Russell's marvelous *History of Western Philosophy* (1946).

5. Darwin (1872) p. 525. The German biologist Ernst Haeckel (1834–1919) was one of the first (Darwin preceded him) to make much of the relations between development and evolution, although his views are in many ways too simple and lent themselves to the "ontogeny recapitulates phylogeny" parody. Gould (1977) has summarized the modern view as follows: "Evolution occurs when ontogeny is altered in one of two ways: when new characters are introduced at any stage of development with varying effects upon subsequent stages, or when characters already present undergo changes in developmental timing" (p. 4). The classic work on the development of morphology is D'Arcy Wentworth Thompson's *On Growth and Form* (1917), a wide-ranging book with a strong esthetic element, written at a time when original scientific writing was meant for the general reader. A definitive account of quantitative issues is Julian Huxley's *Problems of Relative Growth* (1932). Important approaches to the development of behavior are due to Kuo (1967, 1970) and Lehrman (1970).

For an excellent account of current views on the mechanisms of evolution and development see the description of a recent Dahlem conference on the subject by Lewin (1981) in *Science*.

6. The problem here is an instance of the well-known difficulty of deciding whether characters are *homologous* (reflect common ancestry) or *analogous* (reflect convergent selection pressures). The human hand and the bat's wing are homologous structures, but the eyes of mammals and octopi reflect convergent selection pressures. A behavior such as suckling is obviously adaptive only at a certain time during a mammal's life; hence its early appearance and subsequent disappearance need not reflect similar behavior in nonmammalian ancestors. As Darwin pointed out, the more useless and apparently irrelevant the character, the more useful it is as a guide to evolutionary origins.

7. Genetic assimilation at first achieved notoriety because of its obvious resemblance to Lamarckian inheritance of acquired characteristics: The venation phenocopies are environmentally induced, yet selection of individual animals showing them eventually leads to spontaneous appearance of the effect. In fact, no direct effect of environment on genotype need be assumed. What is assumed is that the subset of animals showing the venation effect of heat stress have genotypes genetically "closer" to the genotype necessary for its spontaneous appearance than the rest. This assumption, plus the assumption of essentially random genetic variation about the mean, is sufficient to account for the effect of selection in shifting the mean genotype to the point where venation effects appear spontaneously.

Waddington's "epigenetic landscape" and his genetic assimilation experiments are described in a 1956 paper and in a number of general accounts (e.g., 1962).

8. The allusion to Mozart is originally due to Darwin, who pointed out that "if Mozart, instead of playing the pianoforte at three years old with wonderfully little practice, had played a tune with no practice at all, he might truly be said to have done so instinctively" (1872, p. 267). The phenomenon, however, is not uncommon. Consider the great mathematician Blaise Pascal (1623–1662), who because he was weakly as a child was protected from the excessive intellectual stimulation provided by Euclid – and reinvented much of geometry. His sister Gilberte wrote ". . . Since my father has been so careful to conceal all these things [mathematics] from him that he [Blaise] was forced to invent his own names. Thus, he called a circle a 'round,' a line a rod and similarly for all the rest. Using these names he set up axioms and finally complete proofs. And since, in these matters, one proceeds from one thing to another, he continued to make progress and pushed his investigations to the point where he reached the 32nd proposition of Book I of Euclid. . ." (Meschowski, 1964, p. 34). Pascal obviously did not need a great deal of specific experience to learn the primitives of geometry.

The key issue in the nature–nurture controversy is where the relevant information comes from. There are only two alternatives: If an organism shows some organized structure or behavior, the information must have been communicated either during phylogeny (when we are inclined to term the behavior *instinctive*) or during ontogeny (when it is termed *learned*) or, more probably, both. By any reasonable criterion it seems that Pascal's knowledge of geometry was largely instinctive.

See Staddon (1981a) for a speculative extension of these ideas to the relation between cultural and genetical evolution.

9. Calls made in the egg serve the vital function of synchronizing hatching – which is beneficial to all because it simplifies the logistics of feeding versus incubation by the parents. Gottlieb's work is described in his book (1971) and papers (e.g., 1974).

10. The term *template* has been most extensively used by Peter Marler and his associates in their elegant studies of the ontogeny of song learning in the white-crowned sparrow (*Zonotrichia leucophrys*; see e.g., Marler & Hamilton, 1966). They used it to refer to the "model" set up in the male by early experience of the adult song to which its own singing will later be matched. Strictly speaking, a template is a simple but literal model of some stimulus feature that can be acquired by experience, as in the sparrows, or is more or less built in.

The term *template* is an attempt to label the process of *encoding* by which any organism must abstract information about the external world. No representation, even a color photograph or movie, is complete; moreover, the more literal and complete a representation, the more memory storage it requires. Hence, natural selection has strongly favored the elimination of *redundancy* in the formation of templates: Features that are not reliable predictors of the object to be identified are eliminated; reliable features, especially if they are simple, are retained. Thus, the male English robin (*Erithacus rubecula*) recognizes other males simply as red fluffy objects of a certain size; for the male stickleback (*Gasterosteus aculeatus*) another male is any red object subtending a certain visual angle.

Information on imprinting is widely available. Good secondary sources are Hinde (1970) and Bateson (1974). The original work was done by Heinroth (1911) and his student Konrad Lorenz (1935, reprinted and translated in Lorenz, 1970). Related effects of early experience are discussed in Chapter 13.

2

VARIATION AND SELECTION OF BEHAVIOR

Orb-web spiders have devised a most efficient net for catching flying insects, yet we can trace no history of trial and error in the life of an individual spider that could explain the excellence of the web's design. Spiders don't learn how to weave good webs; no spider tries different designs and discards all but the most efficient. It is this instant perfection that sustains mystical beliefs in the power of instinct and fortifies disbelievers in evolution. Yet there surely was trial and error, in the form of ancestral spiders that built webs with varying efficiencies. Those that build most efficiently were better nourished and had most progeny. Granted that variation in web-building ability is, to some degree, inherited, then the ability to build better webs evolved, not by selection of good webs by individual spiders but by selection of spiders who made better webs. Web building depends on history, but more on the history of the race than on the history of the individual spider.

In "higher" animals, however, more and more behavior is selected in the first way, in the life of the individual, not by differential reproduction of individuals with different innate talents. This book is about the ways that individual history determines adaptive behavior.[1]

The simpler the animal, the more we know about the processes that select one behavior over others. This chapter has two objectives: First, to illustrate the processes of variation and selection of individual behavior in some simple cases – orientation mechanisms in plants and protozoa. Second, to describe elementary processes such as habituation, adaptation, and the "laws" of reflex action, that are some of the ingredients of adaptive behavior in higher animals. The next chapter introduces the notion of *feedback* in connection with direct (taxic) orientation.

SIMPLE ORIENTATION MECHANISMS

Finding the proper habitat, a place not too hot or cold, nor too dry or wet, safe from predators and with a supply of food, is a major behavioral problem for all animals. For simple animals such as protozoa and primitive invertebrates it is the main problem. Since these organisms possess either no, or only the most rudimentary, nervous system, they must solve it in an exceedingly economical way. Hence, simple orientation mechanisms exhibit the properties of adaptive behavior in their clearest form.

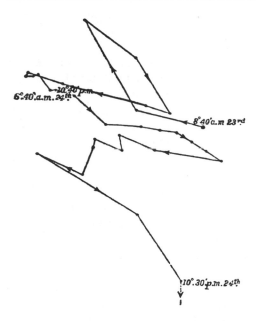

Figure 2.1. Tracing of circumnutation of the plant *Trifolium subterraneum* on horizontal glass, illuminated from above, from 8:40 A.M., July 23, to 10:30 P.M., July 24. (From Darwin, 1880, Figure 92.)

The movement of climbing plants provides a good illustration of the orientation problem, and of a simple solution to it. An important objective for any green plant is to gain maximum access to sunlight. Obstacles are presented by other plants with similar requirements. Where the need for light is not outweighed by other considerations, such as avoidance of predation or extremes of temperature or the effects of wind, plants therefore grow vertically and seek the highest, best-lighted point.

Darwin identified rotation of the growing plant tip (he called it *circumnutation*) as a key element in direct plant growth. He describes circumnutation as:

When the shoot of a hop (*Humulus lupulus*) rises from the ground, the two or three first-formed, whilst very young, may be seen to bend to one side and to travel slowly round towards all points of the compass...From seven observations made during August...the average rate during hot weather and during the day is 2 hrs. 8 m. for each revolution.... The revolving movement continues as long as the plant continues to grow; but each separate internode, as it becomes old, ceases to move. (1875b, pp. 2-3)

Figure 2.1 shows Darwin's record of the changing position of the tip of a *Trifolium* plant, with points plotted every half hour.

Two forces are at work here, one directed and one undirected. The upward growth of the plant is a directed movement, a taxis; the direction is opposite to the force of gravity (negative *geotropism*). The way taxes work is described in the next chapter. But the turning movement is undirected – until the moving tip encounters a vertical obstacle such as a stick or the stem of another plant. Once such an obstacle is encountered, the turning movement, together with upward growth, ensures that the plant will twine about it and be lifted to a higher (and presumably more satisfactory) location.

Most plants also possess a second taxic mechanism, *phototropism*. This combines with upward growth to direct the tip of the plant (on the average) toward the lightest part of the sky. The efficient light seeking of the plant can, therefore, be explained by the combined effects of three separate adaptive mechanisms: negative geotropism, circumnutation, and positive phototropism.

This example is very simple. There is no real need to talk about variation and selection here because all the ingredients are open to view. Nevertheless, circumnutation plays the role of variation: The plant tip sweeps out an arc "searching" for some vertical support. The environment, in the form of vertical obstacles, then plays the role of selection by blocking the moving tip and constraining it to a particular place. As we get to more complicated behavior, the mechanisms that underlie variation and selection are less well understood, although both variation in behavior, and the selective elimination of variants, can always be observed.

Light seeking by a plant illustrates two aspects to the study of adaptive behavior: analysis as the route to understanding, and the link between physiology and behavior. The analysis of behavior into component processes, such as the three just described, is a major theme of this book. In chemistry the aim is to describe all compounds by the rules of combination of a few elements. In similar fashion, the aim of behavioral analysis is to explain all the behavior of an organism in terms of a limited number of fundamental processes and their rules of interaction. We are closest to achieving this with simple orientation mechanisms. We still have a way to go in understanding the adaptive behavior of most vertebrates.

The second aspect of the study of adaptive behavior is its relation to physiology: Once an adaptive mechanism has been identified at the purely behavioral level, it becomes profitable to look for its physiological basis; that is, for the structures and chemical processes that underlie it. In the behavior of higher animals, the best understood processes tend to be those that involve the peripheral parts of the nervous system: sensory processes, simple perception, and some aspects of the organization of motor behavior, such as simple reflexes. It is no surprise, therefore, that it is in just these areas that our understanding of the physiological basis for behavior is most advanced.[2]

Following this pattern, the study of plant movement by Darwin and others led eventually to the discovery of *auxins*, powerful regulators of the differential growth that underlies movement in plants.[3] The movement of bacteria, the next example, is also being traced to complexes of reversible chemical reactions.

Indirect orientation (kineses)

All living creatures behave; not even a virus is completely inert. Quite primitive animals often show extraordinarily clever behavior. For example, in the 1880s

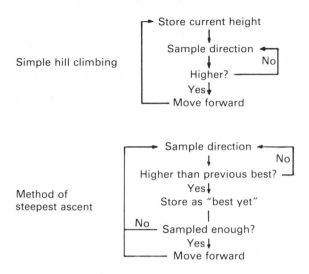

Figure 2.2. Flowcharts defining the logic of two kinds of hill-climbing process.

the German biologists Engelmann and Pfeffer discovered bacterial *chemotaxis* (actually a *kinesis* in the terminology I am using[4]): the ability of bacteria such as *Salmonella* to move up or down a chemical gradient. By inserting a capillary tube containing a suspension of attractive substance into a medium containing bacteria they showed that more bacteria entered the capillary than would be expected if they simply moved randomly. This is extraordinary behavior for an organism only about 2 μm long, with only the most primitive sensory and motor apparatus, and no nervous system at all.

Bacteria have no *distance receptors;* they cannot detect the source of a chemical in the way that we can detect an illumination source – from a distance, by looking at it. (We have no distance receptors for chemical gradients either, of course, nor are such things conceivable.) The only way to find the source of a diffusing chemical is by means of a process called *hill climbing*, where the hill to be climbed is the chemical gradient. Hill climbing can be explained by the simile of a blind man trying to find his way to the top of a hill. His simplest course is just to sample a direction: If it is downhill or level, sample again, but if it is uphill, follow it for one step, then sample again, and so on. A more efficient type of hill climbing is the *method of steepest ascent*: Here the man samples different directions from his current position and picks the one that gives the biggest increment. The logic of these two strategies is illustrated by flowcharts in Figure 2.2. Both require a small *memory*, sufficient to store a single slope value, but in addition the steepest-ascent method requires a *stopping rule* to tell the man when to cease sampling at each step. Still more complex, and more efficient, variants of hill climbing are possible (indeed, all adaptive behavior can be thought of as a sort of hill climbing).

The essential process in hill climbing is comparison of two heights, which defines the gradient. There are only two ways a bacterium can detect a chemical gradient. One is by a comparison of chemical concentrations across the minuscule length of the body (simultaneous comparison), which would involve detection of concentration differences of as little as one part in 10,000, a formidable task. This mechanism would allow for *directed movement* up the gradient, a *taxis* in the classification of Fraenkel and Gunn, whereas the observed movements look almost random with respect to the gradient. The second possibility is for comparison *across time* (successive comparison), allowing modulation of a largely undirected pattern of movement by increases and decreases in concentration. This kind of orientation mechanism is termed a *kinesis* in the Fraenkel and Gunn scheme.

Macnab and Koshland (1972) showed that temporal comparison was the key in an ingenious experiment that exposed the bacteria (they used two species: *Salmonella typhimurium*, the agent of common food poisoning, and *Escherichia coli*, a resident of the human gut much used in molecular biology) to temporal rather than spatial concentration gradients. First, they showed that the movement pattern of the bacteria was the same at different absolute concentrations of an attractant. Then they subjected the bacteria to a sudden, "step" change in concentration, and looked at them immediately after mixing was complete. If the organism's behavior is guided by comparison between the concentrations at its "head" and "tail," there should be no effect of the sudden change. But if their movement was determined by a difference across time, then a sudden drop in concentration should produce the same effect as swimming down a gradient.

And so it proved. These organisms show essentially only two modes of movement: straight-line swimming, and "tumbling." Bacterial tumbling increased dramatically if attractant concentration decreased, as it would when swimming down a gradient; tumbling decreased when attractant concentration increased, as when swimming up a gradient. The bacteria have reduced a most complex problem in three-dimensional orientation to simple on–off control of a random pattern: When things are getting worse, tumble; when they begin to improve, swim straight. Bacteria find their way up the attractant gradient by means of the simple hill-climbing strategy in Figure 2.2: Straight swimming corresponds to "move forward" in Figure 2.2, and tumbling corresponds to "sample direction." The tumbling performs the role of a random, slow search of the environment; when the search turns up a slight improvement, the organism follows that direction: "By taking giant steps in the right direction and small steps in the wrong direction, it biases its walk very effectively in the direction which aids its survival" (Koshland, 1977, p. 1057).

These experiments show that the bacteria are able to detect a gradient by making successive, rather than simultaneous, comparisons. Over how much distance can they compare? This is determined by a process of adaptation that is, in effect, a simple memory (recall that some memory is essential to hill climbing). It can be measured as follows: In the attractant-increase experiment the

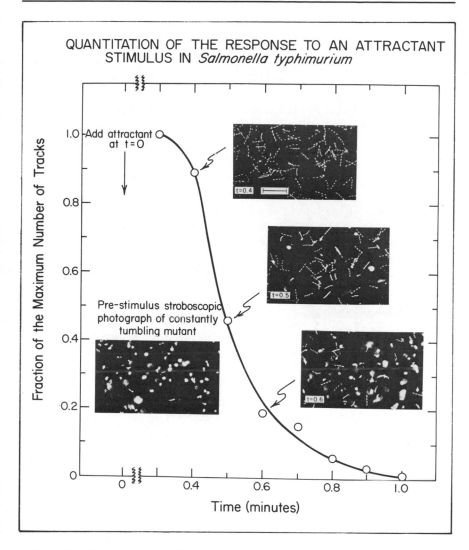

Figure 2.3. The panels show the appearance in the dark-field microscope as an increasing proportion of bacteria cease tumbling and begin swimming in a straight line: The field is illuminated stroboscopically, so that when the camera shutter is left open for a brief period a smooth swimming bacterium shows up as a series of white dots. The graph shows the proportion of organisms tumbling as a function of time since the attractant (serine) was added at time zero. (From Koshland, 1977.)

initial effect is an increase in smooth swimming, and a decrease in tumbling. As time elapses, however, fewer and fewer bacteria swim smoothly and eventually all resume tumbling. This almost-constant tumbling is consistent with the initial observation that under constant conditions, whether attractant concentration is

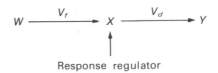

Figure 2.4. Hypothetical biochemical model of bacterial orientation mechanism. (Adapted from Koshland, 1977.)

high or low, the bacteria show the same pattern of movement. Figure 2.3 shows a graph of the adaptation process. It shows the proportion of bacteria swimming smoothly at different times after mixing; the numbers were obtained from photographs like those in the figure. The curve is declining, showing that fewer and fewer bacteria are swimming smoothly as time goes by. The form of the curve is exponential, following an initial period of slower decline.[5]

The adaptation mechanism allows these organisms to make comparisons across a distance of between 20 and 100 body lengths, which reduces the analytical problem from 10,000 to 1 to between 100 and 1,000 to 1: difficult, but better than simultaneous comparison.

The hill-climbing process used by these bacteria is so simple that Koshland has proposed to derive it from a relatively simple chemical process. His proposal is shown in Figure 2.4. X is a "response regulator," a substance whose concentration determines the type of response: For example, if X suppresses tumbling, then when its concentration reaches a threshold value, tumbling is suppressed and the bacterium swims smoothly; but when the level of X is below threshold, the organism tumbles. Thus X is like the temperature input to an on–off thermostat. The rate of adaptation is determined by the rate, V_f, at which X is formed from the precursor W relative to the rate at which it is decomposed, V_d. In this chemical system, V_d is directly related to the concentration of X, so that the system is inherently stable: An increase in X caused by a rise in V_f will eventually be compensated for by an increase in V_d, so that the level of X will return to its initial level.

If V_f increases more than V_d when the bacterium travels up a gradient, more X will be produced, further suppressing tumbling and allowing further progress. If the level of attractant ceases to increase, however, V_d will catch up with V_f, the level of X will decrease, and the organism will begin to tumble, slowing its forward progress. On going down a gradient the opposite would occur: V_d will gain on V_f, tumbling will occur, and progress will be arrested.

The quantitative and biochemical details of this system remain to be worked out.[6] Key quantitative elements are the value of the threshold and the rate of adaptation, since these will determine the range of gradient slopes where the system will function adaptively: For any choice of threshold and adaptation rate there will be some gradients where the system will fail to climb or may even descend the gradient.

Kinetic responses of the type shown so clearly by bacteria are common in

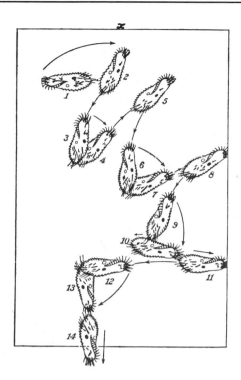

Figure 2.5. Reaction of *Oxytricha fallax* to heat applied at the top of the figure (at *x*). The animal moves backward from position 2 to position 3, then swivels its posterior end to position 4, and then goes forward to position 5. The set of movements 2 to 5 constitutes one turn. This turn did not take the animal out of the warmed area and turning was repeated in (5–8), (8–11), and (11–14). Evidently it was cooler at position 14 and the animal then went straight on. (From Jennings, 1906, Figure 84.)

simple animals that need to orient to stimuli that (a) cannot be sensed at a distance, such as temperature, chemical and humidity gradients (direct [taxic] orientation is more usual in illumination gradients); and (b) are likely to be gradual. For an organism the size of a bacterium or a *Paramecium*, or even a planarian, sharp spatial changes in temperature or chemical concentration are rarely encountered.

Tumbling in bacteria responds symmetrically to attractants and repellants: An increase in attractant concentration tends to suppress tumbling, as does a decrease in the concentration of a repellant. This relatively simple process thus provides the animal with a complete motivational system, able to deal adaptively with both positive and negative events.

As we move from bacteria to more complex single-celled animals, kinetic orientation becomes correspondingly more complex. For example, Jennings (1906) in a classic account described how ciliates such as *Paramecium* and *Oxytricha* avoid a region that is unfavorable (e.g., because of its temperature or pH). Figure 2.5 is taken from his book and shows the reaction of a single animal to a heated zone at the top of the figure. When it first enters the unfavorable region the animal stops, backs, then turns to the right and starts to move in a new direction; if the bad area is encountered once again, the process is repeated until a successful direction is found. The direction of the turn each time is unrelated to the actual direction of the boundary. This trial-and-error mechanism is obviously similar in many ways to the initiation of tumbling of bacteria when circumstances change for the worse; but just as obviously it is more efficient, because the first reaction of the animal is actually to reverse its direction of movement. It would be even more efficient if the turn were reliably away from the boundary, but this requires localization of the point of maximal stimulation on the animal's body and is presumably beyond the sensory capacities of *Oxytricha*.

Paramecium and many other ciliates show essentially the same pattern as *Oxytricha*, reacting to negative gradients by retreating and turning. As with the bacteria, this process serves both to avoid bad regions and seek good ones. If a bad region surrounds a good one, the process keeps the animal within the good region.

REFLEX MECHANISMS

The beauty of simple orienting mechanisms is that we can see all together the entire behavioral repertoire of the animal. The familiar psychological categories of motivation, cognition, sensation, and perception are all fused in a simple set of processes that initiate action, respond to stimulation, and guide the animal toward things that will aid its survival and reproduction. Behavior such as the *avoiding reaction* of *Paramecium* – reversal of movement in response to a sudden "change for the worse" – makes sense as part of a harmonious system of reactions that combine to produce adaptive behavior. With animals like birds or mammals, on the other hand, it is more difficult to see how things fit together. Their behavior is so varied and dependent on individual history, and their niches are so complex that we cannot grasp as a whole the functional relations among the various elements in their repertoire. There is a natural tendency to look for some particular behavior as the key to behavior in general, when the real key is no behavior in particular, but rather how all behaviors fit together to yield adaptation to a niche.

Paramecium can vary only its rate and direction of movement. Its rather rich sensitivity to a variety of chemical, photic, and thermal stimuli must, therefore, be funneled into these limited modes of action. "Higher" animals are not so limited: not only can they sense even more aspects of the physical environment, they can do many more things in response. The avoiding reaction of *Paramecium* is one extreme of a continuum of reactions that serve the function of bringing the animal into a congenial environment. The *reflexes* of higher animals serve a similar function – most reflexes avoid, escape from, or minimize the effect of, noxious stimuli – but rather than being modulated by the environment, like the avoiding reaction, they are simply replaced by quite different kinds of behavior when circumstances demand it. The boxer reflexly blinks and ducks as his opponent's fist approaches his face, but if he avoids successfully, his next reaction is likely to be a planned offensive strategy, not an automatic reaction.

Thus reflexes are a part, actually a rather small part, of the adaptive repertoire of higher organisms. Nevertheless, they have had historically a special role because of the apparent simplicity of the relation between stimulus and response. The immediacy of reflexes, and their relative independence of the animal's past history, make them easy to study. Moreover, the obvious adaptiveness of reflex properties, and the similarity between reflex properties and some aspects of the properties of "higher" behaviors, suggested to many that a complete psychology might be built on deep understanding of reflexes. This is almost certainly a mistaken view. Nevertheless, for historical reasons and reasons of simplicity, a discussion of reflexes is a natural preliminary to discussion of more complex kinds of behavior.

The reflex idea has a long history, beginning in the modern era with the French philosopher René Descartes (1596–1650), who was perhaps the first to propose that all animal, and much human, behavior can be explained mechanistically, rather than by reference to a "soul" that directs the animal from within. This idea was elaborated by the Russian physiologist I. M. Sechenov (1829–1905) who also carried out experimental work of the modern type on "spinal reflexes," that is, the built-in protective and integrative responses that can be demonstrated in animals in which the higher brain centers have been severed from the spinal cord (so-called "spinal preparations"). This work was brought to its highest pitch by the British physiologist C. S. Sherrington and is summarized in his influential book *The Integrative Action of the Nervous System* (1906); based on a course of lectures delivered at Yale University.[7]

Sherrington's work followed a familiar pattern. Rather than attempt to study sensory and motor integration as a whole, a confusing and impossible task, he attempted, instead, to reduce it to its simplest elements. Instead of working with an intact dog or cat, he operated surgically on the animal so as to disable the higher brain centers. These "spinal" animals cannot learn and behave in an automatic way that can be easily studied. In his characteristic, rather literary style, Sherrington describes the properties of his experimental preparation as follows:

Experiment today put[s] within reach of the observer a puppet-animal which conforms largely with Descartes' assumptions. In the more organized animals of the vertebrate type the shape of the central nerve-organ [i.e., the brain] allows a simple operation to reduce the animals to the Descartes condition. An overlying outgrowth of the central nerve-organ in the head can be removed under anaesthesia, and on the narcosis passing off the animal is found to be a Cartesian puppet: it can execute certain acts but is devoid of mind. . . . Thoughts, feeling, memory, percepts, conations, etc.; of these no evidence is forthcoming or to be elicited. Yet the animal remains a motor mechanism which can be touched into action in certain ways so as to exhibit pieces of its behavior. (1947, p. xi)

With this reduced, decerebrate animal it is possible to study in full quantitative detail the simplest level of reflex, sensorimotor integration, free of the complications introduced by spontaneity and "volition."[8]

Reflexes are automatic but not crude or undifferentiated, or without a clear function in the normal life of the animal. Sherrington continues with a number of examples:

The movements are not meaningless; they carry each of them an obvious meaning. The scope commonly agrees with some act which the normal animal under like circumstances would do. Thus the cat set upright...on a "floor" moving backward under its feet walks, runs or gallops according to the speed given the floorway. Again in the dog a feeble electric current ("electric flea") applied by a minute entomological pin set lightly in the hair-bulb layer of the skin of the shoulder brings the hind paw of that side to the place, and with unsheathed claws the foot performs a rhythmic grooming of the hairy coat there. If the point lie forward at the ear, the foot is directed thither, if far back in the loin the foot goes thither, and similarly at any intermediate spot. The list of such purposive movements is impressive. If a foot tread on a thorn that foot is held up from the ground while the other legs limp away. Milk placed in the mouth is swallowed; acid solution is rejected. Let fall, inverted, the reflex cat alights on its feet. The dog shakes its coat dry after immersion in water. A fly settling on the ear is instantly flung off by the ear. Water entering the ear is thrown out by violent shaking of the head. An exhaustive list would be much larger than that given here...But when all is said, if we compare such a list with the range of situations to which the normal dog or cat reacts appropriately, the list is extremely poverty stricken....It contains no social reactions. It evidences hunger by restlessness and brisker knee-jerks; but it fails to recognize food as food: it shows no memory, it cannot be trained or learn: it cannot be taught its name. (1947, pp. xi-xiii)

The deficiencies of the reflex animal are in the simplicity of the stimuli to which it can respond and in the absence of any but the briefest memory. It can neither learn new things nor recall past experiences. It is just this absence of history that makes the preparation of such analytical interest. The assurance that what is observed *now* can be traced to causes in the environment that are either present now, or no more than a minute or two in the past, much simplifies the scientist's task.[9] The absence of spontaneous movement completes the picture. How different is the normal animal, whose behavior now may reflect experiences months or years ago and whose ability to adapt to new situations demands the expression of a variety of novel, spontaneous behaviors.

Sherrington was a physiologist and his concern was with the functioning of the nervous system. He defined a reflex as a sensorimotor (stimulus-response) relation involving at least two neurons between receptor and effector, that is, at least one *synapse*, the then hypothetical, now much-studied point of contact between communicating nerve cells. Most of his experiments on reflex function were purely behavioral, however, and the reflex properties that emerged from them turn up even in organisms lacking a nervous system. The same properties – habituation, spatial and temporal summation, thresholds, refractory period, "momentum," and others – are shown by many reactions of intact higher animals. The properties of the reflex seem to reflect evolutionary *convergence;* that is, a set of similar adaptations to similar environmental features.

Sherrington's concept of the reflex is far from the simple, inflexible, push-button caricature sometimes encountered in introductory textbooks. To be sure, there is always a stimulus and a response; but the ability of the stimulus to produce the response depends on the reflex threshold – and the threshold of each reflex depends not only on the state of many other reflexes but also (in the intact animal) on higher centers, which retain the effects of an extensive past history. The function of reflexes is the *integration* of behavior, which would be impossible without well-defined rules of interaction.

Reflex properties can be divided into two classes: the properties of reflexes considered in isolation, and the rules governing interactions between reflexes. I consider each class in turn.

Individual reflexes

A stimulus – a sound, a light, a touch – is presented and, after a brief delay, a response – a movement of the ear, contraction of the pupil, a scratching movement of the paw – ensues. *Reflex* is the name for the properties of this relation between stimulus and response. Stimuli have intensity, quality, duration, location, and perhaps other properties; responses have similar properties and, in addition, follow the stimulus with a certain delay (*latency*) and with a certain reliability (*probability*). The properties of the reflex are the relations between these stimulus and response properties.

Underlying all is the theoretical notion of *reflex strength*, the tendency for the reflex response to occur. Reflex strength cannot be measured directly, but can be estimated by, for example, the intensity of stimulus required to elicit the response, by the degree of variation from the optimal stimulus that is still sufficient to produce the response, and by the latency between stimulus and response: The weaker or more different from optimal the stimulus, and the more rapid the response, the stronger the reflex.

There are seven main reflex properties:

Threshold. The stimulus for a reflex must be above a certain minimum level if it is to elicit a response. This is the *absolute threshold* for the response. The threshold is not fixed, but depends on a number of factors. In the intact animal, the threshold depends on, for example, the stimulus situation, the animal's state of attention, its motivational state, and its past history – both immediate and remote. In the spinal, or decerebrate animal, the threshold is mainly affected by immediate past history and the state of other reflexes.

Thresholds are typically measured in the following way. Weak stimuli (in the neighborhood of the threshold) are presented in a random order, well spaced in time, and occurrences of the reflex response that exceed some criterion are counted. A stimulus close to threshold will sometimes elicit the response and sometimes not. The proportion of times that each stimulus is effective is plotted as a function of stimulus intensity, and the result is usually an S-shaped (sigmoidal) function, as shown in Figure 2.6. The threshold is taken, by convention, to be the stimulus intensity that elicits the response 50% of the time. Since the location of the function, and hence of the threshold, along the stimulus-intensity

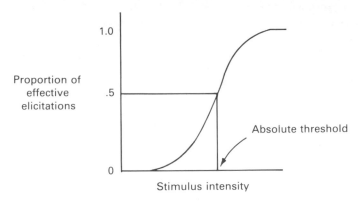

Figure 2.6. A typical threshold function: probability of elicitation as a function of stimulus intensity.

axis depends both on features of the experimental situation and on the criteria chosen, the threshold function is not of much interest in isolation; rather it is used as a *dependent variable*. One is usually interested in the change in the threshold caused by some other experimental manipulation, such as earlier elicitation of the same or an antagonistic reflex, a change in the animal's motivational state, or some other factor that affects reflex strength.

The gradual slope of the threshold function reflects the contribution of *noise* – unpredictable fluctuations in threshold – traceable to lack of perfect constancy in the physiological state of the system. If noise were to be completely eliminated (a physical impossibility), the threshold function would be a straight, vertical line, a "step" change from 0% to 100% response at the threshold intensity. In most cases, threshold variability is "planned" (rather than a reflection of poor construction) in the sense that some variation in threshold is usually adaptive. When variation makes no adaptive sense, the system shows very little variability: Under optimal conditions, human visual sensitivity is close to the limits imposed by the quantal nature of light, for example.

Latency. The time between stimulus onset and the occurrence of the reflex response is one of the most useful measures of the state (strength) of the reflex. Like threshold (and for the same reasons), latency measures show some variability; the same stimulus will produce responses with slightly different latencies on successive presentations, even when temporal interactions (to be discussed subsequently) are well controlled.

A typical latency distribution is shown in Figure 2.7. Latency distributions are usually of the form to be expected from a relatively simple *stochastic*, that is, probabilistic-in-time, process. For example, suppose that immediately following the onset of the stimulus, the probability that the response will occur rises from some very low level to a high level: That is, as soon as the stimulus has occurred,

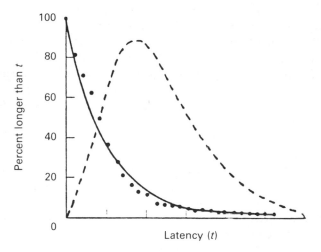

Figure 2.7. Latency distributions. Solid line: exponential frequency distribution: $y = A \exp(-\lambda t)$. Dashed line: sum of two exponential distributions: $y = A \exp(-\lambda_1 t) + B \exp(-\lambda_2 t)$.

the probability, in each brief interval of time Δt, that the response will occur rises from P_0 ($\simeq 0$) to P_1. This model may be compared to tossing a biased coin, whose probability of "heads," P_1, is equal to the probability a response will occur at the end of every time increment Δt. If heads comes up, the response occurs, and the probability drops from P_1 back to its original low level, P_0. It is easy to see intuitively that if we repeat this experiment a large number of times, and count up the number of times heads occurred after zero "tails," one tail, two tails, and so on (equivalent to latencies of 0, Δt, $2\Delta t, \ldots$) that short runs will predominate. Indeed, the theoretical distribution is the exponential (see note 5) illustrated by the solid line in Figure 2.7.

Actual distributions are similar to this prediction; they differ in having their mode (maximum value) displaced from zero: This can be accommodated by adding a second random process to the first, to take account of the unavoidable "dead" time before the physiological processes responsible for the reflex response can act (dashed line in Figure 2.7). In fact, a two-stage model does a creditable job of explaining many latency distributions.

The virtue of these simple stochastic models, in addition to economy of description, is that things that affect latency often have selective effects on just one of the parameters of the stochastic model – just on the "strength" parameter, for example.[10]

There are two general messages from this brief discussion of latency distributions: The first is that a process that is random in time yields an exponential latency distribution (not one that is flat or bell shaped); and second, that quite simple processes may underlie apparently complicated distributions.

Latency depends on the same extrinsic factors as threshold, and the two measures tend to covary: a low threshold being associated with a short latency, a high threshold with a long latency. Latency also depends on stimulus intensity; usually a more intense stimulus elicits a more rapid response. These characteristics are consistent with the protective function of most reflexes. A more intense stimulus is likely to be more dangerous and to justify a more rapid and vigorous response. Thus the strength, as well as the speed, of most reflex responses is directly related to stimulus intensity.

Refractory period. After a reflex response has occurred, the threshold of the reflex may be elevated for a brief refractory period. The refractory period is intrinsic to the physiology of nerve conduction, is common in spinal reflexes, but may or may not occur in stimulus-response relations in the intact organism.

Reflexes are either *tonic* (i.e., maintained by continuous stimulation, as in postural reflexes), or *phasic* (i.e., transient or repetitive, as in the scratch reflex). A refractory period is obviously essential to phasic reflexes, else continued stimulation would lead to a continual excitation of the response, as in tonic reflexes. The function of the scratch reflex, for example, is to produce repeated, back-and-forth limb movements in response to a continuous "tickle" stimulus. Time for limb withdrawal (flexion) is provided by the refractory period of the active (extension) phase of the reflex: After the limb has extended, continued stimulation can produce no further extension. Withdrawal occurs because of reciprocal inhibition between flexors and extensors, which is discussed later: Flexor and extensor muscles are so connected that facilitation of one produces inhibition of the other, and vice versa. Hence, inhibition of the extensor in the refractory period of the scratch reflex automatically yields activation of the flexor, and, thus, withdrawal of the limb ready for the next stroke. The process is cumbersome to describe in words, but can be expressed economically in the formal language of oscillatory networks. Reciprocal inhibition and *successive induction* are properties of such networks.

Temporal summation. Two subthreshold stimuli, spaced closely in time, may excite a reflex when either alone would be ineffective; this is temporal summation. For example, the so-called *orientation reflex* is a set of coordinated movements by which an animal focuses on a novel stimulus, such as an unexpected sound. A cat will turn toward the sound, looking at the direction of the source with ears pricked forward, pupils dilated, and every muscle tensed. The animal is in the optimal state both to receive any new information and to act on it immediately. It is a common observation that one weak stimulus may not elicit this response (i.e., "get the animal's attention," as it is usually expressed), but two in close succession may.

Temporal summation is a widespread property, especially of reflexes related to orientation. For example, the prey-catching response of frogs toward small moving objects can be facilitated by small movements of the same stimulus a few

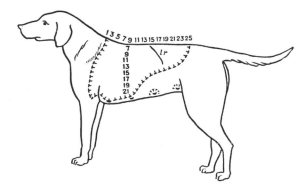

Figure 2.8. The receptive field for the hind-limb scratch reflex. The reflex can be evoked by touch anywhere within a saddle-shaped area. *lr* marks the position of the last rib. (From Sherrington, 1906.)

seconds earlier, even though the initial movements seldom trigger the response (Ingle, 1975).

Spatial summation. Somatosensory reflexes (such as the scratch reflex) each have a specific region of the skin within which a stimulus can excite a response. This region is the *receptive field* of the reflex (see Figure 2.8). Two stimuli, each individually subthreshold, may excite the response if presented together: This is spatial summation. The degree of summation, in general, depends on the degree of proximity of the two stimulated regions within the receptive field; the closer the stimuli, the more they summate.

The temporal and spatial summation interact, as graphically described by Sherrington:

A subliminal [below-threshold] stimulus applied at a point A will render a subliminal stimulus applied at point B near A supraliminal if the second stimulus follow within a short time, e.g., 500 msec. The space of receptive surface across which this can be demonstrated in the scratch-reflex amounts to 5–6 cm. . . . the phenomenon is characteristically and simply illustrated by the difference between the potency as a stimulus of the edge of a card, say 6 in. long, pressed simultaneously over its whole length against the receptive skin field, say for 5 sec., and on the other hand lightly drawing one corner of the card along the same line in the skin field also for 5 sec. The former application simply evokes a reflex of a few beats, which then dies out. The latter evokes a vigorous reflex that continues and outlasts the application of the stimulus. A successive line is more effective as a stimulus than a simultaneous line of equal length and duration. Again, if a light disk 3 cm. in diameter and a fraction of a millimetre thick be freely pivoted in bearings at the end of a handle, so that it turns when pushed by its handle over the skin surface, such a wheel may not, when pushed against a spot of the receptive surface, excite the reflex, but it excites it when it is rolled along it. The same thing is seen with a spur wheel. Even when the points are 2 cm. apart, as the spur wheel is rolled over the surface successive summation occurs, and the reflex is evoked as the progress of the wheel proceeds. If a parasite in its travel produces excitation which is but close below the threshold, its progress is likely to so develop the excitability of the surface whither it passes that the scalptor-reflex [scratch reflex] will be evoked. In the skin and the parasite respectively we have, no doubt, two competing adaptations at work. It is perhaps to avoid the consequences of the spatial spread of the "bahnung" [facilitation] that the hop of the flea has been developed. (Sherrington, 1947, p. 185)

Momentum (after-discharge). The excitation of most phasic reflexes will generally outlast the stimulus that produced it. Sherrington's discussion of the scratch reflex provides an example (". . . a vigorous reflex that continues and outlasts the application of the stimulus"). Under restricted conditions, repeated elicitation of a reflex may become easier and easier, and this may also be related to momentum.

Habituation. Repeated elicitation of any reflex response eventually leads to a decrease in vigor and loss of stability (irregularity and tremor) and finally to cessation of the response. Habituation to a weak stimulus occurs sooner than to a strong one, presumably because the reflex threshold rises above the level where a weak stimulus is effective before it comes to exceed the level of a stronger stimulus. Thus, a reflex fatigued by a weak stimulus can be elicited once again by a stronger one, and does not persist when the stimulus intensity is reduced again. Most important, reflex habituation is not due to muscle fatigue: Flexor muscles that cease to act as a consequence of habituation of the flexion reflex (withdrawal of a limb to painful stimulation) are still readily excited by the scratch reflex; a scratch reflex habituated by repeated stimulation at a point in its receptive field is readily excited again by stimulation at a new place within the field a little distance away from the old. Reflex habituation, like habituation in the intact animal, is stimulus specific. Habituation dissipates with time: If a habituated reflex is left unstimulated for a time, it will recover so that the same stimulus applied after a delay is once again effective.[11]

Reflex interaction

Very little of the behavior of intact, higher animals is purely reflexive: The same stimulus doesn't invariably elicit the same response; the *effective stimulus* (i.e., the set of all physical stimuli adequate to produce the response) is often complex and hard to define in physical terms, and the organism's history (for more than just a few minutes past) affects responsiveness. The intact animal, unlike the spinal preparation, cannot be restored to some constant, "ground" state simply by the lapse of time; stimulus effects are not completely reversible.

Nevertheless, the properties of reflexes are of interest to students of the normal, intact animal because reflexes exhibit in clear and relatively simple form the essential features of any behavioral unit: temporal properties, the effects of stimulus intensity and frequency, self-exciting (momentum), and self-limiting (habituation) effects. These properties of reflexes are useful dimensions for the analysis of more complex behaviors.

Some of the reflex properties already described are not characteristic even of all reflexes, much less of the behavior of intact animals. For example, learned stimulus-response relations, or *operants*, do not always show the direct relation between stimulus intensity and response vigor that is shown by protective reflexes: Yelled instructions aren't necessarily more effective than instructions delivered in more normal tones. The latency measure is not appropriate for many operants where time is part of the effective stimulus. On the other hand, many phasic operants, such as pressing a bar or pecking at a fixed stimulus, have properties akin to habituation and momentum. The phenomenon of *stimulus generalization* parallels temporal and spatial summation.

The constraints that determine reflex interaction apply to all behavior, however. There are two: (a) The constraint that the organism can do only one, or at most a few, things at a time (this corresponds to Sherrington's *final common path* for reflex action);[12] and (b) The adaptiveness, or lack of adaptiveness, of intermediate courses of action. Most commonly, perhaps, "he who hesitates is lost" – a state of activity intermediate between two opposed courses of action is less adaptive than either. When confronted with a stimulus that has both attractive and fearsome features, flight or fight are both likely to be better than hesitation; simultaneous excitation of flexors and extensors results in a crippling rigidity useful only to believers in isometric exercises; orientation to either of two simultaneous novel stimuli coming from different quarters is better than looking between them. On the other hand, conflicting postural demands are best resolved by an intermediate solution, rather than either extreme.

There are three principles of reflex interaction.

Reciprocal inhibition (competition). Since incompatible reflexes must compete for the same final common path to the effectors, the primary principle of reflex interaction is reciprocal inhibition: Facilitation of reflex A inhibits reflex B; and, conversely, inhibition of A facilitates B. As we have seen, this rule holds for incompatible behavioral units at any level of complexity.

This principle was particularly obvious to reflexologists because of the *reciprocal innervation* of flexor and extensor muscles that is essential to the proper coordination of movement. The reflex principle of reciprocal antagonism applies to individual responses, but a similar *principle of antithesis* seems also to apply to whole systems of muscles that are affected by positive or negative "moods" of an animal.

Darwin (1872) noticed that each emotional "state of mind" tends to be associated with a stereotyped set of reactions. For example, a person who is perplexed may scratch his head, roll his eyes, or cough; someone who disagrees with a proposition is likely to avert or close his eyes and shake his head, whereas if he agrees, he will look straight with open eyes, and nod his head. In greeting his master a dog typically takes up the submissive posture shown in the upper panel of Figure 2.9: The forelegs are bent, the coat smooth, and the tail wagging. The dog's mood when confronting a potential adversary is in a sense opposite to his mood when facing a well-loved individual. Darwin pointed out that the active

Figure 2.9. Upper panel: posture of a dog greeting its master (submissive). Lower panel: posture of a dog with hostile intentions (aggressive). (From Darwin, 1872.)

muscles are also opposite: if a flexor is tensed in the first case, the extensor will be in the second. The bottom panel of Figure 2.9 illustrates such a threat posture: The dog's head is erect, his ears pricked, his legs are straight, and his fur and tail raised – all opposite actions to the submissive posture shown in the upper panel.

Many expressions of emotion function as communications, and are, therefore, subject to special constraints. In particular, messages requiring very different actions of the receiver should also be very different in form, so that the inevitable confusions are more likely to occur between messages similar in meaning than between messages of very different meaning. Thus the property of reciprocal inhibition may have quite a different functional basis in reflexes and in emotional expression, although in both it can, ultimately, be traced to the adaptive importance of decisive action.

Most reflexes compete, but some combine synergistically. The second property of reflex interaction may therefore be termed:

Cooperation. Reflexes that share response components may facilitate one another ("allied" reflexes), or when simultaneously excited may yield a response whose character is intermediate (blending). For example, subthreshold stimuli for the flexor and scratch reflexes may summate to elicit limb withdrawal. This is an example of mutual facilitation (also termed *immediate induction*) by two reflexes sharing a common response. The same reflex excited by stimuli from widely different parts of its receptive field behaves in many ways like two separate reflexes (for example, habituation of the reflex to one stimulus may have little effect in the threshold for the other); thus, spatial summation is a phenomenon closely related to cooperation.

Because posture must be maintained at all times, postural reflexes always combine synergistically with other reflexes to reach a compromise that is usually adaptive. Blending is also a common mode of interaction between action tendencies guided by some of the orientation mechanisms discussed in Chapter 3.

Successive induction. An important effect that derives from reciprocal inhibition (competition) and the temporal properties of reflexes is *successive induction* (also termed "spinal contrast"). In Sherrington's words: ". . . the extension-reflex predisposes to and may actually induce a flexion-reflex, and conversely the flexion-reflex predisposes to and may actually induce an extension-reflex" (Sherrington, 1947, p. 213). The role of this mechanism in recurrent reflexes (e.g., scratch reflex) has already been touched on, and its usefulness in alternating movements, such as those involved in walking, is obvious.

An example of successive induction is shown in Figure 2.10, which is taken from Sherrington's (1906) book. It shows the magnitude of the extension reflex of the hind limb of a spinal dog (measured by means of a pen arrangement connected to the limb and writing on a rotating cylinder) both before (panel A) and after (panels B–F) strong excitation of the opposed flexor reflex. The extensor reflex is strongly facilitated by the prior excitation of its opponent (panel B). This effect depends both on the reciprocal inhibition between flexor and extensor reflexes, and on the time between stimulation of the flexor and subsequent elicitation of the extensor: the longer the time between, the less the successive induction effect. Most of the decrease in the extensor reflex shown in panels B–F is simply due to the passage of time, rather than to habituation of the extensor reflex due to repeated elicitation.

The term *contrast* is often used for successive induction because of the obvious resemblance to perceptual effects: An illuminated area surrounded by a dark ring appears brighter than the same area surrounded by a light ring (simultaneous brightness contrast); similarly, an illuminated area viewed after a period of darkness appears brighter than following a period of illumination (successive brightness contrast).

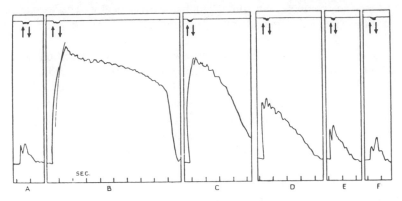

Figure 2.10. Successive induction in the cross-extension reflex of the hind limb of a dog. The brief eliciting shocks are shown at the top of the figure. The displacement of the limb appears below, and the ordinate is time in seconds. Between A and B a strong flexion reflex of the limb was produced and maintained for 55 sec. The next crossed-extension reflex, B, is greatly augmented in magnitude and after discharge. The augmentation has dissipated by the 5th min after the intercalated flexion reflex (F). (From Sherrington, 1947, Figure 57.)

Contrast effects turn up in all sensory modalities, in the behavior of most, if not all, animals, and even in experiments on reward and punishment (*behavioral contrast*; see Chapter 11). Several interesting examples are discussed by the entomologist Kennedy (1965) who has repeatedly emphasized the importance of temporal interaction between behaviors as a source of so-called motivated behavior.

Kennedy's most extensive series of experiments involves the aphid *Aphis fabae*. This animal has two main modes of behavior, "settling," which comprises several related actions (probing the leaf, feeding, excretion, and reproduction), and "flight," which involves locomotion (by walking or flight) between feeding sites. There is reciprocal inhibition between these two systems, and independent stimulus control over each: The stimulus that tends to excite flight is light, and the stimulus for settling is the surface of the host plant. This system of interactions can conveniently be represented as in Figure 2.11, which shows the two activities, S and F, their facilitating stimuli, P and L, and the reciprocal inhibition between activities.

The temporal interactions between settling and flight in this animal can become quite complicated. However, it is possible to distinguish two relatively simple modes. Because each activity has its own controlling stimulus, each can be elicited separately. If the aphid has been flying for some time (and this can be arranged with the aid of a vertical wind tunnel, so that the animal maintains a constant air speed without moving relative to the surroundings), presentation of an appropriate surface will elicit settling, which, therefore, inhibits flight. Conversely, if the animal is settled, turning on the overhead light will usually elicit upward flight, which, therefore, inhibits settling.

Figure 2.11. Interaction between reciprocally inhibitory activities.

In addition to these direct inhibitory effects there are also aftereffects. The details have been best worked out for flight as it is affected by prior settling: If the tendency to flight is already strong (i.e., the animal shows a high air speed) when it is suppressed by settling, then removal of the settling stimulus causes a rebound of the tendency to fly above its level when the settling stimulus came on. This is an example of successive induction of excitation. Conversely, if the tendency to fly is low when the settling stimulus is presented (and by implication, the tendency to settle is relatively high), then flying may be depressed (relative to the preinhibition level) after the settling stimulus is withdrawn: This is successive induction of inhibition. The balance between these two effects, poststimulus excitation versus poststimulus inhibition, is unstable and shifts from moment to moment. This instability is probably not accidental, but represents a useful source of variability in behavior. Since adaptive behavior involves the selection of adaptive variants by an unpredictable environment, a residue of behavioral variation is often essential if an animal is not to become "locked in" to maladaptive modes of behavior.

A second example, from a study of walking in locusts by Moorhouse, Fosbrooke, and Kennedy (1978), shows how periodic changes in the stimulating environment can interact with the successive-induction process to produce paradoxical effects. Young locusts were positioned in the treadmill apparatus shown in Figure 2.12, which allowed their walking and turning movements to be monitored automatically. Moorhouse et al. divided the behavior of the insects into two mutually exclusive categories: "walking," which entails forward progression at or above a minimum speed, and "nonwalking," which is everything else (backward progression, slow forward progression, grooming, biting, "peering," and complete immobility). The animals were subjected to three experimental treatments: continuous light, continuous darkness, and alternating light and darkness. The pattern of behavior following light onset is shown in the three-dimensional graph in Figure 2.13. The graph shows the range of walking speeds in 10-min segments following light onset. For the first 10 min, the animal walks little, but the amount of walking gradually picks up over time, reaching a peak after about 60 min and declining slowly thereafter. The pattern of walking in continuous darkness was different in two main respects: The onset of walking was almost immediate, so that the animals walked substantially even during the first 10 min, but the rate of walking peaked sooner, after about 30 min, and declined thereafter to a low level. Overall, the animals walked more in the light than in the dark.

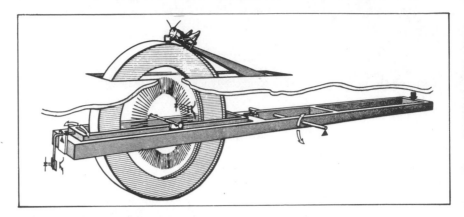

Figure 2.12. Apparatus used to monitor locust activity in the experiment by Moorhouse et al. (1978). Main structure is made of balsa wood, pivoted in the center with a counter-weight at the arrowed end. Center disk of wheel has 500 radial stripes at intervals equivalent to 1 mm at circumference. Electronic symbols represent infrared sensors for rotational and lateral movements of the wheel.

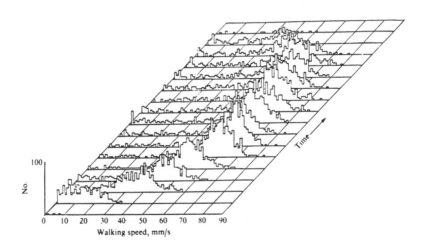

Figure 2.13. Running speed distributions at 10-min intervals following light onset at time zero in the Moorhouse et al. (1978) experiment.

The third treatment involved continuous darkness followed by alternating 10-min periods of light and dark. Here, surprisingly, the animals walked more in dark periods than in light, walked much less in the light than in continuous light (the first treatment), and walked more in the alternating periods even than in continuous light. The results of the three treatments are compared in Figure 2.14.

These results are readily explained as successive induction of excitation. The slow buildup of running in the light suggests that following light onset, activities

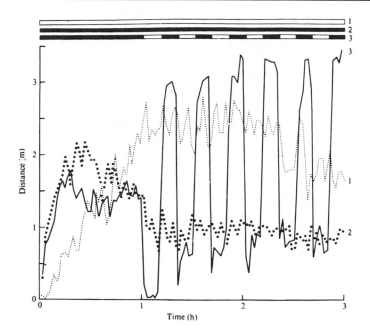

Figure 2.14. Summarized results of three treatments – constant light (. . . .), constant dark (••••), and alternating light and dark (———) – in the Moorhouse et al. (1978) experiment. Ordinate shows distance run per 100 sec; abscissa is time.

antagonistic to running are strongly, but transiently, excited. When light and dark periods alternate, only these nonwalking activities have time to occur during each light period. However, since these activities inhibit walking, and since they are not excited by dark (indicated by the immediate onset of walking in the dark), their offset in the dark allows walking to rebound to a level above its level in continuous light.

Although simple in principle, it is obvious that the interactions among time-varying excitatory and inhibitory processes involved in successive induction can, in practice, become quite complicated. Measurement of the quantitative properties of these processes remains one of the major unsolved problems in the experimental study of behavioral integration.

INHIBITION AND REFLEX STRENGTH

Inhibition and reflex strength have already appeared in earlier discussion, but they deserve separate treatment. *Strength* is a useful but hard-to-define term that refers to the various measures associated with the "elicitability" of a reflex: threshold, latency, vigor, fatigability (i.e., susceptibility to habituation), and probability of elicitation, all considered in relation to the adequacy of the

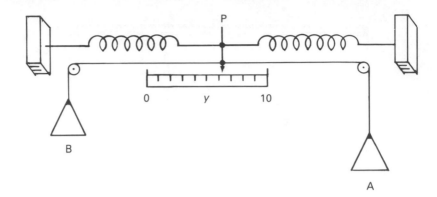

Figure 2.15. Physical model of additive behavioral interaction.

stimulus, a weak response to a poor stimulus having similar significance to a strong response to a good stimulus. For most protective reflexes, under most conditions, these measures are correlated: A short latency goes along with a low threshold and a vigorous response elicited with high probability. The correlation tends to break down for other (e.g., postural, autonomic) reflexes and for learned behaviors. The concept of strength is, nevertheless, indispensable, even though some discretion must be exercised in picking appropriate measures in each particular case. For many recurrent habitual behaviors (e.g., lever pressing, running, pecking, etc.) the rate of occurrence of the activity (i.e., instances per unit time), or the proportion of time spent engaging in it, are useful measures of strength (more on this in later chapters).

Part of the difficulty in defining strength comes from inhibitory effects. For example, an animal may have a relatively strong tendency to engage in an activity such as walking, but the activity may, nevertheless, fail to occur because of *inhibition* by some other activity, as in the locust experiments just described: The high strength of walking was apparent as soon as the inhibiting activity ceased, since the inhibited activity then rebounded (successive induction). Evidently, a suppressed response may have considerable strength even though it is not occurring.

The first direct demonstration of inhibition in the nervous system was Eduard and Ernst Weber's discovery in 1845 that stimulation of the vagus nerve suppresses the heartbeat. This is straightforward enough: An ongoing activity is weakened by the addition of some factor. Nevertheless, the concept of inhibition has a history of muddle and mystification. Perhaps the simplest way to think about it is that inhibition refers to *one of the ways in which causal factors may combine* to produce an effect. In the simplest case (and even quite complex cases can be reduced to this one by appropriate scale changes and other manipulations) causal factors combine *additively*, so that inhibitory factors are just those that add in with negative sign. A concrete illustration is provided by the physical arrangement shown in Figure 2.15. The dependent variable ("level of response": y) is the pointer reading, which can range from 0–10. (Note that there are no negative numbers, since the measured level of a response must be nonnegative.) The causal factors affecting y are simply weights that may be added to pans A and B. If there is no friction and the spring is perfectly elastic, then obviously, y, the pointer displacement, is proportional to the net weight acting against the restoring spring, so that:

$$y = C(a_1 + a_2 + \ldots - b_1 - b_2 - \ldots),$$

or, using summation notation:

$$y = C(\sum_{i=1}^{m} a_i - \sum_{j=1}^{n} b_j), \tag{2.1}$$

where a_i ($i = 1, 2, \ldots, m$) denotes one of m weights added to pan A, b_j ($j = 1, 2, \ldots, n$) denotes one of n weights added to pan B, and C is a constant of proportionality that depends on the strength of the spring. Thus the B-factors are inhibitory with respect to y and the A-factors are excitatory.

Figure 2.15 also illustrates three other characteristics of "real" behaviors. (a) If the B-weights are large, the pointer will be off the scale to the left (the behavior will not be occurring). Yet obviously the number of A-weights necessary to bring the pointer on-scale will not be constant but will depend on the number of B-weights (this is sometimes called *inhibition below zero;* if there are always a certain number of B-weights, the resulting minimum A-weight necessary to bring the pointer on-scale corresponds to an *absolute threshold*). Thus, the fact that a behavior is not occurring says nothing about what its "real" strength may be. (b) Similarly, if there are many weights in pan A, the pointer may be off the scale to the right (we can consider this to be a pointer reading of 10). This corresponds to the physiological factors that limit the actual level (force, rate of occurrence) of any behavior (this is often called a *ceiling effect*). (c) Real balances are subject to friction, which means that small changes in the level of the A- or B-weights may have no effect. Friction is analogous to a *differential threshold*, and can produce *hysteresis*, that is, a dependence of the pointer reading not just on the levels of the A- and B-weights, but also on the way in which those levels were arrived at.[13] These factors may limit the proportionality between causal factors and the observed level of a behavior.

These three factors, ceiling effects and differential and absolute thresholds, mean that a given response measure may not be *linearly related* to response strength. There are two main types of nonlinearity: One tends to limit the effects of causal factors, the other to exaggerate them. The model in Figure 2.15 illustrates the limiting type, in which extreme causal-factor values are limited by a ceiling. This effect is usually gradual, rather than abrupt: After the absolute threshold is exceeded, fixed increments in causal factors produce smaller and smaller increments in the level of behavior (this is called a *negatively accelerated* function. The solid line in Figure 2.16 shows an example). The effects on

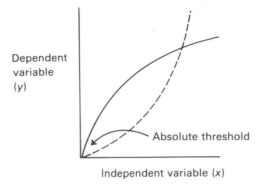

Figure 2.16. Monotonic functions: negatively accelerated (solid line) and positively accelerated (dashed line). Equations often used to fit the negatively accelerated function are $y = A[1 - \exp(-\lambda x)]$ and $y = Ax^m$, $m < 1$, and for the positively accelerated function, $y = Ax^m$, $m > 1$.

behavior of external factors, such as stimuli and reinforcement, generally follow negatively accelerated functions; in this way the unlimited range of physical stimuli is funneled into the limited range of physiological response.

The inhibition between competing reflexes usually follows the exaggerating, *positively accelerated*, kind of nonlinearity (the dashed line in Figure 2.16), however, because this ensures that either one or other, but not both, can occur. Elimination of the spring from Figure 2.15 leaves a model of this kind of nonlinearity in its most extreme form. Without the spring, the pointer has only two stable positions, 0 or 10, depending which pan is heavier.

SUMMARY

Reflexes and kineses are both limited in a critical way: They depend only on events in the present or the immediate past. Reflexes are limited in an additional way, in that they are *ballistic*, that is, they occur in response to an eliciting stimulus and are not readily modified by their consequences: they are almost independent of *feedback*.

This difference between reflexes and orienting mechanisms simply reflects their different functions. Kinetic orientation mechanisms have a large random component because simple organisms are in no position to predict important features of their environment. Consequently, a large pool of behavioral variation is necessary if the right variant (location) is to be found. Conversely, reflexes have evolved to deal with highly predictable contingencies: Limb withdrawal is almost certainly the best response to painful stimulation of an extremity; limb flexion is certainly best accompanied by relaxation of the extensor, and so on.

Later chapters deal with behavior that shows progressively more and more dependence on past history. The next chapter takes up the general topic of feedback and its role in orientation.

NOTES

1. Of course the distinction between racial vs. individual history as determinants of behavior is arbitrary, as all such distinctions between nature and nurture must be. For example, web building in general may be innate, but the construction of any particular web clearly is not. The spider searches for and selects a particular place for its web, adapts the web design to the local topography, and can cope, to some degree, with damage to the web. These aspects of its behavior depend on its individual history, and are obviously not encoded directly in its nervous system. What is innate is the developmental specification for a particular *program* for action that can accept as inputs the relevant topographic details.

This same argument applies to learning: The *capacity* to learn is innate, but the particular things learned are obviously not. The attempt to understand how a particular animal learns can be thought of as a search for whatever it is that is inherited that permits, and limits, learning – the analog to the web-building program.

2. *Neurophysiology and behavior.* There are two contrasting philosophies about the relations between physiology (especially neurophysiology) and the study of behavior. At one extreme is an implicit reductionism that devalues any purely behavioral explanation. This view is not actively proposed by any particular writer, but tends to be accepted in an unquestioning way by many neuroscientists, especially those concerned with anatomical questions or with the identification of the biochemical substrates of behavior. The reductionist position comes in two varieties. The more tolerant concedes that explanations at the purely behavioral level are possible (although not very interesting), but argues that, in practice, the route to their discovery is through study of the internal structure and function of the organism (i.e., through neuroanatomy and physiology). Extreme reductionism holds that a behavioral explanation is no explanation at all, and that the only valid account is in terms of physiological substrates. A common fallacy here is the assumption that once some phenomenon, such as ring-dove courtship or human depression, is traced to the action of a hormone (say) it has been completely explained. Discovery of such a link may be of great practical value, of course: The link between diabetes and insulin has provided relief to millions of sufferers from this condition. But just as we still know rather little about the causes of the insulin deficiency in diabetes, so the simple finding of a link between a chemical and a behavior neither explains what causes the chemical to vary in level, nor explains why this chemical, rather than some other, produces this particular behavior. The only satisfying explanation is one that relates the behavior to others and to the environment – in other words, an explanation of how the system as a whole functions.

Students of the integrative functions of behavior, whether trained as physiologists, psychologists, or zoologists, rarely take a reductionist position. The reason is not so much a philosophical as a practical one: Even in organisms as simple as arthropods, the neural machinery is too complicated to be studied in isolation from behavior. Without some understanding of behavioral organization it is hard even to ask sensible questions of the neural material. For example, Kennedy (1967) writes:

Neuropile that looks hopelessly complicated when peeped at through the tip of a microelectrode nevertheless produces a limited number of orderly behavioural actions in orderly sequence. So the functional relations of their neural mechanisms (localizable or not) must themselves be rather simple and discoverable – provided methods appropriate to this level are used. The problem at any level of analysis is to see how the units work together to produce the emergent or resultant properties they lack singly. To discover this the units must actually be working together when studied and for the highest levels, behaviour, that means using the whole animal. This is not the baffling impasse it looks like to the neurophysiologist itching to get in there with his powerful but inappropriate tools. (p. 261)

In psychology the view that behavioral explanations are necessary precursors to physiological ones has been most forcefully argued by B. F. Skinner. In his 1938 book *The Behavior of Organisms*, Skinner wrote that the letters CNS (central nervous system) as used by most neurophysiologists could as well stand for "conceptual nervous system," since many supposedly physiological concepts (most notably Sherrington's reflex) derive not from direct manipulations of the nervous system, but from behavioral studies. Unfortunately, Skinner went beyond a justification of purely behavioral explanations to argue against the value of physiological ones. He felt (unjustly, I believe) that there is something inherently improper in using physiological terminology for concepts derived purely from the analysis of behavior. Although this practice can certainly lead to error, there are, nevertheless, many successful precedents in which inferences from behavioral studies have subsequently been confirmed physiologically: for example, in studies of the mechanisms of hearing, in the identification of visual pigments, and in the properties of the synapse itself.

3. See Bell (1959) for an excellent summary of subsequent work on plants deriving from Darwin's researches.

4. The terminology in this area is confusing. Fraenkel and Gunn (1940), in an influential book based in part on the earlier, German, work of Kuhn, clearly separated direct and indirect orientation by the terms *taxis* and *kinesis*, but other workers have not been as particular. *Taxis* is often used to mean any kind of simple orientation: Koshland's *chemotaxis* is actually a kinesis in the Fraenkel and Gunn scheme, for example, as are some of the *tropisms*, described by Loeb (e.g., 1918).

5. We will encounter the exponential function frequently. The curve in Figure 2.3 approximates the function

$$y = K \exp(-at), \qquad (N2.1)$$

where exp( ) denotes the exponential function raised to the power in ( ), K is a scale constant, a is the reciprocal of the *time constant* (the larger the time constant, the slower the decay of y with respect to t), y is the dependent variable (percent bacteria tumbling in Figure 2.3) and t is the independent variable, here *time*. The exponential function decays with negative acceleration, so that the shallow, positively accelerated portion between the first two points of the curve in Figure 2.3 is a departure from the exponential form.

6. Koshland's chemical model and the experimental work on which it is based are described in more detail in his book *Bacterial Chemotaxis as a Model Behavioral System* (1980).

7. For a more extensive history of the reflex concept than that given here see Boring (1942), and Brazier (1959). Useful summaries are also given in Swazey (1969), Granit (1966), and in an excellent integrative book by Gallistel (1980).

8. *The mind–body problem.* The set of philosophical problems raised by the commonsense notion that mental phenomena such as volitions, thoughts, attitudes, and the like cause behavior is known collectively as the mind–body problem. Like many old philosophical riddles, the mind–body problem has acquired an air of permanence and insolubility that tends to produce a defeatist attitude – if so many great thinkers of the past have failed to solve this deep problem, then how can we mere worms expect to do so? This attitude is disturbing to thoughtful students, but is often a comfort to their instructors, who can take satisfaction in demonstrating the impotence of youth in this matter if not others.

The problem has two main aspects. One is the practical one of measuring "mental" phenomena and demonstrating their causal efficacy. The difficulty here is simply the logical one that correlation does not imply causation: Even if a given mental event is invariably associated with some action, we cannot thereby prove the one causes the other, simply because we cannot directly manipulate the mental event. This is not a major difficulty, since invariable association under a variety of circumstances would be interesting enough no matter what the logical deficiencies. The second aspect is more serious and less soluble and concerns the role of mental events as extraphysical causes of behavior. This idea can only be disproved by exclusion: When we can explain every conceivable behavior of animal or human in purely physical terms, then there will obviously be nothing left for nonphysical causes to account for. Clearly the nonphysical-cause idea will be safe for a while.

Many researchers have followed Sherrington in a Cartesian dualism (Descartes proposed that everything in human behavior is automatic, except a residue determined by the soul which, as every schoolboy knows, is located in the pineal gland) that nevertheless allows them to do their own work. It may be epitomized as follows: "Well, clearly there *are* mental effects not totally traceable to physical causes, but the part of the nervous system (or type of behavior, or species of animal, or...) that I happen to work on is, in fact, perfectly deterministic and can be understood by the usual scientific methods."

Some take this position for religious reasons, others because the application of determinism to our own behavior seems to pose problems. The difficulty arises, in part, because science must assume determinism. If this determinism applies to the lower animals, considered as subjects for psychology, then evolutionary continuity requires that it also apply to man, the experimenter. But this seems to pose a problem, since the capacity to discover the truth seems to presuppose the *capacity* for error; that is, for "spontaneity," "free will," "choice," or something of the sort. It makes no sense (for many people) to ask if a conclusion is true if it can also be shown to follow from a completely deterministic process.

Yet there is nothing essentially implausible about the idea. For example, quite early in the history of artificial intelligence, computers were programmed to discover and prove simple theorems in mathematical logic. It was necessary for this purpose only that the machine be programmed to generate logic statements (i.e., that it have some mechanism of behavioral variation) and then apply to them a set of logical tests (i.e., rules of selection), rejecting statements that failed (false propositions) and retaining those that passed (true propositions). To be sure there is often in such programs a random element in the generation process, as there is in the behavior of organisms; yet the process as a whole is perfectly comprehensible from a deterministic point of view. Consequently, there is nothing paradoxical in the notion that animal, and human, behavior (including consciousness and "mental life") is determined by the physical activity of the nervous system and follows comprehensible, deterministic laws. The human scientist may differ from an "intelligent" computer only quantitatively, in the range of different hypotheses he can entertain, and in the subtlety of the tests for truth that he can apply.

This view precludes the possibility of "absolute truth" (although it doesn't thereby justify the kind of feeble relativism that argues for the abolition of the terms "black" and "white" just because some grays are hard to distinguish), since truth is always relative to the set of alternatives that have been considered, and the criteria used for selecting from them. As the human species evolves and our cultural heritage develops, better criteria and more powerful hypotheses may appear, so that deeper truths become possible. But this kind of relativism is widely accepted by scientific workers of all philosophical persuasions.

For views of the mind–body problem different from the one I have sketched see Sperry (1969) and Popper and Eccles (1977). For discussions of evolutionary epistemology see Campbell (1975), Popper (1974), and Toulmin (1967). For a view of the mind–body relation similar to the one assumed here see Griffin (1976).

9. It is important not to underestimate the difficulties for causal analysis posed by time delays. For example, in a few primitive cultures, the relation between conception and pregnancy was not known – presumably because of the long time delay involved. Recent medical research suggests that some degenerative diseases characteristic of adulthood are caused by dormant, "slow" viruses that infect in childhood, many years before symptoms appear. It may be that many causal agents of this sort remain to be found because of the difficulty of identification caused by long time delays.

Even quite short time delays pose problems. For example, in an unpublished experiment in my laboratory, subjects were confronted with several push buttons and told that "by pushing the buttons you can make the counter advance" (counts signaled the amount of money reward the subject was to receive at the end of the experiment). Most subjects developed elaborate sequences of button pushes and verbalized correspondingly complicated hypotheses about the reinforcement rule in effect. The actual rule was a schedule in which pressing any button produced a count, providing presses were separated by more than 10 sec. Most subjects failed at first to detect the temporal contingency, coming up instead with "superstitions" about the efficacy of sequences of button pushes that took up the necessary amount of time. I return to this issue later in connection with the problem of reinforcement mechanisms (Chapters 13 and 14).

10. A model that represents latency as the sum of two random processes is

$$P(R > t) = A \exp(-\lambda_1 t) + B \exp(-\lambda_2 t), \tag{N2.2}$$

where A and B are constants representing the relative contribution of delay and reflex strength to $P(R > t)$, probability of response at poststimulus times greater than t. λ_1 and λ_2 are the time constants of the delay and strength processes, respectively. More commonly, perhaps, the delay has a fixed minimum and so must be modeled by something a bit more complicated than an exponential distribution; the first term in the sum might then be a normal or a Poisson distribution. If the delay represents a minimum processing time, then variables assumed to affect reflex strength affect only the strength parameter, λ_2.

11. There is little consistency in the terms used for the waning of responses under repeated stimulation. I use *habituation* (rather than *fatigue*, which was Sherrington's term) for stimulus-specific waning that is not attributable to changes in the effector – for which the term (muscle) *fatigue* is preferable – or the receptor, for which the term *adaptation* is preferable. Habituation is thus reserved for a process that by demonstration or inference is central (see Thompson & Spencer, 1966).

12. This limitation is sometimes termed the "Ford effect," after the derisive (and doubtless unjustified) comment of Lyndon Johnson about another president that "He can't walk and chew gum at the same time."

13. For example, suppose that the B-weights are fixed, and the A-weights are gradually increased by small increments until the pointer just moves to a new position, where the values of the A- and B-weights are recorded. Let some more A-weights be added to move the pointer a bit further. Suppose now that B-weights be added until the pointer returns to the original position again. The ratio of the A- and B-weights will be different on this second occasion than they were at first, because of the effects of friction. Hysteresis is the name given to this effect where the value of a dependent variable (pointer position) "lags behind" the value of the independent variable (the ratio of A- and B-weights) as it is progressively changed. Hysteresis is a common effect in threshold experiments.

3

DIRECT ORIENTATION AND FEEDBACK

Kinetic (indirect) orientation works by means of successive comparisons. Animals orient in this way either when the gradient to be detected is too shallow to permit simultaneous comparison, or when the source of stimulation cannot be sensed at a distance because the direction of local gradients is not perfectly related to the direction of the source. Visual and auditory stimuli allow for a more efficient strategy: The steepest local gradient always points to the source, and gradients are usually quite steep. Thus, simultaneous comparisons can provide immediate feedback about the proper direction of movement, allowing direct orientation with no wasted motion.

In this chapter I discuss simple mechanisms of direct orientation, first in a descriptive way, and then in terms of feedback mechanisms. The discussion of feedback serves several purposes: It shows how feedback is involved in orientation, and how a mathematical model can account for experimental results. It shows how apparently different orientation reactions can be considered as variants of the same process. The discussion of feedback models allows me to explain the difference between *static* and *dynamic* behavior theories. The remainder of the chapter discusses the type of explanation given by feedback accounts and shows how the concept of feedback control provides a mechanistic explanation for motive and purpose. The idea of feedback is essential to an understanding of motivation (Chapter 6) and reinforcement schedules (Chapters 5 and 7).

TAXES

Fraenkel and Gunn classify direct orientation (taxic) reactions into four main types. These do not, in fact, correspond to completely different mechanisms, but the mechanisms will be easier to understand after I describe the different types and the experiments used to tell one from the other.

Klinotaxis

The first taxic reaction, *klinotaxis*, is really an intermediate case, since it involves both direct orientation and successive comparisons. Figure 3.1 shows an

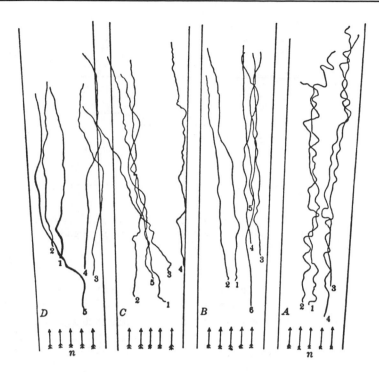

Figure 3.1. Tracks of four maggots (A to D) in a beam of horizontal light (arrows). Each maggot was tested 4–6 times. (From Mast, 1911.)

example of the relevant behavior: Mast tested each of four maggots (*Lucilia sericata*) four to six times for its orientation to a beam of horizontal light. Movement was always directly away from the light, although the tracks are not perfectly smooth. Substantial head movements (wiggles in the record) are apparent particularly in the track of maggot A. These head movements provide a clue to the underlying process. Fraenkel and Gunn write: "During steady forward crawling the head is sometimes put down symmetrically in line with the axis of the body, but from time to time it comes down alternately to the right and to the left. When the maggot first begins to crawl, these lateral deviations of the head are usually considerable, and may even result in the body assuming a U-shape" (1961, p. 60). The animals look as if they are searching in some way for the direction of the light. Once they find it, they move in the opposite direction.

The animals find the direction of the light by successive comparison. This can be shown by tricking the animal: turning on an overhead light when the animal is turning in one direction and turning it off when it faces the other direction. Under these conditions a negatively phototactic animal moves in the direction associated with the low light level.

This process is similar to so-called trial-and-error learning, but is much sim-

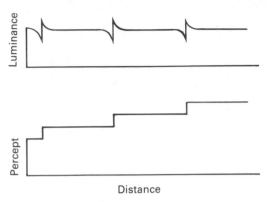

Figure 3.2. Top: gradient of luminance across a striped array. Bottom: perceptual effect of this luminance distribution – a staircase pattern of progressively increasing brightness.

pler than the learning of higher organisms. No long-term memory is involved; the maggot does not remember where the light was or anything of the sort. The maggot has fewer modes of reaction than an animal such as a rat, and can, therefore, be led into maladaptive behavior by relatively simple means. For example, the animals may move away from a directional source even if it means moving into a region of higher, but nondirectional (e.g., overhead) illumination. They can also be induced to move up (rather than down) a gradient of illumination by turning on a second light only when the animal is facing down the gradient. In the absence of a long-term memory, the animal has no way of discovering that its situation while improving on a moment-by-moment basis is getting worse overall.[1]

Lest we begin to feel arrogant, I should point out that very similar tricks can be played on the human visual system. Figure 3.2 shows at the top a sawtooth gradient of illumination: from left to right, the illumination repeatedly grows slowly dimmer, and then rapidly brighter. The bottom part of the figure shows what this pattern looks like to the eye. Because of inhibitory, rate-sensitive mechanisms, the rapid increases in illumination have a greater perceptual effect than the slow decreases. Consequently what is seen is a staircase pattern of progressively increasing brightness.

These effects arise because sensory systems, and the orientation mechanisms that depend on them, are usually more sensitive to the *rate of change* of the relevant stimulus than to its absolute value: The important variables are the *higher derivatives* of the stimulus value, with respect to space and time, rather than the stimulus value itself.[2] *Sensory adaptation* is one of many labels for this sensitivity to rate of change, since it refers to the waning effect of a stimulus with time following its onset. This mechanism is enormously useful to animals, and represents in most cases a great improvement over dependence on absolute stimulus values. As we have seen, however, in the absence of other, longer-term memory processes it can sometimes lead to maladaptive behavior.

The details of the mechanism of klinotaxis, in *Lucilia* or any of the other species that show the pattern (e.g., the protozoan *Euglena*, earthworms, the ascidian *Amaroucium*) have not been fully worked out. It is likely that the behavior of head swinging (which also guides the animal) occurs only when the rate of change of illumination of the receptor(s) exceeds some threshold value. In a stable, uniform light field, the animal moves little and swings its head from side to side in a relatively narrow arc; but if the light falling on the receptors increases greatly (which might be because the animal has entered a lighter region, or because the ambient illumination has changed) a large head excursion occurs, followed by a smaller return swing. If any of these subsequent swings again produces an increase in receptor illumination, further swings are produced. As time elapses, the effect of increases in receptor illumination diminishes because of adaptation, so that successive head swings are likely to be of smaller and smaller amplitude. If the source of illumination is fixed, the animal will, in the meantime, have become oriented away from it, so that tracks similar to those in Figure 3.1 result. If the source of illumination is moving, then obviously the animal's ability to orient away from it will depend critically on quantitative details involving its rate of movement, of head swinging, and of sensory adaptation. I return to quantitative questions in the discussion of feedback later in the chapter.

The remaining taxic orientation reactions, tropo- and telotaxis and the light-compass reaction, all involve simultaneous comparison of the stimulation of two or more bilaterally symmetrical receptors. One clue to the type of reaction is provided by the track of an animal orienting in a uniform gradient: The track is convoluted if the mechanism is a kinesis, wavy for klinotaxis, but straight for the taxes that use simultaneous comparison. Tropotaxis and the light-compass reaction differ chiefly in their quantitative properties.

Tropotaxis

The two-light experiment, illustrated in Figure 3.3, is critical to the identification of tropotaxis. The figure shows the tracks of several pill-bugs (*Armadillium* sp.), a terrestrial crustacean that lives under rocks and decaying wood, placed some distance from a pair of equal lights. *Armadillium* behaves in a positive phototactic fashion in this experiment: It approaches the two lights. But when placed equidistant between the two lights the animal will often follow a path in between them, rather than going straight to one or other of the two.

Tropotaxis is the outcome of a balancing process: The animal turns until the two eyes are stimulated equally and then proceeds forward. As the animal approaches a point directly in between the two lights, it will go directly to one or other of them, as the tracks show. At that point, a head swing that faces the animal directly toward one of the lights will also eliminate the influence of the other one.

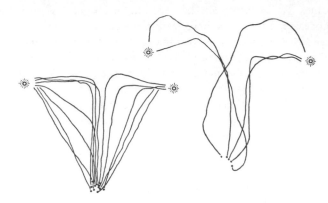

Figure 3.3. Tracks of photo-positive *Armadillium* toward two equal lights. (After Müller, 1925.)

A second means of identifying tropotaxis is to eliminate the stimulation on one side of the animal, either by blinding it in one eye or painting over the eye if this is possible. The result is *circus movements*: In a uniform light field, such as an overhead light, elimination of one eye means that stimulation appears to come from only one side. If the animal is positively phototactic it then turns continuously toward the illuminated side; if it is negatively phototactic, it turns continuously toward the blinded side. In the normal animal with an eye-level light source, these reactions would lead eventually to equal stimulation of both eyes and leave the animal facing either away from, or toward, the light source. Figure 3.4 shows an example of circus movements produced in unilaterally blinded *Ephestia* larvae.

Light-compass reaction

The third taxis in Fraenkel and Gunn's basic list is the *light-compass reaction*. This pattern is very common in animal orientation. For example, bees use it in returning to the hive, and ants, although often able to rely on odor trails, can also orient with respect to the sun, finding their way back to the nest by keeping a fixed angle between their path and the direction of the sun. Moreover, many species are able to compensate for the movement of the sun by a change in this light-compass angle, so that a fixed direction is maintained (this is sometimes termed *sun-compass orientation*, and it appears to be one of the main components in the feats of navigation performed by migrating birds). This compensation depends on an "internal clock," tied to the 24-hour cycle (a *circadian rhythm*). I show in a moment that light-compass orientation can be derived from the same feedback mechanism as tropotaxis.

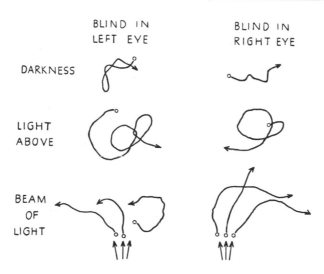

Figure 3.4. Tracks of photo-negative *Ephestia* larvae, blinded on one side, under three illumination conditions. (After Brandt, 1934.)

Telotaxis

In tropotaxis, animals usually orient between two equal, symmetrically disposed lights, and only directly approach one of them when the angle subtended by the lights is large (see Figure 3.3). In telotaxis, the animals generally head straight for one of the two lights (although they may switch from one to the other as they approach) even if the lights are quite far away. Fraenkel and Gunn give a graphic, natural-history description of *telotaxis* as follows:

Large numbers of [the little mysid crustacean, *Hemimysis lamornei*] are to be found in the aquarium tanks of the Marine Biological Stations at Plymouth and Naples. When a single light is placed at the side of a glass tank containing *Hemimysis*, the animals swim to and fro continually, always keeping in line with the beam of the light. They swim about 10 cm towards the lamp, then turn sharply through 180° and cover about 10 cm again before turning back towards the lamp, and so on. If an additional light is arranged so that the two beams cross at right angles, some of the mysids are quite unaffected in their behaviour while others switch over to this second light and behave as if the first one were non-existent. The result is that the mysids form two streams which, so to speak, flow through one another, crossing at right angles and not interfering with one another. (1961, p. 90)

Examples of telotactic tracks of hermit crabs and an isopod (*Aega*) in a two-light experiment are shown in Figure 3.5.

The mechanisms involved in telotaxis depend on the type of receptor the animal possesses. If the eye is capable of forming a rudimentary image, that is, of immediately identifying the bearing of a source of stimulation, then the animal

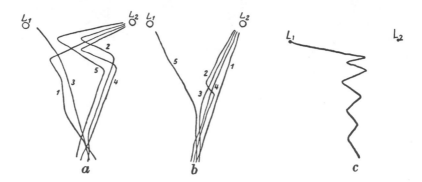

Figure 3.5. Tracks of hermit crabs (a) and (b) and an isopod (c) in a two-light experiment. Each part of the track is directed toward one light only. (a and b after Buddenbrock, 1922; c, Fraenkel, 1931.)

has the necessary information available on the receptor surface. The animal can orient correctly by placing its walking apparatus under the control of one image exclusively. This is not a trivial problem, since the image-identification system must compensate for changes in retinal position caused by the animal's own movement. Both the lensed eyes of vertebrates and the compound eyes of insects and other arthropods provide the necessary directional information.

Many animals with image-forming eyes prefer to orient to a light using the balance mechanism of tropotaxis – shown both by orientation between two lights, and by circus movements when unilaterally blinded – even though the same animal may respond telotactically under other conditions. For example, at first a unilaterally blinded bee will show circus movements, but after a while they cease and the animal will head straight toward a light source – a telotactic response. The preference for tropotaxic orientation may be a reflection of the complex computations required for telotaxis.

Animals show many other simple orientation reactions. For example, so-called skototaxis (approach to dark areas or objects); a variety of reactions to temperature gradients (usually favoring a particular zone of temperature, either because it is appropriate for the organism's metabolism, or because it is a signal of the animal's prey, as when ticks prefer the skin temperature of their host); postural reactions to light and gravity – geotaxis and the dorsal light reaction; reactions to physical contact – many small and vulnerable animals seek out and conform to crevices and corners; reactions to fluid flow (rheotaxis) – fish generally face upstream, flying insects orient into the wind, some tidal crustaceans orient using cues from wave currents; reactions to chemical and humidity gradients such as those discussed in Chapter 2; orientation in sound fields (sound localization). Similar principles – simultaneous or successive comparison, feedback, time-dependent effects – are involved in all these reactions. Despite the relative

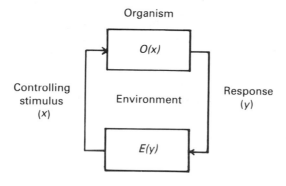

Figure 3.6. Organism–environment relations in adaptive behavior.

simplicity of each reaction analyzed in isolation, in combination they can lead to quite complicated, "intelligent" behavior. More on this later.

Feedback analysis

Tropotaxis depends upon a balance between stimulation of receptors on opposite sides of the body. The animal turns until stimulation of both receptors is equal and then approaches (positive phototaxis) or moves away from (negative phototaxis) the source of illumination. The process can be represented as a simple feedback mechanism, as shown in Figure 3.6. The relations between organism and environment are here represented as a *feedback loop*, which shows the mutual relations between some aspect of the environment and some aspect of behavior that is linked to it.

The picture in Figure 3.6 applies to many things of interest to psychologists and ethologists. It will be helpful to have standard terms for its elements. Variable *x*, the aspect of the environment to which the system is sensitive (the *input*) corresponds to different things in different experimental situations. Here it is simply the receptor disparity (measured in illuminance units) associated with a given orientation of the animal in relation to the light source. The *output* or *response*, *y*, is the aspect of the organism's behavior that affects, and is affected by, feedback; here it is just the angle of the animal in relation to "north" (i.e., a standard direction), or in relation to the source, if the source is fixed. The two boxes labeled "organism" and "environment" in Figure 3.6 contain two functions, $O(x)$ and $E(y)$, that describe how the input (*x*: receptor disparity) is related to the output (*y*: direction) by the animal (function $O(x)$), and how the output is related to the input by the environment (function $E(y)$). In most experiments $E(y)$, called a *feedback function*, is known, and the objective is to discover the *control function*, $O(x)$, imposed by the organism.

In tropotaxis, the feedback function, $E(y)$, depends on the shape of the bilateral receptors, their sensitivity over their surface, and the geometry of the situation. In the simplest case, the receptors can be considered as points or spheres, and the stimulation falling on each receptor will, therefore, be determined solely by its distance from a source of illumination, according to the inverse square law. If the distance from the source is great, it is easy to show that the difference in stimulation of two bilateral receptors, y, is directly proportional to their separation, and to the sine of the angle between the midline of the animal and a line drawn from the source to the center of the animal (i.e., the animal's *heading* with respect to the source).[3]

Consider now how the control function, $O(x)$, might work from moment to moment. The animal is presented with a certain disparity of stimulation, x, which will, in turn, be a function of its heading with respect to the fixed light source (call that angle θ). Ignoring adaptation for the moment, the disparity must have an effect on the animal's tendency to turn. Perhaps the simplest possibility is just that the animal's *rate of turning* (measured in angular units per second) is proportional to the receptor disparity, with a sign such that the turns are toward the more illuminated side (for positive phototaxis). This assumption can be expressed formally by saying that the *first derivative* of the heading angle, $d\theta/dt$, is proportional to the receptor disparity. (This is termed *integral control*, because the controlled variable, θ, is determined by the time integral of the controlling variable, x.) We thus arrive at a simple formal model of this situation:

$$x = A \sin \theta, \tag{3.1}$$

giving the receptor disparity as a function of heading angle (the feedback function), and

$$y = d\theta/dt = -Bx, \tag{3.2}$$

giving the relation between rate of turning and disparity (the control function), where A and B are constants (A is proportional to receptor separation, and incorporates a scale factor; B represents the *gain* of the system: how fast the animal turns for how much disparity).

Two aspects of this little model are of interest. One is the *steady-state* solution: Assuming the light source is fixed, what will the animal's final heading be? A second is the dynamics of the situation: How does the animal's heading change with time?

The static problem is simple. When things have settled down, we know that the animal will have ceased turning; that is, $d\theta/dt$ will equal zero, which means that $x = 0$ (from Equation 3.2), which means that $\theta = 0$ (from Equation 3.1). Hence the animal will eventually point toward the light source. The dynamic solution is a little more difficult to arrive at, and requires that we know the *initial conditions* (i.e., the animal's initial heading) and, if the source is moving, the way in which it is moving as a function of time. I return to the dynamic problem in a moment.

This model illustrates the distinction between closed- and open-loop control. The stability of the system is ensured under normal conditions by the immediate effect of the response, turning, on the feedback input, receptor disparity. But suppose receptor disparity were independent of turning (this is known as "opening the loop")? This could be done by blinding the animal unilaterally and leaving it in uniform light. Under these conditions, $x = C$ (constant). You can see at once from Equation 3.2 that $d\theta/dt$ must then equal BC, which is also a constant. Thus, the animal must turn constantly. This is, of course, exactly what is observed, in the form of the "circus movements" already described. In a slightly more elaborate experiment, fixed amounts of light might be presented to each eye, allowing $d\theta/dt$ (turning rate) to be measured as a function of C (receptor disparity). The slope of the resulting line then gives the value of parameter B, the gain of the system. Opening the loop is obviously a useful technique for getting at the properties of feedback mechanisms.

The virtue for the animal of a negative-feedback mechanism is that it minimizes the effect of changes in the control function, $O(x)$. For example, suppose that because of age or injury the animal is no longer able to turn as rapidly as before. In terms of our model this change might be represented by a decrease in the value of parameter B in Equation 3.2. The steady-state solution is unaffected by this change: Even with impaired movement, the animal must still eventually orient toward a fixed source of illumination. This is just what is observed when the locomotor apparatus of a tropotactic animal is surgically interfered with (e.g., by immobilizing or amputating legs): The response may be slowed, but the animal continues to orient appropriately. Surprisingly, perhaps, not all directed movement involves feedback.

The light-compass reaction seems more complicated than the mechanisms discussed so far, but it need not be. Consider, for example, a simple modification of Equation 3.2, our illustrative model for tropotaxis: Instead of $d\theta/dt = -Bx$, where x represents receptor disparity and θ the heading angle, suppose we add a constant, so that

$$d\theta/dt = -Bx - C \qquad (3.3)$$

When $d\theta/dt = 0$, $x = -C/B$, rather than 0, the previous steady-state solution. Consequently, the resting value of θ, $\hat{\theta}$, from Equation 3.1, is $\hat{\theta} = \sin^{-1}(-C/AB)$, where A, B, and C are constants. Thus, the system represented by Equation 3.3 maintains a constant angle to a light source (light-compass orientation), rather than going directly toward it as in tropotaxis.

The dependence of light-compass angle on time (sun-compass orientation) can be incorporated simply by making C a function of time, so that the model becomes:

$$d\theta/dt = -Bx - C(t), \qquad (3.4)$$

where $C(t)$ is a function with a 24-hour period chosen so that light-compass angle shows the appropriate temporal shift.

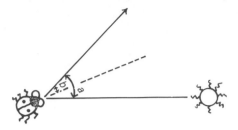

Figure 3.7.

Thus, feedback analysis shows that three apparently different orientation reactions can be considered just as variants on the same integral-control system.

Dynamic analysis

Equations 3.1 and 3.2 describe rate of turning as a function of time; they are a *dynamic model* for behavior. These equations illustrate two properties of dynamic models: First, such models generally express the effect of an independent variable on the *rate of change* in behavior, that is, as a differential or difference equation. In the tropotaxis example, equation 3.2 shows how a given interocular disparity changes the animal's rate of change of direction. But we are generally interested not so much in the change in behavior as in its actual value: We want to know where the animal is heading at a given time, not its rate of turning. The differential equations must be *solved* before we can make predictions about behavior as a function of time.

The second feature of dynamic models is that they have a steady-state solution that can often be deduced without having to solve the differential equations of the model. Much can be learned from the static analysis: Does the animal orient to the light? Does it show circus movements or not? Where does it come to rest? Most of the theories described in this book are static theories. Nevertheless, static theories omit important system properties. They give few clues to the *stability* of the system, for example. An organism that under frequently encountered conditions shows uncontrollable oscillations, such as the tremor of Parkinsonism or the continual turning of circus movements, is at a severe evolutionary disadvantage. Organisms must be stable in response to normal perturbations. Identification of the necessary and sufficient conditions for stability is one of the main objectives of the dynamic analysis of feedback systems.[4]

Solving Equations 3.1 and 3.2 is actually quite simple and can serve as a model for much more complicated analyses. The solution requires some familiarity with elementary calculus; readers lacking that background can skip the math and get on to the conclusions, which are all that is needed for later chapters.

First, it will be convenient to measure the bearing of the light source (angle *a* in Figure 3.7) and the heading of the animal (angle *b*) with respect to a fixed reference direction. Rewriting Equation 3.2 with these new designations yields:

$$d(a-b)/dt = -Bx, \qquad (3.5)$$

and Equation 3.1 becomes

$$x = A \sin(a - b). \qquad (3.6)$$

It will greatly simplify later discussion if we consider only small values of $a - b$ (i.e., headings close to the correct one) for which $\sin(a - b) \doteq a - b$. Equation 3.6 then becomes

$$x = A(a - b). \qquad (3.7)$$

We wish to solve Equations 3.5 and 3.7 so as to obtain the change in the animal's heading angle, b, as a function of time, given that it starts from some initial heading b_0. If the source is stationary ($a = $ constant), this is easily done by substituting for x in Equation 3.5 and integrating: $d(a - b)/dt = -AB(a - b)$. Since a is a constant, this reduces to $-db/dt = -AB(a - b)$, or

$$dt = db/AB(a - b). \qquad (3.8)$$

Integrating the left hand side of Equation 3.8 from 0 to the present time, t, and the right hand side from the initial heading, b_0 to the present heading, b, yields:

$$\int_0^t dt = (1/AB) \int_{b_0}^b db/(a-b).$$

Evaluating the integrals yields:

$$ABt = {}_{b_0}^{b}[- \ln(a-b)],$$

putting in the limits of integration and exponentiating yields:

$$\exp(-ABt) = (a - b)/(a - b_0), \qquad (3.9)$$

which is the desired function of time. Since a and b_0 are both constants, this reduces to an expression of the form:

$$a - b = \theta = \theta_0 \exp(-ABt), \qquad (3.10)$$

where $\theta_0 = a - b_0$, which is the familiar exponential decay function illustrated in Figure 3.8.

Notice that the exponential function in Equation 3.10 has the properties it should: At $t = 0$ the bearing of the animal is θ_0; as $t \to \infty$, θ approaches 0; and the rate of change of θ with respect to time (the slope of the curve) is at first large, when θ is large, but slows down progressively as $\theta \to 0$. The rate at which the angle θ is reduced is determined by the quantity AB; the time that it takes for θ to be reduced to 0.368 ($= 1/e$) of θ_0 is equal to $1/AB$, which is termed the *time constant* of the system: the smaller AB, the more slowly θ is reduced.

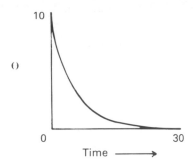

Figure 3.8. Output variable (θ: heading of tropotactic animal) of a simple integral-controlled servosystem following a "step" input (i.e., system starts with a fixed displacement between initial and desired orientation). Curve is a plot of Equation 3.10 in the text with $\theta_o = 10°$ and $AB = 5$.

Frequency analysis. So far I have assumed that the light source is fixed. This may be realistic for light sources, but it certainly isn't true of prey items, for example, for the tracking of which predators possess comparable feedback mechanisms. What if the controlling (feedback) input varies with time?

Before looking at more theory, let's consider the kind of experimental results we want to explain. We will probably be interested in the limits of the system: Can the animal track any moving target, or just slowly moving targets? If so, how slowly must they move? If it fails to track, how does the behavior break down? Does the animal just stop tracking completely (remain motionless); does it track *out of phase* (i.e., look left when the light is on the right and vice versa); or does it begin to oscillate wildly from side to side? Does the tracking break down gradually or all at once?

To answer these questions we need to present the animal with a target that moves from side to side at different frequencies and in different time patterns. For each input time pattern ($x(t)$ in Figure 3.6) we will obtain a comparable output pattern of angular variation ($y(t)$). A useful explanation should summarize the results of all possible such experiments – give us $y(t)$ for any $x(t)$.

The theoretical problem can be handled in two main ways. One is simply an extension of the analysis already described: The constant term, a (the bearing of the light source), can be replaced by a time function, $a(t)$, and the resulting differential equation solved directly for particular forms of $a(t)$. This is tedious and may be difficult in particular cases. The second way is simpler, providing the system is *linear*, that is, describable by a linear differential equation (I explain the implications of this in a moment).

For those with an interest in the mathematical basis for the ensuing arguments, I present a summary of linear-systems analysis in the Notes to this chapter.[5] The basic notions can be explained without mathematics in the following way. Problem: To describe the input–output behavior of a system to all possible time-varying inputs. How can this be done without having to list every possible input and its output? Suppose that the following three conditions are met:

1. The response of the system to a simple periodic input can be described by two quantities: the *phase* relation between input and output, that is, how much does the output lag behind the input; and the relative amplitude of the output and the input (system *gain*), that is, is the output smaller (or larger) than the input, and by how much?

2. Any periodic waveform (or any aperiodic waveform of finite duration) can be built up by adding together simple periodic waveforms.

3. When two or more simple waveforms are added together and presented as input, the resulting output is what would be expected if the two had been presented separately and their separate outputs added together; that is, the effect of a simple waveform on the output is the same whether it is presented by itself, or summed with others. This property is known as *superposition*; together with the first property it means that the system is *linear*.

I have not defined what I mean by "simple periodic input," but for the purposes of this argument it is any kind of cyclic variation that passes through the system without distortion – the shape of the output is exactly the same as the shape of the input. Thus, phase and amplitude are the only two properties of the output that can be different from the input.

The point of these three conditions is that when they hold, we can summarize the properties of the system by two functions of *frequency*: a function showing how the ratio of input amplitude to output amplitude varies with frequency (this is termed a *gain plot*); and a function showing how phase lag depends on frequency (this is termed a *phase plot*). The two curves together are termed a *Bode plot*; I show an example in a moment. If we know the Bode plot, and the component waveforms in the input, we can predict the output.

It turns out that many real systems obey these three conditions well enough to be useful. The first condition is satisfied by the sine wave (the pattern traced out on a moving paper by a swinging pendulum): a sine wave passes through a linear system without distortion. Examples of sine waves, changed in phase and amplitude, are shown in Figure 3.9. The second condition was shown to be analytically true by the French physicist Fourier: Any periodic waveform can be built up out of sine waves of frequencies N, $2N$, $3N$, etc., where N is the *fundamental* frequency and the others are *harmonics*. Figure 3.10 shows how a complex waveform can be built up out of (or analyzed into) two sine-wave components. The third, superposition, condition is satisfied by many natural systems, providing the input amplitude is relatively small, since the independence of input components depends on their sum not being too large. If the input amplitude is too large, real systems often cannot follow, so that the output signal is *clipped*. This is another example of a ceiling effect (see Chapter 2).

Figure 3.11 shows an example of a Bode plot. The upper panel shows gain in decibels (dB), a logarithmic unit, so that equal *differences* represent equal *ratios* of output to input. The lower panel shows the phase lag, in degrees (360° repre-

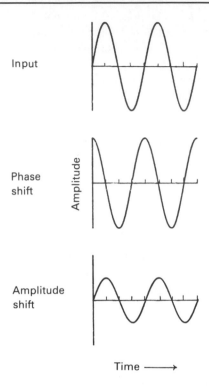

Figure 3.9 A sinusoidal input (top panel) shifted in phase only or amplitude only (bottom panel). In real control systems amplitude change is almost always accompanied by a phase shift.

sents one cycle at the fundamental frequency), as a function of frequency, also in logarithmic units. The behavior shown here is typical of many real systems: At low frequencies the output faithfully matches the input, gain is close to unity (zero on the log scale) and phase lag is small. As frequency increases, gain decreases and phase lag increases (indeed, in so-called minimum-phase systems, the two are interrelated).

The phase-amplitude-frequency information in a Bode plot is summarized mathematically by the *system transfer function*, which is described in more detail in note 5. In most respects, the transfer function can be treated as a simple multiplier. Its usefulness can be illustrated in the following way. The first step is to modify the generalized feedback diagram seen earlier in Figure 3.6 to incorporate the notion of a set point.

The modified diagram is shown in Figure 3.12. It has three elements: a transfer (control) function for the organism, G; a transfer (feedback) function for the environment, E; and a comparator, which subtracts the feedback input, x, from a *set point*, x_0. The difference between x and x_0 is the input to G. The comparator and G are both inside the organism, as indicated by the dashed line;

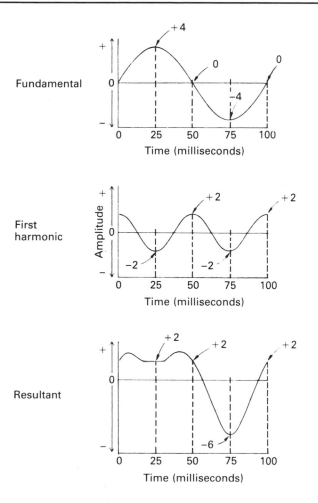

Figure 3.10. Graphical illustration of Fourier analysis of a period signal (bottom panel) decomposable into two sinusoidal components in harmonic relation to one another (top two panels: fundamental at top, first harmonic in the middle). At any point on the time axis, the amplitude of the compound wave is equal to the sum of the amplitude values of the two component waves.

x_0 may be a purely internal reference, as in homeostatic mechanisms such as temperature regulation – in which case the environment enters in as perturbations of input, x – or it may be an external input, as in orientation, where x_0 is given by the location of the (moving) object to be tracked. (In the first case the system is termed a *regulator*; in the second a *servomechanism*.) The input and output variables x and y are *frequency functions*, as explained more fully in note 5. Because the feedback term, x, is subtracted from the set point, x_0, the feedback is

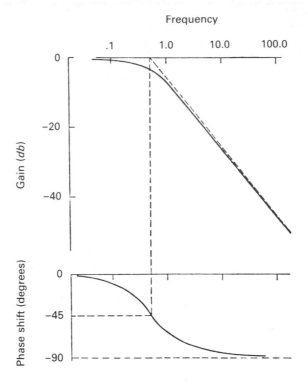

Figure 3.11. Bode plot for the exponential lag: $G(s) = a/(s + a)$. Top panel: gain as a function of input frequency (both in log units). Bottom panel: phase as a function of input frequency.

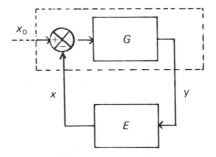

Figure 3.12. Conventional diagram of a control system: G is the transfer function of the organism, E the transfer function in the environment (feedback function), x_0 is the set point, which may be external, as in a servomechanism, or internal, as in a regulator. The environmental input, x, is subtracted from the set point at the comparator, indicated by the circle and cross. The net input then goes to transfer function, G.

negative and the function of the system is to diminish the difference between x and x_0.

G and E can be treated just as as multipliers (*gain* functions). Thus, if the input to E is y, its output will be $x = Ey$; similarly the input for G is $x_0 - x$ and its output is $G(x_0 - x)$. The two simple, linear equations describing the relations around the loop are therefore

$$x = Ey \quad \text{(feedback function)}, \qquad (3.11)$$

and

$$y = G(x_0 - x) \quad \text{(control function)}. \qquad (3.12)$$

Eliminating y between the two equations and rearranging yields

$$x = x_0 GE/(1 + GE), \qquad (3.13)$$

which is the fundamental feedback equation.

There are three things to notice about Equation 3.13. First, there will always be some discrepancy between the set point, x_0 (zero retinal disparity in the tropotaxis example) and the actual input, x (actual retinal disparity will always be greater than zero for a moving target). The size of this error is inversely related to the product of the two gains, EG: the larger the loop gain, the smaller the error. Second, the gain terms (transfer functions) around the loop combine multiplicatively. Third, negative feedback reduces the net gain of the system, but at the same time ensures that the relation between x and x_0 is relatively insensitive to the absolute value of loop gain, providing it is fairly large: as $EG \rightarrow \infty$, $x \rightarrow x_0$. In any real control system, components such as muscles or motors are liable to deteriorate with time, so that an arrangement that protects performance from these effects is obviously of great value. Hence feedback systems are widespread in engineering and biology.

Equation 3.13 is a general description of any control system. Even if E and G are interpreted just as static multipliers (not transfer functions) this equation provides a passable model for feeding regulation that can explain otherwise puzzling properties of food intake and its relation to the taste of food and the effort necessary to obtain food (see Chapter 6). Many properties of reinforcement schedules can be elucidated by looking for the properties of reward and punishment that act as feedback to the animal (see Chapters 5 and 7).

Three simple feedback responses. A linear system is completely defined by its Bode plot, which gives its response to all frequencies. This suggests two ways to discover the properties of a linear system: systematically present sine waves across the whole range of frequencies and measure the system output; or present a brief signal whose *spectrum* contains all frequencies (recall Fourier's theorem that any signal of finite length can be broken down into a series of sine-wave components). The first method is called *frequency analysis*, and the second (which is much more useful) is known as *transient analysis*. Signals particularly

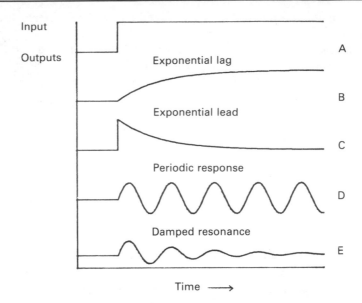

Figure 3.13. Simple responses to a step-function input. Panel A: input as a function of time. Panel B: exponential lag response. Panel C: exponential lead response. Panel D: oscillatory response. Panel E: combination exponential lead and oscillatory response (damped oscillation).

useful for transient analysis are the *unit impulse* (a very brief "spike" input) and the *step* input, since both these signals are made up of all frequencies, from the lowest (since the signal occurs only once its "fundamental" is at zero) to the highest (since both signals change level at an infinite rate, they must have components of infinite frequency). It turns out that a system's response to a step or an impulse is a unique "signature" that defines all its properties (this is why the quality of the "click" you can get by touching the stylus of a stereo system is such a good indicator of the fidelity of the system).

There are three simple responses to a step input, and many natural systems correspond to one of these or to some combination. They are shown in Figure 3.13. The tropotactic system corresponds to either the exponential lag or the exponential lead, depending on how we define the dependent variable: If we are measuring θ, the angle between the animal and the target, then the system is an exponential lead – θ decreases with time after the initial target displacement. If we are measuring the angle between the animal's current and initial positions, then the system is an exponential lag – the angle begins at zero and increases in negatively accelerated fashion to a maximum. The third response, simple oscillation, would be produced by a step input if our tracker behaved like a pendulum, that is, no friction, a restoring force proportional to displacement, and (a key element) some inertia. Since a real animal will have some inertia we might expect some oscillation in response to rapidly moving targets, and a step response similar to the damped oscillation in the bottom panel of Figure 3.13.

The exponential lag and lead both correspond to first-order systems (i.e., differential equations containing only first derivatives); these responses are derived formally in note 5. Oscillatory responses require at least a second-order system.

Numerous behavioral effects look like exponential lags. For example, many experimental situations show "warm-up" effects in which the organism takes some time at the beginning of each experimental session to come up to its stable level of performance. This effect is common in shock-postponement schedules, where an animal such as a rat is pressing a lever to avoid periodic, brief electric shocks. Motivational variables often show lagged effects: A sudden reduction in daily food intake shows up only gradually in a reduction of body weight. Adaptation and habituation are examples of exponential leads: A step change in stimulation produces an initial response that wanes with time. Pure oscillatory responses to stimulation are rare in the animal world, because they imply instability and a loss of contact between the organism and its environment; damped oscillation is commonplace, however, in systems where rapidly moving objects must be tracked and the physical mass of the animal's head and body enters into the feedback equations.

These three linear processes, plus a few nonlinear processes, such as delay (see the discussion of reflex latency in Chapter 2), account for most of the simple orientation mechanisms already discussed, for many physiological processes, such as aspects of temperature regulation and motor performance, for prey tracking in mantids and other animals, for simple reflexes such as pupillary contraction to light, for aspects of food and water regulation and even copulatory behavior, and for many aspects of operant behavior to be described in later chapters.

When is feedback useful? Feedback mechanisms are found whenever the animal must cope with unpredictable variation, either in the external world, or in the properties of its own muscular system. Most motor skills involve feedback, because hand–eye coordination must change to accommodate growth and variation in muscular strength. On the other hand, feedback slows the response of any system (I show this formally in note 5, but it is pretty obvious intuitively). Consequently, activities that must be carried out rapidly usually do not involve feedback. For example, Mittelstaedt has shown that the praying mantis aligns itself with its prey using visual feedback but its strike is entirely ballistic and unguided. Thus, if the animal's "prewired" setting is wrong (and the feedback that ensures that it is not is at the phylogenetic, not the ontogenetic level: Evolution selects efficient strikers, not efficient strikes, as in the orb-weaving example with which I began Chapter 2), the strike shows a constant error that never improves. Even a vertebrate such as the frog shows persistent errors of this sort following surgical intervention. In a series of classic experiments, Sperry[6]

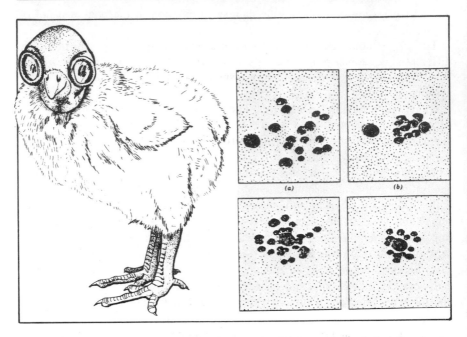

Figure 3.14. Left panel: chick with prism goggles. Right panel: clustering of pecks at a nailhead embedded in a soft surface as a function of age in normal chicks and chicks reared with prism goggles. (From Hess, 1956.)

interchanged the optic nerves between the two eyes of immature frogs. When the frogs matured, they struck to the left when a fly was on the right, and vice versa, and these errors were not corrected with practice. Baby chicks with prism lenses on their eyes that shift the world over a few degrees peck to one side of the grain; their pecks cluster more tightly as they mature, but the constant error remains (Figure 3.14).

Much human behavior can be either ballistic or feedback-guided, at the option of the individual. The Harvard physiological psychologist Karl Lashley raised the problem of feedback in connection with behavioral sequences, such as skilled typing or the arpeggios of the expert pianist, which occur too fast for tactile sensations from one keystroke to affect the next:[7] How are they organized, if not by feedback about the success or otherwise of each movement? What is being executed is not individual movements, one after the other, each guided by feedback, but a preprogrammed pattern of successive responses. There is feedback (the pianist will eventually notice if he has made a mistake), but not on a response-by-response basis: keystroke N might be affected not by the outcome of keystroke N-1, but by the effect of some earlier keystroke, say N-4. By allowing feedback to be delayed, the rapidity of a movement is limited only by muscle speed. On the other hand, feedback is not omitted entirely.

The difference between this system and the moment-by-moment one is in what is affected by feedback. In a moment-by-moment system, each response as it occurs is guided by feedback – as in threading a needle, or when first learning to type, for example. In the arpeggio and prism-adaptation systems, the parameters of a ballistic motor program are adjusted on the basis of feedback from successive responses. This is a form of learning and allows the animal to anticipate future stimulus-response relations, because it is adjusting its "strike" for future action based on a history of past hits and misses – implicitly assuming that the world in the future will be essentially the same as the world in the past. A feedback system of this sort is sometimes termed a *mesh* system.

Thus, both mantids and people respond ballistically when speed of response is critical; but people (and some other mammals) can both alter the parameters of the ballistic system (as in their adaptation to prism lenses) and deal with the inevitable delays when feedback is allowed to lag behind action (as in fast typing).

A ballistic system may itself be the result of learning: Ballistic typing only develops after a hunt-and-peck stage in which each keystroke is feedback-guided.

Feedback increases phase lag, which may, in turn, contribute to instability. Notice that in the Bode plot in Figure 3.11 phase lag increases with frequency. This is almost universal in real systems. The lag in Figure 3.11 does not go beyond 90° (a quarter of a cycle), but in many systems can go to 180° or more. At a lag of 180°, the output of the system is at its minimum when the input (to the set point) is at its maximum, and vice versa. Consequently, negative feedback, which is a stabilizing effect, has been turned into positive feedback, a destabilizing, "vicious circle" effect. It is not surprising, therefore, that many feedback systems get into difficulties at high frequencies. The ability to adjust system parameters, and to anticipate regularities in the input, are important protections against this kind of instability.

THE INTEGRATION OF BEHAVIOR

H. G. Wells once remarked that "The end of all intelligent analysis is to clear the way for synthesis." There is little point in breaking down the behavior of animals into simple units, be they reflexes, servomechanisms or "elementary information processes," if we cannot use these units to understand the behavior of the whole animal in its natural environment. The only way to tell if we have all the pieces of the puzzle is to put them together and see if we have reconstructed the natural behavior of the animal. I first discuss the role of variability in behavioral integration, then give examples of how a few simple mechanisms, acting together, can produce quite flexible and adaptive behavior.

The initial problem is to account for *variability* in behavior: If we have a well-defined set of elementary processes, it is natural to expect that they should lead to unique predictions about behavior – yet behavior is often unpredictable. What is the adaptive utility of behavioral variation, and on what does it depend?[8]

In some cases, the simple kineses discussed in Chapter 2, for example, the variability is intrinsic to the mechanism and serves the function of random sampling of an unknown environment. In other cases, a degree of variability even in an otherwise determinate mechanism can prevent the animal from getting trapped by particular circumstances. For example, a tropotactic animal that adjusts its orientation so as to balance the stimulation from bilateral receptors would pass completely between two equal lights and never reach either, if its mode of movement were completely regular. A little variability allows it to find the other balance point that involves going straight to an adjacent light even though it started out going between the two lights.

Variability is also helpful in getting around obstacles. Many insects are photopositive in dim light and everyone has seen flies and other insects trapped in a room buzzing at the window. The fly does not search systematically for a way out, but if a gap exists the animal's random movement will eventually find it. As wasteful as it appears, the fly's random buzzing is much more successful than would be a precise homing mechanism that simply settled down at the lightest part of the window pane.

Different individuals of the same species often behave differently under similar conditions. For example, photo-negative *Ephestia* (meal-moth) larvae move generally away from a source of light, but different individuals show different amounts of error and the distribution of heading angles of a group of larvae is approximately normal (i.e., the bell-shape characteristic of a process determined by many independent causes). The nauplius larvae of the barnacle *Balanus* are a more interesting case: Some are photo-negative and some photo-positive, so that the distribution of heading angles to light is bimodal. Such bimodal distributions, an example of *behavioral polymorphism*, are quite common, and quite puzzling at first sight.

The functional basis for behavioral polymorphism seems to be *frequency-dependent selection*, that is, selection where the Darwinian fitness of one type of behavior is directly related to its frequency relative to the other. For example, suppose that a particular prey species can find food in two different types of habitat that are not spatially contiguous; predators must decide to hunt in one or the other. Clearly it pays the predator to spend most of its time in the more popular habitat. Hence a prey individual that prefers the less popular habitat will be at an advantage. This system ensures that approximately equal numbers of prey individuals will be found in both types of habitat.[9]

Behavioral polymorphism can obviously take two extreme forms: either the same individual can show two or more modes of action at different times, or there may be two or more *types* of individual, each showing one mode. Both kinds of variation are found, but the first kind is potentially more flexible, as it allows for the evolution of *systematic* variation: that is, selection by the animal of one mode or the other, depending upon circumstances. Systematic variation tends to be the strategy increasingly employed by mammals and other "higher" animals. In a simple form it is common even in invertebrates. For example, the protozoan *Euglena* is photo-positive in weak light and photo-negative in strong light. Consequently the animals congregate in intermediate levels of illumination that may represent an adaptive compromise between the bright light that provides maximum energy from their photosynthetic organ and the dim light that provides greatest safety from predators. Some individual cockroaches, *Blatta orientalis*, are hygro-negative (avoid high humidity), but after some time in dry air they lose water by evaporation and then become hygro-positive. In mammals and birds a change of this sort would be termed a *motivational* one, caused by an altered physiological water balance. Reactions also change as a function of time: habituation, warm-up effects, and circadian and other rhythms are examples that have already been discussed. A taxis may even reverse periodically, as in the *Mysis* crustaceans in the Naples aquaria.

The sense of a taxis may change as a function of some quite different variable. For example, under suitable conditions, *Paramecium* is geo-positive (it descends) in light, and geo-negative in darkness. Many marine invertebrates show a similar reaction, which may be designed to vary the animal's depth as a function of the time of day. The water-flea, *Daphnia*, shows an interesting adaptive response to an increase in the carbon-dioxide concentration: It becomes photo-positive, a reaction that would normally take it to the water surface where the CO_2 concentration is likely to be lower and the oxygen tension higher. Exposure to dry air tends to make the woodlouse (sowbug), *Porcellio*, photo-positive rather than photo-negative. This case is more puzzling because it seems very unlikely that light is a reliable cue to humidity for this animal; indeed, in general, the opposite is more probable: Dark places such as crevices and holes are much more likely to be damp than bright, open areas. However, evidently the woodlouse, like a good politician, is a pragmatist: Ideology may say that dark means damp, but if one is in the dark, and dry, then maybe light is worth a try. This mechanism is a primitive, built-in, version of the "win stay, lose shift" strategy of trial-and-error learning in higher animals.

A simple set of mechanisms may, in combination, yield remarkably intelligent behavior. Consider the behavior of the sea-snail *Littorina neritoides*, as analyzed by Fraenkel:

This animal is found several metres above high-water marks of European seas. It is usually geo-negative and never geo-positive. When out of water it is always photo-negative. In the water it is too, except when it is upside down, and then it is photo-positive . . . These reactions may be expected to guide the animal from the bottom of the water to rocks (dark) and then up the rock face (geo-negative); if the light is very bright, the animal stops and settles down at the water surface. In crawling up under water, if it gets into a deep horizontal cleft, negative photo-taxis takes it inwards on the floor, negative geo-taxis upwards on the end wall, and positive photo-taxis outwards on the roof in the inverted position. Upward progress is therefore not barred by such a cleft. Above the water surface, the sign of photo-taxis does not reverse in this way, so the animal comes to rest in such a cleft. Under the influence of dryness and other unfavorable conditions the animal closes up; it may then fall back into the sea, but if conditions are not very unfavorable it may live for months out of water. This behavior provides examples of taxes in opposition

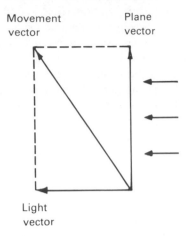

Figure 3.15. Vector model of the combined effect of negative geotaxis (tendency to move directly up the slope of a plane) and negative phototaxis (tendency to move directly away from a light). (Adapted from Crozier & Cole, 1929.)

(gravity and light) and of one stimulus (the presence of surrounding water) affecting the reponse to another (light). (Fraenkel & Gunn, 1940, pp. 297–298)

In these examples the animal's response to one variable, such as light, is modulated by another, *contextual*, variable, such as gravity or whether the animal is in or out of water. Often, however, two or more variables may each have an independent taxic effect; for example, both geotaxis and phototaxis may operate simultaneously. What happens when these variables act in opposition, if negative geotaxis dictates upward movement, say, and negative phototaxis urges downward movement, away from the light? Here, as in the case of opposed reflexes, there are only two possibilities – competition or cooperation. Most taxic mechanisms appear to interact cooperatively. For example, a slug placed on an inclined plane turns so that it travels straight up the plane, along the line of greatest slope (it appears to accomplish this by equalizing muscle tension on both sides of the body). Slugs are also negatively phototactic, and if a light is shone on the animal from one side as it moves up the plane, it will turn away and adopt an intermediate orientation; the stronger the light, the greater the turning away and the larger the angular deviation from a straight upward path. The animal's direction of movement can be modeled here as the resultant of two vectors, one representing the slope of the plane and the other the direction and intensity of the light, as illustrated in Figure 3.15. Under other circumstances one reaction may totally dominate the other, particularly if one is a protective reaction of some sort such as the withdrawal reaction shown by *Stentor* and other small, sessile invertebrates to a passing shadow.

These two kinds of combination rule, cooperation and competition, are not, of course, perfectly distinct, but merely extremes along a continuum. At one end the effects are additive (linear); this is cooperation. At the other end is total competition: The response with greatest strength occurs exclusively and the other is completely suppressed. As in reflex interaction, which of these possibilities actually occurs no doubt depends on the Darwinian fitness of intermediates: When a compromise is less adaptive than either alternative alone, as in the feeding versus flight choice for *Stentor*, one or other will occur exclusively.

There seems to be a trade-off across animal species between systematic and unsystematic variation. Systematic variation means that the rules the animal plays by depend upon the situation, on contextual variables such as time of day, condition of light, presence of another animal, and so on. In each situation, the animal's behavior may be very predictable, yet the animal may be sensitive to many different situations (not all of them, perhaps, known to an outside observer) so that its overall repertoire of behavior may be large. Unsystematic variation means that the animal always plays by the same set of rules, but these include a substantial unpredictable element – kineses are the obvious example. There seems to be a trade-off in the sense that the more predictable the animal in any particular situation, the richer the variety of situations to which it is sensitive. Human beings can carry out particular tasks with enormous precision, for example, yet there is no denying their unpredictability in general, or its dependence upon the individual's perception of situation.[10]

THE NATURE OF EXPLANATION

The emphasis of this book is on explanations for behavior in purely environmental terms: I ask, How does the animal's past history interact with rules built in to the animal by its evolutionary history and individual development to produce future behavior? I am not directly concerned with the animal's internal structure, or with intentional states we can relate to our own introspections (see Chapter 1). Since there is often confusion about the meaning of explanations at this level, I conclude the chapter with brief discussions, first of "black-box" explanations, and then of the relation between feedback theory and explanations of behavior in terms of motive and purpose.

The meaning of "black-box" analysis

Scientific explanations can be at many levels. The behavior of a moth circling a flame might be explained in a number of ways: (a) as an "instinct"; (b) as a form of taxic reaction (e.g., light-compass reaction); (c) as a particular kind of control system; or (d) as the response of a particular neural network connected to a set of receptors and effectors.

The instinct account says nothing about the process involved; it is really just a kind of classification. It places the moth's behavior in a group containing all those behaviors that seem not to depend on experience. Although this is a relatively primitive kind of explanation, it is not empty; it may in fact be essential to go through such a "natural history" stage at the beginning of any science.

Explaining the behavior as a taxis is also classificatory, but now in terms of the results of experiments – unilateral blinding, the two-light experiment – and begins to get much closer to the underlying process.

The control-system account rests on more experiments and more quantitative results and constitutes an explanation in terms of process. It gives the "rules of the game," those aspects of the behavior that are invariant, that is, independent of particular inputs and outputs (recall the constancy of the transfer function of a linear system). However, the control-system model is a black-box account: The equations of the model, and the boxes and connecting arrows that make them easier to understand, may have *nothing whatever* to do with the elements of the fourth level of explanation: an account in terms of neural and other structures. Nevertheless, the control-system account says what the neural structures must do, and gives hints on how they might do it. But block diagrams cannot be interpreted literally as connections among neural circuits.

The lack of relation between block diagrams and the physical constitution of the system is especially obvious if we are modeling a real physical system. For example, the exponential lead describes the relation between current (output) and voltage (input) across an electrical capacitor (essentially a pair of parallel plates capable of storing a quantity of electrical charge): If the voltage across the capacitor is suddenly increased, there is an initial large current flow that subsequently decays to zero. Where is the "loop" here? There is no physical link corresponding to the feedback path in Figure 3.12, nor is there anything corresponding to a comparator or a set point. There is negative feedback, but it is in the form of the increased repulsion among charges as the number of charges on each plate increases. Many other physical and biological systems behave like exponential leads: a spring-loaded dashpot, a pendulum in glue, the inflow of organisms into a suddenly denuded habitat. The negative feedback (restoring force) in the first case is the spring; in the second, gravity; and in the third, competition among individuals for limited resources.

Black-box accounts, the third level of explanation, provide a natural preliminary to accounts at the fourth level, in terms of neural structures, but cannot be directly interpreted in neural terms (see note 2, Chapter 2).

Each level of explanation has its uses. Explaining the moth's circling in terms of instinct is a useful starting point for the student of behavioral development and it may have implications for the evolution of behavior. The taxis account relates the behavior to other orientation reactions, and may be useful for understanding species differences. The control-system account is useful for making quantitative predictions, and as an element in accounts of the integration of behavior, that is, in theories which bring together a number of mechanisms so as to account for something approaching the organism's total behavioral repertoire.

Purpose, teleology, and mechanism

The concept of *purpose* has occupied a prominent place in the history of psychology. Although slightly discredited by the behaviorists, it is covertly retained even by them in the form of so-called "control of behavior by its consequences." It will not have escaped the alert reader that the set point of a feedback mechanism (the *sollwert* or "should-be value" in German) corresponds pretty closely to the intuitive idea of a "goal" or "motive." This connection was pointed out some years ago in a classic philosophical paper by Rosenblueth, Wiener, and Bigelow (1943) in which they showed that the idea of feedback provides the conceptual link between mechanistic accounts of behavior, that is, explanations in terms of antecedent events (proximal causes), and teleological accounts, that is, explanations in terms of goals or motives (final causes). The superiority of a feedback account rests in its self-contained ability to account both for those cases where the goal is attained, and those where it is not. The former simply represent the domain of stability and negligible phase lag of the system, the latter its region of large phase lag and instability.

Commonsense explanation in terms of purpose does not easily account for failures to achieve a goal. For example, a student may do well on an organic chemistry examination, and this might be attributed to his goal of getting into medical school (say). But suppose he had failed the exam, despite his strong motivation to succeed? Explanation in terms of goals or motives must then postulate some competing motivation – he may have spent too much time with his girl friend, perhaps – or else say something about the student's capabilities and his methods of study, that is, resort to a mechanistic account. In short, conventional motivational accounts offer only two explanations for failures to achieve an objective: either competing motives, or an unsuspected structural limitation.

Both these escape routes have an ad hoc look to them, yet they roughly correspond to respectable explanatory alternatives: an optimality account, that provides the laws by which motives compete, and a mechanistic account, that dispenses with motives entirely and explains behavior by antecedent conditions and structural properties of the subject (see Chapter 1).

Different fields have inclined more to one or the other type of explanation. Classical economics, for example, explains people's allocation of time and money between different commodities, or between work and leisure, in terms of a balance of motives, a teleological account. The individual is assumed to optimize his total utility by allocating his resources so that the marginal gain from switching from one thing to any other is constant; that is, he spends so that his last nickel will buy him the same benefit, no matter where he spends it. In this way, total benefit will usually be maximized. Optimality analyses (about which I shall have much more to say in later chapters) are the ultimate form of teleological account. On the other hand, the field of ecology (which has much in common with economics as the etymology implies) has generally favored mechanistic accounts: The distribution of species within and between habitats is usually explained in terms of their relative efficiencies and reproductive rates.

Psychology and behavioral biology have at different times favored both approaches. Reinforcement theory is basically teleological in the sense that behavior is assumed to be guided by access to a key event, the reinforcer, which in effect functions as a goal: The animal works for food or to avoid electric shock. Purposive terminology is studiously avoided (the phrase "works for" is frowned on in the young, and is only employed in the privacy of the laboratory), but the lack of any generally accepted mechanism to account for the effectiveness of reinforcers means that the term *reinforcement* is actually used as if it meant "purpose." For example, when several experiments showed a few years ago that hungry rats would press a lever for food even if free food was available, a popular response was that the lever-pressing behavior was "self-reinforcing." The difference between this and the naive explanation that the animal "likes" to press the lever may be too subtle for the naive reader to detect. As we shall see in later chapters the essence of reinforcement, both as an effective procedure and an explanation of behavior, is restriction of access, so that the notion of self-reinforcement (as a sufficient explanation for behavior) is a contradiction in terms.

Other areas of psychology have looked for mechanistic accounts. For example, the now largely discredited Hullian theories of learning were a praiseworthy attempt to specify the causal links between the stimulus, as cause, and the response, as ultimate effect.[11] More recent mathematical theories of classical conditioning and the effects of food and other hedonic stimuli on general activity (arousal) are strictly mechanistic, looking to antecedent, rather than consequential, events for the explanation of behavior. Theories that explain choice and the distribution of behavior in terms of competition between the tendencies to engage in different activities are also, like their ecological counterparts, mechanistic in spirit.

Both mechanistic and teleological accounts have their uses, although mechanistic theories are obviously desirable where they are possible. But when the means an organism, or an economy, can use to attain some end state are many and poorly understood, but the goal is relatively clear, then a teleological theory may be the best that is available (see Chapter 1). An adequate teleological explanation (like the empirical principle of reinforcement) is certainly better than a premature mechanistic theory, as the early followers of Hull found to their cost.

SUMMARY

The mechanisms discussed in this chapter and the preceding one represent the best that animals can do when limited to what might be termed *local memory*, that is, when their behavior is affected only by present events and events in the immediate past. Although simple, and relatively easy to analyze, these mechanisms can nevertheless produce remarkably flexible behavior. In the rest of the book, the emphasis is on learned behavior, that is, behavior that depends upon more remote past history.

Learned behavior can be studied in two ways: as learning, that is, as a *change* in behavior with time and experience; or as *habit*, that is, as a more or less fixed pattern, built up after much experience. The study of habit has the advantage that like the mesh feedback systems discussed earlier, once set up, habits may show rather little dependence on past history. They can be studied like orienting mechanisms, in individual animals, using reversible procedures. The next several chapters look at learning from this point of view. Later chapters deal with learning as change.

NOTES

1. This limitation is a general feature of all hill-climbing mechanisms: They find their way to the top of *a* hill, but not necessarily the *highest* hill. They are *local* rather than *global* optimizers.

2. *Contrast mechanisms.* The initial advance in understanding these rate-sensitive effects of sensory mechanisms was made by the Austrian physicist and philosopher Ernst Mach (1838–1916). Floyd Ratliff, in his book *Mach Bands* (1965), has provided a fascinating account of Mach's sensory work and translations of relevant papers, as well as a summary of more recent work on the problem. See also Arend, Buehler, & Lockhead, 1971; von Békésy, 1967; and Ratliff, 1974.

The intimate relation between the perceptual effect of a luminance gradient and its higher derivatives can be seen from Figure 3.16. The top curve is a gradient of luminance as a function of distance, $V(s)$, such as would be obtained by measuring at right angles to a set of progressively lighter bars (*ab, cd, ef,* etc.) separated by regions of gradually increasing lightness (*bc, de,* etc.). The next curve, labeled "response" shows the appearance of this luminance gradient. There are two features that are of special interest: First, the progressive change in lightness in regions *bc, de,* etc. is barely perceptible, and the successive bars *ab, cd,* etc. increase but little in perceived brightness. Second, narrow dark and light bars (*Mach bands*) are seen at the points of inflection *b, c, d,* etc. The third curve shows the first derivative of the luminance curve and the bottom curve is the second derivative. Mach saw at once that the light and dark bands are derivable (to an approximation) from the negative of the second derivative of luminance, and that the perception as a whole can be represented as the weighted sum of the original luminance distribution (transformed via a compressive function as described in Chapter 2) and the second derivative of that distribution:

$$R(s) = A \log [V(s) - (B/V)d^2V/ds^2] + C, \qquad (N3.1)$$

where A, B, and C are constants, and the second derivative is multiplied by $1/V$ because the effect of an inflection is inversely related to the brightness level.

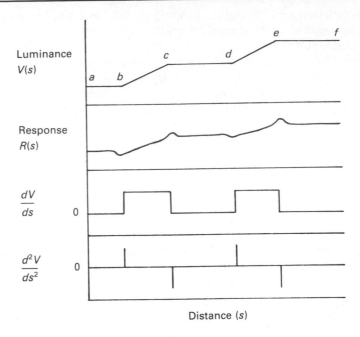

Figure 3.16. A luminance gradient (top curve), the perceived brightness gradient (second curve), and the first and second derivatives of the luminance gradients (third and fourth curves).

Figure 3.17 is redrawn from Figure 3.2 in the text, with the addition of the first and second derivatives. It is clear that it poses difficulties for Mach's scheme (which is itself equivalent to several more recent models, as Ratliff shows), since no simple combination of the original luminance distribution with the second derivative will yield the decreasing staircase that represents the perception.

If for simplicity we ignore the compressive transformation (or, alternatively, simply assume that $V(s)$ is measured in logarithmic units), then Equation N3.1 becomes

$$R(s) = aV(s) + bd^2V/ds^2 + c. \tag{N3.2}$$

It is obvious that $R(s)$ cannot be derived from this equation for the luminance distribution shown in Figure 3.17. However, Equation N3.2 can be rewritten in a way that immediately suggests a modification to accommodate Figure N3.2 and related effects. First, we rewrite Equation N3.2 in terms of first and second derivatives only:

$$R(s) = a\int_0^s (dV/ds)ds - b(d^2V/ds^2) + c'. \tag{N3.3}$$

(The integral term reduces to $V(s) - V(0)$, so that $V(0)$ is incorporated in a new value of the additive constant, c'.) Equation N3.3 is equivalent to Equation N3.2

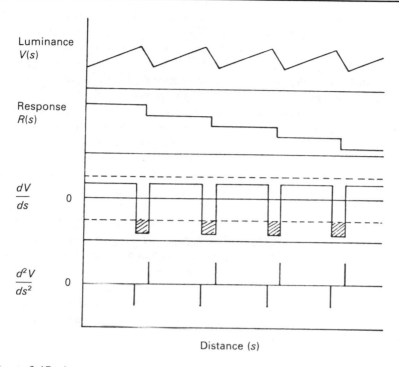

Distance (s)

Figure 3.17. A sawtooth luminance gradient (in logarithmic units: top curve), its perception, and the first and second derivatives of the luminance gradient. Suprathreshold regions of the first-derivative curve are crosshatched.

save for the implication that absolute values of $V(s)$ are not sensed directly, but rather affect sensation via the integration of the first derivative. Whatever the stimulus aspect that is sensed by a sensory system, there will always be a threshold. If rate of change of luminance, rather than absolute luminance, is important, then there will be a rate of change so slow as to have no effect. The value of this threshold will presumably be related to the difference limen (differential threshold) for brightness. Thus, the term dV/ds in Equation N3.3 refers only to suprathreshold values of dV/ds.

The implications of this change can be seen in Figure 3.17 where the threshold is indicated by the two dashed lines (two because the first derivative can take on negative as well as positive values): The small positive values of dV/ds associated with the gradual part of the sawtooth are below threshold and therefore do not contribute to the integral, whereas the large, but briefer, negative values associated with the steep part of the sawtooth add to the integral yielding a fixed increment in the sensation at each cycle. The positive and negative parts of the second derivative are very close together because of the steepness of the falling part of the sawtooth, and so interfere with one another; since their average is zero (and since the contribution of the second derivative to perception is probably much less than the contribution of the first derivative), they have little net effect on perception. Thus, the perceptual effects of luminance gradients can be derived, to a first approximation, from the simple hypothesis that the visual system (and other sensory systems as well) is sensitive only to rates of change.

This analysis applies most easily to temporal variations because here the direction of integration is determined unambiguously by the flow of time: The system has no alternative but to integrate past stimulation up until the present. Spatial interactions are more difficult to handle: Should the system integrate from left to right, from top to bottom, or what? One possibility is that spatial changes are converted at once to temporal ones via eye movements, which occur all the time. Land and McCann (1971) have suggested other possibilities, based upon integration of differences along interconnecting paths. This question remains open.

As we have seen in previous chapters, habituation and adaptation are names for the processes that permit special sensitivity to changes in stimulation, and the curves in Figure 3.17, particularly, are very similar to those encountered earlier. For example, in Koshland's experiments on bacterial orientation a "pulse" of nutrient produced a transient decrease in tumbling, very like the transient increase in sensation level produced by a sudden stimulus change in Figure 3.17. The transience of the stimulus effect in the bacteria was attributed to adaptation, which can, therefore, be seen simply as another name for the rate-sensitivity of many biological systems.

3. This result can be obtained as follows: Let A_0 be the intensity of the light source at O in Figure 3.18, and L and R be the two bilateral point receptors. Y is a point on the midline of the animal on the line joining the receptors and θ is the heading of the animal with respect to the source. Receptor disparity is the difference in light intensity falling on the two receptors; by the inverse square law this is:

$$\Delta I = A_0/OL^2 - A_0/OR^2, \tag{N3.4}$$
$$= A_0(OR^2 - OL^2)/OL^2OR^2.$$

Considering each term in the numerator separately:

$$OR^2 = OX^2 + RX^2 \tag{N3.5}$$
$$OL^2 = LX^2 + OX^2 = (LR + RX)^2 + OX^2$$
$$= LR^2 + LR \cdot RX + RX^2 + OX^2.$$

Substituting in Equation N3.4 yields: $\Delta I = -A_0 \cdot LR(LR + 2RX)/OL^2OR^2$. But by elementary trigonometry, $\sin \theta = YX/OY = (RX + LR/2)/OY$; hence

$$\Delta I = -A_0LR \cdot OY2 \sin \theta/OL^2OR^2, \tag{N3.6}$$

which is the stated relation. If the animal is far away from the source, so that $OY>>LR$, then $OL \simeq OR \simeq OY$, and this relation reduces to

$$\Delta I = -A_0LR \cdot 2 \sin \theta/OY^3, \tag{N3.7}$$

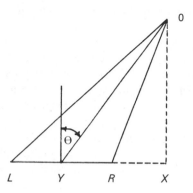

Figure 3.18. Geometric relations between a light source, 0, and two bilateral, point receptors, *L* and *R*.

so that receptor disparity is directly proportional to receptor separation, *LR*, and inversely proportional to the cube of the distance to the source.

4. Regulatory gadgets such as the furnace thermostat and the centrifugal governor used in early clocks have been around for some time. The modern, quantitative study of feedback devices dates from James Clerk Maxwell's paper (1868) on the centrifugal governor used in James Watt's steam engine. Feedback theory received its greatest impetus during the second World War when tracking devices of all sorts – automatic gunsights, radar trackers, homing torpedoes – elicited a level of public interest lacking in more peaceful times. Since then the study of control systems has grown into a highly developed field of largely mathematical inquiry.

There are now so many texts on the theory of control systems that it is hard to give an objective survey. The classical source is MacColl's *Fundamental Theory of Servomechanisms* (1945), but this, although relatively easy to follow, is not explicitly directed at the novice reader, and gives no account of the now-standard Laplace transform method. More recent books or papers aimed at biologists or psychologists are Toates (1975; a clear elementary account), McFarland (1971), Powers (1978), Grodins (1963), and Milsum (1966). Useful summaries of empirical and theoretical studies are to be found in McFarland (1974), the 1964 Symposium of the Society for Experimental Biology, Toates and Archer (1978), and Davis and Levine (1977). A comprehensive account of the mathematics involved is Kreider, Kuller, Ostberg, and Perkins (1966).

My account of control theory as applied to tropotaxis is, of course, greatly simplified. More comprehensive applications of the approach to orienting mechanisms such as the optomotor response (visual following of moving stripes, shown by insects and many other animals), are to be found in the German literature, e.g., Mittelstaedt (an account in English is 1964; see also 1972).

5. *Linear systems analysis.* The three fundamental assumptions of linear systems analysis have been given in the text. Here I describe the Laplace transform method that allows systems of linear differential equations to be solved in an almost automatic way.

Recall that the properties of any linear system are completely described by its Bode plot, which gives the gain and phase lag for all frequencies of sine-wave input. The *Laplace transform* is a function that incorporates all the information in a Bode plot. The transform method is similar to the use of logarithms, a technique already familiar to most people with some mathematical experience, in that both methods change quantities into a form where they are more convenient to work with. In the days before microchips, multiplication or division of large numbers was a tedious chore. The invention of logarithms in the seventeenth century by the Scottish mathematician John Napier allowed numbers to be converted to their logarithms, whereupon the sum of the logs yielded the log of the product, so that multiplication and division were replaced by the much simpler operations of addition and subtraction. The key to the simplicity of the method lay in the laboriously compiled tables of logarithms and antilogarithms that enabled the results of these additions and subtractions to be at once translated back into the raw number domain.

Most time functions that are of practical interest can be similarly converted into their Laplace transforms; so can the differential equations that describe the control function. These transformations allow the operations of differentiation and integration (necessary for the solution of differential equations) to be converted into arithmetic operations of multiplication and division, as I show in a moment. In this way any linear differential equation can be solved just using simple algebra. The solution takes the form of an expression in the complex variable s. By consulting a table of Laplace transforms (analogous to a table of antilogarithms), the time function corresponding to this expression can then be found.

The Laplace transform is denoted thus: $\mathcal{L}F(t) = F(s)$ where s is a complex variable containing frequency, phase and amplitude information. Formally,

$$\mathcal{L}F(t) \quad \int_0^\infty F(t)\exp(st)dt. \tag{N3.8}$$

For details of the derivation and the limitations that allow the integral to exist (i.e., be finite) see (for example) Toates (1975), Milsum (1966), or Kreider et al. (1966). The Laplace transform is closely related to the moment-generating function, familiar to students of probability theory.

The thing that makes the Laplace transform so useful in the analysis of linear systems is that the ratio of the output $F_o(s)$, and input, $F_i(s)$, is *constant* for all inputs; this ratio is the *transfer function*. Thus, the transfer function, $G(s)$, is defined by $F_o(s) = F_i(s) \cdot G(s)$. The transfer function behaves just like a multiplier, when input and output are expressed in terms of the complex variable s.

We can derive $G(s)$ in three separate ways. (a) Analytically, from the differential equations of the system, if these are known. (This is very straightforward because the two operations of differentiation and integration correspond just to s and $1/s$, respectively, so that $\mathscr{L}(d/dt)F(t) = sF(s)$, for example, where $F(t)$ is some function of time.) (b) From the input–output relations derived from certain inputs, such as the "unit step" and "unit impulse," that contain a wide range of frequencies and produce responses with a "signature" characteristic of the system. This is known as *transient analysis*. (c) From the Bode plot obtained by empirically probing the system with a wide range of sinusoidal frequencies; this is termed *frequency analysis*.

As an example of the derivation of $G(s)$ from the differential equations of a system, consider again the simple model of tropotaxis described by Equations 3.5 and 3.6. First, since this is a dynamic analysis, we can replace a and b by $a(t)$ and $b(t)$, indicating that both these variables are now functions of time. Substituting in Equation 3.7 and rearranging yields: $db(t)/dt = AB(a(t) - b(t))$. Taking Laplace transforms of both sides yields:

$$sb(s) = AB[a(s) - b(s)].$$

Rearranging so as to obtain the ratio output/input, $b(s)/a(s)$, yields:

$$b(s)/a(s) = G(s) = AB/(s + AB). \tag{N3.9}$$

The expression on the right-hand side is therefore the transfer function for this system. The output for any given input signal, $a(t)$, can now be obtained simply by substituting the Laplace transform for that signal for $a(s)$ in Equation 3.9 and rearranging to obtain an expression for $b(s)$ in a form suitable for consulting a standard table. The response explored earlier is equivalent to the "unit step," an instantaneous displacement of the target light from a value of 0 (the animal heading straight to the target) to 1. The Laplace transform of the unit step is just $1/s$, so that $b(s)$ is equal to $AB/s(s + AB)$. This expression does not appear in standard tables, but the expressions $1/s$ and $1/(s - a)$ do appear there. $AB/s(s + AB)$ can be written in this form using the method of partial fractions, which yields:

$$b(s) = 1/s - 1/(s + AB).$$

From the tables this means that

$$b(t) = 1 - \exp(-ABt), \tag{N3.10}$$

Which is just Equation 3.8 with $a = 1$, $b_o = 0$, and $b = b(t)$. The response of this system to any other time-varying input can be obtained in the same way.

Equations N3.10 and N3.8 are termed *exponential lags*, and this example shows that an exponential lag in response to a step input is the characteristic signature of a transfer function of the form $a/(s + a)$, that is, of integral control. The same information can be derived from a Bode plot for this system, which has already been shown in Figure 3.11, using the vector properties of the complex variable s.

Exponential lags result when the rate of change of the dependent (controlled) variable is proportional to the level of the independent (controlling) variable – or,

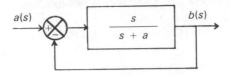

Figure 3.19. Effect of feedback on the exponential lead.

equivalently, when the controlled variable is proportional to the integral of the controlling variable. There is obviously a converse process, when the controlled variable is proportional to the rate of change of the controlling variable (rate control). Its properties can easily be deduced from what we already know about the Laplace transform method. Thus (using the same terms as before), rate control means that

$$b(t) = A \cdot d\,[a(t) - b(t)]/dt,$$

where A is a constant. Taking Laplace transforms and rearranging, as before, yields

$$b(s)/a(s) = s/(s + 1/A).$$

Setting $a(s)$ equal to the Laplace transform for a step input ($1/s$) and rearranging yields

$$b(s) = 1/(s + 1/A),$$

so that (from tables)

$$b(t) = \exp(-t/A),$$

which is again an exponential with time constant A. This function, the exponential lead, should be especially familiar as the temporal change characteristic of effects variously labeled as adaptation, habituation, or fatigue; that is, rapid response to a sudden change in stimulation, such as the onset or offset of a light, followed by slow waning of response back to a low level.

As an example of the use of the transfer function in learning about the temporal properties of a system, consider again the fundamental feedback equation (Equation 3.10). The transfer functions G and E are used just like simple multipliers in that equation, so we can use the equation to predict the effects of adding a feedback loop to one of the elements shown in Figure 3.13. Consider, for example, the exponential lead. Suppose we connect it in a feedback loop as shown in Figure 3.19; what will be the transfer function of the combined system? We can find out at once just by substituting the transfer function, $s/(s + a)$, for G in Equation 3.10, with a feedback gain (E) of unity. The result is the transfer function $s/(s + a/2)$; thus, the time constant, which was originally $1/a$, is now $2/a$ – the system responds twice as slowly. This is a general result: The flexibility conferred by feedback is bought at the cost of a slower response. Consequently, activities that must occur as rapidly as possible often do not involve feedback, as I describe in the text.

6. Sperry (1951) is a good review.

7. Lashley (1951). Any rapid typist can verify immediately that his keystrikes are not guided by immediate feedback by noting how many keys he hits after the end of a line is reached and the carriage will not advance further (assuming that he has already ignored the bell). If he is at the same intermediate stage of incompetence as I, he will type fast enough to hit several keys after the carriage has reached the limit of its travel. This is a measure of the time it takes for feedback to affect the motor program.

8. There is a tendency to equate *unpredictability* with *randomness*. It may be worth pointing out that quite simple deterministic systems can lead to behavior that looks random to an observer who is ignorant of the generating process. Mathematicians have shown recently that a perfectly deterministic set of (nonlinear) differential equations can, nevertheless, lead to apparently chaotic, and certainly unpredictable, behavior (see, for example, May, 1976). Thus, unpredictability, from the point of view of an outside observer, does not imply absence of determinism: A system may be both deterministic and irreducibly unpredictable – unpredictable is not the same as random. (I return to this topic in Chapter 4 in connection with the concept of *observability*.)

9. Behavioral polymorphisms have been analyzed most thoroughly in connection with social behavior. Maynard Smith (e.g., 1976; see Krebs & Davies, 1981, for an elementary discussion) has used game theory to derive the idea of an *evolutionarily stable strategy* (ESS) – a stable mixture of behaviors that coexist in a population because of frequency-dependent effects.

10. The argument here is closely related to W. R. Ashby's (1956) *law of requisite variety*, which states that if an organism is to maintain its internal environment constant in the face of environmental perturbations, the variety of its output must equal the variety of its input. This is easy to understand if you think of a simple tracking mechanism, such as the model of tropo-taxis discussed earlier. If the system is to continue to be aligned accurately with a moving target, then its correcting movements (responses) must exactly match those of the target. The ability of the tracker to do this is, of course, a function of its frequency response in relation to the frequencies in the pattern of movement of the target (which is directly related to the input *variety*, in the communications theory sense).

Ashby's law (which derives from the theory of communication pioneered by Shannon & Weaver in 1949) refers both to the way an animal varies a single response and to its ability to come up with other responses if one fails. Linear systems theory is concerned more with the quantitative properties of a single response, such as shaft angle or voltage, to a single input, whereas real organisms cope with environments that require variation in the type, as well as the quantitative properties, of responding. The principle applies equally to both cases, however, The implication of this principle is that the reciprocity between systematic and unsystematic variation is likely to hold good only as between species that live in environments of comparable richness.

11. The Yale psychologist Clark L. Hull (1884–1952) was one of the most influential figures in the history of behaviorism and stimulus-response psychology. His best known books are the *Principles of Behavior* (1943), a theoretical and experimental account of how rats learn about mazes; the *Mathematico-Deductive Theory of Rote Learning* (1940, with several other authors), an account of how people learn nonsense syllables; and *A Behavior System* (1952), an updated version of the *Principles*. A good secondary account of his work appears in Osgood (1953). Hull, although not a mathematician either by nature or training, was nevertheless inspired by what he took to be Newton's mathematico-deductive method. Unfortunately, he failed to grasp the idea that the power of mathematical theory lies in its economy: Newton needed only three laws to explain the movement of all physical bodies, but Hull and his students, at various times, came up with dozens of axioms and corollaries to explain a set of observations hardly more varied than the theoretical constructs they applied to them. The mass of arduously obtained experimental results that flowed from Hullian theory have yielded very little of permanent value, in large measure just because the experiments were tightly bound to the testing of particular hypotheses most of which are now of only historical interest. Nevertheless, Hull and his followers raised a number of issues that continue to be important, such as the relative contributions of reinforcement (reward and punishment) and practice to be "strength" of a habit, the distinction between learning and performance (see Chapters 13 and 14), and the similarities and differences between different kinds of motivational variables such as deprivation (e.g., time without food) and incentive (e.g., the attractiveness of food).

4

OPERANT BEHAVIOR

Some behavior makes sense in terms of the events that precede it; other behavior makes more sense in terms of the events that follow it. Reflexes are behavior of the first kind. The type and vigor of the reflex response are closely related to the type and intensity of the eliciting stimulus. Kineses are behavior of the second kind. The movement of the orienting bacterium from moment to moment is unrelated to any single stimulus, yet its behavior as a whole can be understood in relation to the prevailing chemical gradient: The organisms aggregate in high concentrations of an attractant and disperse away from high concentrations of a noxious substance.

Behavior of this type is *guided by its consequences*: Under normal conditions, the location of the bacterium is determined by the chemical gradients in its environment. Because the antecedent (proximal) causes of its behavior are many but the final cause is one, we can best describe the behavior by its outcomes: The bug finds the food.

Behavior guided by its consequences was called by Skinner *operant behavior*, and the term has become generally accepted. The word *operant* refers to an essential property of goal-directed behavior: that it have some effect on the environment. If the bacterium cannot move, or if movement has no effect on the organism's chemical environment, then its behavior will not appear to be guided by a goal.

Skinner was interested in learning and restricted his definition to operant behavior that reflects learning. It is useful to have a term for the wider concept, however, and so I use operant behavior to refer to any behavior that is guided by its consequences.[1]

The consequence that guides operant behavior plays the same role as the set point of a feedback system. If the set point is external, as in a tracking system, then the position of the animal will appear to be controlled by the external reference: As the target moves, so does the animal. If tracking is very good, the illusion that the animal is physically attached to the target can be very strong. The term *control* is sometimes used to refer to operant behavior also, as behavior that is "controlled by" its consequences. The meaning is the same. In both cases, the behavior is guided in some way by the discrepancy between the present state of affairs and some ideal state.

In this chapter I provide the outline for a comparative psychology of learning. I discuss the kinds of mechanism an animal must possess for it to be capable of operant behavior and look at differences between the operant behavior of simple animals and the operant behavior of mammals and birds. This comparison suggests that a major difference between "simple" and "complex" animals is in the number of different "states of the world" they can distinguish. The middle part of the chapter is an overview of classical and operant conditioning and related effects, such as sensitization and pseudoconditioning. Conditioning is taken up again in more detail in the next chapter. The last section of the chapter deals with the concepts of *stimulus, response*, and *internal state*, and their mutual relations. This discussion leads to formal definitions for each of these terms. The chapter ends with a discussion of *learning* and *memory* and how they relate to behavioral mechanisms.

CAUSAL AND FUNCTIONAL ANALYSIS OF OPERANT BEHAVIOR

How does the operant behavior of mammals differ from the kinetic behavior of protozoa? There are, of course, trivial differences: Protozoans are small and mammals are big; protozoans can move about but can't do much else, whereas mammals can climb and fly and talk and press levers. These are the same sorts of difference that exist among different mammal species; they have nothing to do with the essential feature of operant behavior, which is its goal-directedness. Yet there are differences. One way to get at them is to look carefully at the operant behavior of a simple animal and consider how it differs from the operant behavior of a mammal or a bird. This comparison can also tell us something about the mechanisms that must be involved in learning.

One of the most beautiful examples of intelligent behavior by protozoa is provided by Jennings' account of how *Stentor*, a single-celled pond animal (see Figure 4.1), copes with a mildly irritating substance introduced into the water in its vicinity:

Let us now examine the behavior under conditions which are harmless when acting for a short time, but which, when continued, do interfere with the normal functions. Such conditions may be produced by bringing a large quantity of fine particles, such as India ink or carmine, by means of a capillary pipette, into the water currents which are carried to the disk of *Stentor* . . .

Under these conditions the normal movements are at first not changed. The particles of carmine are taken into the pouch and into the mouth, whence they pass into the internal protoplasm. If the cloud of particles is very dense, or if it is accompanied by a slight chemical stimulus, as is usually the case with carmine grains, this behavior lasts but a short time; then a definite reaction supervenes. The animal bends to one side. . . . It thus as a rule avoids the cloud of particles, unless the latter is very large. This simple method of reaction turns out to be more effective in getting rid of stimuli of all sorts than might be expected. If the first reaction is not successful, it is usually repeated one or more times. . .

If the repeated turning to one side does not relieve the animal, so that the particles of carmine continue to come in a dense cloud, another reaction is tried. The ciliary movement is suddenly reversed in directions, so that the particles against the disk and in the

Figure 4.1. *Stentor roeselii* attached to its tube and drawing a cloud of carmine particles into the ciliary disk as part of its normal feeding pattern. (From Jennings, 1906,
Figure 111.)

pouch are thrown off. The water current is driven away from the disk instead of toward it.
This lasts but an instant, then the current is continued in the usual way. If the particles
continue to come, the reversal is repeated two or three times in rapid succession. If this
fails to relieve the organism, the next reaction – contraction – usually supervenes.

Sometimes the reversal of the current takes place before the turning away described
first; but usually the two reactions are tried in the order we have given.

If the *Stentor* does not get rid of the stimulation in either of the ways just described, it
contracts into its tube. In this way it, of course, escapes the stimulation completely, but at
the expense of suspending its activity and losing all opportunity to obtain food. The
animal usually remains in the tube about half a minute, then extends. When its body has
reached about two-thirds original length, the ciliary disk begins to unfold and the cilia to
act, causing currents of water to reach the disk, as before.

We have now reached a specially interesting point in the experiment. Suppose that the
water currents again bring the carmine grains. The stimulus and all the external conditions
are the same as they were at the beginning. Will the *Stentor* behave as it did at the
beginning? Will it at first not react, then bend to one side, then reverse the current, then
contract, passing anew through the whole series of reactions? Or shall we find that it has
become changed by the experiences it has passed through, so that it will now contract
again into its tube as soon as stimulated?

We find the latter to be the case. As soon as the carmine again reaches its disk, it at
once contracts again. This may be repeated many times, as often as the particles come to
the disk, for ten or fifteen minutes. Now the animal after each contraction stays a little
longer in the tube than it did at first. Finally it ceases to extend, but contracts repeatedly
and violently while still enclosed in its tube. In this way the attachment of its foot to the
object on which it is situated is broken, and the animal is free. Now it leaves its tube and

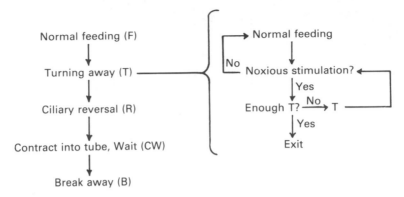

Figure 4.2. Chain-reflex model for *Stentor* avoidance behavior. Left: feeding and the four types of avoidance reaction, in the order they normally occur. Right: a program for each link in the chain.

swims away. In leaving the tube it may swim forward out of the anterior end of the tube; but if this brings it into the region of the cloud of carmine, it often forces its way backward through the substance of the tube, and thus gains the outside. Here it swims away, to form a new tube elsewhere (Jennings, 1906, pp. 174-5).

The behavior of *Stentor* as Jennings describes it is marvellously adaptive. How might we explain it, both mechanistically and from a functional point of view?

The animal has four levels of response to escape from the carmine, each more costly (but also more likely to be effective) than the preceding one: turning away (T) uses little energy and doesn't interfere with feeding (F); ciliary reversal (R) uses little energy, but is an interruption of feeding; contracting into the tube and waiting for a while (CW) is energetic and seriously interferes with feeding; breaking away (B) is most energetic of all, and means abandoning a known feeding site. We don't know what causes the animal to shift from one mode of behavior to another. Jennings is typically cautious: ". . . shall we find that it has become changed by the experiences it has passed through. . . ?" He avoids saying what aspect of the animal's past experience might be responsible for the change from CW to B, the most drastic change in the sequence.

There are two simple ways to explain the succession of avoidance reactions. One way is illustrated in Figure 4.2. The figure shows on the left the sequence of five activities, and on the right the sequence of decisions that allows each activity either to be repeated, or to be followed by the next. The basic idea is that each activity is repeated for a while ("Enough?") and then, if the noxious stimulation persists, the next avoidance response is tried. "Enough" might correspond to number of repetitions, time spent in the activity, or some combination of time and number.

The model in Figure 4.2 is an example of a reflex *chain*: Each activity is the cause of the next one in the sequence. Chains are the first thing to come to mind

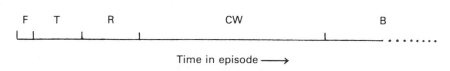

Figure 4.3. A temporal decision rule for *Stentor* avoidance.

when one ponders how a simple animal might put together a sequence of acts. They were very popular with early stimulus-response psychologists. But chains represent a rather rigid form of behavioral organization. If one activity fails to occur or is blocked in some way, the next activity cannot occur, for example. Perhaps for this reason, chain reflexes are rarely found, even in something like walking, where it might seem reasonable that the movement of one limb should precipitate the movement of the next.[2]

An alternative way to organize avoidance behavior is by means of a *temporal program*. The risk to *Stentor* is directly related to the *time* when the carmine is present. The four avoidance reactions are progressively more costly. The animal needs a rule that will match the cost of the avoidance response to the risk. One way to determine which behavior should occur, therefore, is simply to identify three cutoff times, t_T, t_R, and t_{CW}, such that $t_T < t_R < t_{CW}$. Then activity R (say) occurs if $t_T < t < t_R$, and so on, where t is the time since the beginning of the episode. The temporal rule describing the animal's behavior is shown graphically in Figure 4.3. Thus, an internal clock that starts running with the onset of the carmine, with settings corresponding to changes in behavior, $T \rightarrow R$, $R \rightarrow CW$, and so forth, is an alternative method of generating the observed sequence of avoidance reactions.

Noxious substances come in different concentrations, as well as lasting for different times. Risk is directly related to concentration as well as time. The scheme in Figure 4.3 can easily be generalized to two or more risk factors. For example, suppose that risk (H) is proportional to the product of concentration (C) multiplied by time: $H = tC$. Then the animal simply has to identify three values of $H - H_T$, H_R, etc. – that define transitions from one behavior to the next, as before. The whole thing can be represented in a two-dimensional space (rather than the one-dimensional space of Figure 4.3), as shown in Figure 4.4. Obviously, the scheme is not restricted to a product decision rule or to two dimensions: Any function, $H = f(t, C)$, can be used to partition the space into regions associated with different activities, and the space can have as many dimensions as there are risk factors.

The one-dimensional, temporal decision rule suggests an obvious mechanism for the *Stentor* behavior – a clock. But the two-dimensional case makes it clear that an explanation by decision rules is functional, not mechanistic. It is not at all obvious what mechanism the animal uses to compute the relative importance of time and concentration in deciding on an avoidance reaction. Yet in this relatively simple case it is easy to imagine possible mechanisms. For example,

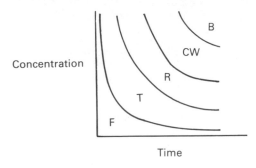

Figure 4.4. A time-concentration decision rule for *Stentor* avoidance.

suppose that the threshold concentrations at which each of the four reactions occur are ranked according to their cost $C_T < C_R < C_{CW} < C_B$, and that the reactions are mutually exclusive (see the discussion of thresholds and reflex competition in Chapter 2). Then reaction T, with the lowest threshold, will occur first, blocking the others. Suppose further that the threshold of any reaction *increases* as the reaction continues to occur (habituation). This mechanism will ensure that eventually it will be supplanted by R, as the threshold for T increases owing to habituation. The same process repeated leads to the transition from R to CW, and so on. Given appropriate choices for time constants and inhibitory relationships, a plausible mechanism for *Stentor* avoidance might be arrived at (see note 2).

This mechanism is not the only possible one. Even if we restrict ourselves just to processes involving time-varying thresholds, several mechanisms for the *Stentor* avoidance reaction might be invented. To discriminate among them would require careful behavioral and perhaps even physiological experiments. To eliminate the chain-reflex possibility, some way would have to be found to prevent one of the reactions. If reaction R can occur without preceding reaction T, then a chain is ruled out. Since this apparently occurs once in a while without any intervention, the chain reflex idea is unlikely from the outset. To discriminate among other possibilities, we need to know the dynamic properties of the system: How quickly do the activities succeed one another in a constant carmine concentration? What is the response to a step change and to alternations between different concentrations? and so on (see the Moorhouse et al. experiment discussed in Chapter 2). Whatever the answer it will have to be compatible with the very simple, static functional account.

The functional account is more general than the particular mechanism that underlies it, because it will apply not just to *Stentor*, but to any small animal similarly situated – even though different species (and perhaps even different individual *Stentors*) might well do the job in different ways. If we can think of more than one way to solve a behavioral problem, there will often be comparable diversity in nature's solutions.

An animal's ability to behave according to the prescriptions of a functional account depends upon particular mechanisms – here of the reflex type, but in other cases, more complicated. For some purposes the functional account is more useful, for others we might like to know more of the details of how the animal does it. For example, if we are interested in how an animal's behavior relates to its niche, the functional account is better; but if we want to get at the physiological basis for behavior, the mechanistic account is more useful.

Obviously, the more intelligent the animal, the more possible mechanisms may underlie a given example of adaptive behavior. Consequently, when we are dealing with mammals and birds, functional or descriptive theories generally do much better than mechanistic ones.

Operant behavior and learning

How does the avoidance behavior of *Stentor* stack up against the operant behavior of higher animals – with the behavior of a cat escaping from a puzzle box, for example, Thorndike's original learning situation?[3] At one level, the similarity is great. In both cases, a change for the worse in the animal's environment causes a change in its behavior. Further changes continue to occur in a systematic way until the irritating condition is relieved (variation and selection). The increasing rapidity of contraction (CW and B) in response to the carmine resembles the increasing efficiency with which the cat escapes on successive trials. The cat's improvement would probably be called "learning": Does *Stentor's* behavior also qualify?

The same external situation – carmine, plus normal feeding behavior by the animal – leads initially just to turning away, but later to breaking away. This means that there has been a change in *Stentor's internal state*. Such changes must have been occurring right along, otherwise the initial avoidance reaction, turning away, would have persisted indefinitely. Thus the animal shows one essential feature of learning, a change in behavior potential as a result of experience. So far our *Stentor* is doing about as well as Thorndike's cat. What is the crucial difference?

The big difference is in the effect of either lapse of time or change in situation on the animal's behavior. The same *Stentor* exposed to carmine after 24 hours will almost certainly behave as it did when first exposed, not as it did at the end of the first exposure. A small lapse of time will have lesser effects on *Stentor's* behavior, but some new experience during that time – a new chemical stimulus, for example – is likely to abolish the effect of the earlier experience. But the cat that escapes efficiently from the box after a few successive trials will not be the same even after a week as it was when first put in the box. It may not escape quite as rapidly as it did on its last trial, but it will certainly do better than it did on its very first trial. The ability of the cat to respond correctly after lapse of time is not immune to the effects of intervening experience, but is likely to be much less affected than the behavior of *Stentor*. What does this difference between cat and protozoan mean?

The obvious answer is that the protozoan cannot remember things for as long as the cat. Leaving aside for a moment the problem of exactly what we mean by "remember," I think that this begs the question. The problem is, *Why* doesn't *Stentor* remember things as long as the cat? The difficulty is unlikely to be some general limitation on the persistence of behavioral changes in very small animals. It is in theory and in fact a simple matter to set a molecular "switch" that affects behavior after long delays. Even precisely timed delays are not rare in nature. Many very simple animals show periodicities that extend over very long times – circumannual rhythms, the periodicity of the 17-year locust, and so on. Nature has no problem in building clocks or retaining changes over long periods.

We can get an idea of the real difference between cat and *Stentor* by looking at what each creature *ought* to do in its respective problem situation – given what we can guess of their different discriminative abilities. The cat faces a much more clear-cut problem. The cat can discriminate puzzle boxes from the many other things with which he is familiar. When he senses the box again, he is unlikely to confuse it with anything else and has no reason to suppose that his previous solution is not appropriate. The *Stentor*, however, is unlikely to be able to discriminate carmine from many other chemical mixtures. Consequently, when it gets the carmine on a later occasion, it may simply be unable to identify it as the same event: Even if the carmine has exactly the same sensory effect on both occasions, the effect may be sufficiently similar to the effect of other chemicals, experienced before and after, that the *Stentor* may still be literally unable to identify its second carmine experience as a repeat of its first. Even a highly intelligent *Stentor* might be well advised to treat a second exposure (especially if much delayed after the first) as a completely new situation.

Stentor learning about carmine may be like a man with very poor vision meeting again someone he has met before: If he has to rely entirely on vision, he will necessarily be unsure whether he has really met the person before or not, quite apart from his intelligence and ability to remember a face. The poorly sighted person (like the hearing-impaired person) may appear dumb not because he cannot remember and understand, but because he cannot discriminate. In other words, *Stentor* probably hasn't sufficient information to justify repetition on the second carmine exposure of the behavior it had developed in response to the first exposure. *Stentor* may fail to learn, therefore, not because it cannot remember, but because it cannot be sufficiently sure when the same situation has recurred.

The abilities of animals are always in harmony with one another; an animal will not develop one ability to a high level if lack of another makes the first useless. An animal that can only move slowly will not usually have highly developed distance vision; a short-lived animal such as a butterfly may have little ability to repair injuries; a fish that lives in caves will often be blind. In similar fashion the persistence of memory is functionally related to the *number of things that can be discriminated*. This relation can be illustrated as follows. Imagine a very simple animal that can classify stimuli in just four ways: good–bad, and two other,

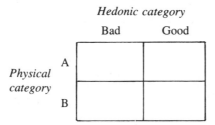

Table 4.1. *Contingency table for a simple animal that divides all stimuli into two categories, A and B.*

"neutral" categories unrelated to good–bad (two categories of physical size, or intensity, for example). Every situation that the animal encounters must then fit into one of the four cells defined by this 2 x 2 table (Table 4.1). In order for it to be worthwhile for an animal to form an association between a given state of the world and its hedonic consequences, there must obviously be a real association. For example, if our animal can only partition neutral events into the two categories of "intense" and "weak," then there must really be some general correlation between the intensity of a physical event and its hedonic consequences; loud noises must be generally associated with bad things and soft noises with good or neutral things.

It is obvious that the more possible categories that animal has available, the higher the real correlation between events in a given category and their hedonic consequences is likely to be. For example, many loud noises are quite harmless, but loud, staccato noises coming from long objects are always potentially dangerous. The more qualifiers (additional categories) the animal has available, the more accurately it can identify the real concomitants of good and bad. The more accurately the animal can identify signals for good and bad, the more worthwhile it becomes to remember them.

I conclude that the poor long-term memory of *Stentor* is much more likely to reflect a limitation on its ability to classify physical events into a number of categories than some limitation on storage or the persistence of physical change. The problem is an information-processing one. In order to react to carmine as it did on the first occasion *Stentor* must be able to identify the relevant features of its environment accurately enough to minimize the chance of the two kinds of possible error: reacting to the wrong stimulus and failing to react to the correct one. If its categories are so few[4] that carmine does not uniquely occupy one of them, then it lacks the ability to tell whether a given chemical stimulus is carmine or something else, and so cannot detect the special properties of carmine presentations. Hence, if we want to show learning in simple organisms, we need to pick stimulus categories so separated that the animal is unlikely to confuse the experimental stimuli with others encountered naturally. The major difference between animals that can, and cannot, learn – between protozoans and pigeons – is in their ability to differentiate among states of nature: The difference is in what they know, rather than in how well they can remember.

Operant and classical conditioning

The basic operation in learning is using something as a signal for something else. The signal can be used in different ways, corresponding to habituation, sensitization, pseudoconditioning, and classical (Pavlovian) and operant (instrumental) conditioning. Sometimes only the latter two are classified as *learning*; this is the narrow sense of the term. For lack of any other general term, I will also refer to all five types as *learning*; this is the broad sense. I define these five types of learning now in a general way. The next chapter goes into more detail about particular operant and classical conditioning procedures. The mechanisms of classical and operant conditioning are discussed at length in the last two chapters.

Every stimulus has two aspects: good–bad (hedonic quality), and other, nonhedonic properties. For example, an electric shock is bad, but it also has a certain "tickling" quality, a certain duration, a limited spatial extent, and so on, and these things have no particular good–bad aspect to them. Similarly, food is good but it also has a certain taste, texture, temperature, and color. Sometimes the hedonic and nonhedonic aspects are distinct, but sometimes they are not. Spatial location rarely has any hedonic quality, for example, but the taste of a food and its value may be inseparable. Separable or not, every stimulus has its hedonic and nonhedonic aspects. In the simplest kinds of learning, the animal uses the nonhedonic characteristics of a stimulus as a cue for its hedonic qualities.

Habituation. As we saw in Chapter 2, habituation is the selective waning in strength of a response to repeated stimulation. For example, anything novel is potentially dangerous. Hence any sudden novel sound tends to be treated as potentially harmful: *Stentor* retreats into its tube; a rabbit pricks up its ears, turns toward the sound, and gets ready to bolt for its burrow. But as the sound is repeated and is not accompanied by any bad consequences, its nonhedonic properties, which at first were taken as potentially alarming, are eventually treated as a "safety" signal and the sound alarms no more. Habituation is, therefore, a kind of learning in which the nonhedonic aspects of a stimulus are treated as a signal for safety or neutrality. The stimulus is reclassified as "harmless."

In higher animals, the effects of habituation may be long-lasting or not, depending upon the stimulus. Wild birds may take alarm at a new feeding station, for example, but their alarm will habituate and will not reappear even after a few days away. On the other hand, the effects of habituation to a series of pistol shots will not persist unless the experience is repeated many times. The more intense the stimulus, the more transient the habituation to it. In lower animals, the effects of habituation rarely persist.

Habituation can often be abolished by some new experience, *dishabituation*. For example, if an animal has been habituated to a series of loud noises and is now placed in a new situation, or presented with a bright light, another animal or some other striking stimulus, presentation of another loud noise will often alarm again.

Sensitization. This and the other three types of learning I will discuss all involve two kinds of stimulus. First, a stimulus with hedonic value and a stimulus with less or no hedonic value: a *reinforcer* or *unconditioned stimulus* (US) and a *conditioned stimulus* (CS). These terms derive from the use of hedonic stimuli such as food and electric shock in Pavlovian and instrumental conditioning situations. A noxious stimulus, such as an electric shock or a loud sound, will usually elicit a startle reaction from a human or an animal. A loud sound will startle, a weak sound will not; but a weak sound (CS) presented after one or two shocks (US) may startle again. This is not because the animal has learned anything about the relation of the shock and the sound: The sound is not a signal for the shock. Sensitization is a sort of *confusion* effect: The animal is anticipating the shock; sudden shock and sudden sound share nonhedonic qualities; so sound after shock is reacted to like shock.

The term *anticipation* here says rather more than necessary. All that is involved is that the shock changes the animal's state in such a way that another shock, or stimulus like shock, will be reacted to more strongly than if the first shock had not occurred. In the same way, *Stentor* after a second or two exposure to carmine reacted differently to further carmine: The initial few seconds of carmine exposure sensitized the animal so that further exposure elicited more and more extreme reactions. Only local memory is involved; a weak sound long delayed after shock will elicit no reaction, just as a second dose of carmine delayed after the first produces the same effect as the first rather than a stronger effect.

Any frightening situation sensitizes the animal; it is not necessary actually to present a noxious stimulus. For example, the following situation provides a good classroom demonstration of the human startle response: A student is invited to participate in front of the class in an experiment to measure skin resistance. He (or perhaps more usually, she) is asked to place two fingers on two metallic contacts protruding from an unfamiliar piece of obviously electrical apparatus. At the instant that the finger touches the (completely harmless) contacts, the demonstrator sounds a loud siren. The siren produces no, or at most a weak, startle when presented without the buildup. When it coincides with what the student fears may be a painful, or at least a novel, experience, however, the reaction is vigorous.

Pseudoconditioning. This is very similar to sensitization. In both cases, prior presentation of a US causes a CS to elicit the same reaction. The effect is called

Shock Onset
(training)

		Gradual	Sudden
Light Onset (testing)	Sudden	2/7	7/9
	Gradual	8/10	1/8

Table 4.2. *Summary results from a pseudoconditioning experiment by Wickens and Wickens (1942). Cells show numbers of animals per group showing the four possible results.*

sensitization if the CS at a stronger intensity can elicit the reaction (shock → startle; loud sound → startle; weak sound ↛ startle unless preceded by shock or loud sound). The effect is called *pseudoconditioning* if the CS never elicits the reaction on its own.

An experiment by Wickens and Wickens (1942) is a neat demonstration of the role of confusion or *generalization* in these effects. They trained two groups of rats in a box with two chambers to run to the second chamber when shocked in the first. For one group the shock came on suddenly, for the other, its onset was gradual. Half the animals in each group were then tested with a light that came on suddenly; the other half with a light that came on slowly. As you can see from Table 4.2, the light nearly always elicited running when its speed of onset matched the training condition, but rarely did so when it did not match.[5]

Sometimes animals habituate to a stimulus, sometimes they are sensitized by it, so that their reaction increases with successive stimulus presentations. We seem to have things both ways here: Since a reaction can only decrease or increase, we can explain everything. Fortunately, the effect to be expected from a given stimulus does seem to depend on measurable stimulus properties, most notably stimulus intensity. Animals seem to habituate to stimuli of low or moderate intensity; but become sensitized by intense stimuli. For example, a sudden, loud tone will elicit a startle response from rats. But after repeated presentations, the response habituates. However, if the same loud tone is presented against a background of loud white noise, the reaction not only fails to habituate, it increases across tone presentations. The critical factor seems to be the damage-potential of the situation.

Classical conditioning. If a hedonic stimulus (US) is reliably preceded (signaled) by a neutral stimulus (CS), many animals can learn to use the CS as a signal for the US. The process differs from sensitization and pseudoconditioning in that the CS must really be a good predictor of the US; it is not sufficient that they occur more or less close together in time. In the standard procedure studied so extensively with dogs by Pavlov, a US such as food is repeatedly preceded by a neutral CS such as a tone. After a few such pairings, the salivation produced originally only by food is now produced by the tone as well. The reaction to the US is called the *unconditioned response* (UR); the reaction to the CS is called the *conditioned response* (CR). A comparable experiment with *Stentor* would involve pairing brief carmine presentations (the US) with some other stimulus, such as a change in illumination or temperature (the potential CS). The carmine elicits turning away; the light-change initially elicits nothing. If after a few pairings the light produces turning, then we *may* have classical conditioning. Additional control experiments in which the order of CS and US is varied (to rule out sensitization and pseudoconditioning) are necessary to be sure – we want to be sure that it is the predictive relation between CS and US that is important. And additional tests with the CS alone, after a delay, are necessary to see if the change is a relatively permanent one – as it usually is in higher animals. In practice, of course, protozoans rarely pass all these tests; most of their learning is habituation, sensitization, or pseudoconditioning.

Classical (Pavlovian) conditioning is the prototype for all signal learning. The CS is a signal for the US, and the animal reacts to the CS as if it were the US – although careful scrutiny usually shows that the reaction to the CS is anticipatory, rather than just a copy of the reaction to the US.

Operant conditioning. Suppose we pair a tone with food a few times, in Pavlovian fashion, but then present the tone alone: What will the animal do? Pavlov knew perfectly well that his dogs would not sit quiet under such conditions (that's one reason he restrained them in a harness). Given the opportunity, the dog does things that might produce food: If the experimenter is in the room, the dog will beg from him. If not, the dog will paw at the food bowl and try to get out of the room – whatever the environment permits by way of exploration. Suppose one of these explorations is, in fact, effective, as it might be if the experimenter had merely hidden the full food bowl, for example. The dog then eats and is removed from the room. If the experiment is repeated on the following day, the tone, the room, and the various other stimuli will not now produce the random searching we saw the day before. Instead, the dog is likely to go rather directly to the place where he previously found the food.

This is a two-phase process: (a) The first phase is unsystematic behavior that eventually leads to something good (or avoids something bad). (b) The second phase is the recurrence of efficient behavior when the animal is later returned to the same situation. The two phases together are called *operant conditioning* (an essentially equivalent term is *instrumental learning*). The first phase fits the control-by-consequences definition of operant behavior. The second phase is referred to as the control of operant behavior by a *discriminative stimulus*; more on this in a moment.

Operant and classical conditioning are closely related. The parallel between the two is easy to see in avoidance or escape conditioning. A situation that signals shock or some other aversive stimulus elicits a range of avoidance and escape reactions of three types: fight (attack any plausible target, such as a conspecific), flight (escape from the situation), or "freeze" (immobility). These reactions tend to occur in a systematic sequence (as in the *Stentor* example), depending upon the severity and duration of the threat and the opportunities offered by the environment. If no reaction is successful in escaping from the threat, then the animal settles on the final reaction in the series, which, in the case of shock, is generally immobility. This would usually be described as classical conditioning: An aversive CS produces conditioned immobility or "helplessness."

But if one of the reactions *is* effective in eliminating the threat, then a recurrence of the situation is likely after a few repetitions to lead to reliable repetition of this effective avoidance response. This is operant conditioning, and the response is an *operant response*.

The relation between classical and operant conditioning is clear: Classical conditioning is the process that permits an animal to detect that the CS predicts the US. In this way the animal is able to identify the hedonic qualities of the situation: Is it good or bad, and what kind of good or bad is it? Given this information the animal, be it rat or *Stentor*, has available a repertoire of potentially useful reactions that nature (in the case of *Stentor*) or nature and past experience (in the case of the rat) has given it to try. If some reaction is effective, it is selected (in ways I discuss later) and recurs when the situation recurs: This control of operant behavior by the situation in which it has developed is termed *discriminative control*. The environmental features that are effective in controlling behavior are collectively termed the *discriminative stimulus* for the response. Classical conditioning obviously sets the stage for discriminative control.[6]

Generalization and discrimination

Adaptive behavior demands as a minimum that animals respond differently in different situations. The different kinds of learning just discussed define "situation" in different ways. In habituation, the period just following a stimulus is treated by the animal as a different situation from other times. Sensitization and pseudoconditioning work in the same way, the time just after a US is different from other times. In operant and classical conditioning, the animal further differentiates the world according to stimuli in addition to the US. The CS and its associated context (discriminative stimulus) define not only a particular hedonic quality, but also, in the operant case, a pattern, perhaps a unique pattern, of behavior.

Just what the term *situation* signifies in the operant and classical conditioning of birds and mammals is the topic of later chapters. For now, just consider the idealized view illustrated in Figure 4.5. The figure shows the various states of the world ("stimuli") that can be discriminated by an animal, arranged along the

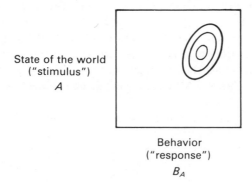

State of the world
("stimulus")
A

Behavior
("response")
B_A

Figure 4.5. Definition of an *operant* as a region in a stimulus-response space. Closed contours are lines of equal probability.

vertical axis; the various modes of behavior of which the beast is capable are arranged similarly along the horizontal axis. Suppose that in the situation defined as "A," behavior B_A develops as a consequence of operant conditioning. Physical situations never recur identically – one day is sunny, the next is overcast or rainy, and so on. Consequently, animals are designed to accept minor variations in stimulus situation. Animals never repeat a response identically either – nor should they, since some variants are more effective than others, and variability itself is sometimes advantageous in avoiding traps (as I noted in Chapter 3). These two sources of variation mean that the pattern of behavior that develops in situation A is better represented as a region or *probability distribution* in the stimulus-response space in Figure 4.5 – rather than as a point where a unique stimulus produces a perfectly stereotyped response. One can think of this relation between situation and action, termed an *operant* (*habit* is the older term), as a hillock in the space centered on the point (A, B_A), where the third dimension, out of the page, corresponds to response strength or probability.

This picture is an enormous simplification, but it enables me to define *generalization* and *discrimination* in a simple way. *Generalization* refers to the effect of systematic variation in the physical stimulus situation on the operant response controlled by that situation. Usually the effect is to reduce the probability of the measured response, as suggested by the hillock picture in Figure 4.5. Thus a pigeon trained to peck a green disk for food reward will peck less frequently on a yellow or red disk. Generalization leads to two kinds of "confusion": A given operant response can occur in situations other than (but similar to) the training situation; and responses other than the given operant can show up in the training situation.

The extent of generalization obviously depends upon the animal's individual and racial experience of variability. Aspects of the environment that never vary can be ignored: Most lizards can move up or down a temperature gradient to find the temperature that suits them best, but some tropical rain-forest species have lost this ability. Because of the constant temperature of their environment they are no longer sensitive to temperature variation. This is an extreme deficit in discrimination and reflects the animals' racial history. An animal reared in an experimental environment where large changes have little significance is likely to generalize more broadly than an animal reared in an environment where large significance routinely attaches to small changes. This is true generalization, an effect of individual history. Measured generalization reflects both individual and racial history.

Discrimination refers to the limits on an animal's ability to treat similar situations differently. Generalization and discrimination are loosely, and asymmetrically, related. Generalization refers to the stimulus variations the animal is willing to accept and still define the situation as essentially unchanged. Discrimination refers to the minimal variations in situation (stimulus) that the animal is capable of treating as different. A pigeon may be able to detect the difference between wavelengths of 500 and 600 nm (they look green and yellow to us), but be willing to treat them alike because in the past they have been associated with the same consequences. On the other hand, wavelengths of 500 and 501 nm can be associated with consequences as different as we like, yet the pigeon will not treat them differently because he cannot tell them apart.

An animal may be able to detect small differences, but be inclined to accept large ones as of no significance. Hence, broad generalization need not imply poor discrimination. But an animal cannot treat similar situations as different if it cannot tell them apart. Hence poor discrimination implies broad generalization.

Learning in the broad sense requires a change in the animal's internal state caused by stimulation. Any change of state is a sort of memory. Habituation, sensitization, and pseudoconditioning require only local memory: a change in state initiated by US presentation that decays with post-US time. Learning in the narrow sense (often termed *associative learning*) requires a change of state of a more subtle kind, one that can be reinstated by only part of the original stimulus complex – the CS alone, without the US. This change is termed *long-term memory*.

Students reading with yellow marker in hand will notice that I have used the terms *memory, stimulus, response*, and *internal state* without defining them precisely. The last section of the chapter plugs this gap.

THE LOGIC OF HISTORICAL SYSTEMS

The eccentric and profound philosopher Ludwig Wittgenstein was discussing the topic of *time* with a colleague. As philosophers often will, the colleague soon asked Wittgenstein to define what he meant by time. Wittgenstein replied along the following lines: "If you know what I mean by *time*, let us continue our discussion. If not, let us discuss something else." I have followed a similar strategy. The terms *stimulus, response, memory*, and so on are understood by everyone at a level sufficient to use them in discussing simple learning phenomena. Nevertheless, unlike "time," the terms can be defined precisely. Definition is helpful because it shows that the role of discriminative and conditioned stimuli must go well beyond their obvious effect in eliciting a response. The analysis also clarifies limitations on our ability to understand *historical systems*, that is, systems whose future behavior is not predictable from what we can discern of their present state, but depends on past events: A computer is a historical system because we can't tell from looking at it what it will do when we type something at the console; but a mechanical clock is not, because we know everything of interest about it from the position of its hands.

The first thing to remember is that when we look at the behavior of an animal and try to understand it, we necessarily deal in *models*. We cannot hope to comprehend the real complexities of any animal, nor would such detailed knowledge be very useful. We are always talking about a simplified "ideal animal" that behaves like the real animal only in respect of the things we happen to be interested in. When we use terms like "stimulus," "response," and "internal state," when we assume that the animal is the same when we repeat earlier treatments, it is understood that we are thinking of our model animal, not the real one. Nothing is ever the same from day to day, least of all a living organism. Every stimulus is physically different from every other, as is every response. Any real animal has an infinity of physically different internal states. Nevertheless, the first step in understanding behavior is to put similar things into classes and look for regularities. The infinity of things that are different about the animal from occasion to occasion is of less interest than the much smaller number of things that are the same, that repeat themselves in a lawful manner.

Let us begin with the concept of *stimulus*. There are two problems: First, how are we to classify stimuli that have no overt effect on the animal? Second, are we to define stimuli physically or functionally?

Imagine a passenger in a bus on a busy city street. The bus halts briefly at a stop and at the same time a blue car pulls up on the other side of the street. Within a minute the bus pulls away and the car is lost to view. The blue car has elicited no reaction, perhaps not even a glance, from the bus passenger. Yet the next day, when questioned by police interested in a robbery, he may be able to recall details about the car such as its color and make. Clearly, the car was a stimulus in some sense. Hence, stimuli need not elicit immediate reactions.

This example suggests an answer to the second question also. The passenger describes his experience not in terms of the physical properties of the car, but in terms of the knowledge categories he shares with the interrogator: The car was a blue Ford sedan. Someone who had never seen a car and did not, therefore, possess the appropriate categories, would have had to resort to a much more cumbersome and detailed, and probably less accurate, description. Hence, stimuli are most usefully defined in terms of what the subject knows, that is, functionally rather than physically.

Finite-state systems[7]

These conclusions can be formalized as three definitions:

1. A *stimulus* (input) is something that either elicits a response, or changes the internal state of the animal.

2. A *response* (output) is something that is jointly determined by a stimulus and an internal state.

3. An *internal state* determines which stimulus will elicit which response and how a stimulus changes the internal state.

Stimulus and *response* correspond approximately to their commonsense meanings. I argue in a moment that *internal state* corresponds to what I have been calling a *situation*. These three definitions formalize something long accepted as a working rule by most psychologists, namely that stimulus, response, and internal state are all defined in terms of one another.

If we simplify things, the relations among these three terms can be illustrated by means of two tables. Suppose that changes inside our model animal are controlled by a clock, and occur only with each clock "tick," rather than continuously as they probably do in real animals. (This is termed a *discrete-time* model.) In between clock ticks, the model animal stays the same. This simplification does not limit the generality of my conclusions, because any continuous system can be mimicked as closely as we pleased by a discrete system with a large number of states.

According to our definitions, there are just two ways that an internal state can affect behavior: (a) It can affect the relation between stimuli and current responses. For example, in state 1, stimulus S_1 might elicit response R_1, whereas in state 2, S_1 might produce R_3. (b) A state can affect the relation between stimuli and future states. For example, in state 1, stimulus 1 might cause a transition to state 2; but in state 2, stimulus 1 might cause a transition to state 3. In other words, if we are going to describe a historical system in terms of stimuli, responses, and internal states, we can describe everything we want to know about it by means of two tables. One table shows the relations between stimuli and current responses, with state as the parameter. Given the state, this *S-R* table shows which stimulus elicits which response. A second table shows the relations between stimuli occurring at one time (tick of the clock) n and the subsequent state at time $n + 1$, with state at time n as the parameter: given the state at time n, table *S-O* (*O* for organism) shows which stimulus causes which state *transition*.

These two tables can be used to illustrate the logic of the earlier example. To make the tables easier to read I adopt the convention that STATES of the model are represented by capital letters, stimuli by lowercase letters, and *responses* by lowercase *italics*. The model obviously does not define all the behavior of the organism: unknown states, stimuli and responses are denoted by x's.

Table 4.3 illustrates the car story: The *S-R* table is on the left and the *S-O* table on the right. The *S-R* table shows the response, *n, y,* or *x,* elicited by the stimulus? (the question) or by some other stimulus, *x,* when the individual is in state *X* ("other") or *S*. The *S-O* table shows the state produced by stimuli *c* (the car) or *x* ("other"). The tables have a lot of x's in them, because all we know about the situation is that if the individual saw the car (i.e., his state was changed by it), then he responds "yes" to the question, but otherwise he does not. The critical entries are in the *S-R* table where state *S* yields a positive response to the

	S - R State			S - O State	
	X	S		X	S
s?	*n*	*y*	c	S	S
x	*x*	*x*	x	X	X

Stimulus

Stimuli	Responses	States
s?: Did you see a car	y: Yes, I saw it	S: having seen the car
c: car	n: no	X: unknown
x: unknown	x: unknown	

Table 4.3. *Finite-state tables for a two-state model. Left: Stimulus → Response relations. Right: Stimulus → State-change relations.*

question (s? → y), and in the *S-O* table, where the sight of the car changes any state (X) to S (c → S). The other entries are essentially "don't knows" – we don't know the effects of stimuli other than the ones whose effects we are describing.

The essential information in Table 4.3 is shown in another way in the upper part of Figure 4.6. Here states are represented as nodes in a directed graph. Each arrow in the graph begins with the stimulus received and ends with the response made, and the arrow connects the initial and final states. The lower part of Figure 4.6 shows a slightly more realistic representation: Instead of just two states, "having seen the car" (S) and "not having seen the car" (X), it shows three, the third one being "recalling the car" (S*). The idea here is that the question "Did you see it?" changes the individual's state into one where he is actually recalling the car. The evidence for this additional state is that the first question about the car is usually answered more slowly than later ones, suggesting that the first question somehow reinstates or makes more readily available other information about the original experience. Obviously still more complex representations could be devised to accommodate additional experimental facts.[8]

State S* corresponds to the *recall situation*; it is the "state of mind" created by the interrogator's question. Discriminative and conditioned stimuli seem to have a similar effect, in the sense that they reinstate a certain set of behavior potentialities. The term *expectancy* is sometimes used but, like the term *anticipation* I used earlier, it may say more than necessary. I return to this question in the last two chapters.

Equivalent histories. The common feature of all these representations is that "seeing the car" produces a persistent change that can be detected by the different

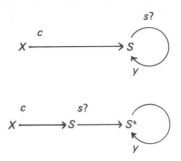

Figure 4.6. Two finite-state models for recall.

responses made to a later question. All the models are oversimplified in an obvious way. The schemes in Figure 4.6 make no provision for our passenger to remember anything other than the car. Clearly state S must be further differentiated into states S_1, S_2, and so forth, to take account of the effects of later experiences. In the same way, state X must also be differentiated into numerous states to accommodate the varied possibilities for prior experience. Nevertheless, if we ask our passenger only the one question, everything we know about him can be represented by the upper diagram in Figure 4.6. The states of our model must be inferred from the questions we ask of the subject, and they are limited by those questions.

A series of questions, or stimuli, is a *history*. The *states* in our models (any models) are *equivalent histories*. This idea can be explained in the following way. Imagine that you have an infinite number of *replicas* of the historical system (animal, human, machine) that you wish to study. Imagine further that you know in advance what things are stimuli (inputs) for the system and what things constitute its responses (outputs). Then take replica number one, and subject it to a sequence of inputs of infinite length. Then take replica number two and do the same thing, but with a different, infinite, sequence of inputs; and so on for all the replicas and all possible sequences (this is a long experiment!). Now sort all the stimulus-response sequences you have obtained. Suppose you find, for example, that if the first three inputs are the sequences abc, acb, abb, or acc then the response to the fourth input, whatever it may be, is the same; and similarly for the fifth, sixth, and so on. Apparently these four histories are equivalent in the sense that the system's behavior afterward is the same for all. It is natural to summarize this list of experimental results by saying that these histories all put the system into the same *state*. Thus from the experimenter's point of view, postulation of a state is just a way of summarizing the effect of various past histories that are equivalent in terms of the system's future behavior.

The experimenter's hope is that by classifying the various histories he has obtained, he can reduce the infinity of input–output relations to a finite number of states. Since a real experimenter has less than infinite time and replicas available, he must proceed by shortcuts, making guesses about possible states based on a preliminary classification, and then choosing "test histories" that will check out his guesses. The classifying process is called *induction* or the *inductive method*; the guess-and-test is called the *hypothetico-deductive method*. All science involves both, although the proportions are different for different sciences: Paleontology and sociology, for example, are largely inductive; physics is largely hypothetico-deductive.

It should be clear that understanding a system with very many states is a formidable task. Moreover, the logic of historical systems imposes impossible conditions on the conscientious experimenter – an infinity of replicas, and infinite time with each. These conditions are never satisfied. In the matter of replicas, there are only two options: either take different animals and assume they are essentially the same, or take the same animal at different times and assume that the intervening history has had negligible effect. The first option corresponds to the *between-group* method of experiment. Different histories, the experimental treatment and one or more control treatments, are given to each group and the average results are compared. We can never be certain here whether the animals are really equivalent to each other, or whether it is legitimate to average the results.[9] The second option is the *within-animal* method. Treatments to be compared are presented successively to the same beast. To check that no systematic changes are taking place, "control" treatments are often repeated in between different "experimental" treatments; but we can't be certain here whether similar behavior on successive occasions means that the animal returns to essentially the same state, or not.

Despite these logical difficulties, both within- and between-animal methods are widely, and successfully, used, albeit for rather different kinds of problems. We have no real choice. When the causes of present behavior can lie in the remote past, the problem of experimental analysis becomes essentially insoluble, in the sense that there is no practicable method that will *guarantee* a solution. Hence, the experimenter has no choice but to set up situations where he hopes he can rule out the effects of remote past history. He tries to ensure that the animal is in the same state when he makes the manipulations whose effects he is comparing, which means either assuming that different animals are equivalent, or that certain kinds of intervening experience have negligible effect, so that "behavior" (really, the *state* of the animal) is *reversible*.[10]

Memory. Memory in the broad sense is simply an effect of a past event on future behavior. Any system with more than one state has memory of a sort. As we have seen, there are two main types of memory: habituation, sensitization, and pseudoconditioning require only local memory; classical and operant conditioning require long-term memory. The adjective "long-term" is unfortunate in its emphasis on the persistence of memory. The distinctive thing about long-term memory is not so much its persistence as its *context sensitivity*, that is, the power of a stimulus situation to recreate a given behavior potential.

Long-term memory can be further subdivided into two types, synchronous and temporal. *Synchronous* refers to control of operant behavior by the current physical stimulus situation: the rat pressing a lever in a particular Skinner box, the porpoise doing its tricks on the trainer's signal. *Temporal* refers to control of operant behavior by an event in the past. For example, if fed at fixed time periods, most animals will learn to anticipate food delivery and will nearly always be close to the feeder as feeding time approaches. Each food delivery is a temporal signal for the next and controls future food anticipation. Animals can also learn to choose one stimulus rather than another on the basis of the most-recent, or least-recent, stimulus they have seen. Temporal control differs from local memory in two main ways: (a) It is context-sensitive – an animal can be controlled by one past event in one context, by another in another. Apart from dishabituation, local-memory effects are largely independent of context. (b) It is modifiable by training: The animal can learn to do different things under the control of different stimuli after different times, in more or less arbitrary combinations.

Temporal and synchronous control are not mutually exclusive; all temporal control is context-dependent, and the effects of current stimuli are almost always modified by events in the animal's immediate past. For example, an animal once fed for pressing a lever and now not will eventually cease to press; but a single food delivery in the familiar context will at once reinstate the old pattern.

Endogenous rhythms. These rhythms, such as *circadian* (approximately 24-hour) rhythms shown by almost every organism, are intermediate between synchronous and temporal control. Rhythms have no single temporal stimulus; external stimuli are involved in the form of *zeitgebern* or "Timekeepers" such as the light–dark cycle that advance or retard the natural rhythm so as to synchronize it with an environmental cycle.

SUMMARY

The most important difference between kinetic behavior and the operant behavior of mammals and birds is in its sensitivity to context. The internal state of the kinetic animal depends only on its current environment and the environment in the immediate past (local memory). The bacterium sees the world through a brief, moving time-window, with no memory of the past and no prevision of the future. Mammals, birds, and other intelligent animals can behave in a variety of ways, however; and the ways that they behave depend on both their current environment and their past history (long-term memory).

This chapter has dealt with three related topics: (a) the relation between mechanistic and functional explanations of behavior; that is, between explanations in terms of antecedent events (stimuli past and present) and in terms of consequences (goals, rewards and punishments); (b) the differences between the operant behavior of simple and higher animals; and (c) the logic of explaining behavior in terms of antecedent events.

There are three main conclusions: (a) Mechanistic and functional accounts answer different questions: The mechanistic account is specific and leads naturally to physiological questions. If we are curious about how *this animal* works, this is the kind of explanation we want. The functional account is more general and tells us something about the animal in relation to its niche – not as an individual, but as a member of a species. (b) The difference between the operant behavior of simple and complex animals seems to be in the richness with which the animal's environment is represented internally. The world of the bacterium has few categories, the world of the rat or chicken has many. This difference allows the "higher" animal to detect predictive relations between neutral and valued events – and makes it worth remembering them. (c) Stimulus, response, and internal state are all defined in terms of one another. From the observer's point of view, an internal state is a set of equivalent histories. The term *context* is functionally equivalent to *internal state*. Any system with more than one state shows "memory"; the special feature of operant and classical conditioning (as opposed to other forms of learning) is the ability of different contexts to reinstate previously acquired patterns of behavior.

The operant behavior described in this and the preceding two chapters can be arranged in three levels, defined by the type of behavioral variation: (a) innate unsystematic variation, as in the kinetic behavior of bacteria; (b) innate systematic variation, as in the different reactions of the snail *Littorina* depending upon whether it was in or out of water (this is context-sensitive behavior, but of a built-in sort); and (c) learned systematic variation, as in operant and classical conditioning, in which varied reactions (which may be built in, or may themselves have been built up by past selection) are selected and come to occur in varied contexts. The remainder of the book is concerned with learned systematic variation. I first discuss the *steady state*, that is, the ways that animals adapt to particular learning situations. The last three chapters look at the processes by which these steady states are arrived at.

NOTES

1. *Skinner's definition of operant behavior.* Skinner in his various writings has used the term *operant behavior* in three independent senses: (a) By exclusion, as behavior that is not tied to an eliciting stimulus. It is, if not spontaneous, at least behavior for which ". . .no correlated stimulus can be detected upon occasions when it is observed to occur" (Skinner, 1938, p. 21). (b) As behavior involving units, operants, defined by the correlation between stimulus and response. Skinner argued that both stimulus and response are classes of events, and the definition of each class is an empirical matter: Each should be defined in terms of the other so that the elements show "orderly changes" as a function of relevant variables (see Staddon, 1967, for a discussion of the philosophical basis for this position). A stimulus becomes "that which causes a response," and a response "that which is caused by a stimulus." This kind of bootstrap definition may seem peculiar but is, in fact, quite characteristic of scientific theory generally (see Staddon, 1973). I provide a formal definition along these lines for stimulus and response later in the chapter. (c) As behavior that is "controlled" by its consequences. The third definition is the one that has achieved widest currency and carries the fewest theoretical overtones. This is the sense in which I use the term.

There is a degree of contradiction between Skinner's first two characteristics of operant behavior: How can a behavior occur in the absence of any stimulus and yet be defined by the closeness of association of stimulus and response? Skinner was aware of this problem and has always insisted on using the term "occasioned by," rather than "elicited by" when referring to the stimuli that cause operant behavior. The paradox arises from his considering only *learned* behavior. In most higher animals, behavior that is effective in securing a reward or escaping a punishment becomes associated with the circumstances under which the animal learned to make the effective response. Features of the environment become *discriminative stimuli*, having the power to reinstate this behavior on later occasions: The dog learns to avoid shock by pressing a lever when the buzzer comes on so that in due time the buzzer becomes a discriminative stimulus for lever pressing. These are the stimuli to which Skinner refers. There is nothing in the notion of behavior guided by consequences that requires either learning or (which is implied) the existence of discriminative stimuli, however: The behavior of bacteria aggregating in a nutrient medium, or the escape behavior of *Stentor*, are both examples of operant behavior in the third sense, although neither involves either learning (in this sense) or discriminative stimuli. There is no compelling reason to reserve the term operant behavior just to the more complex case, and I use it here in the widest sense.

The first and third of Skinner's defining properties, spontaneity and control by consequences, are not contradictory but complementary. If behavior is to be determined by its consequences, it must first occur in sufficient variety that effective variants are available to be selected. Unfortunately, Skinner took this process of generation for granted and concentrated almost entirely on the fact of selection.

The phrase "control by consequences" is, of course, an oblique reference to feedback control, as I note in the text.

2. *Parallel models.* Chain reflexes are historically the oldest kind of mechanism proposed for adaptive behavior. The most implausible thing about chain reflexes is that they are *serial* processes: Each activity depends on the preceding one. Serial processes are vulnerable to disruption and imply a degree of stereotype rarely found in behavior.

Obviously many other mechanisms could be invented for *Stentor* avoidance. Parallel models explain behavior as the outcome of competitive interactions among a set of possible activities, only one of which can occur at a time. Parallel models adapt readily if one activity is blocked and derive behavioral variability from variations in thresholds, a well-known phenomenon. The behavior of Jennings' *Stentor* is most plausibly explained in this way.

The reasoning is as follows. (a) Since the animal engages in one or other of the five possible activities shown in Figure 4.2, never in more than one at a time, we assume that they *compete* for access to the "behavioral final common path." (b) Since the activities occur in a fairly rigid order, we assume that their *thresholds* differ, "turning away" occurring at low values of an irritant stimulus, "breaking away" at high concentrations. (c) Under constant stimulation, all activities habituate after a while. Hence, "turning away" will tend to weaken after some time. As it weakens, its inhibitory effect on the other four possible activities must also diminish. The activity with the next-highest threshold is presumably "ciliary reversal," which, accordingly, supplants "turning away" after a while. Soon, it also habituates, to be supplanted by "contraction," and so on. In this way the entire avoidance sequence can be generated.

The key features of this kind of model are (a) that the tendency for any activity to occur depends both on the current stimulus and the state of other activities; and (b) that activities tend to habituate (which can be thought of as self-inhibition), thus effectively reducing the threshold for those activities they inhibit.

Parallel models of this sort are simple in principle. In practice, they give rise to behavior that depends in complex ways on the quantitative properties of habituation and reciprocal inhibition. Nevertheless their simplicity, and consistency with what we know of the "wiring diagrams" of simple animals, suggests that they accurately represent an essential feature of much adaptive behavior (see Grossberg, 1982; Kennedy, 1967; Ludlow, 1976, 1980).

3. Edward L. Thorndike is the inventor of the *law of effect*, the basis of the modern principle of reinforcement. His original experiments were done with cats and other animals escaping from cages that could be opened from the inside by pressing a lever, pulling a string, or some other simple but relatively "unnatural" activity on the part of the animal. I return to Thorndike and the law of effect in Chapter 5. For a review of the work that has been done on learning and habituation in protozoa see Corning, Dyal, and Willows (1973).

4. *Information theory*. I write here and later as if the animals have a set of watertight compartments into which physical stimuli must fit. In reality, of course, things are not so clear-cut. In the case of an animal with two categories, for example, it is not that every stimulus that it ever encounters fits into category A or category B, with no overlap. Rather, some stimuli are always in category A, some always in B, some classified sometimes as A and sometimes as B, and perhaps others are sometimes C and sometimes B, and so on. The critical limitation is in the degree of *information* we have about the stimulus, knowing into what category the animal has classified it (or vice versa). If the information is one *bit*, then we can treat the animal as if it had two categories, with every stimulus falling uniquely into one, even though, in fact, it may have N categories, with considerable uncertainty about how a given stimulus is classified.

The term *bit* comes from the theory of communication (Shannon & Weaver, 1949; see Cherry, 1961, for an exposition for psychologists) and is defined as the information necessary to decide between two equiprobable alternatives. If you are curious about the sex of my cousin and I tell you she is female, I have transmitted, and you have received, one bit of information. On the other hand, if you know my cousin's name is Gail and I tell you she is female, I may have transmitted less than one bit, since the name Gail is more often a female than a male name (i.e., you already know that the alternatives are not equiprobable). The general formula is

$$I = -\sum_1^N p_i \log_2 p_i.$$

Thus, if you know I live in the 489 telephone-exchange area and I tell you my phone number, I have transmitted one set of four decimal digits out of a possible 10,000, that is, 13.29 bits of information $-\sum_1^{10,000}(1/10,000)\log_2(1/10,000))$.

Protozoa seem to be limited as to the amount of information they can process about their environment. This corresponds to a limitation on the number of internal states (in the sense defined later in the chapter) that they have.

The limited information-processing capacity of simple animals accounts for the importance of innate behavior to these animals. Even the most primitive animal must be able to identify a few things with high accuracy: a potential mate, common dangers, food. Given limited computational capacity, it is more efficient to build in sensitivity to a few specific signals (usually chemical) for these predictable and necessary tasks, than to use limited general learning capability that may not be able to discriminate with the necessary precision.

5. The standard treatment of the properties of habituation is by Thompson and Spencer (1966). Sensitization and pseudoconditioning are discussed at length in many of the older learning texts, such as Hilgard and Marquis (1940) and Kimble's update (1961).

6. The word *control* means quite different things in the two usages, *discriminative control* and *control by consequences*. Discriminative control is closer to the direct meaning – in many cases, discriminative control cannot be distinguished from simple elicitation of an operant response by its stimulus. I argue later that discriminative control is better thought of as a determination of the internal state of the animal by its present environment and past history, rather than as elicitation. Control by consequences means feedback control, as we saw earlier. It is unfortunate that the same word has these two senses, both in the rather restricted context of operant behavior.

7. The following discussion of finite-state systems follows the excellent account in M. Minsky's (1967) book: *Computation: Finite and Infinite Machines*. For an application of this approach to the causal analysis of operant behavior see Staddon (1973).

8. State X is the *set* of all states other than S; equally obviously, S is also a set, the set of all states having the common property that the individual responds "yes" to the question. States in these representations are always sets of real states. This is just another way of saying that we necessarily always deal with models, simplifications of the enormously complex real animal.

Time-to-respond, *latency*, is a much-used dependent variable in numerous experiments on human cognition. These experiments show that questions about, or exposure to elements from, previously experienced situations much facilitate responses to additional questions or elements. See, for example, Posner (1978) for a review of this work.

9. *Averaging of data*. The problem of the relation between averaged data and individual performance has been an enduring theme in methodological discussions over the years. For example, in the heyday of Hull and his students, numerous experiments studied the process of learning by looking at the improvement in performance across trials of groups of animals set to running mazes or solving discrimination problems. The resulting average curves are usually negatively accelerated, with each trial producing a substantial increment at first, and smaller and smaller ones as training progresses. The results can be fitted by a variety of negatively accelerated functions, of which the negative exponential is the most popular:

$$y = A(1 - \exp(-Bx)), \tag{N4.1}$$

where y is a performance measure such as "percent correct," x is the number of trials, and A and B are fitted constants.

How representative is Equation N4.1 of the performance of individual animals? This question usually cannot be answered in the obvious way, by looking at the learning curve of each animal, because these curves are typically very variable. If we take the variability at face value, then there is no point to averaging, since each animal is different. If we take the variability to be "noise" – random variation – then the legitimacy of averaging depends on what is assumed to be varying. What form of *individual* relation between y and x is consistent with the relation between averages given by Equation N4.1?

I'll just look at two possibilities. One suggests that averaging is legitimate, the other, equally (some might say more) plausible, suggests that it is not.

If we take the "noise" assumption at face value, then each measured performance value, y, can be represented as the sum of the true value, plus a noise component, ϵ, which is a sample from a population with zero mean. We also assume that each individual is identical, so that the true relation between y and the independent variable x, $y = g(x)$, is the same for all. For a group of N animals, therefore:

$$y_1 = g(x) + \epsilon_1$$
$$\vdots$$
$$\tag{N4.2}$$
$$y_N = g(x) + \epsilon_N.$$

Averaging performance across animals therefore yields:

$$1/N\sum_1^N y_i = \bar{y} = g(x) + \bar{\epsilon}. \tag{N4.3}$$

By hypothesis, ϵ_i is drawn from a population with zero mean, so that $\bar{\epsilon}$ will tend to zero as N increases. To an approximation, therefore, the average result will be:

$$y = g(x),$$

which can be compared with the obtained relation,

$$y = F(x).$$

Hence the "noise" assumption allows us to equate $F(x)$, the average curve, with $g(x)$, the individual function.

This model has two rather implausible assumptions: that all animals are identical, and that all variability is "measurement error" of the same sort as the error in reading a meter in a physics experiment. All animals are almost certainly not identical; and there is nothing corresponding to measurement error in most psychological experiments. It is at least as reasonable to assume that even if each animal obeys the same general law, the parameters of the function will differ from individual to individual, and that this is the main source of individual variation. With this change, Equation 4.2 can be rewritten as:

$$y_1 = g(x, a_{1,1}, \ldots, a_{1,M}) + \epsilon_1$$

$$\cdot$$
$$\cdot \tag{N4.4}$$
$$\cdot$$

$$y_N = g(x, a_{N,1}, \ldots, a_{N,M}) + \epsilon_N,$$

where y depends on x according to a function with M parameters that may differ in value from individual to individual. By the same argument as before it follows that

$$F(x) = 1/N\sum_1^N g(x, a_{i,1}, \ldots, a_{i,M}).$$

That is, the average function, $F(x)$, is equal to the average of each individual function of x. For what class of functions, $g(x)$, will the average function, $F(x)$, be of the same form as the functions averaged? The short answer is: not many. It is true if $g(x)$ is linear, as can be seen at once by averaging two instances:

$$y_1 = m_1 x + c_1$$
$$y_2 = m_2 x + c_2,$$

where m and c are constants. Hence,

$$y_1 + y_2 = x(m_1 + m_2) + c_1 + c_2,$$

or

$$y = x(m_1 + m_2)/2 + (c_1 + c_2)/2,$$

which is still a linear function.

It is not true for nonlinear relations such as Equation N4.1. For example, suppose that $y_1 = m_1 x^{a_1}$, and $y_2 = m_2 x^{a_2}$, which are power functions, with m and a as constants. Then

$$y = m_1 x^{a_1}/2 + m_2 x^{a_2}/2,$$

which is not a power function.

Thus, rather little can be learned about the performance of individuals by looking at group averages. For a more extensive discussion of this issue see Estes (1956), Stevens (1955), and Sidman (1960).

10. *Unobservability.* In addition to these difficulties, there are unavoidable uncertainties in the study of black-box systems. Because a stimulus can change the state of the system, some states may be literally unobservable by an experimenter. For example, even under ideal conditions, we can never be sure that a given event is *not* a stimulus. Failure to find any effect is always inconclusive. This limitation holds because the tests themselves can eliminate the evidence they are intended to elicit (see E. F. Moore, 1956). Stated formally, Moore's "uncertainty-principle" theorem is that "there exists a [finite-state] machine such that any pair of its states are distinguishable, but there is no simple experiment which can determine what state the machine was in at the beginning of the experiment" (1956, p. 138). It is easy to see the logic of this limitation in the case where there are three states to be distinguished, call them A, B, and C. Let us suppose that A and B are distinguished by the fact that input 1 elicits output 1 from state A and output 2 from state B; B and C are distinguished by the fact that input 2 produces output 2 from B and 3 from C. Input 1 fails to distinguish B from C because it elicits 2 from both; similarly, input 2 fails to distinguish A from B because it elicits 2 from both. Finally, suppose that inputs 1 and 2 both send the system into state C.

To discover the initial state of the system we have only two choices, either input 1 or input 2. The first choice will distinguish between initial states A and B, but will fail to distinguish between states B and C. Moreover, additional inputs will not help, because the system will now be in state C no matter what its initial state; a similar problem results if the initial input is 2. There is no theoretical solution to this problem, of course, other than having replicas known to be in the same state at the beginning of the experiment, so that different, noninterfering tests can be carried out on each replica. Unfortunately, real animals are replicas only in respect to restricted aspects of their behavior, and it is often far from clear what these aspects are. It is reasonable to assume equivalence in such matters as the properties of cells and organs, and possibly some psychological functions of sensory systems. It is much less clear that the processes involved in learning are essentially identical from individual to individual.

Everyday experience shows that Moore's theorem is relevant to psychology. For example, asking someone to recall an event may actually cause him to forget it, although some less direct test might have shown that he had, in some sense, remembered it. Many people have had the experience of being asked to recall verbally a familiar telephone number and being unable to do so; yet they could have dialed the number without difficulty. Sometimes the ability to dial correctly is retained even after failure of verbal recall, but more commonly failure to recall also abolishes (often only temporarily) the ability to dial correctly.

5

REWARD AND PUNISHMENT

All functional explanations of behavior depend on some notion of what is good and bad. If we are talking in terms of evolutionary adaptation, good and bad boil down to values of inclusive Darwinian fitness, a measure reasonably clear in principle, but often elusive in practice. If we are talking in terms of the operant behavior of individual animals, good and bad correspond to reward and punishment, to situations better or worse than the current situation. This chapter is about reward and punishment: about the concept of *reinforcement* that includes them both, about how it is defined, and the procedures used to study its effects.

There are two kinds of question to ask about reward and punishment: (a) What makes situations good and bad? Can we define good and bad independently of the behavior of the animal – do all good situations share common features? Or must we always see the effect of a situation on an animal before we can be sure of its hedonic value? (b) Granted that we know the hedonic properties of a situation, how does it affect behavior? What are the mechanisms, what are the rules?

Generations of philosophically minded students of human affairs have labored over questions of the first kind, which Aristotle termed the definition of "the good." The early twentieth-century philosopher G. E. Moore summed up the modern consensus in dry but exact fashion as follows: "I have maintained that very many things are good and evil in themselves, and that neither class of things possesses any other property which is both common to all its members and peculiar to them" (Moore, 1903, p. x). All that good things have in common is that they are good; all that bad things have in common is that they are bad.

The early behaviorists were undeterred by the philosophers' failure to find an independent yardstick for value. Deceived by the apparent simplicity of the white rat, they tried to reduce motivated behavior to a small set of "primary drives": *Hunger, thirst*, and *sex* were on everybody's list. For the rat, at least, the definition of a good thing was that it led to the reduction of one or more of these three drives. But opinions differed about what should be done when rats sometimes did things that could not be explained by one of the three. Rats in a new environment will eventually explore it, for example; given a weak electric shock for pressing a lever, they are likely to press it again rather than avoid the lever after their first press, and so on. One school proposed new drives, like "curiosity," "habitat preference," "freedom," "sleep," and "aggression." The other school, more parsimonious, held to the original trinity and proposed to solve the problem of additional motives by linking them to the basic three. For example, exploratory behavior might be explained not by a "curiosity drive," but by a past history in which exploration had led to food, water, or sexual activity.

Neither course was wholly satisfactory. Among those willing to entertain additional drives, there was no general agreement beyond the basic three. Indeed, someone acquainted with desert animals might question even *thirst* as a primary drive, since many rarely drink in nature, obtaining the water they need from their food.. The fate of the fundamentalists was no better. Although some activities could be plausibly linked to primary drives, attempts to include others appeared strained; here also there was no consensus.

As I explain in Chapter 7, the modern version of the multiple-drive view is the economic concept of a *preference structure*. This idea is both more and less ambitious than the earlier view. It is more ambitious in that it proposes to accommodate not only more than one drive (or desirable thing), but also shows how competing drives are to be reconciled. Drive theory could not easily explain how an animal both hungry and thirsty should choose between food and water, for example. It is less ambitious because it leaves on one side the issue of completeness: It is content to take situations one at a time, propose a preference structure for each, and test the implications for choice – leaving for later the issue of whether there is a global preference structure from which all operant behavior can be derived.[1]

The view that there are primary and secondary motives, that all behavior can be derived from a small primary set, has no real contemporary descendant, but recent attempts to derive a universal preference structure from a limited set of motivational *characteristics* is clearly in the same tradition (see Chapter 7).

The question of what makes a situation good or bad from an animal's point of view, interesting as it is, has not led to conclusive answers. Psychologists have been driven back to Moore's conclusion that there is nothing beyond the animal's reaction that marks a situation as good or bad. This has led to the animal-defined concept of *reinforcement*, as I describe in a moment.

The question of how situations with known hedonic properties affect behavior has attracted much more attention, for two reasons. First, it suggests many more experiments. If we have something an animal wants, like food, and we know what the animal can do in a gross sense, then we can require all sorts of things of him as a condition for giving him the food. The way in which the animal copes with the problems we set tells us something about the machinery he has for improving his situation. Second, and most important, the more we know about the machinery, the closer we are to answering the first question. In science, as in education, it is as well to begin with simple, extreme cases. If we understand how animals avoid and escape from electric shock, how hungry animals cope with food problems, then we will be in a better position to understand their behavior in situations where the rewards and punishments are less drastic and their behavior accordingly more subtle.

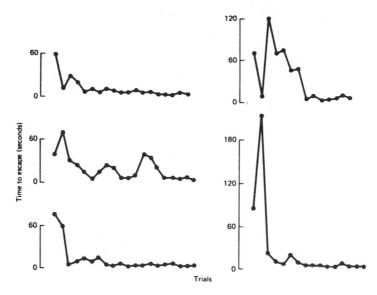

Figure 5.1. Time taken to escape from a puzzle box on successive trials by five different cats. (From Thorndike, 1898.)

This chapter discusses the main experimental arrangements that have been used to limit animals' access to food or to permit them to escape or avoid electric shock. I emphasize procedures and methods of data analysis, but also say something about how animals adapt to the procedures – later chapters expand on this. I begin with the concept of reinforcement.

REINFORCEMENT AND THE LAW OF EFFECT

The modern, experimental study of reward and punishment is usually dated from the work of Edward L. Thorndike.[2] During the last years of the nineteenth century, while a graduate student, first at Harvard University and then at Columbia, Thorndike studied the behavior of cats and other animals escaping from puzzle boxes. The cats could escape from the box by clawing on a wire loop or a bobbin, or by making some other response of this sort to unlatch the door. After each successful escape (trial) Thorndike gave the animal a brief rest, then put it in the box once again. This process was repeated until the animal mastered the task. Thorndike measured the time the animal took to escape on successive trials, producing for each a *learning curve* like the ones shown in Figure 5.1.

The learning curves in Figure 5.1 are quite variable. This is because they just measure times, and not activities. What seems to be happening is that on early trials the cat tries various things – such as pawing at the door, scratching the walls of the box, mewing, rubbing against parts of the apparatus, and so on. Most of these are ineffective in operating the latch. Because these activities occur in an unpredictable sequence from trial to trial, the effective response occurs at variable times after the beginning of a trial. Trial times improve because the ineffective acts gradually drop out.

Thorndike concentrated on trying to find the selection rule that determines how the effective act is favored over ineffective ones. He decided that *temporal contiguity* is the critical factor, together with the hedonic value of the outcome. He stated his conclusion as the well-known *law of effect*:

> Of several responses made to the same situation, those which are *accompanied or closely followed by satisfaction* to the animal...will, other things being equal, be more firmly connected with the situation...; those which are accompanied or closely followed by discomfort...will have their connections with the situation weakened....The greater the satisfaction or discomfort, the greater the *strengthening or weakening of the bond*. (Thorndike, 1911, p. 244, my italics)

This principle provided a framework for American studies of learning for the next sixty years. The first phrase in italics identified as critical the close temporal relation between reward (or punishment) and subsequent behavior. The term *satisfaction* identified reward and punishment as necessary for learning and raised the issue of the definition of what was subsequently to be termed *reinforcement*. The term *bond* led to the view that learning involves the formation of links or associations between particular *responses* and particular *stimuli* (situations). These three ideas have been more or less severely modified by later work. Let's look at each of them.

Obviously the law of effect would be of little use without some independent measure of what is meant by "satisfaction." If we want to train an animal according to Thorndike's law, we must know what constitutes satisfaction for it; otherwise the principle is no help. Thorndike solved this problem by making use of the fact that animals can do more than one thing to get something: A satisfying state of affairs is anything the animal "does nothing to avoid, often doing such things as to attain and preserve it." A little later, this idea was formalized as the *transsituationality* of the law of effect: If something such as food will serve as a *reinforcer* for one response, it should also serve for others. Another way to deal with the same problem is to take *approach* and *withdrawal* as reference responses. A "satisfier" (*positive reinforcer*, in modern terminology) is something the animal will approach; a "discomforter" (*punisher, aversive stimulus*, or *negative reinforcer*[3]) is something it will withdraw from. *Reinforcers* are the *unconditioned stimuli* (USs) of the previous chapter.

Thorndike's definition of a reinforcer is the aspect of his law that has been least altered by later work. As I point out in Chapter 7, it has been extended somewhat by the notion of a preference structure, but its essential feature – that hedonic quality is revealed by the animal's own behavior – has been retained.

Comparison of Thorndike's law with the discussion of learning in the previous chapter shows that Thorndike made no distinction between local and long-term memory. Learning to escape from a puzzle box is one thing; recalling the effective response after a delay, or after being removed from the situation, is quite another. A cat may learn to escape from a puzzle box, just as *Stentor* may "learn" to escape from carmine, without being able to repeat the feat 24 hours later. The availability of a rewarding consequence is certainly necessary for adaptive behavior like this, but its role in enabling the animal to remember what it learned is not obvious. We certainly cannot assume, as Thorndike did, that reward is necessary for memory (i.e., the formation of "bonds," in his terminology). (We will see in Chapter 12 that valued events seem to be better remembered than neutral ones, but that is another matter.)

The third element in Thorndike's law is his assumption that the effective response is directly "strengthened" by its temporal contiguity to reward. It turns out that contiguity is terribly important; but it is not the only thing that is important, and Thorndike's emphasis on the single effective response at the expense of the many ineffective responses has been misleading in some ways. Michelangelo is reported to have responded to a question about how he was able to sculpt so beautifully by saying: No, it is really quite easy: I just take away all the marble that is *not* the statue, and leave the rest. Thorndike's law does not sufficiently emphasize that reinforcers act by a process of selective elimination.

Later exeriments have shown that response-reinforcer contiguity is not sufficient for a reinforcer to be effective, and may not always be necessary. The logic of the thing shows that strengthening-by-contiguity cannot be sufficient by itself to explain operant behavior. There are two ways to deal with this problem: One is to consider what additional processes may be necessary. The second is to look in more detail at the functional properties of operant behavior: To what procedural properties is it sensitive? In what sense do animals maximize amount of reward? The second question is much easier than the first. Moreover, the more we know about the functional properties of operant behavior, the more educated our guesses about the underlying processes can be. I follow up the question of mechanism in the last two chapters. The functional properties of reinforcement and reinforcement schedules are taken up in a preliminary way in this chapter.

Experimental methods

All science begins with taxonomy. If we want to understand the properties of reward and punishment (i.e., reinforcement), the first step is to gather some examples of how they act, and then begin classifying. But how are examples to be gathered? One could collect anecdotes: "Little Freddie used to pick his nose, but when I thrashed him soundly for doing it, he soon stopped." But this is obviously unsatisfactory: We don't know how soundly Freddie was thrashed or how soon the thrashing followed the offense or how quickly Freddie desisted. We have no precise measure of the response, the punishment, or the frequency with which one followed the other. We don't know Freddie's past history. Useful data on the effects of reward and punishment cannot be gathered like bugs at a picnic, without planning or design. We must do experiments – but what kind of experiments?

Experiments on reinforcement are of two general kinds: experiments in which the animal can improve his situation by moving about; and experiments where movement is irrelevant, but the animal can improve things by making a spatially localized response. The first category includes studies in which the animal must find food in a maze or runway or avoid electric shock in a shuttle box. Mazes are of two main sorts: the Hampton-Court variety where there is one goal box and many blind alleys, and the animal's task is to learn the one path to the goal; and the newer, radial maze, where every goal box contains food and the animal's task is to visit each goal box without repetition. Early studies of learning all used situations where locomotion was essential to reward. The second category comprises mainly so-called free-operant or Skinner-box experiments, in which the animal must press a lever or peck a lighted key for reinforcement, which is delivered in a fixed place by an automatic mechanism.[4]

Maze-type experiments are useful if spatial behavior is of special interest, or if one wants to make use of animals' natural tendency to approach some things and withdraw from others. No special training is required for an animal to go from the part of a shuttle box where it has just been shocked to a part where it has never been shocked, for example. Rats will explore the goal boxes of an eight-arm radial maze without having to be led down each arm. Spatial tasks are less useful if one is interested in time relations – between reward and response, between successive responses, or between stimuli and responses. Skinner-box experiments usually require that the animal be first trained to make the instrumental response, but they are ideal for the study of time, because the experimenter can measure exactly when a particular response occurs and arrange for reward or punishment to occur in a precise temporal relation to it. The Skinner box also lends itself easily to automation: Given a food-pellet dispenser, a transducer for measuring specified aspects of the animal's behavior, and a computer to record response information, present stimuli, and operate the feeder according to a rule specified by the experimenter, human intervention is required only to place the animal in the apparatus and type "GO" at the console. Since temporal relations are very important to operant behavior, and repetitive labor is irksome to most people, Skinner box experiments have become very popular.

Thorndike looked at changes in behavior across learning trials; in contemporary terms, he studied the *acquisition* of behavior. If he had persisted in running his cats even after they had mastered the task, he would have been studying *steady-state* behavior, the properties of a developed *habit*. Steady-state behavior is more interesting if reinforcement does not follow every response (*intermittent reinforcement*). It is easier to study if the animal need not be reintroduced into the apparatus after each occurrence of the reinforcer. Both these requirements favor Skinner's method over Thorndike's, and the Skinner box has become the preferred apparatus for studying steady-state operant behavior. Let's look at some common arrangements.

The Skinner box. Skinner boxes come in many varieties. The standard version, for a rat, is a small, metal-and-Plexiglas box about 20 cm on a side. On one wall

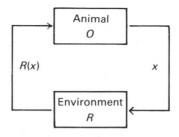

Figure 5.2. Feedback relations in an operant conditioning experiment. x = response measure; $R(x)$ = reinforcement produced by x; R — feedback function (reinforcement schedule); O = control function (behavior laws).

is a lever or two, often retractable, so it can be presented or withdrawn under remote control. A feeder, for either pellets or liquids, dispenses food at an aperture in the middle of the wall. Stimulus sources, in the form of a loudspeaker or buzzer, and lights above the levers, are also on the wall. The version for a pigeon is a little larger, food comes from a grain hopper, and stimuli and response transducer are combined in the form of a translucent pecking key, on which colored lights or other visual stimuli can be projected. Pecks on the key, or presses on the lever, go to a controlling apparatus (originally a tangled mess of wires, timers, and electromagnetic relays, nowadays usually a computer) that operates the feeder and turns stimuli on or off according to the experimenter's program.

This basic plan can be modified in several ways. Additional transducers (for the same or different responses) can be added or the transducers might be modified for different species: Ethologists, for example, have studied Great Tits (small European perching birds) in a Skinner box with a pair of perches to record hops and a conveyor belt to present mealworms as reinforcers. In my own laboratory we have built hexagonal or octagonal enclosures for rats with transducers for different responses – such as wheel running, gnawing, licking, and others – in each segment. Complex stimuli can be presented on a computer display or with an automated slide projector.

The essential features of all these arrangements are represented in Figure 5.2, which has two parts: the programming computer, labeled as R, which provides reinforcers ($R(x)$) for the animal (I ignore computer-controlled stimulus changes for the moment), and the animal, labeled O, which provides responses (x) for the computer. Animal and apparatus constitute a *feedback system*; anything we measure about steady-state operant behavior, such as the animal's rate of lever pressing, or the rate at which he gets fed, will generally reflect properties of both halves of the system: the animal *and* the programming computer. Figure 5.2 is a model for all interaction between an animal and its environment (compare it with Figure 3.6); the Skinner box just represents a highly controllable environment.

R and O are *functions*: R defines how the response the animal makes (x) will be

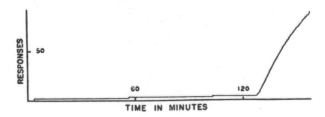

Figure 5.3. Cumulative record of the acquisition of lever pressing by a rat reinforced with food on a fixed-ratio 1 schedule. The first three feedings had little effect; the fourth is followed by a rapid increase in lever-press rate. (From Skinner, 1938, Figure 3. Reprinted by permission of Prentice-Hall, Englewood Cliffs, N.J.)

translated into the reinforcers it receives ($R(x)$). R is of course known, since the experimenter determines the program for delivering reinforcers. Program R is termed a *feedback function* (or *schedule function*). Other names for R are *contingencies of reinforcement*, or *reinforcement schedule*. O, the *control function*, defines how the reinforcers the animal receives will be translated into responses. Another name for O is the *laws* or *principles of behavior*. O is not known in general, and the aim of experiment is to help refine our understanding of it.

Figure 5.2 can be converted from an illustration to a formal model once we decide on the proper way to measure x (responding) and $R(x)$ (reinforcer presentation: *reinforcement*, for short).[5] Chapters 6 and 7 go into this in more detail. For now let's just consider some commonly used programs (reinforcement schedules) and how the animal adapts to them.

Response- and time-based schedules of reinforcement

The simplest feedback function is when every lever press yields a food pellet. This is also the simplest *ratio schedule: fixed-ratio 1* (FR 1), also known as *continuous reinforcement*. Figure 5.3 shows how one individual, hungry (i.e., food-deprived) rat first learned to respond for food pellets delivered on a fixed-ratio 1. The rat had previously been exposed to the Skinner box and given occasional opportunities to eat from the automatic feeder, but responses to the lever had no effect. This is known as *magazine training* and just gets the animal used to eating from the feeder. On the day shown in the figure, the lever was connected to the feeder for the first time. The rat's lever presses are shown as a *cumulative record*: Time is on the horizontal axis, and each lever press increments the record on the vertical axis (each response produces only a small vertical increment, so that cumulative records appear quite smooth so long as response rate changes gradually). The first three lever presses (at time zero, close to 60 min, and about 95 min) produce food but not additional lever pressing. But at the fourth response, the animal evidently "catches on" and presses rapidly thereafter, the rapidity of his presses showing in the steepness of the record. The

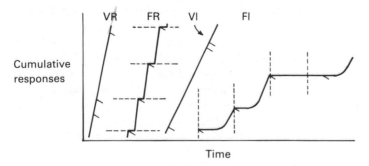

Figure 5.4. Stylized cumulative records of steady-state performance on fixed- and variable-ratio and interval schedules. Rate of responding (responses/time) is represented by the slope of these curves. Dashed lines show when a reinforcer is available for the next response.

record begins to tip over at the extreme right of the figure, as the animal's rate of pressing slows – presumably because he is getting less and less hungry.

Skinner discovered that a hungry rat will continue to press a lever even if food doesn't follow every lever press. When the number of presses required for each food delivery is constant, the resulting arrangement is termed a *fixed-ratio schedule*; when the number varies from food delivery to food delivery, it is termed a *variable-ratio schedule*. The ratio value is the ratio of responses made to food deliveries received, over some period of time. When the time interval involved is a single experimental session (typically 30 min to 3 hr), the relation between responses made and reinforcers received (i.e., between response and reinforcement *rates*) is known as the *molar feedback function*. For ratio schedules it takes a uniquely simple form:

$$R(x) = x/M, \tag{5.1}$$

where M is the ratio value, $R(x)$ the frequency of feeder presentations per unit time (food rate), and x the rate of lever pressing. Molar feedback functions are important for the regulation of feeding by operant behavior and for understanding the different effects of different schedules (see Chapter 7).

Fixed- and variable-ratio schedules have the same molar feedback function, but differ, of course, in their local, *molecular* properties. This difference shows up in cumulative records of steady-state (i.e., well-learned) behavior, which are shown in stylized form in Figure 5.4. The diagonal "blips" on the record indicate food (reinforcer) deliveries. The dashed horizontal lines through the FR record are separated by a constant vertical distance, to indicate that each reinforcer occurs after a fixed number of responses. Records like this have been produced by pigeons, rats, people, monkeys, and numerous other animals – performance on simple reinforcement schedules differs little across a range of mammal and bird species. Both FR and VR schedules generate high rates of responding, as shown by the steep cumulative records in the figure, but the local structure of behavior is different: Animals typically pause briefly after each food delivery on fixed-ratio schedules, but respond steadily on variable-ratio. This difference is a reaction to the fact that food never immediately follows food on FR, but sometimes does so on VR. Food *predicts* a period of no-food on FR, but if on VR the number of responses required varies randomly from one interfood interval to the next, food predicts nothing and there is no reason for the animal to deviate from a more or less steady rate of responding.

The feedback rule for ratio schedules is that reinforcer occurrence depends upon number of responses. There are obviously two other simple possibilities: dependence on *time*, or joint dependence on time and number. Pure dependence on time is an open-loop procedure, in the sense that reinforcer occurrence is then independent of the animal's behavior, so that the response input (labeled x in Figure 5.2) doesn't exist. Other names for open-loop procedures are *classical* or *Pavlovian conditioning* (see Chapter 4). I return to them in a moment. The only remaining operant possibility, therefore, is joint control by time and number. The most frequently used procedures of this type are *fixed-* and *variable-interval* schedules. Both require the passage of a certain amount of time followed by a single response for the delivery of the reinforcer. The sequence is important: A response that occurs too early is ineffective; it must occur after the time interval has elapsed.

I showed in Figure 5.3 how a rat learns to respond on an FR 1 schedule. How might an animal learn to respond on a fixed-interval (FI) schedule? The process takes much longer than on the simple FR 1, because the animal has to learn about two things: the response contingency – the fact that a response is necessary for each crack at the food; and the minimum interval between food deliveries (i.e., the FI value). He has to learn only the first of these on FR 1. Let's begin with a magazine-trained pigeon maintained at about 80% of its normal body weight (i.e., very hungry!), with the controlling computer set to limit food deliveries to no more than sixty within a single-day session (so as to prevent the animal from gaining weight from day to day). We have trained him to peck (using one of a variety of methods discussed later) but he has so far received food for every effective key peck. Now we introduce him to the fixed-interval procedure, with the interval set to perhaps 60 sec.

Figure 5.5 shows in stylized form the stages that the pigeon's keypecking goes through as it converges on the final steady-state performance. These stages are by no means clear-cut, nor are the transitions between them perfectly sharp, but we nearly always see these four patterns succeed each other in this order. Each stage takes up many interfood intervals, perhaps 100 or more (i.e., several daily experimental sessions). At first (stage I) each peck-produced food delivery (indicated by the blips in the cumulative record) produces a burst of further pecks, which slowly dies away; the animal pecks slower and slower when these pecks do not result in food. After a time greater than 60 sec has elapsed, and the response rate is now very low, an isolated response immediately produces food and this at once elicits a further pecking burst.

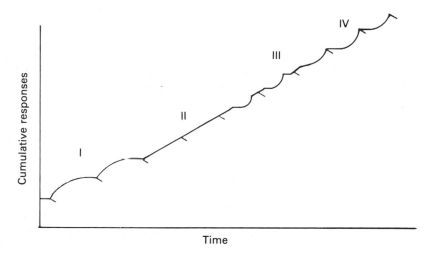

Figure 5.5. Schematic cumulative record of the changing patterns of responding as a pigeon adapts to a fixed-interval schedule. (Adapted from Ferster & Skinner, 1957, p. 117.)

In the second stage, the temporal pattern of pecks between food deliveries changes from negatively accelerated to approximately constant. The pigeon now responds at an approximately steady rate for many intervals. This pattern is succeeded by breaks in the steady responding that take the form of brief periods of acceleration followed by returns to a lower rate of response. This is stage III. This pattern shifts gradually to the final form, which involves a pause in responding after each food delivery, followed by accelerating responding (stage IV). This is the so-called fixed-interval "scallop," a highly reliable pattern shown by many mammals and birds. As Figure 5.4 shows, the steady-state FI pattern is quite similar to the FR; differences are the lower "running" rate (rate after the postreinforcement pause is over) in FI, the slightly shorter pause (in relation to the typical interfood interval), and the "scalloped" pattern of the FI record, indicating gradual acceleration in responding, rather than the "break-and-run" pattern characteristic of FR.

Each stage of FI acquisition (as this process is termed) makes good adaptive sense. The first stage – a burst of rapid responding after each food delivery – seems to be an innate adaptation to the fact that food often occurs in patches. Finding some food after a lull strongly suggests that there is more where that came from, so that foraging efforts should be stepped up. The spatial equivalent is *area-restricted search*: When a pigeon foraging in nature finds some grain after a period of unsuccessful search, his rate of turning increases (that is, he continues to look in the vicinity) and his rate of movement may increase as well. The process is a sort of kinesis, adapted to keep the animal in the "hot" area. An animal on an FI schedule is restricted to the same "food patch," but it can follow the temporal part of this rule by looking especially hard for more food right after it gets some.

The final stage is also adaptive: The animal pauses after food because it has learned that no more food is likely for a while and it is free to occupy that time in some other way. The two intervening stages represent the transition period when the animal is gradually giving up its initial, "default" rule (area-restricted search) in favor of a new rule (the FI scallop) adapted to changed circumstances. Since the animal cannot be certain from day to day that the new pattern of food delivery will persist, it makes sense that he should change only gradually from one behavior to another.

The pause after food on FI is obviously adaptive, but it is nevertheless usually too short: The efficiency of steady-state FI performance is surprisingly low. Strictly speaking, only a single response need be made for each food delivery, namely, the first response after 60 sec. Yet a pigeon might make 30–40 key pecks in an average interval, only one of which is essential. Part of the explanation for this inefficiency lies in limitations on the animal's ability to estimate the passage of time, but that is not the whole story. I suggest a possible explanation later.

The difference between FI and VI parallels that between FR and VR: On VI (if the interreinforcement intervals are chosen randomly) the probability that a response will produce food is constant from moment to moment. Food delivery has no special predictive significance, so that animals tend to respond at a more or less steady rate that is a bit slower than the rate on a comparable VR schedule (I discuss the reason for this in a later chapter).

Equilibrium states

The four patterns shown in Figure 5.4 are the outcome of a converging process. In the acquisition phase, the animal at first shows more or less innate responses to occasional, unpredictable (from its point of view) food deliveries. As food continues to be produced by this interaction, food deliveries begin to take on a predictable pattern; this pattern, in turn, guides future behavior until the process converges to produce the steady-state pattern of behavior shown in the figure. On fixed-interval schedules, for example, the major regularity, the fixed time between successive food deliveries, *depends on the animal's behavior* as well as on the schedule. If the pigeon pecked slowly or erratically, so that many food deliveries were obtained well after the time at which the apparatus had "set up" (i.e., well after the dashed vertical lines in Figure 5.4), then the time between food deliveries would not be fixed – although the *minimum* interfood interval might still approximate the fixed-interval value. By varying its behavior early in training, the animal is able to detect invariant properties of its hedonic environment: On fixed-interval schedules, the FI value is detected and controls behavior; on FR the ratio value; on VI the mean, minimum interfood interval, and so on.

These examples illustrate a general characteristic of operant behavior: The stimuli that come to control behavior are often themselves dependent on behavior. This kind of interaction is not restricted to the somewhat artificial conditions of the Skinner box. For example, a young squirrel learns about the tastiness of various nuts by first opening them in an exploratory way and sampling the contents. This allows it to learn that some nuts are better than others, so that it will seek out and respond to particularly tasty types that it ignored previously. With additional experience, the animal may come to learn that hazelnuts (say) are to be found in the vicinity of hazel trees, or under hazel leaves. Thus at each step, the animal's initial explorations reveal correlations – between the appearance of a nut and its taste, between a habitat and the occurrence of desirable nuts – that guide future behavior. The fixed-interval case is simpler only because the situation is artificially constrained so that the only relevant explorations are along the single dimension of time. The animal varies its distribution of pecks in time and the invariant that emerges is the fixed minimum time between food deliveries. This then guides the future distribution of pecks, which conforms more and more closely to the periodicity of the schedule and, in turn, sharpens the periodicity of food deliveries.

The fixed-interval scallop, and other properties of such stable performances, are aspects of the *equilibrium state* reached by the feedback system illustrated in Figure 5.2. This equilibrium is dependent both on the fixed-interval feedback function, and on the mechanisms that underlie the organism's operant behavior. The examples I have given should make it clear that a particular equilibrium *need not be unique*, however. The final equilibrium depends on two things: the range of *sampling* – the variability in the animal's initial response to the situation, the number of different things it tries; and the *speed of convergence* – the rapidity with which the animal detects emergent regularities and is guided by them. Too little sampling, or too rapid convergence, may mean an equilibrium far from the best possible one. In later chapters I discuss phenomena such as "learned helplessness" and electric-shock-maintained behavior that represent maladaptive equilibria.

Equilibria can be *stable, unstable, neutral*, or *metastable* in response to environmental changes. Stability is not an absolute property of a state but a label for the observed effect of a perturbation. A state may be recoverable following a small perturbation but not after a large one. For example, the physics of soap bubbles shows that their form is the one with the lowest free energy for the number of surfaces: Thus, a spherical bubble returns to its original, efficient shape after a slight deformation. The spherical shape is a stable equilibrium under moderate perturbations. A more drastic deformation will puncture the surface, however, whereupon the bubble collapses to a drop and the original state cannot be recovered. The spherical shape is stable under slight deformation, but metastable in response to a severe deformation.

The four types of equilibrium can be illustrated by a visual metaphor. Imagine that the state of our hypothetical system is represented by the position of a ball in a terrain of hills, valleys, and plains. On the plain, the ball stays where it is placed: This is neutral equilibrium. In a valley, if the ball isn't moved too far, it returns to the valley floor: This is stable equilibrium. On a hill, even a small displacement lets the ball run down to the valley: This is unstable equilibrium. If the ball starts in a valley it returns to the valley only if it isn't moved too far up the hill; too big a displacement, and it rolls into the next valley: This is metastable equilibrium.

The valley metaphor is realistic in one other way, in that it implies oscillatory behavior as the ball rolls from one valley wall to another after a perturbation. Many natural systems show persistent oscillations in the steady state; motor tracking and predator–prey interactions are well-known examples.

Equilibria on simple reinforcement schedules are generally stable: A given schedule usually yields the same pattern of behavior, and this pattern can usually be recovered after some intervening procedure. Exceptions are of two kinds. Occasionally a pattern is unstable in the sense that it persists for a brief period (which may be several days, or even weeks), but then without any change in the schedule it alters in an irreversible way. For example, occasionally an animal will show instead of the typical scallop pattern a more or less steady rate of responding on a fixed-interval schedule. Indeed, all animals pass through such a period. However, once this pattern changes to the scalloped one, the original pattern never reappears. It, therefore, represents an unstable equilibrium.

A more interesting exception is *metastability*. In this case, the pattern of behavior under a given schedule remains stable, but it is not *recoverable* when the schedule is reimposed after an intervening treatment. This effect is quite common. So-called *spaced responding* provides an example. Hungry pigeons can be trained to space their key pecks in time by delivering food only if a key-peck is separated from the preceding one by at least t sec, where t is on the order of 10 or 20. They adapt to this procedure with difficulty because the better they adapt, the more frequently they receive food, and the more frequently they get food the more inclined they are to peck. Since more frequent pecking reduces the rate of access to food, the stage is set for a very slow and oscillatory process of adaptation. Pigeons do settle down eventually, however, but at first their performance is far from optimal. For example, an animal initially exposed to a 10-sec timing requirement may after several weeks still space most of its pecks less than 5 sec apart, and the mode of the interpeck interval distribution may be at only 1 or 2 sec. If the spacing requirement is changed, then, on returning to the original requirement, the pigeon will do much better than before: The average spacing between pecks will be greater and the modal peck closer to the timing requirement. As the animal is exposed to different timing requirements, the performance at each requirement gets better, until the mean and modal interpeck times come to approximate the spacing requirement. This pattern represents the stable equilibrium. The earlier patterns, in which mean and modal interpeck time were much shorter than the timing requirement, are metastable equilibria.[6]

The procedures of fixed-ratio, fixed-interval, and variable-interval schedules are summarized in the form of event diagrams in the bottom half of Figure 5.6.

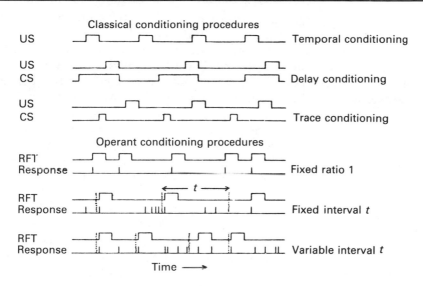

Figure 5.6. Event diagrams of common classical- and operant-conditioning procedures. Time is on the horizontal axis and the occurrence of the labeled event (CS, US, response, RFT = reinforcement) is indicated by upward deflection of the line. In the classical procedures, the occurrence of the US (unconditioned stimulus) is independent of responding, but systematically related to a CS (conditioned stimulus), post-US time or post-CS time. In the operant procedures, the reinforcer depends on a response and (on all schedules other than fixed-ratio 1) on some other time, response or stimulus condition. Thus, in fixed interval, a certain time, *t*, must elapse before a response is reinforced; on fixed-ratio N, N-1 responses must occur before the Nth response is effective.

The top half shows open-loop (classical conditioning) procedures, which I discuss next.

Classical conditioning

The study of classical conditioning begins with the Russian I. P. Pavlov, whose work made its major impact in the West with the publication in 1927 of an English translation of his lectures on conditioned reflexes. The lectures had been given three years earlier to the Petrograd Military Medical Academy and summarized several decades of active work by a large research group. The subtitle of *Conditioned Reflexes* is "An Investigation of the Physiological Activity of the Cerebral Cortex," which gives a clue to Pavlov's objectives. As a physiologist he was interested in behavior as a tool for understanding the functioning of the brain. Like Sherrington, however, his experiments involved little surgical intervention. Most were purely behavioral, and though, like Sherrington, he often interpreted his results physiologically – inferring waves of excitation and inhibition spreading across the cortex, for example – unlike Sherrington, later physiological work has not supported his conjectures. Nevertheless, Pavlov's observations, and his theoretical terms, continue to be influential.

Pavlov's basic procedure was the one labeled *delayed conditioning* in Figure 5.6: A brief stimulus, of a few seconds' duration, such as a tone or a bell or a flashing light, is periodically presented to a dog restrained in a harness. The dog has recovered from a minor operation in which the duct of its salivary gland has been brought to the outside of the cheek, so that the saliva can be collected and measured. At the end of this brief stimulus (which is to become the *conditioned stimulus*, CS for short) some food powder (the *unconditioned stimulus*: US) is placed in the animal's mouth. The food powder, of course, induces salivation; this is the *unconditioned response* (UR). This sequence of operations and effects, tone → food → salivation, is repeated several times and soon a new effect is noticed. Salivation now begins to occur when the tone comes on, and before food has actually been placed in the animal's mouth. This is termed the *conditioned* or *conditional response* (CR) – conditional because it depends upon the relation between CS and US during prior training.

This effect can hardly be called startling and must have been observed by anyone who had kept animals and fed them on some kind of schedule. Bernard Shaw in his satire "The Adventures of the Black Girl in Her Search for God," parodies Pavlov thus:

"This remarkable discovery cost me twenty-five years of devoted research, during which I cut out the brains of innumerable dogs, and observed their spittle by making holes in their cheeks for them to salivate through. . . . The whole scientific world is prostrate at my feet in admiration of this colossal achievement and gratitude for the light it has shed on the great problem of human conduct."

"Why didn't you ask me?" said the black girl. "I could have told you in twenty-five seconds without hurting those poor dogs."

"Your ignorance and presumption are unspeakable" said the old myop. "The fact was known of course to every child; but it had never been proved experimentally in the laboratory; and therefore it was not scientifically known at all. It reached me as an unskilled conjecture: I handed it on as science." (1946, p. 36)

Shaw's parable is a reminder that selling as behavioral "science" the commonplace embedded in jargon is nothing new. But in this case his criticism is not just. Pavlov's contribution was not the discovery of anticipatory salivation, but its measurement and use as a tool to study behavioral processes (the "physiology of the cerebral cortex" in his terminology). For example, he found that if the duration of the CS is longer than a few seconds, the saliva does not begin to flow at once, but is delayed until just before the delivery of food. (Pavlov called this *inhibition of delay*.) This period of delay is fragile, however, in the sense that any unexpected change in the situation immediately causes a copious flow of saliva. This phenomenon, a sort of dishabituation (see Chapter 4), is found in operant as well as classical conditioning; and its study has revealed important things about memory and the organization of action. Inhibition and the related phenomenon of successive induction, both of which Pavlov defined and demonstrated, have already been mentioned. By looking at the effects of varying the type of CS and the time between CS and US, by pairing one stimulus with the US and another with its absence, by looking at what happened when the CS was presented without the US, and by numerous other similar manipulations, Pavlov was able to develop for learned behavior a set of principles comparable to those earlier derived by Sherrington in his studies of spinal reflexes.

Thorndike looked at instrumental learning and decided that its essential feature was temporal contiguity between response and reinforcer ("satisfier"). Pavlov came to the same conclusion about classical conditioning: The necessary and sufficient condition for conditioning is temporal contiguity between CS and US. Both men had good reasons for their belief – although both were wrong. The first experiments that showed why, and point to a sort of alternative, are classical-conditioning experiments.

Pavlov could point to much data suggesting the importance of contiguity. For example, compare delay conditioning with so-called *trace conditioning* (the second and third panels in Figure 5.6). Both procedures involve a delay between CS onset and occurrence of the US, but in the trace case, the CS ends some time before the US begins. Trace conditioning is much more difficult to get than delayed conditioning, which suggests that any delay between CS and US is detrimental to conditioning. Many other experiments demonstrated the bad effects on conditioning of CS–US delays. The only discordant note was provided by *temporal conditioning*, the top panel in Figure 5.6, which is just periodic presentation of the US (like a fixed-interval schedule, but without the response requirement: Temporal conditioning is also known as a *fixed-time schedule*). In temporal conditioning, the US is also a (temporal) CS, like the neutral CS in trace conditioning. Despite the delay between CS (US) and US, temporal conditioning is very effective, even with long delays. This difference between temporal and trace conditioning seems to depend on the properties of memory – which also account for other examples of long-delay conditioning discovered subsequently (see Chapters 12 and 13). But temporal conditioning attracted little attention until relatively recently, and the main attack on contiguity came from another quarter.

CONTINGENCY AND FEEDBACK FUNCTIONS

A seminal paper in 1967 by R. A. Rescorla made a major advance. Rescorla used a procedure, invented by W. K. Estes and B. F. Skinner in 1943, that does not involve salivation at all. The Estes–Skinner procedure is used to study classical conditioning, but it is a mixture of both operant- and classical-conditioning procedures. The key ingredient is a variable-interval schedule of food reinforcement. Responding on VI schedules is an admirable *baseline* with which to study the effects of other independent variables: Other things being equal, the animal responds at a steady rate; hence, any change in rate associated with the presentation of a stimulus can safely be attributed to the stimulus, rather than to accidental variation.

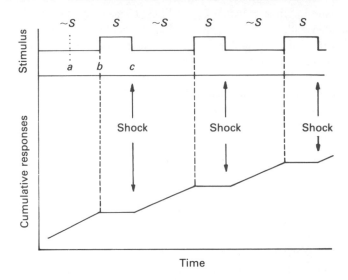

Figure 5.7. The suppression of a response maintained by a variable-interval schedule of reinforcement during a stimulus, S, ending with a brief electric shock. The degree of suppression is measured by comparing response rate during the stimulus (period bc) with responding during the same period before stimulus onset (period ab).

Estes and Skinner made use of a VI baseline to study the effect of occasionally presenting a relatively brief, neutral stimulus of the type used by Pavlov. After an initial "novelty" effect, the animal (usually a rat) continues to respond when the stimulus is present at about the same rate as when it is absent. This is the control condition, which establishes that the stimulus by itself has no effect. In the next phase, the rat is briefly shocked (through the metal grid floor) at some time during the stimulus presentation. After a few such stimulus-shock pairings, the stimulus produces a clearly recognizable suppression of lever pressing, as shown in Figure 5.7. This *conditioned suppression* (also termed the *conditioned emotional response*, or CER) can be measured by the relative rate of responding in the presence of the CS, the *suppression ratio*: $S = N_s/(N_s + N_{ns})$, where S is the suppression ratio, N_s is the number of lever presses when the stimulus is present, and N_{ns} is the number of lever presses during a comparable period when the stimulus is absent. $S = 0$ if the animal stops pressing completely in the presence of the CS, and .5 if the CS has no effect.

Conditioned suppression behaves in essentially the same way as the salivation of Pavlov's dogs, but has numerous practical advantages. Rats are cheaper than dogs; no operation is required; there are fewer physiological limitations on lever pressing than on salivation; it is less messy. Although other classical conditioning methods are sometimes used in Western laboratories (e.g., measurement of skin resistance, of blinking, or of the nictitating-membrane response in rabbits),

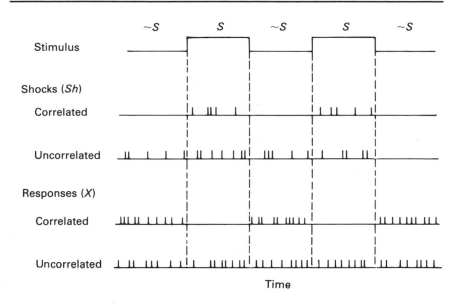

Figure 5.8. Stimulus contingencies in classical conditioning. The top panel shows the alternation of two stimuli. The next two panels show correlated and uncorrelated stimulus contingencies. The last two panels show the effects of each on responding maintained by a VI food schedule.

salivation is hardly studied at all, and the conditioned-suppression method is widely favored.[7]

A simplified version of Rescorla's procedure is shown in Figure 5.8. There are two stimuli (e.g., a tone vs. a light, or buzzer vs. absence of buzzer), labeled S and $\tilde{\ }S$ ("not-S," the absence of S). The stimuli are perhaps 60 sec in duration and, in this simplified version, occur in strict alternation, with about 50 such cycles making up a daily experimental session (only 2 cycles are shown in the figure). In the *correlated* condition (second row in Figure 5.8), brief, randomly spaced shocks occur only during stimulus S. In the *uncorrelated* condition, the shocks occur throughout the experimental session, that is, indiscriminately in the presence of both S and $\tilde{\ }S$. (The uncorrelated condition is sometimes also called the *truly random control* condition.)

The effect of these two procedures on lever-pressing maintained by the variable-interval schedule is shown in the bottom two rows of the figure. In the *correlated* condition animals typically respond for food only in the presence of stimulus $\tilde{\ }S$, which is not associated with shock: This is conditioned-suppression, just discussed. The interesting result is obtained in the *uncorrelated* condition (bottom row): In this condition animals respond indiscriminately in the presence of both stimuli, although at a somewhat lower rate than in stimulus $\tilde{\ }S$ in the correlated condition.

This result completely rules out CS–US contiguity as an adequate explanation

	US		
	~ Sh	Sh	
S	0	2	$p(Sh|S) = 1.0$
~S	2	0	$p(Sh|{\sim}S) = 0$

CS

Table 5.1. *Correlated condition.*

for classical conditioning. Simple pairing of US (shock) and CS (stimulus S) cannot be sufficient for conditioning, since this pairing holds in both the *correlated* and *uncorrelated* conditions of Rescorla's experiment; yet conditioning occurred only in the *correlated* condition. What, then, are the necessary and sufficient conditions for classical conditioning?

Intuitively, the answer is clear. The animals show conditioning to a stimulus only when it *predicts* the US: CS and US must therefore be correlated for conditioning to occur. This conclusion is appealing and widely accepted, but deceptive: One might say of the concept of "predictability," as of the bikini: What it reveals is suggestive, but what it conceals is vital. "Predictability" is something quite different from "contiguity." Contiguity is a perfectly unambiguous, quantitative time relation between two events. But the predictability of something depends upon the *knowledge* of the observer: Since Newton, we can predict when Halley's comet will return, whereas before none could do so; we know something Newton's predecessors did not. The situations used to study conditioning in rats are so simple, and our understanding of them so intuitive, that it is hard to define just what is involved in detecting the kinds of correlation depicted in Figure 5.8 – hard even to realize that correlation is not a simple property like weight or duration. Nevertheless, an explanation of conditioning in terms of correlation or predictability is a functional explanation, and therefore in many ways less powerful than Pavlov's completely mechanistic contiguity account. Rescorla's explanation is better than Pavlov's, but its success carries a cost: A gain in comprehensiveness has also meant a loss in precision.

The idea of correlation can be made more precise with the aid of a common device, the *contingency table*. Tables 5.1 and 5.2 are computed for the *correlated* and *uncorrelated* conditions in Figure 5.8. The rows correspond to stimuli (S and $\sim S$) and the columns to the occurrence or nonoccurrence of shock (Sh and $\sim Sh$). Thus, the entry in the upper right cell in Table 5.1 is the number of occurrences of stimulus S when at least one shock occurred (both presentations of S are accompanied by shock in Figure 5.8). The bottom right cell gives the number of times when $\sim S$ occurred and was accompanied by shock (zero),

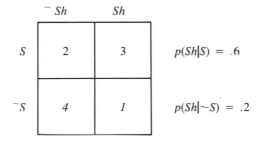

	~ Sh	Sh	
S	0	2	$p(Sh|S) = 1.0$
~S	0	2	$p(Sh|{\sim}S) = 1.0$

Table 5.2. *Random condition.*

	~ Sh	Sh	
S	2	3	$p(Sh|S) = .6$
~S	4	1	$p(Sh|{\sim}S) = .2$

Table 5.3. *Partially correlated condition.*

and so on. The concept of a *stimulus contingency* is obvious from the comparison of Tables 5.1 and 5.2: When the presence or absence of the stimulus is a predictor of shock, entries in the major diagonal of the table are high and entries elsewhere are low, as in Table 5.1. When the presence or absence of the stimulus is uncorrelated with the presence or absence of shock, the rows are the same, as in Table 5.2. The entries to the right of the tables ($P(Sh|S)$, etc.) are conditional probabilities, which are defined in a moment.

Tables 5.1 and 5.2 present "pure cases": The contingency between S and shock is perfect in Table 5.1 and zero in Table 5.2. Intermediate cases are also possible, and one is illustrated in Table 5.3. There is some correlation between *S* and shock in Table 5.3, but it is clearly less than perfect. The degree of association between shock and stimulus in these tables can be quantified using the χ^2 statistic, or the contingency coefficient, ϕ, which is just (χ^2/N), where *N* is the total number of cell entries.[8] I describe a graphical method for representing the degree of contingency in a moment.

The contingency measures represented by Tables 5.1–5.3 are simplified in an important way: They take no account of the relative durations of *S* and ~S. For example, even if shocks occur independently of *S*, if *S* is on for, say, 2/3 of the time and ~S for only 1/3, then the column entries for *Sh* in the table would be

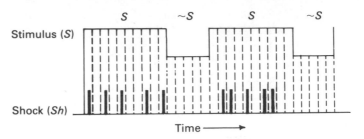

Figure 5.9. Analysis of stimulus contingencies by time bins.

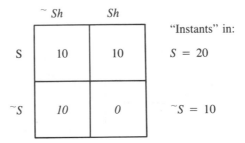

	~ Sh	Sh	"Instants" in:
S	10	10	$S = 20$
~S	10	0	$~S = 10$

Table 5.4. *Correlated condition–time bins.*

	~ Sh	Sh	
S	.5	.5	$p(Sh|S) = .5$
~S	1.0	0	$p(Sh|~S) = 0$

Table 5.5. *Correlated condition–conditional probabilities.*

higher for S than ~S. This gives the appearance of a contingency between S and Sh even though none exists.

One way of handling this difficulty is shown in Figure 5.9. Time is divided up into discrete intervals small enough so that no more than one shock can occur in an interval. The total number of entries in the contingency table, N, is then the total number of such "instants." Cell entries are the number of instances in S and ~S when Sh or ~Sh occurred. Table 5.4 shows some hypothetical numbers for the *correlated* condition in an experiment in which S was twice as long as ~S, and shock occurred in just half the instants in S. Table 5.5 is Table 5.4

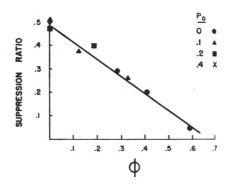

Figure 5.10. Suppression ratio as a function of ϕ for experimental results from Rescorla (1968). Suppression ratio is response rate in the CS divided by rate in the CS plus rate in its absence ("no suppression" = ratio of .5; which is the value for both noncontingent points, where ϕ = 0). Increasing suppression is indicated by smaller suppression values. (From Gibbon, Berryman, & Thompson, 1974, Figure 3.)

reduced to a standard form that takes account of the different durations of S and $\tilde{S}$. Each cell entry is the *conditional probability* of shock, given that S or $\tilde{S}$ is occurring (i.e., $p(Sh|S)$ and $p(Sh|\tilde{S})$); that is, each cell entry in Table 5.5 is just the entries in Table 5.4 divided by the row totals.[9]

The amount of conditioning to a stimulus in a classical-conditioning situation is directly related to the degree of contingency between the conditioned stimulus and the unconditioned stimulus. Figure 5.10 shows a particularly neat example of this relation: Data from an experiment by Rescorla show that the suppression ratio is linearly related, with negative slope, to the value of ϕ, the coefficient of contingency between CS and US. Since low suppression ratios mean high suppression, these data show the direct relation between the effect of a stimulus and its correlation with shock.[10]

Contingency space

Table 5.5 is the basis for the general, two-dimensional, *contingency space* shown as Figure 5.11. Since the row totals always add to 1.0, a given contingency table can be represented by just one of the two columns. By convention the righthand column is usually chosen, that is, $p(Sh|S)$ and $p(Sh|\tilde{S})$. Thus, each such contingency table defines one point in contingency space. (Table 5.5 is represented by the point labeled "X" on the ordinate.)

The contingency space is divided into three regions: (a) Above and to the left of the major diagonal is the region of *positive contingencies*, where $p(Sh|S) > p(Sh|\tilde{S})$ (shocks are more likely in S). (b) Below and to the right of the major diagonal is the region of *negative contingencies*, where $p(Sh|S) < p(Sh|\tilde{S})$ (shocks are less likely in S). (c) The major diagonal itself defines the *absence* of contingency between Sh and S, where $p(Sh|S) = p(Sh|\tilde{S})$

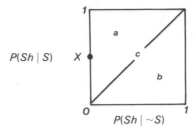

Figure 5.11. Stimulus-stimulus (CS-US) contingency space.

(this is the uncorrelated condition: Shocks are equally likely in S and $\sim S$). Positive contingencies generally produce *excitatory conditioning*, that is, the contingent stimulus produces an effect of the same sort as the US – suppression of food-reinforced responding in the CER experiment. Negative contingencies generally produce *inhibitory conditioning*, that is, effects of a sort opposite to those of the unconditioned stimulus. For example, imagine a CER experiment in which shocks occasionally occur on the baseline (i.e., in the absence of any CS). We could present two kinds of CS: a stimulus in which no shocks occur (a "safety signal" – inhibitory CS), and a signal in which the shock rate is higher than baseline (a "warning signal" – excitatory CS). The safety signal will produce an *increase* in lever pressing (suppression ratio > .5, an inhibitory effect in this context), whereas the shock-correlated CS will produce suppression relative to baseline (suppression ratio < .5, an excitatory effect in this context).

Temporal and trace conditioning

For simplicity, and historical reasons, I have introduced the notion of contingency in connection with the CER procedure and synchronous stimuli – stimuli that occur at the same time as the shocks they signal. But a similar analysis can be applied to temporal stimuli, as in temporal and trace conditioning. These temporal procedures raise two questions: (a) What is the effect of the trace CS on the conditioned response (CR)? Is the CR more or less likely to occur when the CS occurs compared to when the CS is omitted? (This is the question just discussed in connection with the CER procedure.) (b) When does the CR occur? Is it during the CS, immediately after, or after a delay? (This is a question specific to temporal procedures.)

The answer to the how-likely-is-the-CR question is "it depends" – on the temporal relations among CS, US, and other stimuli in the situation. The timing of the CR also depends on these factors to some extent, although there are some general things one can say: The CR almost never occurs during a trace CS, always afterward. It is also often delayed, more or less in proportion to the CS-US interval: the longer the CS-US interval, the longer the trace CR is delayed after CS offset. Figure 5.12 shows in stylized form the typical time course of a trace-conditioned response.

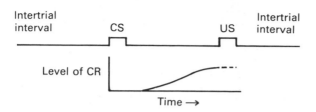

Figure 5.12. Typical time relations between conditioned response (CR) and trace-conditioned stimulus (CS) in salivary conditioning.

I've already mentioned that trace conditioning is hard to get. To begin to see why, look at the four conditioning sequences diagrammed in Figure 5.13. The sequences are typical classical-conditioning procedures, consisting of an *intertrial interval*, the period between the end of the US and the onset of the next CS (period US–CS) and the *trial period* between the CS and the ensuing US (period CS–US). In the diagram these two periods add to a constant, the US–US interval. I'll consider this case first, then look at what should happen if the intertrial interval is allowed to vary, while the CS–US period remains constant. These two periods are formally analogous to the ¯S and S periods in the CER procedure we just looked at. They may not be analogous from the animal's point of view, of course, because they depend upon his ability to *remember* the event initiating the period – the CER procedure imposes no such memory requirement, because the stimuli are continuously present. The behavior to be expected from these four procedures depends entirely on how the animal uses the temporal information available to him.

What is the effect of the CS on the probability of a CR in these procedures? When does trace conditioning occur and when does it fail? Consider two possibilities: (a) The animal has perfect memory (i.e., can use either the CS, the US, or both as time markers). (b) The animal's memory is imperfect.

In the first case, since the CS is always closer than anything else to the next US, the CS might be said to predict the US, and trace conditioning should occur. Since the times between US and US and between CS and US are both fixed, however, the animal might use either or both as time markers. We know that temporal conditioning, sequence D in the figure, yields excellent conditioning, so we know that animals can use the information provided by a fixed US–US interval. If he uses the US, then the CS will have no special effect, and trace conditioning will not occur. But since the accuracy with which an animal can tell time is roughly proportional to the time interval involved (Weber's law for time; see Chapter 12) the animal will obviously do better to use the CS, the stimulus closest to the next US. Moreover, this advantage will be greater the closer the CS to the US: Sequence A should therefore produce the most reliable trace conditioning, B next, and C the worst. If the animal can use either CS or US as a trace stimulus, he should always use the CS.

If the animal's memory is imperfect, however, then some events may make

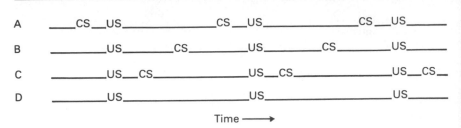

Figure 5.13. Three CS placements in trace conditioning (A, B, and C), compared with temporal conditioning (D).

better time markers than others. In particular, for reasons I discuss in Chapter 12, the US may be a much better time marker than a "neutral" CS. In this case we have two factors that act in opposite directions: The CS is always closer than anything else to the US. Hence, trace conditioning to the CS is favored over trace (temporal) conditioning to the US. But if the US is better remembered (makes a better time marker) than the CS, then, other things being equal, temporal conditioning will be favored over trace conditioning. Trace conditioning should, therefore, occur only under two conditions: (a) When the CS–US interval is much shorter than the US–US interval; or (b) when the US–US interval is *variable*, so that post-US time cannot be used to predict US occurrence. Both these predictions have generally been supported.

Obviously animals behave adaptively in classical-conditioning experiments, much more so than the earlier notion of automatic conditioning-by-contiguity suggests. For the most part, animals become conditioned to the stimulus that predicts the US, and apparent failures to do so in trace conditioning seem to reflect special attention to the US, which is undoubtedly adaptive in other contexts. The subtlety of this behavior poses considerable problems for theory, however. So long as simple pairing seemed to be the critical operation for classical conditioning, attention could be focused on procedural details – the relation between conditioned and unconditioned responses, the effects of CS–US delays, the effects of CS salience, and so on – with the assurance that the basic mechanism was known. It was not. We still know little about the computational process that allows animals to identify and react just to those aspects of their environment that predict hedonic events. I return to this problem in Chapter 13.

Response contingencies and feedback functions

Both Pavlov and Thorndike thought that contiguity was the only, or at least the major, factor in learning. Later experiments have shown that contiguity is not sufficient, that animals are able to detect correlations between CS and US in classical-conditioning situations. As one might expect, the same thing has turned out to be true of operant conditioning – although the theoretical analysis is a bit more complicated.

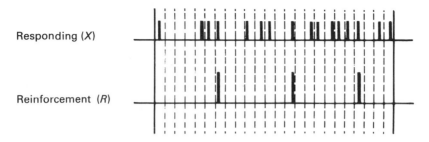

Responding (*X*)

Reinforcement (*R*)

Figure 5.14. Analysis of response-reinforcement contingency by time bins.

The problem for an animal in a classical-conditioning experiment is to detect which stimuli predict the US, and how well they predict it. The problem in an operant-conditioning experiment is similar, namely, what aspects of the animal's *behavior* (i.e., what responses) predict the US (reinforcer), and how well do they predict it. The only difference between the problems of stimulus and response selection is that the animal can control its own behavior, whereas it cannot control the CS in a classical-conditioning experiment. Hence, an animal in an operant-conditioning experiment need not take at face value a given correlation between its behavior and reinforcer occurrence: It can increase or decrease its response rate, or vary responding in other ways, and see if the correlation still holds up. This difference means that the animal can hope to arrive at the *cause(s)* of the reinforcer in an operant-conditioning experiment, but is limited to detecting *correlations* between CS and US in classical conditioning experiments.

We can approach the response-selection problem in two ways: One is by an extension of the discrete-time analysis just applied to CER conditioning, the other by an extension of the memory analysis applied to trace conditioning. I discuss the discrete-time method here, leaving more complex analysis to a later chapter.

The discrete-time representation of contingency in Figure 5.9 can be applied directly to response selection in operant conditioning in the following way. Figure 5.14 shows two event records: responding at the top, and associated variable-interval reinforcement on the bottom. As before, time is divided into discrete "instants" small enough that no more than one response, or reinforcer, can occur in each. To find out if the response predicts the reinforcer we can ask: How many times is a reinforcer accompanied by a response (i.e., both in the same instant)? How many times does it occur without a response? How many times does a response occur unaccompanied by a reinforcer? How many times does neither response nor reinforcer occur?

The contingency Table 5.6 shows the answers for the sample of responses and reinforcements shown in Figure 5.14. Table 5.7 shows the same data in conditional-probability form: The entries in Table 5.7 are the entries in Table 5.6 divided by the column totals. They show the conditional probabilities that *R* or *~R* will

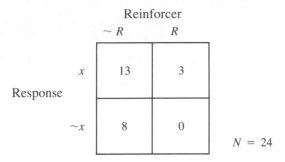

Table 5.6. *Response–contingent reinforcement.*

	~ R	R
x	.81	.19
~x	1.0	0

Table 5.7. *Response-contingent reinforcement–conditional probabilities.*

occur, given x or $\tilde{}x$. An obvious difference between the contingency tables for response and stimulus selection is that the response table is partly under the animal's control. For example, if the animal responds more slowly, the entries in the upper-left cell (unsuccessful responses) will decrease much more rapidly than entries in the upper right cell (reinforced responses); that is, responding predicts reinforcement much better – the contingency between the two improves – as response rate decreases. This follows from the properties of a VI schedule: The more slowly the animal responds, the more likely that each response will produce a reinforcer. With further decreases in response rate, the entries in the upper right cell will also begin to decrease. As response rate decreases to zero, entries in all save the lower left cell vanish.

The lower right cell in Table 5.6 is always zero for the standard reinforcement schedules – the animal doesn't get the reinforcer unless he responds. Consequently variations in the animal's rate of responding simply move the point representing the matrix up and down the vertical axis of the $p(R|\tilde{}x)$ versus $p(R|x)$ contingency space (See Figure 5.16). Molar *feedback functions*, which show the relation between the rates of responding and reinforcement, are therefore more useful than contingency tables as representations of standard schedules.

The relationship between the feedback function and a contingency table such as Table 5.6 can be easily seen by considering the number of responses, x, and

	~ R	R
x	x – R(x)	R(x)
~x	1 – x	0

N = 24

Table 5.8. *Response-contingent reinforcement–analysis by rates.*

reinforcements for those responses, $R(x)$, over a unit time period (i.e., a period chosen so that x and $R(x)$ are just rates). Table 5.6 can then be rewritten entirely in terms of x and $R(x)$, as shown in Table 5.8. The feedback function is simply the systematic relation between x and $R(x)$ enforced by the schedule. Each point on the feedback function (i.e., each pair of [x, $R(x)$] values), therefore, defines a separate contingency table.

Feedback functions for common schedules

Molar feedback functions are important for topics discussed later in the book. To make the concept as clear as possible, and as a way of introducing some new procedures, I now show how to derive feedback functions for some common schedules.

Ratio schedules. Ratio schedules prescribe either a fixed (FR) or variable (but with fixed mean: VR) number of responses per reinforcement. The molar feedback function makes no distinction betwen FR and VR, since it simply relates aggregate responding to aggregate reinforcement. This function was derived earlier and is obviously a simple proportion. Using the previous symbols, the ratio feedback function is given by $R(x) = x/M$, where $R(x)$ is reinforcement rate, x is response rate, and M is the ratio value. The ratio feedback function is therefore a straight line through the origin (see Figure 5.15).

Interval schedules. Fixed-interval schedules must be treated differently from variable-interval, because the animal can predict when food will be available on FI almost perfectly, whereas VI schedules are explicitly designed to prevent this. I just consider VI here. For a VI schedule with random interreinforcement intervals, the average time between reinforcements is made up of two times: the prescribed minimum interreinforcement interval (the VI value), which can be written as $1/a$, where a is the maximum possible reinforcement *rate*, and d, which is the delay between the time when reinforcement is available for a response and the time when the next response actually occurs. Thus,

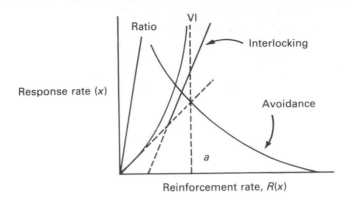

Figure 5.15. Molar feedback functions for four operant schedules: ratio, variable-interval, interlocking, and avoidance (shock-postponement). These curves show how the schedule enforces a relation between response rate (independent variable: x) and reinforcement rate (dependent variable: $R(x)$). The usual convention would put $R(x)$ on the vertical axis and x on the horizontal. The axes are reversed here because in all subsequent discussion we will be much more interested in the reverse relation, where $R(x)$ is the independent variable and x the dependent variable. For consistency, I adopt throughout the latter arrangement.

$$D(x) = 1/R(x) = 1/a + d, \tag{5.2}$$

where $D(x)$ is the actual mean time between reinforcements and $R(x)$ the obtained rate of reinforcement, as before. If we specify the temporal pattern of responding, then d can be expressed as a function of x, the average rate of responding. In the simplest case, if responding is random in time, then

$$d = 1/x, \tag{5.3}$$

the expected time from reinforcement setup to a response is just the reciprocal of the average response rate. Combining Equations 5.2 and 5.3 yields the actual feedback function, which is therefore

$$R(x) = ax/(a + x), \tag{5.4}$$

a negatively accelerated function which tends to a as $x \to \infty$, and to x as $x \to 0$; that is, it has asymptotes at $R(x) = a$, and $R(x) = x$, as shown in Figure 5.15.

If responding is not random, or if it depends on postreinforcement time (as in fixed-interval schedules), the delay, d, may depend on properties of responding in addition to rate. In this case the properties of the schedule cannot be captured by a simple response rate versus reinforcement rate feedback function; something more complicated is required. This difficulty simply emphasizes that although the concept of a feedback function is perfectly general, to get one in a form simple enough to be useful may mean making quite drastic simplifying assumptions about behavior.

Interlocking schedules. Interlocking schedules combine ratio and interval features. Reinforcement is delivered for a response once a weighted sum of time and number of responses exceeds a fixed value. If we neglect the final response requirement, this schedule can be represented by the relation

$$aN(x) + bt = 1, \qquad (5.5)$$

where $N(x)$ is the number of responses made since the last reinforcement, t is the time since the last reinforcement, and a and b are positive constants. But $t = 1/R(x)$ and $N(x)/t = x$, the rate of responding; Hence Equation 5.5 reduces to the feedback function

$$R(x) = ax + b, \qquad (5.6)$$

which is linear, like the ratio function. The line doesn't go all the way to the $R(x)$ axis because at least one response is required for each reinforcer: Hence the interlocking schedule feedback function must have the FR 1 line as an asymptote. Figure 5.15 shows the interlocking schedule function as two line segments: Equation 5.6, and the FR 1 function $R(x) = x$.

These three schedules, ratio, VI, and interlocking, are all examples of positive contingencies: Reinforcement rate increases with response rate. Positive contingencies are obviously appropriate for use with positive reinforcers. Conversely, escape, avoidance, and omission schedules all involve negative contingencies: Reinforcement rate decreases as response rate increases. Negative contingencies are appropriate for negative reinforcers (as earlier defined in terms of approach and withdrawal). In general, the product of the signs of contingency and reinforcer must be positive if behavior is to be sustained. This rule is true of "strong" reinforcers like food (for a highly deprived animal) or electric shock, which usually affect behavior in a consistent way. Animals will cease to respond if responding produces shock (this is the procedure of *punishment*) or if responding prevents food. There are some exceptions, however. As I explain in Chapter 7, it is possible to have too much of a good thing, and some things are reinforcing only in certain quantities (in the limit, this is even true of food, though not, perhaps, of money and other human goodies). Even if the reinforcer maintains its value, paradoxical effects can be produced under special conditions: Animals will respond to produce electric shock; and hungry pigeons can be induced to peck even if pecking prevents food delivery. These exceptions are of interest for the light they shed on the mechanisms of operant behavior, and I deal with them in more detail later.

Escape, avoidance, and omission schedules. None of these negative contingencies lends itself easily to algebraic treatment.[11] In escape, omission, and so-called discriminated avoidance procedures, the opportunity to respond is usually restricted to discrete trials, separated by an intertrial interval. For example, omission training and discriminated avoidance resemble classical delay conditioning, with the added feature that a response during the CS eliminates the US on that trial. (The only difference between omission training and discriminated avoidance is that the former term is used when the US is a positive reinforcer, the latter when it is a negative reinforcer.) Escape is also a discrete-trial procedure: An aversive stimulus such as shock or a loud noise is continuously present during a trial and a response eliminates it. The contingencies in all these cases are obviously negative ones, but the discrete nature of the procedures, and the addition of a trial stimulus, complicates a feedback-function analysis.

Shock postponement (also termed *unsignaled* or *Sidman avoidance*) is a continuous procedure that is very effective in maintaining operant behavior and does allow a relatively simple analysis. In its standard form, the procedure presents brief shocks to the animal at fixed intervals of time, the shock-shock ($S*S$) interval. If the animal responds at any time, the next shock is delayed by a fixed interval, the response-shock ($R*S$) interval. For example, the $S*S$ interval might be 20 sec and the $R*S$ interval 10 sec. Obviously if the animal is careful to respond at least once every 10 sec, it need never receive shock. As response rate declines more shocks are received, their number depending upon the distribution of interresponse times (IRTs). If the distribution is sharply peaked, with most IRTs close to the mean value, then even small shifts in the mean will produce a substantial increase in the frequency of shock. Conversely, if the distribution of IRTs is quite broad, a decrease in response rate may result in quite a gradual increase in shock rate. The feedback function therefore depends on the IRT *distribution* as well as the rate of responding.

Most animals that respond at all on shock-postponement schedules respond at a substantial rate. Consequently, of the few shocks they receive, most are determined by the R*S interval, very few by the S*S interval. (Animals usually respond immediately after a shock, which also tends to eliminate the S*S interval as a factor in the maintenance of this behavior – although not, of course, in its acquisition.) If the IRT distribution is known, it is therefore possible to compute the number of IRTs longer than the R*S interval as a function of the mean IRT. Converting these values to rates of responding and shock then yields the feedback function.

On first exposure to shock postponement (as to any schedule), responding is likely to be approximately random in time. Later the interresponse-time distribution is likely to become more sharply peaked, approximately at the postponement ($R*S$) value, T. For the random case, and $R*S = S*S = T$, the feedback function is

$$R(x) = x \cdot \exp(-xT)/(1 - \exp(-xT)), \tag{5.7}$$

which is shown in Figure 5.15.[12] The function has the expected properties: As response rate increases, reinforcement rate decreases, and when response rate is zero, reinforcement (shock) rate equals $1/T$.

These four examples were chosen for their simplicity, to illustrate the concept of a feedback function, to give some idea of the variety of possible functions, and

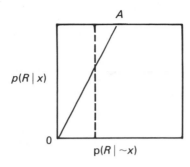

Figure 5.16. Response-reinforcement contingency space.

to make clear the difference between positive (increasing) and negative (decreasing) functions.

The detection of response contingency

So far I have considered only conventional reinforcement schedules, in which the reinforcement, when it occurs, always requires a response. What if we relax the assumption that reinforcement must always follow a response? For example, suppose we take a pigeon already trained to respond for food on a VI schedule, and arrange that half the reinforcers are delivered as soon as the VI timer "sets up," independently of responding (this is known as a *mixed variable-interval, variable-time* [mix VI VT] schedule): What effect will this have? Since pigeons peck fast for food on VI schedules, this change in procedure will have little effect on the overall food rate; all we have done is degrade the contingency between food and pecking: The lower-right cell in Tables 5.6–5.8 will not now be zero. Consequently, the pigeon should now peck more slowly than before, and indeed this is the usual result in numerous such experiments.[13]

How does the animal detect the degradation in contingency? The answer to this is not fully known, but we can get some idea of what is involved by looking at changes in molar contingencies, and changes in temporal (contiguity) relations.

The change in molar response contingency can be represented in a contingency space, as shown in Figure 5.16. The location of the point representing degree of contingency depends on response rate: When rate is high, the point will be close to the diagonal, because many response-reinforcer conjunctions will be accidental. Since half the reinforcers are response dependent, $p(R|x)$ can never be less than twice $p(R|{\sim}x)$, however. At the other extreme, when response rate is very low, almost every response is reinforced, so that $p(R|x) \rightarrow 1$. $p(R|{\sim}x)$ is fixed by the variable-interval value, hence response rate variation can only move the point up and down the vertical line corresponding to $p(R|{\sim}x)$ – where that line lies on the horizontal axis depends upon the duration of the "instants" (see Figure 5.14) we have chosen. Thus, response-rate

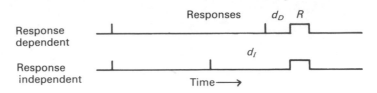

Figure 5.17. Time-delay differences in response-dependent (top) and response-independent (bottom) reinforcement.

variation shifts the contingency up and down the vertical line segment between the upper bound and the line $p(R|x) = 2p(R|^\sim x)$ (line OA), as shown in the figure.

Variation in absolute frequency of reinforcement (variable-interval value) and the proportion of reinforcers that are response independent both have identifiable effects in the contingency space. Variation in interval value shifts the location of the $p(R|^\sim x)$ line along the horizontal axis; variation in the proportion of response-independent reinforcers varies the slope of the radial constraint line OA. Thus, the pigeon has ample molar information to detect a change in response contingency.

He also has molecular information, in the form of time differences. The situation is illustrated in Figure 5.17, which shows the differences to be expected when reinforcement is response-independent versus response dependent. Response-reinforcer delay will vary from one reinforcer to the next, but the variation will be less, and the mean value smaller, when the reinforcement is response dependent. When the distributions of d_D and d_I are reduced to a standard form, in which the variances (spreads) are equal, the separation of the means is a measure of the *detectability* of the difference between them. The significant thing about these two distributions is that the animal can separate them as much as he wishes by slowing his response rate: Evidently the best test for response contingency is to slow response rate and see what happens to the distribution of response-reinforcer delays.

Animals are very sensitive to these time-distribution differences. I describe in a later chapter an experiment in which a pigeon had to make a choice depending on whether a just-preceding event had occurred independently of its peck or not – the birds were able to detect very small time differences. Animals seem also to be sensitive to the contingency-detecting properties of a decrease in response rate, and this may account for their problems with spaced-responding schedules – which mimic VI schedules in showing an increasing probability of payoff as response rate decreases. When the spacing requirement is long (> 30 sec or so), pigeons treat spaced-responding schedules just like long VI schedules: They respond at a relatively high rate, and the peak of the IRT distribution is at 5 sec or less.

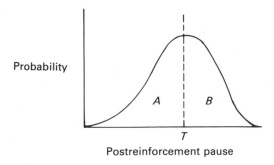

Figure 5.18. Theoretical pause (time-estimation) distribution of a pigeon on a fixed-interval *T* sec schedule.

Sampling always carries some cost. Responding more slowly helps the pigeon estimate the degree of response contingency, but it also delays food because many response-contingent reinforcers will then be "set up" for some time before they are collected. How should the animal weigh this cost? Figure 5.18 illustrates the problem posed by a fixed-interval schedule. The distribution shows the spread in the animal's estimate of time-to-food, as measured by his postfood pause. Area *A*, to the left of *T* (the FI value), represents unnecessary anticipatory responses, responses made before food is made available by the FI programmer. Area *B*, to the right of *T*, represents unnecessary delays, interfood intervals longer than necessary because the animal waited to respond until after reinforcement had set up. If the animal shifts *t* (the mode of his pause distribution) to the left, he will reduce delays (*B*), but at the cost of increasing the number of unnecessary responses (*A*), and conversely. As we have already seen, most animals respond too early on FI, which suggests that they weigh unnecessary delays much more than unnecessary responses. Or, to put the same thing in a slightly different way: Pigeons act as if they weigh potential losses in food much more than wasted key pecks.

SUMMARY

The effects of reward and punishment on behavior are obvious to all. Unaided by experimental psychologists, the human race managed long ago to discover that children and animals desist from punished behavior, and persist in rewarded behavior. Science has added to this rude knowledge in at least three ways. One is to define reward and punishment precisely. The result is the concept of a *reinforcement contingency*. Another has been to emphasize the role of time delays between response and reinforcer. And a third has been to design simple situations that allow us to explore the limits of animals' ability to detect reinforcement contingencies. We have begun to understand the intricate mechanisms that allow mammals and birds to detect subtle correlations among external stimuli (including time), their own behavior, and events of value.

In this chapter I have presented a brief history of operant and classical conditioning and described some common conditioning situations. I have explained the difference between *acquisition*, the process by which animals detect contingencies, and the *steady state*, the fixed pattern that finally emerges. The chapter discussed the concepts of *stability* and *equilibrium*, and how operant behavior, while usually stable, is occasionally unstable or metastable in response to changing conditions. I spent some time on *molar feedback functions*, because these seem to determine molar adjustments to many simple reinforcement schedules, and also shed light on motivational mechanisms, which are the topic of the next chapter.

NOTES

1. A drive implies "push" rather than "pull"; it sounds more like a mechanistic than a functional (teleological) explanation. Perhaps this is why the notion of drive was so appealing to some early behaviorists. The concept of preference structure is clearly functional, however. A preference structure, like the decision rules for *Stentor* discussed in the last chapter, implies no particular mechanism. The distinction is more semantic than real, however. Drives were treated very much like goals by their advocates, so I do the concept no violence by treating drives as a primitive form of preference structure.

For good accounts of theories of primary and secondary drives (theories of which were developed most extensively by Hull and his followers) see Osgood (1953), Kimble (1961), and Bower and Hilgard (1981 5th ed.).

2. *Recent history of reinforcement theory.* The theoretical foundations for Thorndike's work were laid by the English philosopher Herbert Spencer (1820–1903), from whom Darwin borrowed the phrase "survival of the fittest." Spencer applied to trial-and-error learning a version of the Darwinian theory of adaptation via variation and selection, which is essentially the position I am advancing here. Similar views were outlined by the British comparative psychologist Lloyd Morgan (1852–1936), who is perhaps best known for his application of the principle of parsimony to explanations of animal behavior: Never invoke a higher faculty if a lower one will do (Lloyd Morgan's canon).

American comparative psychology was distracted from these rational, biologically based doctrines by the strident tones of J. B. Watson's behaviorism, which was no more objective or parsimonious than Thorndike's views, but was easier to understand. Truth must be understood to be believed, but ease of understanding is a poor guide to validity. Nevertheless, Watsonian behaviorism captured the public imagination, and left American animal psychology with a taste for simplism from which it has yet to fully recover. After his initial contributions, Thorndike moved away from animal psychology, and for some years the field was left largely to three men and their followers: Edwin Guthrie, E. C. Tolman, and Clark Hull. The first two were eclectic in spirit and neither founded a school. But Hull, at Yale, an intellectual descendant of Watson, organized his laboratory along Stakhanovite lines and produced a stream of energetic and dedicated disciples who colonized several other departments of psychology. Soon his students, most notably Kenneth Spence at Iowa, were producing disciples of their own and Hullian neobehaviorism, as it came to be known, became the dominant approach to experimental psychology.

In the mid-1930s B. F. Skinner at Harvard proposed an alternative to Hullian stimulus-response theory. In some ways almost as simplistic as Watson's views, Skinner's avowedly atheoretical approach had the advantage over Hull's far-from-elegant theory of a simple and effective new experimental method, the Skinner box. The social momentum of the Hullian movement ensured its continued dominance until the late 1950s, after which it was superseded by Skinner's operant-conditioning method. Skinner's theoretical views also gained many adherents, although they have never dominated the field in the same way as the disciples of Hullian theory.

Hull's fall and Skinner's rise are probably attributable to growing impatience with the unwieldiness of Hull's elaborating system, and the collective sense of liberation felt by those who turned away from it to Skinner's new method. Skinner wrote persuasively of the power of an "experimental analysis" of behavior and converts to *operant conditioning*, as the field came to be called, found exciting things to do exploring the novel world of reinforcement schedules. It took some time before others began to question the relevance to behavior-in-general of the elaborations of reinforcement-schedule research which, in its heyday, became a sort of latter-day experimental counterpoint to the Baroque fugue of Hullian axioms and corollaries.

Skinner's theoretical position is an intriguing one. On its face it is simple to the point of absurdity. It is a theory and an epistemology, yet claims to be neither. This ambiguous status has conferred on it a measure of invulnerability. Its basis is the principle of reinforcement: All operant behavior (there is another class, *respondent* behavior, that is excluded from this) is determined by the reinforcement contingent upon it. Taxed to identify the reinforcers for different kinds of behavior, Skinner responds that the principle of reinforcement is a definition, not an explanation, thus begging the question. Pressed for the reinforcement for behavior on which nothing external depends, recourse is often had to "self-reinforcement," which is self-contradictory. By giving the appearance of a theory, but disclaiming the responsibilities of one; by providing a method which promises to find a reinforcer for everything – and by invoking a mass of data on schedules of reinforcement exceeding in reliability and orderliness anything previously seen, Skinner was for many years able to overwhelm all argument and present to the world the vision of a powerful and consistent system that promised to remake human society.

Aristotle distinguished four kinds of causes, two of which are of contemporary interest: the efficient cause and the final cause. Efficient cause is *the* cause of mechanistic science, which considers no other an adequate explanation. The stimulus for a reflex response is a cause in this sense. Final causes have not been totally dispensed with, however. The economic concept of *utility*, which is another name for *subjective value*, is a final cause in the Aristotelian sense, because consumers are presumed to adjust their behavior so as to maximize their utility: The consumer's market basket is explained not by the causal factors underlying each purchasing decision, but by reference to the utility of the final bundle. It is obvious that reinforcement is a final cause in the Aristotelian sense, and plays the same role in Skinnerian theory as utility does in economic theory. The parallels are close. For Aristotle, everything has its final cause; for radical behaviorists (Skinner's term), every operant behavior has its reinforcer. In popular detective fiction, the first rule is to look for a motive, which usually involves a woman. Skinner will have nothing to do with motives, but his views can, nevertheless, be epitomized as *cherchez la renforcer . . .*

The intimate conceptual relation between utility and reinforcement theory has led recently to increasing application of economic ideas to behavioral psychology, and vice versa; more on this in Chapter 7.

In addition to the references given earlier, fuller discussions of behaviorism and the law of effect can be found in Boring (1957), Hearst (1979), Postman (1947), Wilcoxon (1969), the volumes edited by Koch (e.g., 1959), and a collection of theoretical reviews by Estes et al. (1954).

3. The terminology in this area is confusing. The terms I have used are probably the simplest and most intuitively obvious. Skinner used the same terms in a slightly different sense, however, and many follow him. Skinner attended to the fact that a reinforcing effect may be produced either by presenting a stimulus (e.g., food) or by removing one (e.g., electric shock). The first case he called *positive reinforcement*; the second, *negative reinforcement*. The symmetry breaks down when applied to suppressive effects, however. Suppression of behavior by the presentation of a stimulus contingent on it is termed *punishment* – I have termed this *negative reinforcement*; but there is no simple term for the suppression of behavior by the response-contingent removal of positive stimulus.

The only really precise way to represent these various cases is by means of feedback functions, which are taken up in connection with the concept of a reinforcement contingency. In the meantime, the terminology I have adopted is probably less confusing than the alternatives because it focuses on the direction of change of the behavior, rather than the kind of stimulus change necessary to produce that change. For example, is *heat* a positive or a negative reinforcer in Skinner's sense? It is difficult to answer this without getting into unprofitable discussions about whether it is the presence of heat or the absence of cold that is reinforcing. The recognition that reinforcement is equivalent to feedback renders such disputes unnecessary.

4. The method of studying reward and punishment via its effects on an "arbitrary" response was independently invented by B. F. Skinner (1932) in the

United States and G. C. Grindley (1932) in England. Grindley used the response of headturning by restrained guinea pigs; Skinner used lever pressing by freely moving rats. Skinner's approach has dominated subsequent research, for at least two reasons: His method was obviously easier to implement than Grindley's; he also discovered that food need not follow every response. This finding led to the concept of a *reinforcement schedule* and the extensive study of the properties of reinforcement schedules and their implications for the mechanisms of operant behavior.

5. I will use the term *reinforcement* in a purely descriptive sense, to refer to the operation of presenting a reinforcer to an animal. Unfortunately it is also often used as an explanation, based on theories (now largely discredited) derived from the original law of effect that assumed that presenting a reinforcer contiguous with a response automatically "strengthened" (reinforced) that response or its connection with current stimuli.

6. See Staddon (1965) for a fuller discussion of this example.

7. Pavlov focused almost entirely on the salivary conditioned response. Recent experiments using the Estes–Skinner procedure tend to look just at the suppression measure. Other recent work has emphasized the earlier conclusion of Zener (1937) that these simple changes are perhaps the least important of the many effects of classical-conditioning procedures, however. I return to a discussion of these other effects in Chapter 13 in connection with the acquisition of behavior.

The question of *why* a stimulus associated with shock should suppress food-reinforced operant behavior is also taken up later.

8. If the entries in the four cells of the contingency table are labeled a, b, c, and d (reading from left to right, top to bottom), then χ^2 is given by $\chi^2 = (ad - bc)^2/(a + c)(c + d)(b + d)(a + c)$; $N = a + b + c + d$. Other measures of contingency are information transmission and the d' measure of signal detection theory (see Luce, 1963; Green & Swets, 1966). See Gibbon, Berryman, and Thompson (1974) for an extensive (though sometimes hard to follow) discussion of the concept of contingency as applied to both classical and operant conditioning procedures.

9. A contingency table says nothing about the direction of causation; it is completely symmetrical. Consequently, we could equally well have divided the entries in Table 5.4 by the column, rather than row, totals, to obtain the conditional probabilities of S and $\sim S$ given Sh and $\sim S$, rather than the reverse. Since the animal's interest is obviously in predicting Sh from S, rather than the reverse, the row totals are appropriate divisors.

10. ϕ here was computed using the time-based method (see Gibbon et al., 1974, for a fuller account).

11. The meaning of "negative contingency" is obvious when an exact feedback function can be derived. It is less clear when no analytic function is available. Perhaps the most general definition is in terms of the sign of the first derivative,

$dR(x)/dx$. The sign of a contingency then refers to the effect on the direction of change of reinforcement rate of small changes in response rate. For complex schedules, the sign of $dR(x)/dx$ may not be the same across the whole range of x. This will always be the case for any schedule designed to restrict x to a limited set of values. For example, on spaced-responding schedules, a response is reinforced only if it follows the preceding response by t sec or more. Thus, when the animal is responding rapidly, so that the typical interresponse time (IRT) is less than t, the contingency is a negative one: Decreases in response rate bring increases in reinforcement rate. But when the animal is responding slowly (IRT$>t$), the contingency is positive: Increases in response rate bring increases in reinforcement rate.

12. This function was derived as follows. If responding is random in time, then the probability that an IRT will fall between zero and T, the postponement value, is given by

$$P(< T) = \int x \cdot \exp(-xt)dt, \tag{N5.1}$$

which equals

$$= 1 - \exp(-xT). \tag{N5.2}$$

The time taken up by IRTs in this range is given by

$$T_s = \int_0^T tx \cdot \exp(-xt)dt \tag{N5.3}$$
$$= 1/x - (T + 1/x)\exp(-xT).$$

Similarly, the probability that an IRT will be greater than T, hence fail to avoid shock, is

$$P(>T) = \exp(xT). \tag{N5.4}$$

The time taken up by IRTs in this range is then just

$$T_u = T\exp(-xT), \tag{N5.5}$$

because of the assumption that the animal responds immediately after shock. The shock rate, $R(x)$, is just the number of shocks received divided by the total time, $T_s + T_u$. From Equations N5.4, N5.3, and N5.5:

$$R(x) = x \cdot \exp(-xT)/(1 - \exp(-xT)),$$

which is Equation 5.7 in the text.

The variable x in these equations is not the actual response rate, but the corresponding parameter of the exponential distribution. When $x>>1/T$, x is approximately equal to x_{ac}, the actual response rate. Otherwise Equation 5.7 must be corrected by the relation

$$x_{ac} = x/(1 - \exp(-xT)), \tag{N5.6}$$

which is derived from the relation

$$x_{ac} = 1/(T_s + T_u).$$

13. For a recent review see Catania (1981) and several of the other articles in the book edited by Harzem and Zeiler.

6

FEEDING REGULATION:
A MODEL MOTIVATIONAL SYSTEM

The operant behavior of higher animals depends on their motivation, the value of the reward or punishment. If we ignore for the moment the problem of competition among motivational systems (hunger vs. sex, vs. thirst, etc.) there are two aspects to motivation: "How hungry am I?" and "How hard should I work to get food?" The answer to the first question depends entirely on the animal's internal state; it is a *regulatory* question. But the answer to the "How hard should I work?" question depends both on the animal's state *and* on the opportunities offered by the environment: It's worth working hard for food if much food is to be got. It's not worth working hard if the probability of payoff is low. This aspect of motivation is known as *incentive*.

This chapter deals with regulation and incentive. I take feeding as a typical motivational system and show how taste, body weight, and even drugs and brain lesions, combine to affect how much rats will eat and how hard they will work for food. The aim of the discussion is to dissect a single motivational system, and show how a few relatively simple principles underlie diverse effects. The next chapter deals with interactions among motivational systems.

REINFORCEMENT AND HOMEOSTASIS

Animals must eat to live and, on an operant food schedule, must work to eat. It would be astonishing, therefore, if schedule performance had nothing to do with the regulation of nutritional balance.[1] The molar (as opposed to molecular) properties of behavior on reinforcement schedules – for example, the average rate of responding over a period of several minutes – are intimately related to energy regulation.

Eating can be looked at from two points of view: diet selection and energy balance. Diet selection deals with the mechanisms that allow omnivores such as rats, dogs, and people under normal circumstances to pick out a diet not deficient in essential vitamins or trace minerals. Energy balance deals with how much animals eat in relation to their energy expenditures: If they eat more than they expend, their body weight goes up; if they eat less, their weight goes down. Performance on simple reinforcement schedules relates most obviously to energy balance. Performance on more complex procedures involving choice between different foods also brings in mechanisms of diet selection. I deal here with energy balance; later chapters say something about diet selection.

The great French physiologist Claude Bernard more than 100 years ago pointed out the importance to organisms of a constant *milieu interne* ("internal environment"): constant or narrowly varying core temperature, adequate cellular oxygen and energy supply, constant blood pH, and so on. In this century, Walter Cannon coined the term *homeostasis* for the processes by which this internal constancy is maintained. Homeostatic mechanisms are of two kinds: internal, acting only within the body; and external, acting on the environment. Examples of internal mechanisms are vasoconstriction and vasodilation, to conserve or shed heat or direct blood flow to active organs; the release by the liver into the blood of glucose and ketone bodies, which maintain the nutritional needs of other organs; the release to the liver of fatty acids and glycerol from cellular fat stores, which provide the raw materials from which glucose and ketone bodies are obtained.

Operant behavior has several functions, but in all species a major and essential one is to serve as the external process of homeostasis. In simple animals such as *Paramecium* and *Stentor* all operant behavior can perhaps be explained in regulatory terms, as Jennings proposed. For these animals, which have limited sexual behavior and usually reproduce by fission, fitness is the same as growth and survival. In more intelligent, sexually reproducing animals, fitness requires more than mere survival, and operant behavior subserves a wider range of social, sexual, communicative, and exploratory functions. Regulation is a prerequisite to everything else, however. One does not compose a sonnet, solve an equation, or even seek a mate, on an empty stomach.

If operant behavior is essential to energy regulation, then to understand reinforcement, which is the guide of operant behavior, we need to know something about how energy regulation works. Regulation implies feedback. Accordingly, in recent years much attention has been devoted to so-called set-point theories, which attempt – either informally, or using the quantitative apparatus of control theory – to explain various aspects of eating, drinking, and even sexual behavior, in terms of feedback loops guided by internal reference levels.

A key feature in all such accounts is the critical variable with respect to which behavior is regulated. In the study of hunger, the spotlight has historically been on physiological measures such as body-fat stores, blood glucose, or free fatty acids and the like. "Lipostats" or "glucostats," sensitive to the levels of these substances, were postulated, and deviations from optimal levels were hypothesized to act as the feedback signals that urge the organism into action, to seek food. Diligent search for these various "-stats" failed to find them, however. Perhaps more important, no single physiological measure has been found adequate to predict the full range of behavior related to eating. Quite apart from the complications associated with specific hungers for particular dietary ingredients such as salt or vitamin B, there appear to be serious difficulties in the way of a simple feedback model of eating with a single physiological variable as its set point.[2]

Alfred Hitchcock used to refer to the thing everybody is looking for in a mystery story as the "McGuffin" – good McGuffins are the Ark of the Covenant, the secret formula, Harry's corpse, and so on. It seems that there is no physiological McGuffin for feeding. Failure to find one has led some to urge the abandonment of the set-point idea. This suggestion misses the point of theoretical models in general, and feedback models in particular. It is an instance of the dangers (discussed in Chapter 3) of interpreting feedback diagrams literally in terms of neural structures. The equations generated by a set-point model may provide a useful description of a system such as a capacitor even though no physical entity corresponds to any of the elements on the diagram, including the comparator and set point. All sorts of expedients, including block diagrams, can be helpful in arriving at a useful description of a system. But these are to thought as scaffolding to a building, and can often be discarded once the objective – a working model – is attained. Any theoretical account contains nonessential features, as well as ingredients that are critical – not that it is ever easy to tell the difference. The block diagrams in feedback accounts are often superfluous.

There are, in fact, good reasons why operant behavior should not be tightly determined by physiological variables, even if maintaining the constancy of the internal environment is, in an evolutionary sense, the function of the behavior. As a familiar analogy, consider the strategy of a traveler driving across country who must keep enough gas in the tank to carry him where he needs to go. Obviously he will not wait until the fuel needle hovers on empty before looking for more. Nor will he pass up a bargain, even if his tank is still half full. His decision to look for gas will be determined by what he knows, has learned, or guesses, of the typical spacing between gas stations along his route, as well as the *variance* in spacing. If stations are far apart, or are spaced unpredictably, he is likely to seize any opportunity, even if the price is high or the mixture not quite right. Conversely, if stations are close together and predictable, he is likely to look for a good price and let his tank run down low.

The general point is that an adaptive organism will be guided in such matters by the best predictors available. Deciduous trees shed their leaves because the benefit from weak winter sunlight is outweighed by the dangers of frost. But the tree does not wait until the first frost to begin to shed. Instead it responds to a weighted average of temperature and day length as a better and safer predictor than the first frost. In arboreal evolution we must presume that trees that waited for frost were more often damaged than those that shed at the appropriate time of year, even if they thereby passed up a few warm and sunny days in some years. Tissue need is the "bottom line" for food-motivated behavior, but it is a poor creature that hesitates to look for food until it is starving.

The distribution of gas stations for the traveler is analogous to the distribution in time of meal availability for an organism. The evolutionary history of each species provides it with some information on this distribution and, presumably, provides each with a rough clock or clocks that partially determine the spacing between meals and the times of day when eating is likely. The period of the intermeal clock will depend both on the expected spacing of food availability, and on the size of the gas tank; that is, the internal food stores in fat, stomach, intestine, and cheek pouch. Many species also have external stores of hoarded food that can extend the time over which they can survive without fresh food sources. The size of internal stores will be constrained by factors such as predator-avoidance that may limit the animal's size or the proportion of its body weight devoted to fat versus muscle and other nonstorage tissues. The size of external stores is limited by the availability of free time for foraging and the pattern of fluctuations in available food density.

In simple animals, the temporal pattern of food seeking is presumably determined almost entirely by ancestral information; that is, by the seasonal, daily, and hourly regularities in the availability of food encountered by successful ancestors. In higher animals, in addition, discrimination learning allows each individual to take some advantage of the peculiarities of its own particular environment to learn the times or places when food is more or less likely to be available. In all species there is likely to be a preferred tempo of eating – of meal spacing – attuned to the animal's storage capacity, size, and metabolic expenditure.

Taste (food palatability) is like the price of gas in the automotive analogy. A satiated animal can be induced to eat by a tasty food, just as most motorists will stop to tank up during a "gas war," whether they "need" more fuel or not. If the price is low enough, additional containers may be found and filled. So also, normal animals will become obese if given ad libidum access to a highly palatable diet.[3]

Overriding all these local considerations is the veto power exercised by tissue needs. The longer the time period we are considering, the more closely must the animal's rate of eating, in calories per unit time, conform to the minimum needed to sustain life.

Organisms are thus designed by their evolution to cope with an uncertain food supply that nevertheless has predictable spatiotemporal features. The level of fat stores, hence body weight, is presumably determined by the maximum periods of deprivation that can be anticipated. The frequency with which the animal seeks food is presumably jointly determined by the current level of fat stores and the clocks that determine the initiation of feeding episodes. These clocks are themselves determined by the temporal distribution (current and ancestral) of food availability and the overall metabolic needs of the animal.

There are adequate functional reasons, therefore, why eating in most species is unlikely to be determined by any single internal or external factor. Nevertheless, despite the probable complexity of the system, it is not necessary either to despair, or to give up the possibility of a behavioral analysis until the physiologists have mapped every internal feedback loop. There are four factors likely to be important no matter what the physiological details, namely, *body weight* (the size of the energy store), *activity level* (energy expenditure), *eating frequency* (energy income), and *taste* (the evolutionary predictor of food quality). I next discuss a simple model of how these four factors combine to guide operant behavior.

OBESITY AND SCHEDULE PERFORMANCE: A STATIC ANALYSIS

It is rare to find something as dear to the hearts of the trendy as it is interesting to science, but obesity is just such a topic. The study of obesity, like research on cancer, death rays, and the internal combustion engine, is one of those areas that needs no defense before the general public. Its interest is manifest, its benefits palpable. In matters of design, "less" may or may not be "more," as the Bauhaus decreed, but in fashion there is no room for doubt: Thin is in and fat is out.

Experimental studies of obesity provide a convenient arena where the interactions among body weight, rate of eating, activity level, and taste have been extensively explored, at least in rats. More limited information is available on other species, including people, but the broad conclusions I will draw probably apply to many omnivores.

Insight into the workings of a system can often be gained from the ways in which it breaks down (its *failure modes*, in engineering jargon). Think of the ways that a TV set can fail. Typical picture defects include various forms of "tearing" and loss of frame hold, impairment of video but not sound, and the reverse, changes in color balance attributable to loss of a single color input, and so on. Even without prior knowledge of the system, or access to its innards, one could nevertheless deduce such things as the separate coding of audio and video channels, the three-color picture code, the line-by-line raster scan, and frame-by-frame synchronization. Experimentally induced aberrations in the normal eating pattern provide similar opportunities to understand the regulation of energy balance.

Animals that eat either less, *aphagia*, or more, *hyperphagia*, than normal can be produced either genetically, or by means of appropriate brain or other lesions. Hyperphagic rats that become chronically obese if given adequate food can be produced by gonadectomy or by electrolytic lesions or knife cuts in the ventromedial region of the hypothalamus (VMH).[4] Lesions in the lateral hypothalamus (LH) produce animals that eat less than normal and settle down to a lower than normal weight even when given unlimited access to food. Comparable results have been produced in cats and monkeys. Clinical data point to a similar role for the hypothalamus in man. Strains of rats and mice are available that are chronically obese. Animals of these different types behave differently when provided with especially tasty or unappetizing food, or when required to work for food on a schedule of reinforcement.

The most revealing effects are those produced by VMH lesions. In the mid-nineteenth century, chronic weight gain in some human patients was traced to tumors of the hypothalamic–pituitary complex. The modern study of this effect begins with a series of classic studies by Hetherington and Ranson (1940, 1942), who showed that electrolytic lesions in the VMH are sufficient to produce voracious postoperative eating and chronic obesity in rats. Subsequent work has identified several reliable differences between VMH and normal rats:[5]

1. VMH animals are "finicky," that is, they eat more of a palatable food than normals, but less of an unpalatable one. This is eventually reflected in their body

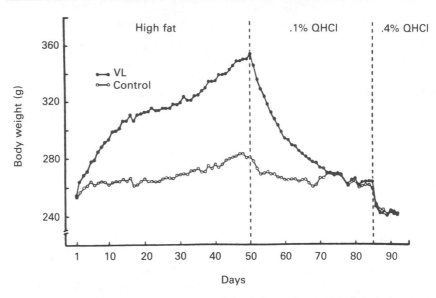

Figure 6.1. Mean body weights of hypothalamic hyperphagic (filled circles) and control rats (open circles) maintained on high-fat, .1% and .4% quinine-hydrochloride-adulterated diets. (From Sclafani, Springer, & Kluge, 1976.)

weight, as illustrated in Figure 6.1, which shows body-weight changes in normal and VMH animals in response to successive adulterations of their diet with .1% and .4% quinine hydrochloride (QHC1) – a harmless, but bitter-tasting substance. The VMH rats eat much less of the adulterated food and show precipitous declines in their body weights, whereas the normals are much less affected. VMH rats become extremely fat on palatable diets (reaching weights two to three times normal), but may show no weight gain or even a loss on an unpalatable diet. This excessive reaction to taste extends to other stimuli as well. VMH rats are more distractible and react more readily to extraneous stimuli than normals. Psychophysical tests show that the difference is one of reactivity, since VMH animals are not more sensitive (in the sense of having lower thresholds) than normals.

The sensitivity to taste shown by VMH animals is of the same type as that displayed by normal animals, but quantitatively greater. Even normal rats will become obese if given unlimited access to a varied and highly palatable diet, and normals sustain some loss of body weight if restricted to a severely adulterated diet. VMH animals show comparable, but much larger, effects.[6]

2. VMH animals are less able than normals to adjust their food intake to compensate for dilution of their food. For example, if 50% or 75% of the diet is nonnutritive kaolin, normal rats soon increase their total intake so that they show little permanent loss in body weight. VMH animals take much longer to show

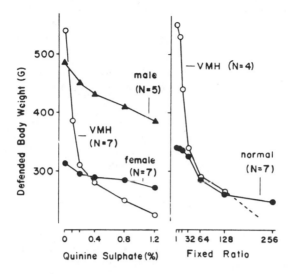

Figure 6.2. Mean stable body weights successively reached by rats exposed to diets adulterated with quinine sulphate (left panel), or available only on fixed-ratio schedules of reinforcement (right panel). VMH rats were females. Both food intake and body weight were stable for at least eight days for each schedule or diet plotted. (From Peck, 1976.)

any compensatory response, and sustain a considerable loss below their table (obese) weight.[7]

3. VMH animals weakly defend their obese settling weight. For example, immediately after the operation, in the so-called dynamic phase, they eat voraciously and are relatively insensitive to taste. As their weight increases, however, they become more finicky and, eventually, maintain the higher weight "indefinitely on a hardly detectable excess of food intake" (Kennedy, 1950, p. 541). If the animals' weight is reduced by food deprivation, when ad lib. food is once again available they go through a second dynamic phase and recover their obese weight. A similar recovery occurs when the normal diet is restored following a period of weight loss induced by quinine adulteration. If obese VMH animals are made even fatter by force-feeding, they lose weight when normal feeding is resumed. If normal rats are made obese by force-feeding before receiving VMH lesions, they show no postoperative dynamic phase, but rather maintain the increased weight by an intake that is only slightly greater than normal.

4. VMH animals are less finicky, and will work harder for food, if their body weight is reduced, particularly if it is reduced below their preoperative weight. For example, the left panel of Figure 6.2 shows the stable weights attained by groups of female, normal and VMH rats on diets adulterated with increasing percentages of quinine sulphate. Initially, the VMH animals ate much less of the adulterated diet than normals, but as their weight dropped they became less

sensitive to the adulteration. Over the range of concentrations from .2% to 1.2% the VMH animals defended their lowered weight about as well as the normals. A similar result can be seen on the far right of Figure 6.1. The right-hand panel of Figure 6.2 shows the results of a similar experiment in which food was available only on a ratio schedule. Once again, the VMH animals at first sustained a substantial weight loss, but then defended their lowered weight almost as well as the normals.

5. Obese VMH animals work much less hard for food than normal rats deprived of food for the same period of time. If food is available on a ratio schedule, VMH rats respond as rapidly as normals at low ratios, but fail to show compensatory increases in response rate at high ratios.

A regulatory model

The effects of VMH lesions on sensitivity to taste, diet dilution, and work requirements, as well as other effects to be described in a moment, can be explained by three simple feedback assumptions:

1. That *eating rate* is a regulated variable, defended by operant means.
2. That incentive factors (palatability) have an additive effect (positive or negative) on operant behavior related to eating.
3. That eating-rate regulation is inversely related to body weight; that is, animals tend to regulate better as their weight declines.

There is empirical evidence for each of these assumptions. For example, several studies have looked at the behavior of rats or guinea pigs responding for food or water on ratio schedules with different amounts of access to each reinforcer (Allison, Miller, & Wozny, 1979; Hirsch & Collier, 1974; Kelsey & Allison, 1976; and Timberlake & Allison, 1974). The results show that the relation between the amount of access to food or water (e.g., seconds of access, or grams, per hour of exposure to the procedure) and the rate of instrumental responding is largely independent of the size of the "meal" the animal receives at the completion of each ratio: if the animal will respond 20 times a minute on FR 30 for a 45-mg pellet, he will respond at approximately the same rate for a 90-mg pellet on FR 60 or a 180-mg pellet on FR 120. These findings imply that eating rate (rather than meal rate, say) is a regulated variable.

It is hardly surprising that animals try and maintain their eating rate when they must obtain all their food or water via the schedule – if not they would soon lose weight under some schedules. However, they also seem to defend eating rate even for short experimental sessions where weight loss is prevented by postsession feeding. Perhaps they don't trust the experimenter! More likely, their evolution has programmed them for a short *time horizon*: present, certain, benefits are weighted much more than distant (and perhaps uncertain) ones. Henry Ford, in his usual sensitive fashion, once commented "In the long run, we're all dead." Rats seem to agree with the sentiment behind his remark.

The second assumption has two parts: that taste affects food intake, and that the effect is additive. The first part is almost self-evident: Common experience, and numerous experiments, show that animals eat more of a palatable diet than an unpalatable one. Davis and his associates have shown that meal size is proportional to the concentration of a palatable substance such as glucose.

Many experiments have shown that taste has an independent effect and may sometimes override regulatory considerations. For example, even normal animals learn only with great difficulty to feed themselves through an intragastric fistula, that is, by a method that bypasses the usual taste inputs (Holman, 1969; Snowdon, 1969). Learning is much aided by giving the animals something pleasant to taste as the intragastric meal is delivered. Rats can learn to maintain themselves on food delivered intravenously, but long-term intake is much below normal and they lose 20–40% of their body weight (Nicolaïdis & Rowland, 1977). VMH animals are even more sensitive to the loss of taste cues than normals: Many fail to feed themselves at all on an intragastric regimen, and all achieve settling (equilibrium) weights much below their weights when feeding normally (McGinty, Epstein, & Teitelbaum, 1965). VMH animals that fail to feed themselves can be induced to do so if allowed to lick saccharin solution with each gastric load. Taste is obviously important. I show in a moment that it is additive.

The third assumption, that body weight affects degree of regulation, is required to account for the improvement in regulatory performance that is always found as body weight declines. It also makes functional sense: A loss in weight is much more costly to a lean animal than a fat one, hence a fat animal should defend its weight less vigorously than a lean.

A convenient way to represent the defense of eating rate by means of operant responding is shown in Figure 6.3. The abscissa units are eating rate (in seconds-of-access-to-food per unit time) and the ordinate is the rate of the instrumental response (e.g., lever presses/time). A ratio schedule constrains the point representing response and reinforcement rates to a straight line through the origin (ratio feedback function; see Chapter 5). Feedback functions for FR 1 and FR 10 are shown in the figure. On a given schedule the point representing performance will settle down somewhere on the feedback function. The equilibrium point, B_1, on FR 1 is shown in the figure. The two dashed lines show the options available to the animal when the ratio size is increased from one to ten. The vertical dashed line represents the option of perfect regulation: Food rate is maintained at R_1 by a tenfold increase in response rate, from x_1 to x_{10}. The horizontal dashed line is the option of no regulation: Response rate is held constant and food rate declines tenfold from R_1 to R_{10}. The point labeled B_{10} on the FR 10 feedback function represents a possible compromise adaptation to the schedule shift. A set of such points can be obtained for a range of ratio values. The line through these points is termed a *response function*; its shape is a measure of the degree of regulation that the animal achieves. For example, if the response function is vertical, then the animal regulates perfectly, adjusting his

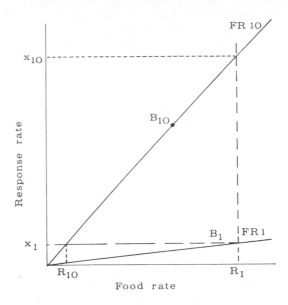

Figure 6.3. Regulatory options on ratio schedules.

response rate exactly in proportion to the demands of the ratio schedule. If the response function is horizontal, the animal does not regulate at all. A function of negative slope represents partial regulation, some negative feedback. A function of positive slope represents the opposite of regulation, positive feedback.

Figure 6.4 shows response functions from two rats that obtained all their food on ratio schedules; the points are the equilibria established on a range of ratio values from 1 through 240. The functions for both rats are approximately linear, with negative slope, representing partial regulation. This result is typical of experiments with rats responding for food. At very high ratio values (low food rates), the instrumental responding itself, and the time taken away from other activities, begin to exert some cost and the function begins to tip over (positive feedback instead of negative). Since this tipover represents a severe drop in food rate at higher ratios, it is not a viable pattern in chronic experiments.[8] Tipover is seen in brief-session experiments only with a wide range of ratio values. In this chapter, I deal just with the regulatory linear segment that is typical of long-term responding and short-term responding with a limited range of ratio values.

It is not possible to see the implications of even the simplest feedback process without making some quantitative assumptions. Fortunately, linear response functions, and many of the other properties of feeding behavior, can be derived from a very simple static model based on the three assumptions stated earlier. The model is illustrated in Figure 6.5. The elements within the dashed lines are inside the animal; the box outside represents the reinforcement schedule. The model assumes that the actual eating rate, $R(x)$, is subtracted from a set point for eating

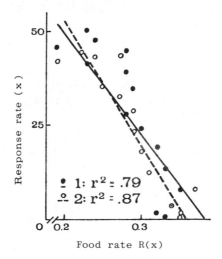

Figure 6.4. Regulatory ratio-schedule responding by two rats that obtained all their food on ratio schedules. The straight lines were fitted by least-squares regression. Response rate is in lever presses/min and food rate in pellets/min. (Redrawn from Collier, Hirsch, & Hamlin, 1972.)

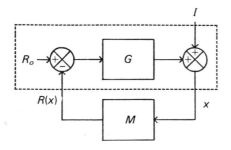

Figure 6.5. Feedback model for the regulation of eating.

rate, R_o. In turn, $R(x)$ is jointly determined by the rate of the instrumental response, x, and the value of the ratio schedule, $m = 1/M$. In most experiments, the animal's state of deprivation is such that $R(x)$ will be less than R_o. G represents the transfer function by which the animal translates a given eating-rate deficit, $R_o - R(x)$, into a given rate of instrumental responding, x. G is assumed to be a function of body weight, W (in which case, it is written as $G(W)$) and perhaps other things as well. I represents incentive factors, which make an additive contribution to response rate. I is assumed to depend upon taste, T (in which case, it is written as $I(T)$), and perhaps other things such as time of day or temperature. There may also be a small effect of body weight on I: When body weight is very high, I may be reduced. R_o, the set point, is assumed to be constant, although this is not essential to the model. Function G may be as complex as we please. For simplicity, and because it is sufficient, I assume G is just a multiplier.

The system in Figure 6.5 can now be analyzed in a similar way to the fundamental feedback equation in Chapter 3 (Figure 3.12, Equations 3.11–3.13). At equilibrium the value of $R(x)$ must be such as to produce a response rate x that, by the terms of the schedule, produces the same eating rate, $R(x)$. In other words, at equilibrium the system is defined by two simultaneous, linear equations: the feedback function, and the hypothetical control function (here a simple multiplier). Thus,

$$R(x) = Mx \quad \text{(the feedback function)} \tag{6.1}$$
$$x = G(R_o - R(x)) + I \quad \text{(the control function)}. \tag{6.2}$$

On a given schedule, G, R_o, and I are constants; hence Equation 6.2 defines a relation between x and $R(x)$ that is independent of the schedule value – the response function – which is here obviously linear, with negative slope. It can be rewritten more conveniently as

$$x = GR_o + I - GR(x), \tag{6.3}$$

which is a straight line with slope $-G$ and x-intercept equal to $GR_o + I$. I shall refer to Equation 6.3 as the *linear model*.[9]

As before, x can be eliminated from Equations 6.1 and 6.2 to yield the obtained food rate, $R(x)$, as a function of the "loop gain," MG:

$$R(x) = MGR_0/(1 + MG) + MI/(1 + MG). \tag{6.4}$$

The first term on the right-hand side of Equation 6.4 is the familiar fundamental feedback equation first encountered in Chapter 3 (Equation 3.13), with set point R_o and gain MG. The second term represents the additive contribution of incentive factors.

Before turning to lesion effects, notice that Equation 6.4 has the properties we might expect from the earlier discussion of normal regulation: The steady-state rate of eating, $R(x)$, although regulated is not absolutely constant. It depends on both incentive, I, and G, regulation. If I is high (as it might be with an especially palatable diet), steady-state eating rate will be higher than if I is low. The effect of I depends on the regulatory factor, G, however: Because I appears only in the second term, divided by $(1 + MG)$, the higher the value of G, the smaller the effect of variation in I. Hence incentive factors will have relatively little effect on the eating rate of normal (high-G) animals.

Body weight appears implicitly in these equations as a factor that affects G (and to a lesser extent, I). There is also a reciprocal relation, of course: between steady-state eating rate and settling body weight. This relation is not well understood. All we are entitled to assume is that a higher eating rate will lead to a higher body weight. Fortunately, no more is needed for the present analysis.

Consider now the possible failure modes of the system in Figure 6.5. There are three simple ones: (a) impairment of the set-point input; (b) alteration of the regulatory function; and (c) alteration of the incentive input. Hypothalamic lesions seem to have effects mainly of the second type. VMH lesions clearly act as if they just impair regulation – a decrease in the value of G in the present model. Other variables, such as a taste or reduced oxygen (hypoxia), have other effects. The VMH effects can be explained as follows:

Finickiness. Obese VMH rats are more sensitive than normal rats to taste factors; that is, they will eat more of a highly palatable food, less of a less palatable one. If one were to graph their eating rate, $R(x)$, as a function of some measure, T, of taste, the function for VMH animals would be steeper than for normals: A given change in taste would produce a larger change in eating rate. The effect on the slope of this function of varying G can be derived as follows. Incentive is directly related to taste: $I(T)$ is an increasing function. If M, G, and R_o are constant, then from Equation 6.4, rate of eating, $R(x)$, is a linear function of I, with slope $M/(1 + MG)$. Clearly if G decreases, this slope will increase: The effects of a change in incentive (taste) will be greater. Hence, if VMH lesions reduce the regulatory parameter G, one effect will be an increase in finickiness.

VMH, and normal, animals become more finicky as they become more obese. But we assume that G is inversely related to body weight; hence the higher the weight, the smaller the value of G, the larger the relative effect of incentive, and the more finicky animals should become.

Response to dilution of diet. VMH animals adjust less well than normals to dilution of their diet with a nonnutritive substance such as kaolin. Rats cannot sense directly that kaolin is nonnutritive. Its effects must therefore be via a small loss in body weight: When kaolin is first added, the animal either maintains or reduces (because the dilution does not improve the taste of the food) its volumetric intake, so that after a few days it loses some weight. Hence the reduced sensitivity of VMH animals to dietary dilution implies that these animals are less sensitive than normals to small changes in body weight.

I assume that weight loss increases G, which, on a neutral or unpalatable diet, causes an increase in eating rate (according to Equation 6.4) that partially compensates for the loss in weight. The steady-state result is a small loss in weight and an increase in eating rate that almost completely compensates for the kaolin dilution. On a highly palatable diet, however, dilution should cause an actual *decrease* in eating rate, according to this model, because the amount by which the animal's actual eating rate exceeds its set point for eating, R_o, is inversely related to G. No one seems to have looked at how the effects of dilution interact with diet palatability, so this prediction remains unchecked. (The predic-

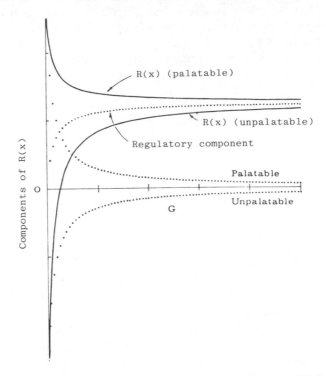

Figure 6.6. The separate and combined effects on eating rate ($R(x)$) of regulatory feedback and taste factors, as a function of gain (G), according to Equation 6.4 in the text. Curves are shown for relatively palatable and relatively unpalatable diets.

tion may not be easy to test, given that a kaolin-diluted diet is probably not very palatable.)

The hypothesized effect of VMH lesions is to reduce G. The first question, therefore, is: How much does a given change in G affect eating rate, $R(x)$? If a given change in G has a larger effect on eating rate in VMH animals than in normals, the differential effect of kaolin dilution is explained.

The answer can be deduced from Figure 6.6, which is a plot of Equation 6.4, showing $R(x)$ as a function of G (i.e., with everything else held constant). The dashed lines represent the two components of Equation 6.4: The upper dashed line is for the first (regulatory) term; the lower two lines represent the second term – the upper one for a palatable diet (I positive), the lower for an unpalatable diet (I negative). The upper solid line shows $R(x)$ as a function of G for the palatable diet; the lower solid line shows $R(x)$ as a function of G for the unpalatable diet.

Just look at the upper solid line (I return to the others in a moment). The *slope* of this line tells us how much $R(x)$ changes for a fixed change in G. The curve is negatively accelerated, and it is easy to see that the amount of change in $R(x)$ for a given change in G *depends on the value of G*: When G is small, a given change in G causes a large change in $R(x)$, because the slope of the line is steep; but when G is large, the same small change in G has little effect, because the line is almost horizontal. Exactly the same thing is true of the lower solid line, except that the changes in $R(x)$ are in the opposite direction. In both cases, the smaller the value of G, the larger the effect on $R(x)$ of a given change in G. Thus, the effect of VMH lesions in reducing sensitivity to diet dilution follows from the assumption that VMH lesions reduce the regulatory parameter G.

Weak defense of settling weight. Given a relatively palatable diet, VMH animals stabilize at a weight that is higher than normal, and return to it when free feeding is resumed after a period of deprivation. The equilibrium rate of eating at a given body weight can be derived directly from Equation 6.4. It is obvious that if G is decreased, the contribution of taste factors, the second term, is increased, and the regulatory component, the first term, is somewhat decreased. The net effect depends on the balance between palatability and regulation, as shown in Figure 6.6. Since normal animals are relatively insensitive to variations in palatability, they can be presumed to lie toward the right, high-G, side of the graph, with stable $R(x)$ values close to the asymptote of these functions. As G decreases, the equilibrium rate of eating, $R(x)$, on the palatable diet rises. Hence the finding that VMH animals eat more than normals, and therefore gain weight, on palatable diets follows from this analysis. Conversely, when I is negative, a decrease in G invariably produces a decrease in eating rate, which is consistent with the finding that VMH animals may actually lose weight on relatively unpalatable diets. Stability of body weight at each palatability level is a consequence of the pre-sumed fixed, monotonic relation between mean eating rate and settling weight: Since each eating rate defines a settling weight, an increase in eating rate entails an increase in body weight; a decrease, a loss of weight.

Improved defense of low settling weights. VMH animals seem to defend their weight about as well as normals, providing body weight is low enough, although systematic tests with different work schedules do not appear to have been done. The present analysis predicts improved defense at low weights because of the presumed inverse relation between G and body weight.

Poor adaptation to work requirements. VMH animals respond for food on low-ratio schedules as well or better than normals, but fail to increase response rate appropriately on larger ratios. This result follows naturally from the linear model. Equation 6.3, the response function, is a straight line with slope $-G$. It is easy to show that as G varies the equation defines a family of straight lines converging on the point $R(x) = R_o$ and $x = T$. Figure 6.7 shows two response functions of

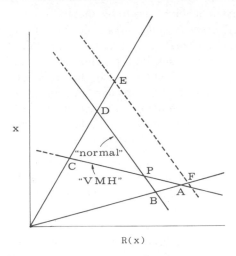

Figure 6.7. Theoretical response functions for VMH, normal, and genetically obese rats exposed to low and high ratio schedules. See text for details.

this sort, together with feedback functions for low- and high-valued ratio schedules. The two response functions that intersect at point P are termed "VMH" and "normal" and differ only in the value of G. The dashed response function is not part of the same family; it has the same slope, hence the same value of G, as the "normal" function, but a higher value for the $R(x)$- and x-intercepts. If all three curves were obtained from animals on the same diet, one could infer that the dashed curve differs from the others in having a higher value for R_o.

This diagram raises some questions and make several testable predictions. Consider first P, the point of intersection of the two solid response functions: Why place P above the equilibrium points, A and B, for responding on the low ratio schedule? (The answer to this question is critical, because the predictions are not right if P is below points A and B.) If for simplicity we let the low ratio equal unity, then the equilibrium value for $R(x)$ can be obtained at once from Equation 6.4. It is

$$R(x) = GR_0/(1 + G) + I/(1 + G). \qquad (6.6)$$

For normal animals G is large ($G >> 1$), hence the first term is approximately equal to R_o. If I is positive (a relatively palatable diet), $R(x)$ on FR 1 will therefore be larger than the $R(x)$ value at point P, which is R_o, the amount of the difference being inversely related to G. The graph illustrates these relations, since the $R(x)$ value of point A (equilibrium for the "VMH" curve on FR 1) is greater than the $R(x)$ value of point B (equilibrium value for the "normal" curve on FR1).

The situation is reversed at high ratio values. Here the "normal" equilibrium at point D is higher than the "VMH" equilibrium at point C. Thus, the assumption the VMH animals have a lower feedback gain than normals accounts both

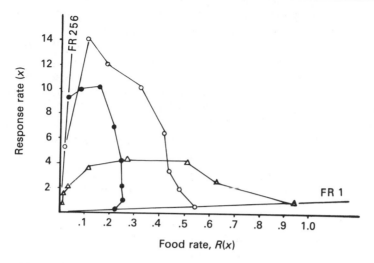

Figure 6.8. Response functions for groups of normal, VMH-lesioned, and genetically obese rats exposed to ratio schedules from FR 1 through FR 128. Response rate is in lever presses/min; food rate in pellets/min. (Redrawn from Greenwood et al., 1974.)

for their excessive eating rate on low ratios and their failure to adapt to high ratios.

The dashed response function in Figure 6.7 illustrates the case of an animal that simply has a higher set point, R_0, or a lower weight, W, but is otherwise normal, that is, has a normal value of G. This animal defends its eating rate as well as the normal animal, but responds faster at all ratio values, hence has a higher eating rate. Some strains of genetically obese rats fit this picture.

Figure 6.8 shows data that conform over much of the ratio range to the predictions in Figure 6.7. The filled circles are the response function for a group of normal animals on fixed-ratio schedules ranging from 1 to 128. The declining limb of the function is approximately linear with a high negative slope, representing good regulation of eating rate. The open triangles represent a group of VMH rats in the dynamic phase. On FR 1 and FR 4 these animals respond, and eat, more than the normals. At higher ratios they fail to increase their response rate appropriately, however, and therefore eat much less than the normal animals, showing very poor regulation of eating rate.

The open circles represent a group of genetically obese rats, also in their dynamic phase. The difference between this response function and that for the VMH animals in striking. The genetically obese animals regulate almost as well as normals (i.e., have a similar value for G), but respond, and eat, more at each schedule value. This difference, as well as their lack of finickiness and normal adaptation to dietary dilution, is consistent with a higher value for R_0 for these animals.

The effects of VMH lesions thus correspond very closely to an effect on regulation alone, with no effect on set point or incentive.

What of lateral hypothalamic lesions? Substantial lesions in the lateral hypo-thalamus (LH) produce a loss in weight accompanied by extreme finickiness.[10] I have not been able to to find satisfactory information either on the finickiness of these animals once they have recovered from the severe initial effects of the lesions, or on their ability to defend body weight against dietary dilution and work requirements. The simplest interpretation of what is known is that LH lesions have an even more severe effect on regulatory feedback, G, than VMH lesions. One objection is that LH animals do not become obese, even on very palatable diets. One possible answer is that severe LH lesions produce a general-ized impairment that interferes with the simple processes of eating and making lever-press responses. There is little doubt that these animals are severely im-paired, and perhaps a model that deals solely with the regulation of feeding should not be expected to deal with a lesion that affects many other functions as well.

Thus, the effects of the classic hypothalamic syndromes, as well as other syndromes associated with obesity, appear to be consequences of a system for the regulation of feeding in which eating rate is affected in an additive way by two factors: a regulatory input that is determined by the difference between the actual eating rate and a "natural" rate for that body weight, and taste factors: by an endogenous regulatory process and an exogenous nonregulatory one. The defects of the VMH animal are traceable to impairment of the endogenous regulatory system, with an inevitable gain in the relative effect of exogenous factors. In consequence the animals become finicky, distractible, and "emo-tional" and cannot adequately defend a body weight that is excessively dependent on the palatability of their diet. The behavior of the classic LH animal is less clearly defined, but may just be a more extreme version of the VMH syndrome. Genetically obese animals are the converse case: Their reference level is elevated and they defend it adequately, without the excessive reactivity to taste of the VMH animal.

This system for feeding regulation has the properties one would expect from the gas-tank analogy. Each body weight defines a certain regulated rate of eating (the first term in Equation 6.4), but that rate can be increased or decreased by incentive (the second term in Equation 6.4). One might also postulate additional additive factors such as temperature and circadian and circumannual periodicities.

Effects of taste and body weight on work schedules

The response functions shown in Figures 6.4 and 6.6 take a long time to obtain: Animals are usually allowed at least one session at each ratio value, and some-times many more. The whole response function may therefore take a week, or even several weeks, to get. This makes it difficult to measure directly the effects of changes in body weight, diet palatability, or other manipulations, such as appetite-affecting drugs, on response functions. It turns out that there is a quicker way to do these experiments. Rats are very good at tracking changes in the

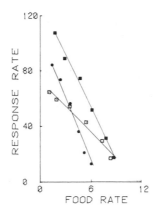

Figure 6.9. The effects of body weight and diet palatability on cyclic-ratio response functions. Points shows data from a group of four rats at either 80% of normal weight (filled squares), 95% of normal weight (open squares), or 80% weight with quinine-adulterated food pellet reward (filled circles). Ordinate is lever presses/min, abscissa is 45 mg pellets/min. Lines were fitted by linear regression. (From Ettinger & Staddon, 1982.)

spacing of food over time – so good, in fact, that they will quickly learn to follow a sequence of ratios that changes in a cyclic fashion. For example, suppose that when the rat is first put in the Skinner box, he gets a pellet. Then after two lever presses (FR 2) he gets another; then after four more (FR 4) he gets the next; the next requires 8, and so on up to 64. Rats very quickly learn to anticipate a *cyclic-ratio* schedule of this form: 2, 4, 8, 16, 32, 64, 64, 32, 16, 8, 4, 2, 4,. . . where they get six or so cycles per session each day. A plot of response rate (ratio size / interfood time) versus food rate (1/interfood time) at each ratio then gives the response function.

Both response functions, and more molecular properties of ratio behavior, correspond quite closely to what is found in the standard procedure where the rat contends with just one ratio value in each experimental session. For example, the effect of ratio size on time-to-first-response (postfood pause) and response rate after the first response (running rate) is similar in both. But the cyclic method makes it easy to look at the effects of such things as body weight and diet palatability on response functions.

Figure 6.9 shows cyclic-ratio response functions obtained from a group of four rats given three treatments: (a) run at 80% of their normal body weight (hungry); (b) at 95% of their normal body weight (not so hungry); and (c) run at 80% weight with quinine-sulphate-adulterated (bad-tasting) food pellets. The effects of the three treatments are just about what we would have anticipated: The effect of the bad taste (filled circles) is to displace the response function inward from the control (80% wt.) function (filled squares); this is an effect on the incentive parameter, I. The effect of satiation (95% wt., open squares) is to decrease the slope of the line. A comparison of response functions across body weights between 80% and 95% showed that only parameter G in Equation 6.3 is affected by changes in body weight. In other experiments (Ettinger & Staddon, 1982) the anorectic (appetite-reducing) drug amphetamine has been shown to affect mainly parameter G, while reduced atmospheric oxygen (hypoxia) affects mainly parameter I – confirming the reports of mountain climbers that food tastes bad at high altitude.

Thus, the effects of both regulatory variables, such as body-weight changes (food deprivation), and incentive variables (taste) seem to affect ratio response functions in the simple way suggested by the linear model: Body weight affects slope, taste affects intercept.

Other motivational effects. The precise quantitative form of the present analysis should not be taken too seriously: The static linear model is not the last word. I introduced some equations only because without them, it is impossible to see how all the factors that affect eating interact. The strong point of the linear model is that it is simple enough to allow one to see pretty clearly how these various factors combine to affect eating rate and work rate on an operant schedule. It suggests simple experiments to check out the effects of physiological manipulations. For example, factors such as taste that enter in additively to the control function (Equation 6.3) appear as terms of the form $MB/(1 + MG)$ in the eating-rate equation (Equation 6.4), where B is the factor in question. Hence the effects of changes in B are inversely related to G. Temperature is known to have a linear effect on eating rate of the sort predicted by this interpretation.[11] VMH animals appear to be characterized by a lower-than-normal value for G; hence their temperature function should also have a lower slope. As far as I am aware, this experiment has not been done.

Another example is sectioning the vagus nerve below the level of the diaphragm.[12] Vagotomy abolishes input to the CNS from the gut and liver and suppresses eating in both normal and VMH rats. The effect on the VMH animals is much greater, however, and generally sufficient to eliminate obesity on the usual diets. Finickiness remains, which suggests that the vagus is involved in a separate additive term in the eating-rate equation. Since such additive effects are inversely related to G, the effect of vagotomy should be, and is, smaller not only in normal animals, but also in genetically obese animals and animals made obese by ovariectomy – because the obesity of these animals does not reflect impaired regulation.

Limitations of the linear model. The defects of the linear model are almost as obvious as its strong points. The most striking is its failure to predict normal satiation. As a starved animal eats, its body weight will slowly rise. But a rise in body weight reduces G; if the diet is palatable, this means that eating rate should *increase* (top curve in Figure 6.6), leading to still more weight increases, and so on. This is an unstable, positive-feedback process that implies indefinite weight increase. Yet starved animals indubitably do eventually slow down to a normal eating rate. There are several ways to deal with this. One possibility is to assume that body weight does have some effect on incentive: Perhaps the taste of any food begins to be aversive at high enough body weights. (The effect must be restricted to high body weights, because weight variation below the normal range seems to have almost no effect on the incentive parameter.) More likely, perhaps, we reach here a limitation of any one-dimensional motivational model. After all, hunger, or any other motivational state, is always relative. If the animal is satiated and its value for G is small, then under normal circumstances it is not likely to look for food or find itself in the presence of food. It will be looking for something else: sex, water, a nest. Equation 6.4 becomes increasingly irrelevant as the animal approaches satiation. We need the multidimensional apparatus discussed in the next chapter to deal with this problem.

It is probably asking too much of the linear analysis to expect it to provide precise quantitative predictions; and, of course, it is a *static* analysis and can give no account of the time course of the various regulatory mechanisms. Response functions are only linear over part of their range, not all as the linear model implies. The curvature in the VMH function in Figure 6.8 is substantial, for example. Most workers in this area have not thought much about quantitative matters, however, so that few data of the precision necessary to test quantitative models are yet available. Moreover, commonly used procedures sometimes involve confounding factors. For example, the data in Figure 6.8 were gathered under conditions where the animals obtained all their food in the experiment. After three days on FR 1, ascending ratios of 4, 8, 16, and so forth, were introduced in succeeding 24-hour periods. Body weight was not controlled, so that the linear model does not in fact predict that the successive daily data points should all lie on the same straight line. If, as seems likely, body weight was increasing over days, the points should lie on a curved line, as illustrated in Figure 6.10. The linear model may therefore do a better quantitative job than is apparent at first sight.

SUMMARY

Feeding, a typical motivational system, involves both internal and external factors. The endogenous factor is regulatory and tends to oppose anything that forces eating rate to be reduced below a preferred value, the set point. The exogenous factors are approximately additive, and cause eating rate to be higher or lower than the set point, accordingly as they are positive or negative. The degree of feeding regulation appears to be inversely related to weight: the more food-deprived the animal (the lower its body weight), the more vigorously it defends its eating rate. Feeding regulation is impaired by ventromedial hypothalamic lesions. Since the endogenous factors are reduced, exogenous factors become relatively more important, and VMH animals are "finicky" and will not work hard for food.[13]

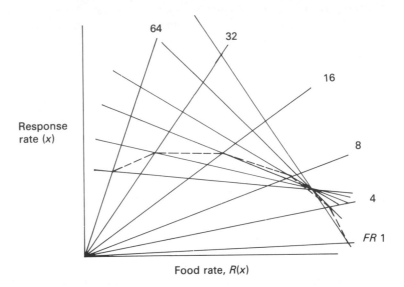

Figure 6.10. Theoretical effects of changing body weight on the ratio-schedule response function. Lines passing through the origin are feedback functions for ratios of 1, 4, 8, 16, etc. used successively in the experiment by Greenwood et al. (1974). Radial lines of negative slope are response functions computed from the linear model on the assumption that body weight is increasing each day. The dots connected by a dashed curve represent the daily equilibria.

Joint determination by exogenous and endogenous factors is just what would be expected from functional considerations: Feeding behavior is guided in sensible fashion by all the information available to the animal, both internal (signaled by body weight), and external (signaled by taste, temperature, and discriminative stimuli).

These regulatory and incentive effects can be described by a linear, static model. The linear model is simple and useful, but it is also limited in a fundamental way: It predicts the same response function (Equation 6.3) no matter what form is taken by the feedback function (Equation 6.1). This is contrary to fact. For example, interval schedules have a nonlinear feedback function, which can be approximated by a hyperbola (see Chapter 5); response functions on such schedules are also nonlinear, with positive, rather than negative, slope over most of their range. When extreme ratio values are considered, the response function on ratio schedules appears to be an inverted-U shape. Response functions on avoidance schedules are often not linear and never of negative slope. The most successful attempts to bring these varied effects together in a unified way involve *optimality analyses*; that is, explanations of behavior as the outcome of processes that act to maximize or minimize some quantity such as utility, cost, or Darwinian fitness. These approaches are the topic of the next chapter.

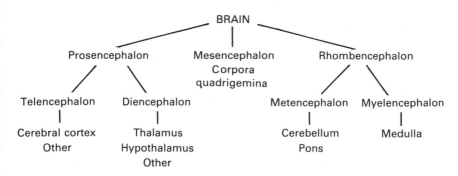

Fig 6.11. Mammalian brain structures, organized in terms of development and anatomy. (Modified from Ranson & Clark, 1958.)

NOTES

1. Astonishing or not, the study of reinforcement schedules was pursued in isolation from regulatory considerations for many years.

2. The difficulties in finding a single critical variable as the trigger for eating are well described in a paper by Friedman and Stricker (1976), which also gives an excellent account of the physiology of hunger. Uses and abuses of the set-point idea are discussed by Mrosovsky and Powley (1977). Wirtshafter and Davis (1977) show by example that regulation implies negative feedback, but does not require any identifiable set point. Chapters by Booth, Bolles and Davis, and Hogan in the book edited by Toates and Halliday (1980) also discuss this issue.

3. The evolutionary significance of taste is not well understood. On general grounds one might expect taste to be an evolutionary predictor of nutritional value: "Sweet" signifies high caloric value, "bitter" potential poison danger, and so on. But the correlations are obviously weak in many cases, and obscure in others. So little is known about the food available to ancestral species that arguments of this sort have few facts to restrain them.

4. The mammalian brain is conventionally classified into a hierarchy of structures, based on anatomical arrangement and pattern of development, as shown in Figure 6.11. The vertical axis of the figure corresponds to the pattern of development: The primitive brain first differentiates into pros-, mes-, and rhombencephalon, then these structures differentiate further into telencephalon, diencephalon, and so on. The left–right axis corresponds to anatomical arrangement: The medulla and cerebellum are closest to the spinal chord, the cerebral cortex further away.

The hypothalamus, thalamus, and epithalamus are the three main divisions of the diencephalon. Experiments in which the hypothalamus has been lesioned, or stimulated electrically or with chemicals, have shown it to be involved in a wide variety of motivational and regulatory functions.

5. *The meaning of lesion experiments.* What is the significance of brain-lesion experiments? It was once assumed that if destruction of a brain area caused loss of some function, than that area was a "center" for that function. A bit more thought soon showed that this inference is unwarranted: Cutting the wire to the speaker will silence a stereo set, but the wire is hardly the source of music. The whole notion of "centers" had to be given up. The brain is a system at least as complicated as a computer, and the effects of interference with it are at least as hard to interpret.

If the interference is sufficiently subtle – the neural equivalent of disabling a single computer-memory element, for example – then very specific effects might be expected. But lesions are not subtle, and even the much more delicate pharmacological and electrophysiological interventions now commonly applied to single neurons or small groups probably represent relatively crude assaults. Lesions therefore probably tell us more about the broad principles of brain organization than about the function of specific neural structures. In other words, these physiological manipulations may tell us more about the functional properties of the brain – the topic of this book – than about structure–function (brain–behavior) relations – the topic of main interest to neuroscientists.

This point is hard to prove, but I can suggest an analogy. Consider the TV-set problem again. Imagine that you have an indefinite number of the things (all the same, or a related, model) and want to find out how they work. But your instruments are limited to a small collection of woodworker's tools: a ball-peen hammer of sturdy design, some screwdrivers, a bit brace, drill press, and a few chisels. Undaunted by the crudeness of the tools, you set to work in methodical fashion. Placing TV set #1 in the drill press (stereotaxic instrument) you drill into it for a precisely measured distance. When the machine has recovered from the surgery, you turn it on and try to tune in a few test patterns. No doubt some drastic defect, from a longer version of the list mentioned earlier, will appear. Encouraged by your success, you take TV set #2, repeat the experiment with some slight variation, and measure the effects. Dogged application of the experimental method continues until you have studied perhaps 1,000 sets with all sorts of assaults, drilling, striking, and chipping in carefully recorded and replicable fashion. What will you find? What can you conclude?

I suggest that you will find much the same kind of thing as from a study of the natural history of defective TV sets: a few common defects, some rare ones, an often-loose correlation between the precise site of the lesion and the exact nature of the defect.

Nevertheless, the defects are likely to fall into a few major groups, defined by the basic functional categories of the machine: video effects, sound effects, effects on picture synchronization, and so on. These will be the common failure modes of the machine. They tell little of the circuit details, but much about functional organization. All is not likely to be perfectly clear, however. Many lesions, as carefully done as the rest, will produce odd defects that fail to fit into any pattern. Perhaps more than one functional subsystem has been damaged, probably because the functionally different structures overlap physically. These effects are much less informative than the common, simple ones.

I suggest that the hypothalamic-lesion studies should be interpreted as this analogy suggests. The common effects are most informative, but there are likely to be many intermediate cases that muddy the picture. It is not surprising that although most VMH-lesion studies report essentially the list of properties described in the text, there are several that present conflicting evidence. A number of these point to the location of the lesion and possible involvement of other structures as the cause. All this is to be expected. Electrolytic lesions, although somewhat more precise than our assault on the TV sets, are still crude when measured by the precision of the neural structures affected. Common, simple effects therefore mean more than rare, complicated ones. The major effects of VMH lesions are the ones I have described.

The experimental literature on the behavioral effects of hypothalamic lesions is voluminous and impossible to more than touch on here. Articles I have found especially helpful are Kennedy (1950), Hoebel and Teitelbaum (1966), Panksepp (1974), Keesey and Powley (1975), and Collier, Hirsch, and Kanarek (1977). The book edited by Novin, Wyrwicka, and Bray (1976) contains a number of useful reviews and theoretical contributions. Particularly relevant to the present topic are chapters by Booth, Toates and Platt, Sclafani, Peck, Panksepp, and Davis, Collins and Levine.

The model I describe, although not identical to any in the literature, is closest to the proposal of Panksepp (1974).

6. Nisbett (1972) in a theoretical review has pointed out that hungry people and rats are in one sense *more* (rather than less) sensitive to taste than nondeprived individuals: When given a choice between tasty and neutral foods, the hungry animal increasingly chooses the more palatable alternative. This apparent finickiness of deprived animals is *not* shown in the single-choice situation, however. When a mediocre food is the only one available, normal rats will eat enough of it to maintain their body weight – but obese VMH animals will not. It is therefore wrong to infer from the choice results that VMH animals resemble food-deprived animals. Moreover, the behavior of the hungry normal animals is perfectly sensible from a regulatory point of view, granted the hypothesis that palatability is an evolutionary predictor of nutritional quality. The undeprived animals can afford to sample alternatives, but the hungry animal, given a choice, would do well to concentrate on the richest payoff.

Theoretical and experimental work in behavioral ecology has expanded this commonsense analysis to include probabilistic food sources. For example, suppose that the animal has a guaranteed, but lean, source of food, and an alternative, chancier, but occasionally richer, source; which should he choose? The answer depends on the animal's estimate of expected food rate: If the rate from the certain, lean, source is high enough to sustain life, then he should concentrate on it (be *risk-averse*, in decision-theory terminology). This corresponds to the Nisbett situation. But if the expected rate is too low to sustain life, then he should go for the rich, but chancy, source, since only that provides any chance of survival. Surprisingly, perhaps, birds seem able to assess the necessary probabilities and behave as theory suggests they should (Caraco, Martindale, & Whitham, 1980).

7. I assume that the effects of kaolin are largely due to its nonnutritive, diluent properties, but later arguments suggest that the relative unpalatability of Kaolin-diluted diets may also be a factor.

8. The data in Figure 6.4 were obtained with an unusual ratio schedule in which the ratio value was fixed, but the meal length was under the control of the animal: At the end of each completed ratio, the feeder remained available until 10 min after the animal removed its head from the feeder opening. This was necessary to ensure that the animal (which obtained all its foods from the schedule in round-the-clock sessions) would not starve, no matter how high the ratio setting. The effect of this procedure is that the animal selects meal lengths appropriate to the ratio value (large meals if the ratio is large, smaller if it is small) and always remains on the linear part of the response function. Similar data can be obtained with the conventional procedure (short sessions and a fixed meal length), providing the ratio values (gram/response or seconds-of-access/response) are kept within the same range. (See Staddon, 1979b, for a fuller discussion.)

In subsequent discussion the term *eating rate* has two meanings: as measured it refers to grams of food (or seconds of access) per hour in relatively brief experimental sessions. However, I assume that eating rate measured in this way is also a valid measure of the free rate of eating in 24-hour sessions, either unrestrained, or with the animal-determined-meal-size ratio-schedule procedure.

9. Equation 6.2, a good approximation to the declining limb of ratio response function, is equivalent to the *conservation model* for schedule performance proposed by Allison (e.g., Allison, Miller, & Wozny, 1979; Staddon, 1979a) – although the number of parameters, and their interpretation, is different.

10. This is true of the classical LH syndrome discussed by Keesey and Powley (1975), Teitelbaum and Epstein (1962), and others. More recently, Peck (1976) has identified two other LH syndromes, types I and III, in addition to the classical one, which he terms type II. For the reasons discussed in the text, the commoner syndrome is likely to provide simpler insights into the nature of the feeding system than exceptional cases, and I do not consider them here.

11. For a recent review of temperature effects see Kraly and Blass (1976). Data from Brobeck (1945) fit linear functions; see also Russek (1976).

12. Studies of the effects of vagotomy have a long history. Recent findings are summarized by Powley and Opsahl (1976).

13. *Human obesity.* What are the implications of animal studies for the "problem" of human obesity? The researcher naturally feels some obligation to comment on a matter of such moment to so many, even though the attempt puts him in dubious company: Those who seek to console, advise, and relieve of a few pounds (or dollars) the fat who would be thin are more numerous than distinguished. Moreover, the innumerable recipes for weight loss may, like swine-flu vaccine, be a cure for which there is no disease. It is easy to get the impression that even moderate obesity is a terminal ailment, even though the evidence for its unhealthiness is mixed at best. Charles Darwin's father died after 82 years of vigorous life at a weight in excess of 300 lbs. A 250-lb. Tongan is regarded as a lightweight, yet the islanders live as long and as well as their slimmer neighbors. Recent medical work seems to concede that "standard" weight tables err on the light side: People thus deemed overweight seem to live as long or longer than their leaner brethren. Thin women are more likely to be infertile than their ampler sisters – jogging may be the best female birth-control device since the pill, as much for its effects on the reproductive system as for its repellent effects on women's shapes. Objective evidence of the virtues of slenderness, at least in middle age and beyond, is hard to come by.

Most of the problems of obesity are probably social rather than physiological. Still, since physiology may be easier to change than fashion, an inquiry into the parallels with animal studies has some appeal. Schachter and his associates (e.g., Schachter & Rodin, 1974) have suggested that obese people are like VMH rats, and both are characterized by excessive reactivity to taste and other stimuli. A number of ingenious experiments with people tends to support this view, although there are some dissenting arguments (e.g., Milich, 1975; Kolata, 1977). My analysis suggests that the reactivity of VMH animals is in fact a secondary effect of dietary dilution and work schedules as well as finickiness. It also accounts for the fact that VMH animals *do* regulate, albeit at a higher-than-normal weight, a fact which, as Schachter points out, is " so troubling to ours as well as all other theories about the obese animal, that they do stop eating" (1974, p. 72). Moreover, there are several possible animal models for human obesity, such as genetic obesity (of several varieties) and obesity caused by gonadectomy. Simple as the linear model is, it nevertheless suggests several ways for obesity or weight loss to occur, and animal examples exist for most of them. The one constant in biological populations is individual variation, and it is unlikely that all human obesity conforms to a single pattern.

The fact of negative feedback means that obesity by itself is uninterpretable. Tests of the sort described in the chapter – dietary dilution and adulteration, work schedules, temperature changes – are essential before the factors maintaining a given body weight can be properly identified.

Unfortunately, even if the causes are identified, altering body weight will still be difficult. Suppose that regulatory feedback is impaired. Body weight will then be more sensitive than usual to dietary palatability and other stimulus factors. But an unpalatable diet tastes bad, and who will insist that the would-be sylph sticks to it after she leaves the spartan regimen of the fat farm? Theory predicts that food intake can also be reduced by imposing a work schedule, but again who will impose it? If someone is obese because of an elevated set point, things are even worse, because regulatory efforts will be vigorous: Sensitivity to taste and work will be relatively low and only enforced deprivation will re-duce eating.

My conclusion is that even for VMH-like obese people, a study of the feeding regulation system by itself provides no real answer to the problem of weight regulation. If there is a solution, it lies in other motivational systems that com-pete with the feeding system for access to the behavioral final common path. Fat people often lose weight when they acquire a new interest, which may be either positive, such as a new job, companion, or religion; or negative, such as a personal or career crisis. Rats housed in activity wheels gain less weight on a palatable diet than animals without this opportunity to run, and the difference is not attributable just to increased energy expenditure – rats decrease, rather than increase, their food consumption when provided with this opportunity for a competing, highly favored activity (Sclafani & Springer, 1976; Premack & Premack, 1963). Opportunities for desirable but drastic changes in life style of this sort are obviously limited in practice. Nevertheless, they may provide the only effective means of producing permanent changes in body weight. The existence of nega-tive feedback means that the only reliable way of controlling one regulatory system is with the aid of another, competing one.

7

THE OPTIMAL ALLOCATION OF BEHAVIOR

UTILITY AND ADAPTATION TO CONSTRAINT

The last chapter took feeding as a typical motivational system and looked at how things like palatability, diet dilution, and brain damage affect the amount eaten and how hard a rat will work for food. These effects were quite well summarized by a simple, static feedback model whose central assumption is that animals regulate[1] eating rate. Chapter 6 did not deal directly with interactions among multiple motivational systems, nor did it deal with the regulation of activities like running or lever pressing that do not fit into the traditional motivational trinity – hunger, thirst, and sex. It turns out that both these omissions – motivational interactions and "weakly" motivated activities – can be handled in the same way; that is, by an extension of the regulatory approach.

The last chapter argued that animals seek to regulate their rate of eating. This is hardly controversial: Food is essential to life and failure to regulate food intake, at least over the long term, does not favor a large posterity. But experimental results suggest that animals do more than this: They regulate even over periods of time too short to pose metabolic dangers – perhaps because of the need to anticipate the possibility of deprivation in the future. Moreover, eating is not the only thing regulated in this way: Essentially everything the animal does tends to occur at a certain rate, and animals will take action to restore this equilibrium rate if it is perturbed in some way. Before I can explain how this works, I need first to explain some basic economic concepts, such as utility (value) and marginal utility, and the idea of a constraint.

I begin by describing a typical experimental situation used to study multiple activities, then describe a fundamental constraint to which all behavior is subject – limitation of time. I go on to recount the pioneering work of David Premack, who first experimented with "weak" reinforcers and proposed a general definition of reinforcement based on this work. Optimality (economic) models of operant behavior derive, in large part, form his proposal. The remainder of the chapter describes the optimality approach and shows how a wide range of experimental results can be accounted for by two principles: that the marginal utility of any activity decreases with the amount available (the *principle of diminishing marginal utility*), and that animals act to maximize utility, subject to constraints of time and reinforcement schedule.

The constraint of time – behavioral competition

Imagine a rat restrained in a seminatural environment that allows it to run in a wheel, eat, drink from a water bottle, and engage in "comfort" activities and "exploration." At a given season and time of day, the proportion of time that the animal spends in these various activities is likely to be quite stable from day to day. A number of experiments[2] have revealed two important properties of this mix of activities: (a) competition among activities for the available time; and (b) regulatory processes that tend to maintain the activities in fixed proportions.

Competition is most easily and directly demonstrated simply by limiting access to one activity and measuring the change in the amount of time spent on others. For independent activities, such as running and drinking, the effect of restricting one is generally to increase the other. For reasons not yet well understood, competition appears to be especially acute under the conditions typical of operant-conditioning experiments, that is, intermittent reinforcement given to a highly motivated animal. The effects of preventing an activity are then especially clear and nearly always facilitatory: Hungry rats pressing a lever for food will press faster if denied the opportunity to run, for example.[3]

In one sense, competition for time is a necessity. Amount of time is always fixed: If activities are measured in time-based units, and if everything is considered to be an activity – the animal is never allowed to "do nothing" – then the total must remain constant. If activities are defined in this way, an increase in one forces a decrease in at least one other. Few situations have been studied carefully enough to meet these conditions: Not everything the animal does is recorded, activities are most often measured as number of occurrences per unit time (rather than in terms of total time taken up), and response topography is not measured in a way that allows one to go back and forth with confidence between rate and time measures.[4] The usual procedure is just to measure two or three "standard" activities (in addition to eating) and look for relations between them.

If less than all activities are recorded, there is no assurance that decrease of one activity will be accompanied by increase of any other. Nevertheless, in experiments with very hungry animals pressing a lever or pecking a key for food, it is safe to assume that the animal does little else but press or peck, and eat. Consequently, such experiments provide good evidence for behavioral competition enforced by the time constraint. For example, Figure 7.1 shows data from a choice experiment by Herrnstein (left panel) in which hungry pigeons could obtain food by pecking at either of two simultaneously available response keys. The delivery of food was scheduled by independent, time-based (variable-interval) schedules associated with each key (a *concurrent VI VI* schedule). Herrnstein varied the average minimum interfood interval (i.e., the VI value) associated with these schedules and determined the equilibrium rate of pecking on each

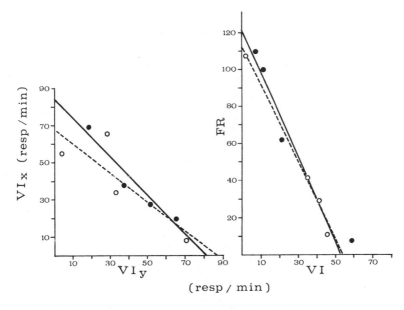

Figure 7.1. Left panel shows response rate on one key plotted against response rate on the other key for several pairs of variable-interval schedules in a two-choice experiment. Open and closed circles refer to data from two different pigeons (Herrnstein, 1961). Right panel shows similar data for a single pigeon in a two-choice variable-interval, fixed-ratio experiment. Open and closed circles show data from conditions with or without a change-over delay (Bacotti, 1977).

alternative at each pair of VI values. Herrnstein was interested in the relation between the proportion of pecks on a key and the proportion of payoffs for pecking that key. But reinforcement for pecking a key can also be regarded as a way of affecting the number of pecks made on that key, independently of other activities; so we can look just at the relation between rates of pecking on the two keys, as the proportion changes under the influence of differential reward. Figure 7.1 shows that pecks on the two keys were reciprocally related – according to a straight line with unit slope – in Herrnstein's experiment: as rate of pecking one key increased, rate of pecking the other decreased by an equal amount.

Data in the right-hand panel of Figure 7.1 show that this reciprocal relation does not depend either on constancy of total reinforcement rate or similarity of the schedules for each key (both of which were conditions of Herrnstein's experiment). Bacotti used a variable-interval schedule for one key and fixed ratio for the other (*concurrent VI FR*), and made no effort to keep total reinforcement rate constant. Nevertheless, his data show the same kind of reciprocal, linear relation between pecks on the two keys.

If we identify the number of pecks per unit time on the two keys as x_1 and x_2, then these two experiments show them to be related by the linear equation

$$ax_1 + x_2 = b, \qquad\qquad (7.1)$$

where a and b are constants: b is a scale constant that depends on the units of measurement, and a is the ratio of durations of the two alternative responses. Thus, animals typically peck at a faster rate on ratio schedules (see note 4), so that in Bacotti's experiment a pigeon will peck more times during a fixed period when it is responding on the ratio key than during a comparable time responding on the interval key (all else being equal). Consequently, if x_1 denotes the ratio schedule, then a is greater than unity; in the symmetrical situation of Herrnstein's experiment, on the other hand, a is close to one.

The functions in Figure 7.1 show the invariant relation between the rates of two activities as the balance between them is altered by external factors: here, different rates of reinforcement for each response, in other situations, variation in the properties of the stimuli controlling one or both activities. In the language of an earlier theoretical era, these functions describe *response-response* (rather than *stimulus-response*) relations.

Behavioral competition is of interest in its own right. It is still imperfectly understood, and experimental work continues. However, the bare fact of competition for available time has implications that can be traced even in the absence of detailed understanding of the process. In Chapter 10 I discuss implications for stimulus generalization and discrimination. Here I show how competition for time can be considered as one of the *constraints* to which animals adapt as they allocate behavior so as to maximize utility.

THE ALLOCATION OF BEHAVIOR

Imagine a rat in a seminatural environment in which it can do several things, each of which we can record and make available or unavailable. There are two main kinds of experiment we can do with such a situation: (a) Add or remove the opportunity to do various things and look at the effect on the amount and temporal order of the remaining activities. (b) Impose *reinforcement contingencies* between pairs of activities and see if the instrumental response increases in rate – make the animal run in order to get access to water, and see if he runs more, for example. David Premack was the first to see a connection between these two kinds of experiment. In studies with rats and monkeys, he showed that the effect of a contingency of this sort depends upon the levels of the activities under free conditions (so-called *paired-baseline* conditions). I next describe one of his experiments and then explain two ways in which the imposition of a reinforcement contingency can be expected to change the levels of the activities involved.

In one experiment, Premack studied the reinforcing relations among the activities of Cebus monkeys. The monkeys were in individual cages and there were four things that they could play with: a lever (L), a plunger (P), a hinged flap (F), and a horizontally operated lever (H). Premack's idea was that "reinforcement"

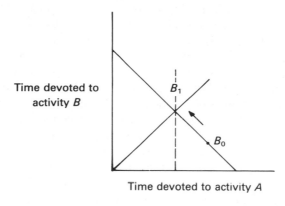

Figure 7.2. Schedule and time-allocation constraints with only two available activities.

is not an absolute property of a "goal activity," but rather is just the relation between a more probable (preferred) activity and a less probable one: An animal should engage in a less probable activity (lever pressing, for example), for the opportunity to engage in a more probable one (say eating). His situation with the monkeys was more convenient for testing this idea than the usual Skinner box, both because more activities were available and because different animals showed different preferences for the different activities. He compared the frequencies of activities under paired-baseline conditions with the effect of imposing a 1:1 contingency between the several possible pairs of activities, that is, a contingency requiring *t*-sec of activity A for *t*-sec access to activity B; A here is termed the *instrumental response* and B the *contingent response*. Premack wrote:

The clearest predictions possible were those for Chicko, who in the first procedure [free access to all activities – paired baseline] showed three reliably different response probabilities [proportions of time spent]. Indeed, Chicko's protocol made possible three kinds of contingencies: contingent response [i.e., reinforcer] higher than, less than, and, in one case, about equal to the free [instrumental] response. . . . the outcomes for the three types of contingencies were as follows: (1) contingent response higher than free response produced. . . an increment in the free response; (2) contingent less probable than free response produced. . . a decrement. . .; (3) the one case in which the responses were about equal produced little or no change, increment or decrement. (Premack, 1965)

Thus, a monkey that spent 10% of its time playing with L, say, and 20% playing with P under free conditions, increased its level of L when L, the instrumental response, had to occur for *t* sec in order for P, the contingent response, to be available for *t* sec. What can we learn from this about the regulatory nature of reinforcement, and about the difference between reinforcers and other events?

Let's look first at the simplest possible theoretical case: The animal can do only two things, and always does one or the other – two mutually exclusive and exhaustive activities. The situation under free conditions is illustrated in Figure 7.2, which shows the time constraint by the line of unit negative slope (sum of times taken up adds to a constant; see Figure 7.1 and note 4). Point B_0, the *free-behavior* point, represents the (proportion of) time taken up by each activity under free conditions.[5] The line of unit slope through the origin represents the added constraint imposed by the contingent relation (recall, t units of instrumental activity were required for t units of contingent activity, hence the unit slope). This is, of course, the feedback function for a fixed-ratio-one (FR 1) schedule. The arrow indicates the forced change (from B_0 to B_1) in B, the point representing the proportions of the two activities. It is clear that the two constraints – time allocation plus schedule – exhaust the degrees of freedom available to the animal in this simple situation. Because, by our initial assumptions, he must engage in one or other activity, the added constraint of the ratio schedule forces the initially low-probability activity (here the instrumental response) to increase above its free value, as the level of the contingent response decreases.

This increase is obviously *not* a "reinforcing" effect of the more probable activity on the less probable. It is not an adaptive response, but merely a forced change, caused by the two constraints: Because the schedule makes instrumental and contingent activities occur equally often, and because the contingent activity occurs more often than the instrumental under free conditions, the first effect of the schedule constraint is to restrict artificially the time devoted to the contingent activity. Since the animal must engage in one activity or the other, restriction of the contingent activity forces an increase in the instrumental response. This change is called a *restriction effect*.[6]

When the number of constraints is not greater than the number of activities, restriction effects are the only possible effects of a reinforcement schedule. When there are more activities than constraints, both restriction and *contingent* effects are possible – more on contingent effects in a moment.

The test for a restriction effect is to ask whether the increase in the instrumental response associated with the imposition of a contingency is greater than, or merely equal to, the increase produced by just restricting the proportion of time the animal can devote to the contingent activity. The effect of restriction is illustrated in Figure 7.2 by the dashed vertical line through point B_1. Clearly, restriction of the contingent response to the level attained under the contingency condition yields the same increase in the instrumental response, despite the lack of any contingent relation between the two activities.

Experimentally, a restriction effect can be demonstrated by what is known as a *yoked-control* procedure. Two animals are involved: Animal A is given contingent access to an activity. In equilibrium, the contingent activity for animal A occurs at a certain rate with a certain temporal pattern. The activity is then permitted to animal B, the yoked animal, at exactly the same time as animal A makes it available for himself. Thus, both animals get the same frequency and temporal pattern of access to the activity, but it is dependent on an instrumental response in one case (A), but independent of the animal's behavior in the other (B): Animal A is on a FR 1 schedule; animal B on a *variable-time* (VT) schedule with interactivity times determined by the behavior of animal A. If animal B shows the same increase in the instrumental response as animal A, we can be fairly sure that the increase is a restriction effect.[7]

Thus, the essence of *reinforcement* is an increase in the level of the instrumental response beyond the increase attributable to restriction. Such an increase is called a *contingent effect*. A contingent effect is possible only if the degrees of freedom in the animal's behavior exceed the constraints imposed by time and schedule: Given two constraints, at least three mutually exclusive and exhaustive activities must be available for an animal to show a contingent effect of the schedule.[8] Even if there are enough degrees of freedom, yoked controls, or comparison with a comparable variable-time schedule, are necessary to be sure that an increase in the instrumental response goes beyond the effects of restriction.

Premack's early experiments lacked such controls. Nevertheless, later work has borne out his thesis that a more probable activity will generally reinforce a less probable one in which it is one-to-one contingent, that is, produce an increase greater than that expected merely from restriction. The one-to-one proportion is critical, for both restriction and contingent effects, however. For example, if we choose a ratio schedule different from 1:1 whose feedback function passes through B_0 in Figure 7.2, then there will be no increase in the instrumental response.

Premack thought of reinforcement as being particularly associated with activities – with eating rather than the intake of food, with running rather than the availability of a running wheel. This is not a necessary part of the regulatory view of reinforcement. Food-in-the-mouth, hence food naturally eaten, is a more effective reinforcer than food delivered intravenously, because of the incentive properties of taste (see Chapter 6). With or without taste, however, it is the rate of access to food that seems to be the critical variable.

Negative reinforcement makes the same point. Receiving electric shock is not an activity. Nevertheless, animals clearly have a preferred rate for such an event – zero. With this sole quantitative difference, electric shock can be treated as just the reverse of positive. Requiring an animal to engage in a low-probability act, or experience an event that is generally avoided, for access to one of higher probability, has a reinforcing effect on the low-probability act, but it has a punishing (negatively reinforcing) effect on the high-probability act. Reinforcement and punishment are thus two sides of the same contingent relation. The only asymmetry is a practical one: Commonly used negative reinforcers, like shock, are usually events that can be delivered to the animal without its cooperation. This is not an essential difference. Animals can be punished by forcing them to engage in more of a low-probability activity than they would like (the schoolboy required to write 100 "lines"), and they can be positively reinforced by hypothalamic brain stimulation requiring as little cooperation as foot shock.

Thus, the essential difference between positive and negative reinforcement is just in the relation between the free levels of the two activities and the direction in which these levels are driven by the contingency. For example, in Figure 7.3 (a repeat of Figure 6.1) the free levels of two mutually exclusive but nonexhaustive

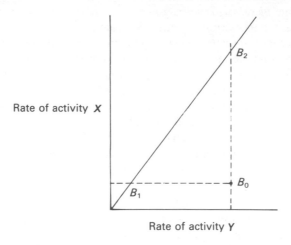

Figure 7.3. Potential adaptations to a ratio-schedule constraint.

activities are indicated by B_0, the free-behavior point. A ratio contingency is illustrated by the ray through the origin. Clearly, if the animal continues to engage in X at its free level, activity Y will be severely curtailed (point B_1). Hence, the contingency is potentially punishing with respect to Y, but potentially reinforcing with respect to X.[9] The *actual* effect of the contingency depends on whether the activities change in the expected way. For example, it is logically possible either for X to show no change (behaviors at B_1: a punishing effect on Y, but no reinforcing effect on X), or for X to increase so much that Y is not reduced (behaviors at B_2: a reinforcing effect on X, but no punishing effect on Y). In practice, either extreme result is rare.

Optimal allocation

What, then, determines the compromise struck by an animal constrained by a reinforcement schedule? Perhaps the simplest possibility is that he adjusts the levels of the instrumental and contingent responses so as to be as close as possible, in some sense, to the mixture under free conditions, the free-behavior point. In the following discussion I will argue that if activities are independent, quite general properties of schedule behavior can tell us how animals must assess the cost of deviations from their preferred mix of activities.

First, a necessary digression on *cost* and *value*.[10] If the free-behavior point represents the most desirable mix of activities, then *deviations* from that point entail some cost. For example, suppose that the animal prefers to run 20% of the time but is prevented from running more than 10% of the time. Increments in running have value because they reduce the deviation of actual from preferred running level, and the amount of value they have depends on the cost of (negative) deviations from the 20% level. Conversely, if the animal is forced to run

30% of the time, reductions in running have value proportional to the cost of (positive) deviations from the preferred level of running. Thus, the value of a given activity level is the same as the cost of the deviation of that level from the preferred level, but with opposite sign.

Two beginning simplifications: First, I will suppose that the values of the instrumental and contingent behaviors are independent; that is, the cost or value associated with a given level of one activity is independent of the level of the other. If so, the total cost to the animal of given levels of the two activities is just the sum of the separate costs. Since B_0, the free-behavior point, represents the most desirable mix, we can guess that the cost of a level of activity different from B_0 can be expressed as some function of the difference. If the coordinates of B_0, for two activities X and Y, are (x_0, y_0) (i.e., X and Y occur with frequencies x_0 and y_0 during a unit time period), then the total cost to the animal of a level of joint activity (x, y) will be some increasing function of the difference, $x_0 - x$ and $y_0 - y$. Formally,

$$C(x, y) = f(x_0 - x) + g(y_0 - y), \tag{7.2}$$

where f and g are increasing functions. $C(x,y)$, is known as the *objective function*, the quantity to be minimized.

Second, I assume that f and g are functions of the same type, for example, both linear, or quadratic, or power functions, or whatever. Thus, Equation 7.2 becomes

$$C(x, y) = f(x_0 - x) + f(y_0 - y). \tag{7.3}$$

Given these two simplifications, we can restrict function f in several ways. The most obvious possibility, that f is just a linear function, that is, $C(x, y) = a(x_0 - x) + b(y_0 - y)$, where a and b are constant multipliers representing the relative costs of X and Y deviations, can be eliminated as follows. First, look at what the linear cost function predicts under free conditions. Obviously C can be made as small as we like by increasing x and y to large values, which is not plausible. Thus, a simple linear function will not do because it does not have a minimum for C at the free-behavior levels, x_0 and y_0.

Perhaps we should look just at the *absolute value* of deviations, so that Equation 7.3 becomes

$$C(x, y) = a|x_0 - x| + b|y_0 - y|. \tag{7.4}$$

This does have a minimum at the free-behavior point under free conditions, but it predicts funny things when we introduce a schedule constraint between x and y, as in Figure 7.3. Although the argument is tedious, it is possible to show that with a ratio-schedule relating x and y ($x = my$, where m is the ratio value) Equation 7.4 predicts one of two possibilities: Depending on the relative values of a, b, and m (the ratio value), the animal should always choose either point B_1 or B_2 in Figure 7.3: if $b/a > m$ he should choose B_2, if $b/a < m$, he should choose B_1 (I show this graphically later in the chapter). That is, he should perfectly regulate either x or y, never compromise. This never occurs.

To make things more realistic, suppose we add a constraint on total time, together with the possibility of a third activity, Z, to add a degree of freedom. The objective function is now

$$C(x, y) = a|x_0 - x| + b|y_0 - y| + c|z_0 - z|, \qquad (7.5)$$

where c is the cost-of-deviation parameter for the third activity, Z. Again, the linear cost function predicts extreme results, depending on the relative values of a, b, c, and m. If $c > a$ or b, the animal should regulate Z perfectly. If not, he should never engage in Z. In either case, in the time remaining he should attempt to regulate either X or Y perfectly, as before, up to a limit imposed by the total time available. If the ratio value is so high that Y cannot be obtained at a rate y_0, even if the animal spends all the available time engaged in X, then rates x and y will be constrained by a line of negative slope (Equation 7.1), representing the time-allocation constraint when all available time is taken up by X and Y. As we saw in Chapter 6, such linear response functions hold only for moderate (not extreme) ratio values.

Thus, the linear cost function fails to predict two basic properties of ratio response functions: that they invariably involve a compromise, and that they are nonmonotonic, turning down at very high ratio values (see Figure 6.8). The cost function must be nonlinear.

The same conclusion follows from a very old economic principle, *diminishing marginal value*. Everyone knows that the appeal of most things declines as one has more of them: An occasional dish of caviar is a treat; one a day may take the edge off one's enthusiasm; one with every meal is a bore. For most goods all the time, and for any good at some time, more and more, in quantity, means less and less, in added satisfaction.

There are some exceptions, of course. To a miser, every additional cent is worth having, no matter how many he already has. Increments in some things become more valuable as we have more of them: To a seamstress, enough material for a dress is worth a great deal more than too little, so the same increment is here worth more when added to a larger base.

These three possibilities are illustrated in Figure 7.4, which shows value plotted as a function of quantity of a good. Curve A fits the miser; value is simply proportional to amount. Hence, marginal value (technically, dV/dx, the slope of the curve) is constant and independent of amount. Curve B corresponds to the seamstress; a large piece of material is of more value than a small, but value increases dramatically when the size is sufficient for some new use: The curve is positively accelerated and illustrates increasing marginal value (dV/dx is a positive function of x). Curve C is negatively accelerated, fits almost everything else, and illustrates diminishing marginal utility (dV/dx is a negative function of x). While many utility (value) functions show regions that are linear or positively accelerated, essentially every good shows diminishing marginal utility at very high values – it is, after all, possible to have too much of practically anything.

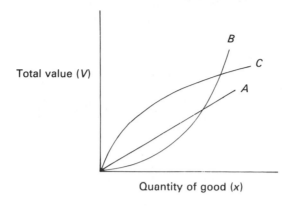

Figure 7.4. Linear (*A*), positively (*B*), and negatively (*C*) accelerated utility functions.

In terms of classical utility theory, cost and value are on the same dimension, differing only in their sign: Values are positive, costs negative, and it is a matter of convenience which we use. Intuition suggests that the cost of an increment in the deviation of x (say) from x_0 (i.e., of a reduction in x) is likely to be greater, the greater the existing deviation: A given reduction in rate of eating will be more costly if subtracted from a starvation than an ad libitum level, for example. This implies that function f in Equation 7.3 is *positively accelerated*. Positively accelerated cost is just the same as negatively accelerated value. Hence a positively accelerated cost function is equivalent to diminishing marginal value in terms of x, for values of x below the preferred level.[11]

We now have all the ingredients for a simple economic analysis of behavioral allocation. Response-independence and a positively accelerated cost function allow us to estimate the value to the animal of a particular mix of activities; time allocation and, most importantly, the reinforcement schedule, constrain the mix to values less than the maximum possible. These are the two components of standard optimality analysis: an *objective function*, and a *constraint set*.

The minimum-distance model

The simplest nonlinear cost function is known as the *minimum-distance hypothesis*, for reasons which will be obvious in a moment. Figure 7.5 illustrates its application to ratio schedules. The figure shows just two activities: x, the rate of the instrumental response, and $R(x)$, the rate of the contingent response (reinforcement) actually obtained (obviously a function of x). The two rays through the origin represent two ratio schedules, I and J. The points B_i and B_j represent the steady-state response and reinforcement rates settled on by the animal as it adapts to these schedules. Compare Figure 7.5 with Figure 7.3; note that both B_i and B_j are compromise adaptations, falling in between the two extreme options. The rule by which they were selected is the simplest one consistent with dimin-

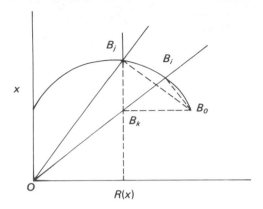

Figure 7.5. Minimum-distance adaptation to ratio constraint.

ishing marginal utility: points B_i and B_j are just the *closest ones* on the ratio-constraint line (feedback function), to the free-behavior point, B_0. The two dashed lines are therefore at right angles to the two ratio lines.

The minimum-distance model predicts a contingent effect of imposing a schedule. Point B_k in Figure 7.5 shows the response rate, x, expected from restriction alone – when reward rate, $R(x)$, is held to the value obtained on the higher ratio schedule. The response rate at B_k is lower than the rate at B_j, showing that the animal responds faster when responding is necessary for reward than when it is not, a result different from the constrained case illustrated earlier in Figure 7.2.

Expressed in the form of Equation 7.3, the objective function for the minimum-distance model is

$$C(x,R(x))) = [(x_0 - x)^2 + (R_0 - R(x))^2]^{1/2}, \tag{7.6}$$

where (x_0, R_0) are the coordinates of the free-behavior point, B_0. Since the extrema (maxima and minima) of a power of any function are the same as those of the function itself, the overall square root can be dispensed with, leaving a quadratic cost function. The minimum of this function, subject to the ratio-schedule constraint, just gives the minimum (Euclidean) distance from B_0 to the constraint line. Hence this particular model, which is just one of many that are consistent with the independence and diminishing marginal value conditions, is termed the *minimum-distance* model. Clearly, the set of points representing adaptations to a range of ratio values will fall on a curve, which is just the ratio-schedule *response function*, discussed in Chapter 6. Elementary geometry shows that the response function for these assumptions is a circle with diameter $0B_0$, as shown.

The minimum distance-model correctly predicts that ratio response functions should be bitonic. Examples of empirical ratio response functions (similar to those already shown in Figure 6.8, but from a different experiment) are shown in

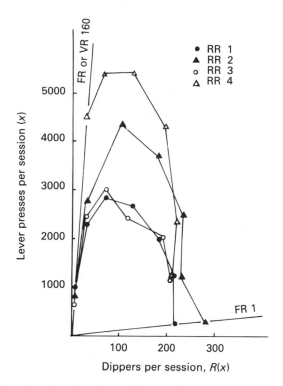

Figure 7.6. Response functions for two individual rats over the indicated range of ratio schedules. Open symbols: VR, closed: FR. Sessions were 1 hr long (Motheral, unpublished).

Figure 7.6. The model fails in one respect, namely, the scales of x and $R(x)$. It is obvious from Figure 7.6 that if the two axes were to the same scale (and even allowing for the fact that x and $R(x)$ are not usually measured in equal-time units), the resulting curve would be very far from a circle, and closer to an ellipse much elongated in the vertical direction. This deficiency of the model is remedied by the very reasonable assumption that the *costs* of deviations in x and $R(x)$ are not equal. In the usual operant-conditioning situation, the reinforcer is something of considerable importance to the animal, such as eating or electric shock. X, on the other hand, will usually be a low-cost activity such as pecking or lever pressing. Indeed, low cost is almost the defining characteristic of most "standard" operant behaviors. Consequently, cost parameters need to be inserted in Equation 7.6, thus:

$$C(x, R(x)) = a(x_0 - x)^2 + b(R_0 - R(x))^2, \qquad (7.7)$$

where a and b are the weighted costs of X and $R(x)$ deviations from B_0: For a contingent response such as eating and an instrumental response such as lever pressing, a is much less than b.

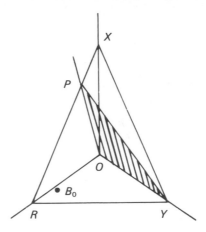

Figure 7.7. Joint effect of ratio constraint (plane *POY*) and time-allocation constraint (plane *XYR(X)*) when three activities are available. B_O, the free-behavior point, lies in the behavior plane *XYR(x)*, but will not generally lie on the line of permissible behavior *PY*.

One major deficiency remains. Ratio-schedule response functions generally have a pronounced declining limb and a relatively abbreviated increasing one (see Figure 7.6): They are not symmetrical as Equation 7.7 requires. Indeed, over a considerable range, the ratio response function can be approximated by a straight line of negative slope, as we saw in the preceding chapter. It turns out that if we include the time-allocation constraint – and permit a third class of activities so as to have sufficient degrees of freedom for a contingent effect – this alters the minimum-distance prediction in the appropriate way. The necessary time-allocation constraint for three mutually exclusive and exhaustive activities is

$$x + R(x) + y = 1, \tag{7.8}$$

where all activities are measured in time units and y is the level of a third activity class. (Equation 7.8 is just an extension of Equation 7.1 to three activities.) In the usual operant-conditioning situation, y might be the time devoted to comfort activities (grooming, etc.), walking around or, for a rat, running in a wheel. This third activity adds a third term to Equation 7.7, thus:

$$C(x, R(x)) = a(x_0 - R(x))^2 + b(R_0 - R(x))^2 + c(y_0 - y)^2, \tag{7.9}$$

and Equation 7.8 imposes a second constraint.[12]

The minimum of Equation 7.9, subject to the constraints of schedule and time allocation, can be found by standard calculus techniques.[13] The predictions of the model are illustrated graphically in Figure 7.7. The level of the third activity, *Y*, is represented by a third axis, at right angles to the *X* and *R* (*R(x)*) axes. The restriction that the three activities must occupy the total time (i.e., add up to a constant, see Equation 7.8) takes the form of a *behavior plane*, *XYR*, on which the point representing the levels of the three activities must lie. The ratio-schedule

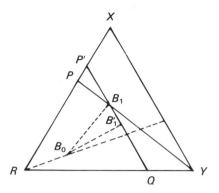

Figure 7.8. Plan view of behavior plane, illustrating restriction vs. contingent effects in the three-behavior case.

constraint is represented by a second plane, POY, the *ratio plane*, passing through the Y axis (since the ratio constraint relates x and $R(x)$, and is independent of y). The intersection of the behavior plane, XYR, and the ratio plane, POY, is the line PY. Clearly, the point representing allowable combinations of x, $R(x)$, and y, the *representative point*, must lie on line PY. In general, B_0, the free-behavior point, will not lie on PY. The quadratic cost function implies that, for equal-cost activities, the representative point will lie at the point on PY closest to B_0.

The minimum-distance prediction for three activities is illustrated in Figure 7.8, which shows the behavior plane, XYR, full on. Line PY is the set of allowable behavior mixes, as before, and B_1 is the minimum-distance solution. The figure shows two things. First, that the minimum-distance assumption predicts (as it should) that x should be higher when $R(x)$ is a positive function of x (e.g., a ratio schedule) than when $R(x)$ is independent of x. In other words, there will be a contingent effect, defined as the difference between the change in x produced by imposing a schedule and the change produced simply by restriction of $R(x)$. Line $P'Q$ represents the control case (this is just the three-dimensional equivalent of the vertical line – now a plane intersecting XYR – through B_1 in Figures 7.2 and 7.5). The minimum-distance adaptation to this condition is at B_1', which represents a lower rate of x than B_1 (higher levels of an activity are represented by points closer to the vertex for that activity[14]).

Second, Figure 7.8 shows that as one activity, such as $R(x)$, is progressively restricted, independently of the others (i.e., presented on a variable-time schedule), the other two do not remain in the same fixed proportion. Thus, as line $P'Q$ is moved progressively closer to R, the proportion of x (x-direction) to y (y-direction) changes as the representative point approaches B_0 along line B_0B_1'. Only in the special case where B_0 lies on the line joining R and B_1 will the proportions of x and y remain constant and independent of the level of $R(x)$. This point is worth making because some theories of choice[15] say that the relative

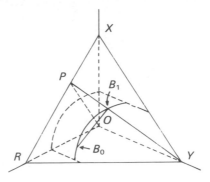

Figure 7.9. Geometrical derivation of the ratio-schedule response function from the minimum-distance assumption.

value of a given pair of alternatives should be independent of other alternatives that may be available.

Prediction of response functions

Figure 7.9 shows in a three-dimensional graph the response function derived from the minimum-distance assumption under both schedule and time-allocation constraints. The response function is generated in the behavior plane; the relation between x and $R(x)$, with y eliminated (which is what is usually measured, as in Figure 7.6) is just the projection of this onto the XOR plane. Figure 7.10 shows theoretical ratio response functions. Parameters a and b are the costs of x- and $R(x)$-deviations, respectively; the five curves represent the effects of different costs of y-deviations. As this cost increases, one effect is to tip the theoretical response functions to the left in the way required by empirical results. The reason for this tipover is not obscure. At high ratio values, the trade-off between x and $R(x)$ is very unfavorable: The increment in $R(x)$ produced by an increment in x is small. Correspondingly, the time taken away from y by the increment in x necessary to yield a fixed increment in $R(x)$ is, at high ratios, substantial. As the cost of y-deviations increases, therefore, the high rates of x associated with high ratios are especially depressed; the time is better spent on y than on x, given the small return for the latter in terms of $R(x)$.

Response functions can be derived in this way for any kind of feedback function, that is, any kind of reinforcement schedule. For example, the feedback function for interval schedules (see Chapter 5) is negatively accelerated rather than linear, with a fixed maximum rate of reinforcement. The minimum-distance response function for interval schedules is consequently different from the corresponding function for ratio schedules: It is lower for a given rate of reinforcement, and tipped to the right rather than the left. Again, the reason is fairly obvious intuitively. At long interval values (low maximum-reinforcement rates),

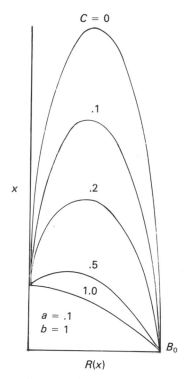

Figure 7.10. Theoretical effects on the ratio response function of increasing the cost of "other" behavior. Cost of instrumental response = a, contingent response = b, other = c.

as response rate increases, the animal very soon approaches the diminishing-returns region of the feedback function where increments in responding produce smaller and smaller increments in reinforcement rate. There is, of course, no such region in the linear ratio feedback function. At the same time (as in the ratio case) more and more time is being taken up by the instrumental response so that other activities are forced further and further from their free-behavior values – at ever-increasing incremental cost, because of the positively accelerated cost function. The increasing cost of these deviations, together with the diminishing payoff entailed by a negatively accelerated feedback function, means that the compensatory increase in response rate as reinforcement rate falls (i.e., as the interval value increases) is soon overwhelmed, so that response rate and reinforcement rate fall together at long interval values. The small increment in reinforcement rate that can be obtained by large increases in response rate is not, at long interval values, worth the added costs – of responding and in time taken away from other activities. Hence, the increasing limb of the interval-schedule response function dominates, whereas the decreasing (regulatory) limb is the dominant feature of the ratio response function (see Figure 7.15).

This discussion of the predictions of economic models has necessarily been

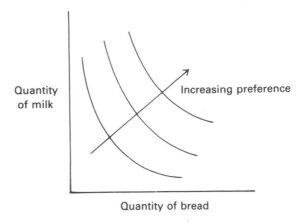

Figure 7.11. Conventional indifference curves.

rather abstract. It is important, therefore, not to lose sight of the essentials, which are that the basic properties of molar adjustments to ratio, interval (and, as we will see, numerous other) schedules are derivable from just three general considerations: The values of activities are independent; activities compete for the available time; and activity values are subject to diminishing marginal utility.

The minimum-distance model is convenient because it lends itself to graphical illustration and because conclusions true of it are also true of almost any model that incorporates diminishing marginal utility. It is one way of generating what in economic parlance may be termed the standard preference structure for partially substitutable goods. Since many of the things I have discussed in terms of costs of deviations and so forth can also be (and often are) described in the economists' language of substitutability, complementarity, and indifference curves, the next section looks at the relations between the two forms of analysis.

Substitutability and indifference curves[16]

Psychologists and philosophers differ on whether it makes sense to give numbers to utilities (some psychologists say you can, most philosophers say you can't), so the analysis I have just described, in which behavior is derived from an assumption about the way the utility depends on the quantity of something, is no longer the mode in economics. No matter; we will get to the same place by a different route. Consider two goods, such as bread and milk. Even if we can't give numbers to the utility of a given quantity of milk and bread, everyone seems to agree that we can always *equate* the value of bundles of goods. For example, 2 liters of milk and 3 loaves of bread may be judged equal to another bundle with 4 liters of milk and 2 loaves, in the sense that we are indifferent as to which bundle we get. There will be a whole set of bundles of this sort, differing in the proportions of bread and milk, but the same in that we are indifferent among

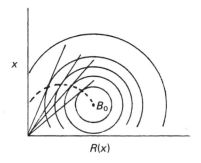

R(x)

Figure 7.12. Indifference curves derived from the minimum-distance model. Rays through the origin are ratio constraints, and the dashed curve is the expected response function.

them. This set defines an individual *indifference curve*: a set of points in a space whose axes are the amounts of the goods being compared. Other names for indifference curve are *preference isocline* and *isopreference curve*.

All agree that bundles can be rank-ordered, A being preferred to B, B to C, and so on. Bundles so ranked obviously cannot lie on the same indifference curve; consequently, to rank-order bundles is also to rank-order indifference curves. The *preference structure* for bread and milk, or any other pair of goods, can therefore be represented by a sort of contour map of indifference curves, such as those illustrated in Figure 7.11. The arrow indicates the direction of increasing preference which, unsurprisingly, is generally in the direction of more of both goods.

It is obvious that a utility or cost function, such as Equation 7.9, defines the form of the corresponding indifference curves (although it is not possible to go in the reverse direction and derive a unique utility function from a given set of curves). For example, suppose that for goods X and Y, the utility functions are $f(x)$ and $g(y)$. An indifference curve is then defined by the relation $f(x) + g(y) =$ constant $= K$, each value of K being associated with a different curve. Thus, for the quadratic cost function, the equation for an indifference curve is $a(x_0 - x)^2 + b(y_0 - y)^2 = K$, which is an ellipse (for $a = b$, a circle) centered on x_0, y_0.

Knowledge of the preference structure is not by itself sufficient to specify behavior – in the absence of any constraint, the organism has no reason to stray from the point of maximum value. Once a constraint is specified, the optimal solution is clearly to settle on the highest indifference curve consistent with it. For linear, or curvilinear, constraints, this implies an equilibrium at the point where the constraint line is tangent to the highest indifference curve. This is illustrated for the circular indifference curves of the equal-cost quadratic model in Figure 7.12. Here the constraint is simply the linear feedback function for a ratio schedule. The dashed line is the ratio-schedule response function. As you can see, indifference-curve analysis gives the same result as the direct minimum-distance method shown earlier in Figure 7.5.

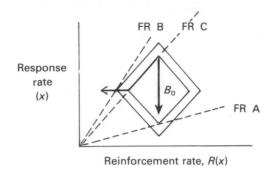

Figure 7.13. Prediction of ratio response functions from indifference curves generated by the absolute-value linear model (Equation 7.4). See text for details.

The indifference-curve method sometimes makes it easy to see things that are otherwise not obvious. For example, Figure 7.13 shows the indifference curves generated by the absolute-value linear model discussed earlier (Equation 7.4). The ratio of height to width of the diamond-shaped indifference contours is equal to the relative costs of the instrumental and contingent responses. The rays through the origin are three ratio feedback functions. Optimal responding is indicated by the point where each feedback function is tangent to an indifference curve. The critical ratio value is the center one, FR C. It is pretty obvious that for any ratio value less than C (such as A), the animal should regulate the contingent response perfectly: The response function lies on the vertical, arrowed line. Whereas for any ratio value greater than C (such as B), he should regulate the instrumental response perfectly: The response function for these ratios lies on the horizontal arrowed line. For ratio C, any point on the feedback function connecting these two lines is optimal. Thus the indifference-curve analysis shows at once the extreme properties of a linear cost function.

Marginal value and substitutability. Economists talk a lot about the *marginal value* of things. The reason is that usually the way to maximize value is to allocate resources (e.g., money) so that the *increment* in value associated with the last little bit of resource is the same no matter to what we allocate it: Given $100 to spend on any mixture of three commodities, bread, milk, and beer, maximum satisfaction is assured if we are indifferent about whether to spend our last dime on more milk, bread, or beer. Indifference means that we have equated the marginal values of the three commodities. Those familiar with differential calculus will recognize that equating marginal value is just the mathematical operation of equating the partial derivatives of the value (or cost) function with respect to each commodity.

The concept of marginal value is useful even without getting into mathematics. For example, it explains why negatively accelerated utility functions (positively accelerated cost functions) turn up a lot. Consider how one should allocate a fixed amount of time between two activities whose values are each simply proportional to time spent: $V_1 = aT_1$; $V_2 = bT_2$. Because the value functions are linear, marginal value is constant in each case. Hence, no pattern of allocation can equate marginal values, and the best course is to devote all the time to one activity or the other, depending on whether $a > b$, or the reverse. When utility functions are negatively accelerated, on the other hand, the more one does of something, the smaller the marginal benefit, so that by adjusting the amounts of different activities it is always possible to equate marginal utilities. The fact that we usually observe *behavioral diversity*, animals doing several things within a given time period, implies negatively accelerated value functions.

Suppose we have a certain amount of one thing and are now offered the possibility of a trade: give up some of good A in exchange for some of new good B. The corresponding behavioral problem is to restrict the level of one activity and ask how the others trade off against one another to fill up the time made available. Another way to look at the same question is in terms of price: If milk becomes more costly, less will generally be bought, and more of other things, like beer, will be bought. If all the reduction in spending on milk is taken up by spending on beer, then beer and milk are acting as perfect *substitutes*. In an operant experiment, a rat may be responding on one lever for cherry cola, and on another for Tom Collins mix;[17] if the price (ratio value) for one commodity increases, by how much will responding for the other increase? The experimental answer tells us to what extent the two drinks are substitutes for one another.

The extent to which one thing is a substitute for another is measured by the *marginal rate of substitution*, or *substitutability*. Given a certain quantity of any good, an individual will be prepared to give up a small amount in return for some amount of another good. The marginal rate of substitution is just the ratio of amount given to amount received, evaluated at a point, that is, as the absolute quantities traded become very small.[18] It is pretty obvious that this ratio will usually vary as a function of the amount of the first good. For example, given a stock of only 1/10 of a liter of milk, and no bread, one might be prepared to give up only 1/20 liter in exchange for a loaf of bread; but given a stock of 5 liters, 1 liter of milk might willingly be exchanged for the same quantity of bread. This can be looked at either as an indication of diminishing marginal value of milk, or of diminishing marginal rate of substitution between milk and bread.

Marginal rate of substitution can also increase or remain constant as a function of amount of a good, corresponding to the cases of increasing and constant marginal utility already described. These three possibilities are illustrated in Figure 7.14: Curve A is the familiar case of decreasing marginal substitution; curve B shows constant marginal rate of substitution – cases of partial and complete substitutability, respectively. Curve C corresponds to increasing marginal rate of substitution.[19]

We have already seen that because of the universal time-allocation constraint, if one of an animal's activities is forced to decrease in frequency, at least one other must increase in frequency. How all other activities change when the

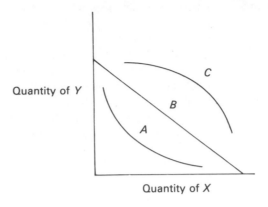

Figure 7.14. Indifference curves showing decreasing (A), constant (B), or increasing (C) marginal rates of substitution.

frequency of one is controlled depends on substitutability relations. For example, with just three activities, if one decreases, at least one of the others must increase; but the remaining activity may either increase or *decrease*. In the latter case, the two activities that change in the same direction are termed *complements*. Everyday examples of complementary goods are meat and meat sauce or cars and gasoline: If consumption of meat goes down (perhaps because of an increase in its price, perhaps because of a diet fad), consumption of meat sauce is also likely to decrease. There are comparable pairs in the behavior of rats on schedules. For example, given periodic, brief access to food (i.e., a controlled food rate, a fixed-time schedule) rats will drink after each food pellet. Over much of the range of food rates, drinking rate and food rate covary – drinking and eating thus act as complements. How can we decide on the direction of change of the free activities when one is restricted?

It turns out that for the quadratic model – indeed for any model of the form $C(x) = a(x_0 - x)^m$, where x_0 is the free-level parameter, a is the cost, and the exponent m is the same for all activities[20] – all activities are partial substitutes, none are complements. For all such models, as one activity is progressively restricted, the levels of any pair of other activities will both increase along a linear function whose slope is the ratio of their costs. If the requirement for equal exponents is relaxed, then restriction of one activity can lead to a decrease in some others. Since complementary activities exist, the equal-exponent simplification proposed during the earlier discussion of Equation 7.2 is not always justified.

The contours of an indifference-curve representation, in as many dimensions as there are independent activities, are a map of an organism's "motivational space" (preference structure, in economic language). All that is assumed by optimality analysis is that such a space exists (preferences are not irreconcilably inconsistent) and that it is *invariant* (preferences do not change capriciously).

This is known as the doctrine of *revealed preference* (see note 10). To go beyond this and attempt to explain the properties of the preference structure, that is, to derive the set of indifferences curves, will generally entail some assumptions about cost or utility. The minimum-distance model is one such assumption. Thus, utility is a theoretical notion but indifference curves are theoretical only in presupposing the stability of preferences.[21]

Goods that are partial or complete substitutes – beer and wine, cars and trucks, for example – generally have some property or properties in common. Similar activities are better substitutes than dissimilar ones. It is intriguing to speculate that all observed activities partake to different degrees of different amounts of a much smaller number of what might be termed (by analogy with ecology) *resource axes*, that are truly independent of each other. Activities are substitutable, then, because they lie on neighboring (nonorthogonal) axes within this hypothetical space. This is an appealing but largely untested possibility (see note 12).

EXPERIMENTAL APPLICATIONS

The power of the optimality approach lies in its ability to predict the differences among the molar response functions associated with different feedback functions (different schedules of reinforcement). There is as yet not much systematic, comparative experimental work on this topic, but the available data from "standard" situations are consistent with the three assumptions I have already discussed: value independence, competition, and diminishing marginal value.[22] For example, the left panel of Figure 7.15 shows the response functions for interval and ratio schedules predicted by the minimum-distance model. The right panel of Figure 7.15 shows response-function data from two similar experiments with rats pressing a lever for food reinforcement. The differences between the curves for ratio and interval schedules are of the sort predicted: The ratio curve is higher at all but the very highest reinforcement rates (corresponding to very short interval values, when an interval schedule cannot be distinguished from FR 1 by the animal); and the ratio response function is tipped to the left whereas the interval function is tipped to the right.

Data from other studies of positive reinforcement, such as interlocking, conjunctive, and tandem schedules (see Chapter 5), seem also to be consistent with these assumptions, to the extent that systematic data are available, and granted the difficulties in arriving at exact, molar feedback functions.[23]

Experimental results from schedules of shock postponement also fit easily into the present analysis. In avoidance schedules, the feedback function is a declining one, that is, as response rate increases, shock rate decreases (see Chapter 5, Figure 5.11). The free-behavior point is close to the origin, since unconstrained animals will choose to receive no shocks and will make few instrumental responses. The response function expected from minimum-distance-type models will be an increasing one: As the free shock rate (i.e., shock rate in the absence of instrumental responding) increases, response rate should also increase, as

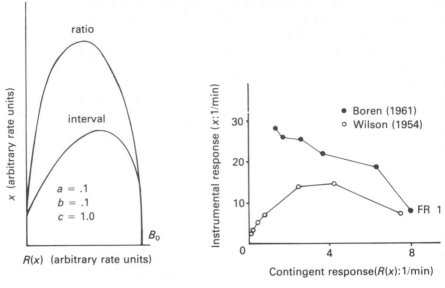

Figure 7.15. Left panel: theoretical response functions for interval and ratio schedules. Right panel: empirical interval and ratio response functions. (From Staddon, 1979a; redrawn from Boren, 1961, and Wilson, 1954.)

shown in the left-hand panel of Figure 7.16. The right-hand panel shows data from an experiment in which rats received shocks at a constant rate if they failed to press a lever. A lever press cancelled the next scheduled shock; additional presses had no effect until the time for the avoided shock had passed. The rats were permitted to establish a stable rate of lever pressing at each shock rate. As expected, equilibrium response rate is a positive function of shock rate. Results from numerous other shock-postponement experiments fit this general pattern.

We saw earlier that if activities other than the instrumental and contingent responses are neglected, the minimum-distance model predicts ratio response functions that are too symmetrical. When these "other" activities are included (that is, their cost-of-deviation and free level are allowed to be greater than zero), the response function tips to the left – the declining (regulatory) limb of the function becomes shallower. Moreover, the greater the cost of "other" activities, the shallower the regulatory limb (see Figure 7.10). We can test this prediction by looking at the effect on the ratio response function of adding a third activity.

Figure 7.17 shows the response functions obtained from a group of four rats responding on a range of ratio schedules. The upper curve was obtained in a standard Skinner box, with no opportunity provided for competing activity. The lower curve shows the response function obtained with a running wheel concurrently available. Not only is it depressed, as would be expected just from competition, but the regulatory limb is also shallower as minimum-distance requires (compare the slopes of the dashed lines drawn through the declining limbs of the two functions). A similar shift can be produced by giving the rat constant access

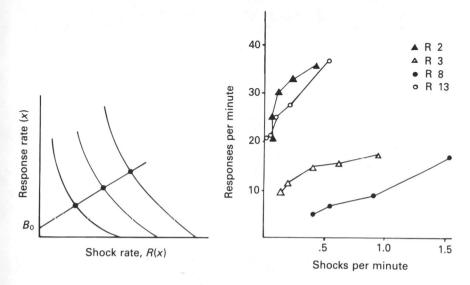

Figure 7.16. Left panel shows hypothetical feedback functions (descending curves), for a shock-postponement schedule, and an estimated response function (ascending curve). The right panel shows empirical response functions from four rats exposed to such schedules; each point is the average of an ascending and a descending run (de Villiers, 1974). See text for details.

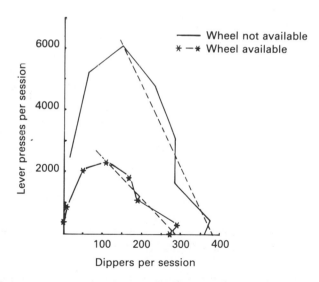

Figure 7.17. Effect on the ratio-response function of introducing a valued competing activity. Upper curve shows the response function for a group of four rats pressing a lever for food with no running wheel available. The lower curve shows the response function when a running wheel was available (Motheral, unpublished data).

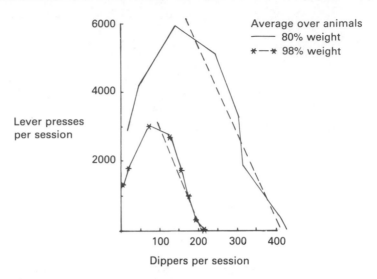

Figure 7.18. Effect on the ratio response function of level of motivation. Upper curve is the ratio response function for four rats maintained at 80% of their normal weight. Lower curve is the response function for the same group maintained at 98% of their normal weight (Motheral, unpublished data).

to an alternative, but less preferred, food (Lea & Roper, 1977). The point is that the slope of the declining limb of the response function depends on the *relative* cost-of-deviation of the contingent response versus other activities. If these activities have high cost-of-deviation, then the rate of the contingent response will be less well regulated (have a shallower declining limb) than if they have low cost-of-deviation. Figure 7.17 shows that the availability of running impairs the regulation of eating. Lever pressing (by which eating is regulated here) and running compete for the available time, so that both cannot, at the same time, be perfectly regulated. If running is a relatively preferred activity (which it is), then adding the opportunity to run must impair the regulation of eating rate.

I argued in Chapter 6 that food deprivation should improve the regulation of eating rate. Data from cyclic-ratio experiments support this contention (Figure 6.9). These results show only the regulatory limb of the response function, and the slope is indeed steeper for the more deprived animals. This result agrees with the commonsense argument that cost-of-deviation in eating rate should be higher for a food-deprived animal than for one close to its normal body weight. This result is not always found. Figure 7.18 shows ratio response functions obtained from a group of rats at 80% and 98% of their free-feeding weights. The difference between the response functions is almost entirely in the value of the free-level of eating (R_0).[24] In this experiment, each ratio was in effect for one session or more (i.e., the cyclic-ratio procedure was not used). Whether this difference in experimental method is responsible for the different effects of food deprivation in the two studies, or whether something else is involved, remains to be determined.

Reinforcement constraints

In the history of reinforcement theory much discussion has been devoted to limitations on the reinforcement principle.[25] Why, for example, is it easy to train pigeons to peck for food, but hard to get them to peck to avoid electric shock? Why is it relatively difficult to sustain treadle pressing with food reinforcement and almost impossible to reinforce wing flapping with food? A popular solution is to invoke physiological "constraints" that prevent the animal from making certain kinds of "associations." Since physiology underlies everything, and associations cannot be observed directly, this kind of account is not satisfying. It serves merely to push the problem back from a region where direct test is possible, observable behavior, to a different level, where, at present, it is not. Always be suspicious of a behavioral scientist invoking physiology: The appeal is likely to be motivated less by faith in physiology than by a reluctance to inquire further.

A different tack is to turn to the ecology of the animal. Pigeons, it is argued, have evolved in an environment where pecking at a potential food item is usually the best way both to test and to eat it. Wing flapping, on the other hand, is useful for avoiding predators and getting from place to place (especially over long distances; it is too expensive in energy cost to waste on short trips, as anyone watching pigeons scuttle out of the way of a car will have noticed). There has been no prior selection for wing flapping to get food. Perhaps, therefore, special links have evolved to "prepare" pigeons to peck for food.

There is much to this argument, although it runs into difficulties in explaining numerous tricks that animals can learn for which they have no obvious evolutionary preparation – seals balancing balls, bears riding bicycles, and so on. It also suffers from the weakness of any purely evolutionary argument about behavior: There is little to restrain it; it makes few testable predictions, and does not relate things that an animal finds easy to learn to things that it finds difficult. The ecological approach gives no clue to the mechanisms that must underlie both constrained and unconstrained learning. We will have to do better than just label things easy to learn "prepared" and things hard to learn "unprepared."

There is an alternative: that most of these effects can be explained by the properties of the animal's preference structure, by differences in the cost-of-deviation (CoD) of different activities. The argument is straightforward: The effectiveness of one activity as reinforcer for another is directly related to their relative CoD values: the greater the CoD of the contingent response relative to the instrumental, the more the instrumental response will change in response to a given deviation of the contingent response from its preferred level. The CoDs of various activities are presumably the outcome of past natural selection, so this account is not in conflict with a functional one; it is just more explicit. Learning (reinforcement) constraints follow from this idea because a low-cost activity makes a poor reinforcer for a high-cost one, both in terms of the absolute rate changes that can be achieved and, more importantly, in terms of the slope of the regulatory limb of the ratio response function. The effects of CoD value can be shown analytically in the following way.

Imagine the animal to be at its free-behavior point; consider the effect on the rate of the instrumental response of an increment in the ratio requirement. If, for a given ratio increment, the increment in instrumental responding is large, the sensitivity of the response to this reinforcer is obviously high; conversely, if an increment in the ratio requirement produces little change in the level of the instrumental response, the reinforcing effect is small – the response is insensitive to this particular reinforcer. Thus, the slope of the response function in the neighborhood of B_0 provides a measure of the *sensitivity* of a particular response to a particular reinforcer. Formally, it is convenient to define the quantity

$$S(u, v) = \partial x / \partial R(x) | u, v, \qquad (7.10)$$

that is, the partial derivative of x (the rate of instrumental response) with respect to $R(x)$ (the reinforcement rate), evaluated at a point u,v (here, the free-behavior point, so that $u = x_0$ and $v = R_0$) as a quantitative measure of sensitivity. When the absolute value of S is high, x is sensitive to reinforcement; when S is low, x is insensitive.

For the minimum-distance model, the expression for sensitivity turns out to be

$$S(x_0, R_0) = -[R_0(b + c) + cx_0]/[x_0(a + c) + cR_0], \qquad (7.11)$$

where a, b, and c are the CoDs of the instrumental, contingent (reinforcing), and other activities, respectively. Since all the terms are positive, the expression as a whole is negative, indicating the regulatory character of the system. If c, the cost of other activities, can be neglected, Equation 7.8 reduces to

$$S(x_0, R_0) = -bR_0/ax_0, \qquad (7.12)$$

so that the sensitivity measure is approximately proportional to the relative costs of the contingent and instrumental responses. Consequently, relative cost is a tolerable measure of the "reinforcibility" of one activity by another.

Hard-to-reinforce activities may, therefore, be high-CoD activities – but in that case they should be good reinforcers themselves. Conversely, low-CoD activities should be easily reinforced by many other activities, but should be poor reinforcers themselves. Presumably activities such as lever pressing, perch hopping, and key pecking, commonly used as instrumental responses, all fall into this category. Indeed, the cost dimension has precisely the properties of the older distinction between *appetitive* or *precurrent* activities and *consummatory* activities advanced by Sherrington and, independently, by the American comparative psychologist Wallace Craig. Appetitive activities are the variable, sometimes exploratory, activities that lead up to consummatory activities such as eating or copulation. Appetitive activities are variable and labile (that is, reinforcible), whereas consummatory activities are stereotyped and intractable. Thus, *appetitive* corresponds to "low CoD" and *consummatory* to "high CoD."

Are all apparent constraints on the effects of reinforcement attributable to an invariant preference structure, or are there indeed special links between some activities and special limitations on others? Such links or limitations should show up as inconsistent relations between sensitivities. For example, consider the sensitivity relations among three different activities, A_1, A_2, A_3. There are six possible relations, since three activities can be paired up in three ways, and each act can serve as both instrumental and contingent response. Thus, in principle, a set of six S values could be obtained, yielding six equations of the form $S(i, j) = -K_{ji}c_j/c_i$, where K_{ji} is the ratio of free levels of activities j and i, and c is the unknown cost of each activity.[26] Thus, there are three unknown costs and six equations. Inconsistencies will be revealed if the costs differ substantially depending on which three equations are solved.

The analysis is particularly simple if instead of sensitivities we use *elasticities*, the ratio of proportional, rather than absolute, changes. The *cross elasticity* for two goods x and y is defined as $(\partial x/\partial y)(y/x)$. Thus, from Equation 7.12, the elasticity of two activities on the ratio response function evaluated at B_0 is just b/a, the ratio of their costs. A reinforcible response, one for which b/a is large, will be highly elastic, showing a large proportional increase in response to an increase in ratio value. Conversely, an inelastic response will be poorly susceptible to reinforcement, showing only small increments. For our three activities, two comparisons, between A_1 and A_2 and A_1 and A_3, yield two elasticities: c_1/c_2 and c_1/c_3. From them, a third, c_2/c_3, can be obtained and compared with the results of a third experiment comparing A_2 and A_3.

Unfortunately, I have not been able to find any experiment in which systematic comparisons of this sort have been made. Nevertheless, there are numerous less-systematic studies showing apparent constraints of a sort quite compatible with a fixed preference structure. For example, male Siamese fighting fish, *Betta splendens*, will regulate their food intake almost perfectly when all their food must be obtained on a ratio schedule for which the response is swimming through a short tunnel.[27] Here, swimming is a highly elastic response. Conversely, *Bettas* increase their frequency of swimming through the tunnel only slightly when the reinforcement is sight of another male *Betta* (a stimulus to which they display): Regulation is poor and here swimming appears to be highly inelastic. Both these results are consistent with a lower CoD for display reinforcement than for food reinforcement. To show real inconsistency, a further experiment is necessary in which some other instrumental response, say fin spreading, is highly elastic for display reinforcement, and inelastic for food. In the absence of such a test, the differences between food and display reinforcement are easily handled by quantitative differences in a cost parameter.[28]

The differences between food and water as reinforcers seem also to be largely in the cost parameter. Lever pressing, as instrumental response, is generally less sensitive to water than to food reinforcement, for example. Heat and access to nest material by mice also behave like low-CoD reinforcers; whereas access to a running wheel is, for a rat, a relatively high-CoD activity almost as effective as food in maintaining lever pressing.

Perhaps the clearest demonstration of real constraints comes from attempts to reinforce "comfort" activities such as grooming. For example, Shettleworth (1975) studied the behavior of golden hamsters (*Mesocricetus auratus*) under free and constrained conditions and found that face washing occurred at a high level under free conditions, yet was easily displaced by other activities. Easy displacement implies low CoD, hence high sensitivity to food reinforcement. Yet another experiment showed face washing to be *in*sensitive to food reinforcement.

There is anecdotal evidence that birds will readily take flight to avoid an electrified grid, even though they can be taught to peck to avoid electric shock only with great difficulty. Yet pecking is easily entrained by food reinforcement, and wing flapping (if not actual flight) is poorly reinforced by food. Thus, although more systematic data would be desirable, there is little doubt that real constraints, not easily handled by a simple economic model based on diminishing marginal utility, do exist. Most effects termed *constraints* are mislabeled, however, being easily explained as cost differences.

The very idea of a constraint – on reinforcement, learning, or whatever – has meaning only in relation to some norm. It is only if a given reinforcer fails to affect a given instrumental response in the expected way that appeal is made to a "constraint." At one time it was thought that all reinforcers should act with equal facility on all responses. This idea was shown to be incorrect almost as soon as people began to look at responses other than the prototypical lever press and key peck, and at reinforcers other than food. We have just seen that many of the constraints thus revealed may reflect nothing more than CoD differences of a sort that might well have been anticipated, had psychologists been thinking in economic terms. Nevertheless, others remain that cannot be so accommodated: They are "real" constraints, relative to the norm of the minimum-deviation model – although they may not be constraints in relation to a more comprehensive economic model.

Still, no matter how fancy the model, some anomalies will probably always remain. Future work will have to decide whether these can usefully be accommodated within economic analyses – or whether a point has been reached where economics must give way to an understanding of the processes involved. The kind of analysis described in this and the preceding chapter is neither dynamic nor causal. It says nothing about the time course of adaptation to a particular procedure, or about the "decision rules" that animals may follow once they have adapted; nor does it account for metastable adaptations or the process of acquisition. For example, the time lag between the delivery of a reinforcer and the behavior on which it is contingent is vital, yet such delays are not adequately represented in a molar feedback function. At long VI values, the feedback function for a variable-interval schedule that delivers food 10 sec after the effective response is almost the same as the function for one that delivers the food at once, yet animals may not learn to respond when there is a delay and, having learned, will respond much more slowly. Observed behavior, whether optimal or not, is always the result of particular, and imperfectly understood, mechanisms for detecting correlations, computing time averages, and assessing stimulus covariation. No matter how subtle such mechanisms may be, there will always be some combination of circumstances for which they will be inappropriate. At some point, probably when they become excessively complex, economic models are no longer useful and must give way to a detailed understanding of the particular dynamic processes that allow reinforcers to exert their effects.

I take up these processes in later chapters, but before leaving this topic some other departures from simple optimal behavior should be mentioned.

Intermittent, brief, predictable feedings often induce food-related activities such as pecking, and even lever pressing. These effects are variously termed *autoshaping* or *superstitious* behavior, depending upon procedural details. They represent departures from a simple economic analysis, in the sense that these activities behave as complements to food, rather than substitutes, since they covary with the frequency of food delivery. They can be accommodated within regulatory models only by assuming that the free levels of some activities are dependent on the levels of others, or by changing other properties of the objective function.

There are several other kinds of activity that are induced or facilitated by intermittent reinforcer delivery, most notably schedule-induced polydipsia (excessive drinking) by rats on food schedules, and schedule-induced attack (on a conspecific or model thereof) by pigeons and other animals. These activities all behave like complements, rather than substitutes, for the reinforcer.

Animals can be induced to behave in maladaptive ways by special training procedures. For example, dogs given recent exposure to severe, inescapable electric shocks are unable without special help to learn a simple response to avoid shock. Restrained squirrel monkeys, trained to avoid shock by pressing a lever, can be brought to actually produce intermittent shocks. These violations of optimality theory depend upon limitations in the learning mechanism. As I have already pointed out, every adaptive mechanism has its limitations. They tend to be more elusive in the case of food reinforcement, perhaps because food is a natural reinforcer. Electric shock is obviously remote from animals' normal experience, and they are accordingly less flexible in dealing with it. Avoidance contingencies ("do this to prevent the future occurrence of that") are also intrinsically more difficult to detect than positive ones ("do this to obtain that"). These issues recur when I discuss the process of acquisition in Chapter 14.

SUMMARY

It used to be customary to discuss *motivation* and *learning* in different books, or at least in separate sections of the same book. This separation is not possible any longer. Regulatory processes are involved in the molar rates even of activities like wheel running and lever pressing, not usually thought of as "motivated" in the same sense as "consummatory" activities like eating and sexual activity. These regulatory processes can be described by relatively simple economic models that embody three major assumptions: that activities are constrained by reinforcement schedules and the availability of time; that the marginal value of an activity diminishes with its rate of occurrence; and that animals allocate time to activities in such a way as to maximize total value.

I have taken the *rate* of an activity as the dimension that is subject to diminishing marginal value, because lever-press or key-peck rate is usually the only thing the animal can vary to affect his eating rate, but this is not essential to the optimality approach. The fundamental idea is that whatever the animal can do that has consequences for him is subject to diminishing returns. For example, in some experiments by Collier and his associates (e.g., Figure 6.4), rats had two ways to affect eating rate: Frequency of access to the feeder was determined by the rate at which the rat pressed a lever on a ratio schedule; and duration of each meal was determined by how long the animal continued to feed once the feeder operated. Granted that both these activities are subject to diminishing returns, one might expect the animal to vary both, according to their relative costs, so as to maintain eating rate relatively constant.

This point, that an optimality analysis must consider *everything* that is under the animal's control, might seem trivial except that its neglect can lead to absurd conclusions. For example, in an experiment by Marwine and Collier,[29] one thirsty rat drank an average of 14.7 ml of water per bout on a FR 250 schedule. Yet on FR 10 the same rat drank only about .4 ml per bout. By so doing, the animal failed to get as much water as it physically could (given the capacity demonstrated on FR 250) and also failed to get as much as its free baseline level. This result can be taken as contrary to an optimality analysis only if the animal is assumed to be indifferent to drinking-bout length. Once we grant that the marginal cost of a drinking bout must increase with its length, the longer bouts on the higher ratio schedules make perfect sense.

Optimality analysis explains reinforcement in terms of relative costs: Activities with low costs-of-deviation from their optimal values are allowed to vary so that activities with high costs-of-deviation can remain approximately constant. Many learning constraints can be explained by CoD differences, without the necessity to invoke special links between activities. Some phenomena, such as schedule-induced behaviors, and easily displaced but unreinforcible activities such as grooming, cannot be explained in this way. No doubt more comprehensive optimality models could be found to accommodate these effects. But since the main virtue of economic models is that they explain numerous facts by a small number of assumptions, complexity diminishes their usefulness. At some point other approaches may prove better.

This chapter has intentionally dealt with a restricted range of situations: no more than three activities, and only one reinforcement schedule. The next two chapters extend the optimality approach to situations with more than one reinforcement schedule and look at applications to natural foraging behavior.

NOTES

1. *Regulation*. Motivational theorists spend much time arguing over the meaning of the term *regulation*, although it's not really clear why. A regulated variable is simply one that is subject to negative feedback. Thus, suppose variable y is a positive function of another variable x, $y = f(x)$, as body weight is a positive function of average eating rate. Suppose y is altered in some way – perhaps by a forced change in x (food deprivation), but perhaps by some other means that affects y directly (a metabolic drug, for example). When the perturbing influence is removed (ad-lib feeding is restored, the drug injections cease), if y is regulated with respect to x, the change in y will produce a change in x in the opposite direction – dy/dx is negative. If body weight has fallen, eating rate will rise; if body weight has risen, eating rate will fall.

The concept of *regulation* is precisely the same as the idea of *stable equilibrium*, described in Chapter 5.

2. For experimental results showing behavioral competition see reports by Dunham (1971), Henton and Iversen (1978), Hinson and Staddon (1978), Premack and Premack (1963), Staddon and Ayres (1975). For a theoretical discussion see Staddon (1977b), and Chapter 10. For theoretical discussions of regulation in the sense of this chapter see McFarland and Houston (1981), Staddon (1979a,b), Timberlake and Allison (1974), Timberlake (1980), and an earlier paper by Premack (1965).

3. See, for example, Staddon and Ayres (1975). Other experiments showing competition are Dunham (1977), Reid and Dale (1983), and Reid and Staddon (1982).

4. *Rate and time measures*. The time constraint for two recurrent activities can be stated formally as follows: Let each instance of activity one take an amount of time t_1, and similarly for activity two. If all the time is taken up with one or other of these activities, then, in an experimental session of unit length, $N_1 t_1 + N_2 t_2 = 1$, where N_1 and N_2 are the number of instances of activities one and two. For m mutually exclusive and exhaustive activities, the constraint is obviously

$$\sum_{i}^{m} N_i t_i = 1. \tag{N7.1}$$

This equation assumes that the time taken up by each instance of a recurrent activity is constant. This is true for many commonly studied activities such as licking, wheel running and lever pressing by rats, and key pecking by pigeons. It breaks down if the activity has different topographies at different rates, as in free locomotion, where animals show different gaits – walking, trotting, cantering, galloping – at different speeds. Some procedures (e.g., ratio schedules) induce pigeons to peck at very high rates, and then the topography of the response differs from pecking at lower rates: Each peck is briefer and less forceful, and movement elements present under normal conditions drop out. Time taken up by an activity is still proportional to its rate, but the constant of proportionality changes with changes in response topography.

If the response topography ("gait") is constant, then time taken up by a response will be proportional to the rate of the response.

5. Activities generally occur in relatively brief bouts, so that even a rare activity (in the sense that it occupies little of the total time) may nevertheless occur quite frequently. A high-probability activity need not be the first to occur, nor does it have to be "exhausted" before lower-probability activities can occur. Activities are generally intercalated; higher-probability activities are simply "on" for a larger fraction of the total time. If activities were not intercalated in this way, the concept of *average rates*, and the notion of a free-behavior point defined in terms of them, would have no meaning. The dynamics of behavioral sequences are discussed at greater length in Chapter 11.

6. This terminology, and the major theoretical arguments in the rest of the chapter, is presented more formally in Staddon (1979a).

7. There are some theoretical problems associated with the yoked-control procedure, because no two animals are identical (see Church, 1964). Even within-animal yoking is open to the standard objections to within-animal experiments in general (see Chapter 4), but as a practical matter yoking is often a useful way to tackle contingency problems (Dunham, 1977).

8. *Income and substitution effects.* The effect of a schedule constraint can always be partitioned into a portion due to restriction and the remainder, which reflects the contingency alone. The restriction component is the change in the rate of the instrumental response, x, that would have taken place had the level of the contingent response, $R(x)$, been artificially held to the value attained under the contingency. The true contingent effect is any increase above that level. If we consider available time as analogous to the *income* of a consumer, and the ratio value as equivalent to the *price* of a commodity, then this distinction between restriction and contingent effects is closely related to the economist's distinction between *income* and *substitution* effects. Other relations between economic concepts and optimal time allocation are taken up later.

9. This argument assumes a bidirectional contingency; that is, x units of B are required for access to y units of A, but these y units must then be used up before access to B is restored: For example, the animal must press a lever five times to get a pellet of food, but then the pellet must be eaten before he again has access to the lever. Under usual conditions only one of these conditions will matter: A starving animal needs no urging to eat the small portions of food delivered by a schedule of food reinforcement, but it will not press the lever much unless lever pressing is required for access to food. In most experiments, therefore, only one half the contingency need be enforced.

10. *Value, reinforcement, and Darwinian fitness.* The term *value* is used here in several senses. To an economist of the old school it meant subjective utility and connoted some kind of pleasurable response in terms of which the value of all "goods" was assessed. It proved impossible to arrive at a satisfactory cardinal scale of utility in this sense. More recently, economists seem to have agreed (one can never be sure with economists) that although magnitudes cannot properly by assigned to utilities, things can at least be rank ordered in terms of value: We can agree that half a loaf is better than no bread, even if not *how much* better. Much ingenuity has gone into showing that most of the predictions of classical utility theory are still derivable from this more parsimonious assumption. Useful reviews of classical utility theory are Hicks (1956), Friedman and Savage (1948), and Stigler (1950).

The modern theory of *revealed preference* treats value as something that can be inferred from (is "revealed" by) the subject's pattern of choices. In other words, value is not measured directly but inferred from behavior, under the general assumption that behavior is guided by a consistent, fixed *preference structure*. I have followed this kind of reasoning in the text, where it led to the notion of a positively accelerated cost function. The approach is scientifically useful when it is possible to derive the preference structure from a limited set of behaviors, and then use it to explain a much larger set. Samuelson (e.g., 1965) is responsible for the concept of revealed preference in economics.

For the biologist, utility can only mean "contribution to Darwinian fitness," but this elegant notion conceals much. Neither organism nor biologist can assess fitness directly. The organism can only evaluate outcomes in terms of some intermediate *currency* selected by past evolution as a good predictor of fitness. For example, in choosing among food grains, a bird cannot know directly which type will most favor future offspring. But if size is correlated with nutritional value, then an inbuilt tendency to select the larger grains will suffice. For such a bird, utility equals grain size. The indirect nature of the process is revealed by allowing the animal to choose between artificial grains, made equal in food value but of different sizes. If the animal retains its size preference, we can be sure of the signal value of size. If he does not, or if he gradually loses his preference, we must conclude that size is not the only signal available – perhaps the animal can learn to correlate the appearance of grain with its nutritional value (as chicks seem to be able to do under some conditions – see Chapter 13). If so, then nutritional value is the fitness signal – prima facie a more reliable one than size, perhaps, but indirect just the same.

The things we identify as reinforcers (more accurately: The preference structure discussed later in the chapter) are just the signals that evolution has identified as the most useful proxies for fitness.

In terms of the regulatory analysis, deviations of different activities from their free-behavior values should be scaled in terms of their probable importance for future reproduction, as estimated by past natural selection. For example, a given reduction in rate of feeding will generally be more of a threat to reproductive success than the same reduction in grooming; this difference might well be reflected in a greater weighted cost for eating-rate deviations.

For further discussion of the strengths and weaknesses of optimality theory applied to evolutionary problems see Maynard Smith (1978) and several chapters in the collection edited by Krebs and Davies (1978).

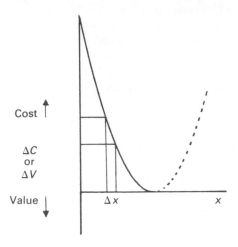

Figure 7.19. The equivalence between a positively accelerated cost function and a negatively accelerated value (utility) function.

11. This conclusion can be stated somewhat more formally as follows: Diminishing marginal value implies that dV/dx is a decreasing function of x or, equivalently, that d^2V/dx^2 is negative over the range of interest. By our intuitive assumption, dC/du is an increasing function of u, where $u = x_0 - x$. By the chain rule, $dC/du = -dC/dx$; therefore, $-dC/dx$ is an increasing function of $x_0 - x$, hence, for $x < x_0$, a decreasing function of x. Since cost is just the negative of value (if you give a man \$10, the cost is added in as -10 to your net worth), dV/dx is, therefore, a decreasing function of x, which corresponds to diminishing marginal value.

For example, let $C(x) = A(x_0 - x)^2$. Since $V(x) = -C(x)$, $dV/dx = 2A(x_0 - x)$, a decreasing function of x for $x < x_0$. These relations are illustrated graphically in Figure 7.19.

12. *Bliss point and characteristic space.* There are at least two possible interpretations of the parameters x_0, R_0, and y_0 in the quadratic cost equation. One is the interpretation assumed in the text, that these are just the free levels of the activities when all three are available; the point (x_0, R_0, y_0) then corresponds to the economists' "bliss" point, the ideal where all wants are satisfied. Unfortunately like all other utopian notions, this cannot be true. There can't be anything special about this point because the addition of a new activity, or replacement of one by another, will generally lead to changes in the free levels of the original activities. The free-behavior parameter must be a property of an activity in a context of other activities, not a property of each activity alone. The effects of changing the behavioral repertoire cannot be predicted from the simple minimum-distance model. An alternative interpretation, therefore, is that x_0, say, is just a property of behavior X that can be estimated from the data. It may correspond to the free level of X in the absence of any competing activity, or under paired-

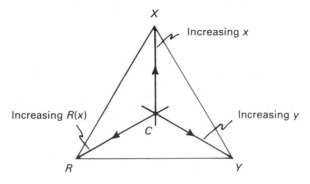

Figure 7.20. Triangular coordinates.

baseline conditions, but it need not – the value of x_0 for some activities may be negative or in excess of the total time available, for example. In this more general view, the true "bliss point" may not be physically realizable.

One way of dealing with the problem of predicting the effects of changes in the set of activities is to postulate a small set of fundamental motivational *character-istics*, V_i – a more sophisticated version of the old notion of primary drives or motives. Any given activity is then assumed to share to some degree in one or more of these characteristics, perhaps in a simple linear way, for example,

$$X_j = AV_1 + BV_2 + \ldots + KV_k,$$

where A K are weighting constants and V_1–V_k are the k characteristics shared in by activity X_j. The animal is then presumed to adjust the levels of the activities available to him so as to minimize deviations from a bliss point in this *character-istic space*. This approach gains generality at considerable cost in complexity, which is why I have elected not to discuss it in the text. The notion of a characteristic space has been explored in economics by Lancaster (1968) and in behavioral psychology by Lea (1981). The ethologists McFarland and Houston (1981) have independently followed a very similar line.

13. See, for example, Chiang (1974); Glaister (1972); and Baumol (1977). The quadratic model is treated explicitly in Staddon (1979a).

14. Points in plane *XYR* in Figure 7.8 are in triangular coordinates, which should be read as shown in Figure 7.20. Thus, points on any line at right angles to line RC represent constant levels of $R(x)$.

15. Most notably the *choice axiom* for independent alternatives proposed by Luce (1959, 1977). The axiom fails here because the alternatives, although independent in terms of the cost function, are not independent in terms of time taken up (the time-allocation constraint). Some results from human choice exper-iments are consistent with Luce's axiom, but data from behavior-allocation ex-periments of the type here discussed have not favored it. Figure 7.8 emphasizes

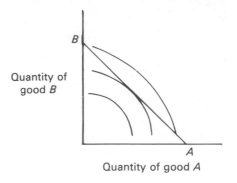

Figure 7.21. Indifference contours and the budget line.

the importance of thinking very carefully about what is meant by "independence" before applying theories that require it.

16. This section gives only the briefest survey of that portion of economic theory most relevant to the analysis of operant behavior. For more comprehensive treatment, see standard economics texts such as Walsh (1970), Baumol (1977), or the most recent edition of Samuelson's popular book. The classical treatment is by Marshall (e.g., 1925).

For recent discussions of the relations between economic theory and behavioral psychology see the volume edited by Staddon (1980), especially the chapters by Rachlin and Kagel et al. Papers by Rachlin, Green, Kagel, and Battalio (1976), and Rachlin and Burkhard (1978) are recent efforts to bring together economics and behavioral psychology.

17. These are the favored substances in a series of studies by Rachlin, Kagel, Battalio and Green. For rats they are imperfect substitutes.

18. Technically the marginal rate of substitution of two commodities is the negative of the slope of the indifference curve at that point. A related but more useful concept is the notion of *elasticity*, which is defined as the ratio of the slope of the indifference curve to the ratio of its absolute values: $\varepsilon_{yx} = (dy/dx)/(y/x)$. For cost functions of the general form $C(x_1, \ldots, x_k) = \Sigma a_i(x_{0i} - x_i)^m$, ($m = 2$ for the minimum-distance model), elasticity of substitution is always unity.

19. This third logical possibility, a curve concave to the origin, is permissible within the present analysis, but cannot occur in ordinary economic theory. The reason is illustrated in Figure 7.21, which shows the income constraint (*budget line, AB*) representing the boundary limiting the consumer's purchases of bundles composed of goods A and B. For example, if A costs \$1 per unit and B \$2 per unit, and income is \$100, then the equation of the budget line is $A + 2B = 100$, where A and B are quantities of A and B. Either preference increases away from the origin (as I have been assuming) or toward the origin. If away from the origin, then points *A* and *B* are both preferable to point *T*, the point of tangency between the budget line and the highest indifference contour (since indifference curves are contours, the assumption that preference increases away from the origin places the origin at the bottom of a utility "bowl," with points *A* and *B* both higher than point *T*). Hence, indifference curves of this form dictate that the consumer shall always purchase one of the two goods exclusively. Conversely, if preference increases toward the origin (the origin is the peak of a utility "hill"), the consumer can maximize utility by purchasing nothing at all.

If we retain the convention that preference increases away from the origin, then indifference curves like those in Figure 7.21 will always dictate exclusive choice. As a practical economic matter, this never occurs: Consumers do not spend all their income on one thing. In the more restricted conditions of behavioral experiments this outcome is still improbable, but is not totally excluded.

There is another important difference between the economics of the human consumer and the economics of behavioral allocation. Money income can be saved or spent, but time, the currency in behavioral allocation, is always expended. Consequently, we can expect that the economic *axiom of non-satiation* – that more is always better, or at least not worse – will regularly be violated, because more of any activity always entails less of others. This means that indifference contours for behavioral allocation will generally be closed curves, and the constraint imposed by a reinforcement schedule (analogous to the budget line) will always restrict behavior to a region away from the region of highest value. For example, the indifference curves for electric shock vs. lever pressing will resemble those of Figure 7.21, with the region of highest value close to the origin. The constraint imposed by an avoidance schedule is of the same general form as the budget line in the figure, but it restricts behavior to the region away from (rather than including) the origin. Optimal performance is, as usual, the point of tangency between this constraint line and the indifference curve closest to the region of maximum value, that is, to the origin.

20. Models of this form are also minimum-distance models, but the space is Euclidean only when $m = 2$.

21. The logical relations between utility and revealed preference are one of the deeper areas of mathematical economics. Questions considered are, for example, under what conditions does a utility function exist? Under what conditions is it unique? Under what conditions are preferences transitive (A preferred to B and B to C implies that A is preferred to C)? The classical treatment of these issues is by Von Neuman and Morgenstern (1947). More recent treatments are Luce and Raiffa (1957), Hicks (1956), Henderson and Quandt (1971), and Intriligator (1971).

22. The alert reader will perhaps have noticed that in a sense only two assumptions, competition and diminishing marginal rate of substitution, are essential for these predictions. After all, many forms of indifference contours are consistent with the diminishing marginal substitution assumption, and by judicious choice among these, quite close matches to the empirical data can be obtained (see Rachlin & Burkhard, 1978). Response independence is a necessary assumption only if we wish to rely on minimum-distance or some comparable additive rule to generate the preference structure.

23. For systematic data on tandem and conjunctive schedules, see Zeiler and Buchman (1979). Susan Motheral, in my laboratory, has obtained systematic data on a schedule related to the interlocking schedule discussed in Chapter 5. Tandem schedules require a fixed number of responses (the ratio requirement) after an interval requirement has been met. An approximation to the feedback function for these schedules can be derived by extending the argument used previously to obtain the feedback function for interval schedules. Thus, on an interval schedule, given random responding, the expected time between occurrences of reinforcement is $D(x) = 1/a + 1/x$, where $1/a$ is the interval value and x is the average response rate. On a tandem schedule, the quantity m/c, where m is the ratio requirement, must be added to this, yielding the feedback function $R(x) = ax/[x + a(1 + m)]$. In accordance with the data of Zeiler and Buchman (1979), this function predicts a bitonic response function; moreover, the minimum-distance model also predicts that response rate (at a given interval value) will be increased by the addition of a moderate ratio requirement, which is also in accord with their results.

A conjunctive schedule imposes a ratio requirement that can be met before the interval requirement has elapsed. It can be considered as a tandem schedule with a variable-ratio requirement smaller than the nominal value (because the animal makes contact with the requirement only when it fails to make the necessary number of responses during a particular interval). This assumption leads one to expect that the peak of the bitonic response function for this schedule should be at a lower ratio requirement than the maximum for the comparable tandem schedule, as Zeiler and Buchman report.

24. Response functions (not shown) were fitted by nonlinear regression, on the assumption that the free level of bar pressing was zero.

25. Several edited books on the general topic of "constraints on learning" have been published in recent years, as well as a number of influential review papers. See, for example, Hinde and Hinde (1973), Seligman and Hager (1972), and papers by Bolles (1970), Breland and Breland (1961), Shettleworth (1972, 1975), and Staddon and Simmelhag (1971).

26. I assume here that all activities are measured in time units. And for Equation 7.12 to be used, it is necessary that in each pairwise comparison the cost of "other" activities be negligible.

27. Hogan, Kleist, and Hutchings (1970). See Hogan and Roper (1978) for a recent review of many similar results.

28. Note that this analysis of reinforcement effectiveness is not the same as the so-called weak reinforcement hypothesis (discussed by Shettleworth, 1972, and others), which states that all reinforcers can be arranged on a continuum of strength, so that a "weak" reinforcer, like access to nest material, will behave just like food for a slightly hungry animal. The minimum-distance model characterizes each activity (reinforcer) by two parameters, a cost and a free-behavior level, not one as the strength view implies. If the equal-form assumption (Equation 7.3) is also relaxed, activities might differ in many other ways. There is no reason why the indifference contours relating one reinforcer and an instrumental response should be of the same form as those relating another reinforcer to the same response.

29. Marwine and Collier (1979). These figures, and the misplaced criticism of optimality theory, are due to Allison (1981).

8

CHOICE AND DECISION RULES

The study of choice is the study of the factors that make animals do one thing rather than another. In this sense all psychology is the study of choice. There is another meaning for choice, *conscious deliberation*, as when we mull over alternatives: Shall we go out and get a pizza, or cook at home? Is it to be medical school, grind, and the big bucks, or genteel poverty and the spiritual satisfactions of renaissance history? No doubt conscious deliberation does occur; but it is inaccessible in animals, and its causal status even in people is by no means clear – some argue that often the reason follows the choice ("rationalization") rather than the reverse, for example. So I'm concerned here just with the first meaning: An animal "chooses" response A only in the sense that it does A, rather than B, C, or some other thing.

The previous chapter was an account of choice in this sense. Two activities were linked by a contingency, and we looked at how their frequencies, and the frequencies of other activities, changed in consequence. There are other, more explicit, procedures for studying choice, however. Instead of rewarding one response and seeing how its frequency changes relative to others, we can reward two (or more), and see how the animal allocates his effort between them. The term *choice* is generally reserved for situations of this special sort.

Choice experiments are done for the same reason as experiments in which only a single response is reinforced: to find the rules by which animals adapt to reward and punishment. Choice experiments simplify things by pitting two or more similar responses against one another, and by rewarding each with the same thing (although not according to the same schedule or in the same amount). When an unrestrained animal shifts from one activity to another, it is choosing between different things: Running is different in kind as well as amount from eating or grooming. These activities are imperfect substitutes for one another. In most conventional choice experiments, on the other hand, the outcomes (usually food) are qualitatively the same. What differs between alternatives is the *means by which* food is obtained.

For example, one of the oldest choice procedures is so-called *probability learning*. An animal such as a rat, a pigeon, or even a goldfish, is confronted with two alternatives, which might be two arms of a T-maze, two keys, or two platforms to which it can jump (goldfish are poor at the latter task, but do quite well at the other two). In the simplest version of this procedure, food is assigned to one or other alternative on every trial with unequal, but complementary, probabilities. Thus, food may be available for choice A on 75% of trials (A is termed the *majority* choice) and choice B on 25% of trials. The animal must choose not between eating and some other activity, but between eating with this or that *probability*.

This is an example of a *discrete-trials* procedure: The animal is not free to respond at any time, but must confine his choices to trials that are specified by the experimenter. Here is another example: A hungry pigeon is confronted with two pecking keys; the keys are only illuminated and effective during a trial. A single peck on a key suffices to turn it off and, if the response is "correct," operate the food hopper. After the response, with its positive or negative consequence, the lights go out or the opportunity to respond is withheld in some other way; this is termed the *intertrial interval* (ITI). After the ITI, the lights come on again, and the next trial begins.

Omitting the ITI, and allowing the animal to respond at any time, converts this to a *free-operant* procedure. In this case, the two alternatives generally differ in the reinforcement schedule associated with them. For example, a popular procedure is to look at the rates of key pecking when one key provides food according to one value of variable-interval (VI) schedule, and the other key according to a different VI; this is termed a *concurrent VI VI* schedule.

There are three ways to look at choice situations: To see whether the changes in performance over time fit learning models; to look at the steady state, in molecular or molar terms; and normatively, to see if animals do what they should to maximize reward or behave optimally in some other way. I briefly describe each of these approaches and then spend the most time on the last two: steady-state molar and molecular performance, and optimality analysis.

Suppose you have a hypothesis about the effect of each eating episode (reinforcement) on the tendency to repeat the response that led to it, a quantitative version of Thorndike's law of effect (see Chapter 5). With an assumption about how the animal *samples* the alternatives before it knows anything about them, you can make predictions about changes in the relative frequency of A and B choices over trials. This is the most profound, and most difficult, kind of analysis, because what is sought is a model of the process by which choices are made and change in frequency with experience – a model of the *acquisition* of behavior. Despite its difficulties, until quite recently this was the dominant theoretical approach to choice.

A second tack is to set aside the problem of acquisition and just look at the steady-state pattern of behavior: What *decision rule* has the animal arrived at? In the simple probability-learning experiment just described, for example, most animals eventually fixate on the majority alternative, choosing A on every trial. Here the decision rule is both obvious and trivial, but in more complex situations, where choice is nonexclusive, the rule may be quite complicated.

Historically, there has been controversy about describing the behavior of animals by decision rules. Objections are of two types: to the implication of conscious deliberation, and to neglect of the learning process that this approach implies.

Some people object to decision rules because they seem to imply conscious deliberation. This objection is heard less frequently as rule-following machines have become commonplace – computer scientists can speak of machines that make decisions without being seriously accused of mentalism or imputing consciousness to silicon chips. An animal may follow a decision rule as a baseball follows Newton's laws of motion, with little reflection.

The second objection is concerned with the process of learning and the properties of steady-state performance. A computer can be programmed to follow a rule, but an organism in a choice experiment must develop the rule through learning. Some have felt that looking at the rule without worrying about how it comes about evades the problem of learning. But if animals show choice patterns that follow rules, realistic learning models must show how these patterns develop. Unfortunately, most learning models so far proposed are stochastic (probabilistic) and are not concerned with moment-by-moment behavior. For example, many assume that on each trial the animal chooses alternative A with probability p, and B with probability $1 - p$. Each reinforced choice is then assumed to increment the probability of responding to that alternative: $p(n + 1) = p(n) + \Delta$, where $p(n)$ is the probability of an A choice on trial n and Δ is the increment produced by reinforcement for that choice. Since probabilities must be less than or equal to one ($0 < p < 1$), Δ is usually defined in a way that limits the maximum value of p to unity, e.g., $\Delta = p(n + 1) - p(n) = w(1 - p(n))$, $0 < w < 1$, where w is a *learning rate* parameter. Correspondingly, on unreinforced trails, the increment is $p(n) = -vp(n)$, $0 < v < 1$, where v is the *extinction rate* parameter.

There is no room for a decision rule in such a stochastic model: The problem of moment-by-moment performance is solved simply by ignoring it. As we will see in Chapter 13, models of this type have been enormously fruitful through their predictions of molar steady-state performance. They are less useful as models of learning.

Lacking adequate learning models, we have little choice but to begin by looking at the end point, the properties of steady-state performance.

A decision rule is a *molecular* notion, it defines what the animal is doing moment by moment. A less ambitious approach to steady-state performance is to look for *molar* rules that describe choice. We may be uncertain about what the animal is doing from moment to moment, but perhaps his average behavior – the proportion of majority choices, for example – follows some simple descriptive rule, an *empirical law* of choice. (This average behavior often follows from a simple stochastic learning model.) *Probability matching*, which states that the steady-state probability of the two choice responses in certain discrete-trial probability learning situations matches the reward probabilities, is one such principle. The *matching law*, which states that in concurrent VI VI schedules the proportion of majority choices equals the proportion of majority reinforcements, is another.

Probability matching, once thought universal, is in fact found rarely; matching does occur, under restricted conditions I discuss in a moment.[1,2]

The third way to look at choice is in terms of the optimal behavior implied by each procedure: we ask, not What does the animal do? but What should he do? For example, if the objective is to maximize average reinforcement probability, then in the simple probability-learning procedure after an initial sampling period there is no point in responding at all to the minority alternative. Hence, the typical steady-state pattern, fixation on the majority choice, is also optimal. Obviously, if most choice behavior is optimal in some straightforward sense, much mental effort can be saved: Instead of remembering the detailed results of innumerable experiments, we need record only the general rule they all follow. The study of optimal choice and the search for decision rules (or molar empirical choice laws) are therefore complementary. Optimality principles can lead us to decision rules, and make sense of decision rules already discovered.

In the remainder of this chapter, I discuss choice from the point of view of decision rules and optimality.

Probability learning[3]

In simple probability-learning experiments, as we have seen, most animal species eventually follow the optimal strategy – exclusive majority choice. This result implies a general principle of enormous simplicity: At any point animals choose the best option available. Many find the simplicity of this rule appealing, but, alas, many others do not. In consequence, most attention has been devoted to procedures that yield nonexclusive choice, where even if this principle still holds, its operation is unobvious. These situations provide a graded rather than all-or-none dependent measure – the choice proportion; and their other features ensure that animals do not respond exclusively to one alternative or the other, but divide their responses between the two.

In what follows I attempt two things: explain what it is about these procedures that leads to nonexclusive choice, and show how the simple rule "at any point, choose the best option" – the hill-climbing rule discussed earlier in Chapter 2 – seems to underlie most of them.

As we have already seen, when animals are allowed to choose, trial-by-trial, between two alternatives rewarded with different probabilities, most end up choosing the majority alternative on every trial. There are two main ways to modify this situation so as to produce nonexclusive choice: *hold* procedures, and *single-assignment* of reward. In *hold* procedures, a reward once made available for a response remains until the response occurs; in *single-assignment* procedures, a reward is made available for one response or the other and no other assignment is made until the assigned reward is collected. In the simple probability-learning situation, where reward is always available for one of the two choices, these amount to the same thing. It turns out that if food once assigned to an alternative remains available until the correct choice is made (and no new assignment is made until that happens), animals will persist in sometimes choosing the minority alternative, that is, nonexclusive choice.

The way in which an animal chooses from trial to trial depends upon the species and the individual. Monkeys and sometimes rats will eventually develop a *lose–shift* pattern: After receiving food for a correct choice, the next choice is always of the majority alternative. If this is correct, the cycle begins again; if not, the next choice is the minority alternative. Since food is always assigned to one of the alternatives on every trial, this minority choice is always reinforced, and the cycle begins again.

This pattern is clearly the most effective one open to the animals: After a rewarded trial, the probability of food is highest on the majority choice, which is the one chosen. But if the choice is unrewarded, food is certainly available for a minority response.

Probability-learning-with-hold illustrates two concepts that will be useful in future discussion: An *initializing event* (IE) is any signal available to the animal that tells him the state of the programming apparatus. The optimal sequence is perfectly defined from that point. The IE in probability-learning-with-hold is food delivery, since the reward probabilities are always p and $1 - p$ on the next trial. The *optimal sequence* is the sequence of choices following an IE that maximizes either momentary or overall payoff (here both). Here the optimal sequence is obviously AB, where A is the majority choice: Food delivery resets the sequence, so that response A always occurs after food; but if no food follows the A choice, the next choice is B, which is always rewarded, and the sequence begins again.

There are two kinds of optimal sequence, *local* and *global*. Local optimizing is just choosing on each trial the alternative with the highest probability of payoff. Other names for local optimizing are hill climbing and *momentary maximizing*. As we saw in Chapter 2, hill climbing doesn't always find the globally best behavior, but in most commonly studied choice situations, it does very well: The difference between the local and global maxima are usually trivial. In probability-learning-with-hold, the local and global optimizing sequences are the same. In most situations, picking the best option on each trial will come close to maximizing overall payoff rate.

While many animals learn to behave optimally in probability-learning-with-hold, others do not. Goldfish, for example, do not develop the precise lose–shift pattern, although they do continue to respond to both alternatives, rather than fixating on one. Why not?

Learning an optimal choice sequence requires that an animal be guided by critical events, both external (IEs), such as food delivery, and associated with its own behavior, such as a particular choice. The decision rule for probability-learning-with-hold is "after food choose A, after an unsuccessful choice [which will always be A if the animal follows the sequence], choose B." Thus, the animal must be capable on trial n of being guided by what happened on trial $n - 1$. The ability to behave optimally therefore depends on the properties of *memory*. It is very likely that the inability of some animals to follow the lose–shift strategy in probability-learning-with-hold reflects memory limitations: The animals are confused about what happened on the previous trial, hence cannot use this information to guide their behavior on the current trial. Memory is a *constraint* (in the sense of the last chapter) that limits their ability to behave optimally.

We are not certain how memory limitations prevent goldfish from following the lose–shift pattern while permitting them to respond to both alternatives in the probability-learning-with-hold procedure. But the following scheme is a reasonable guess (I deal with memory at greater length in Chapter 12): Suppose that the animal is really a hill climber – it always goes to the place where food is most likely. To follow this strategy requires the animal to remember when it last got food: Was it on the previous trial, or an earlier trial? If it ever gets confused about this, it will respond incorrectly. For example, if it *didn't* get fed on the previous trial, but remembers it did, it will choose the majority rather than the minority. Conversely, if it did get fed, but remembers that it didn't it will choose the minority rather than the majority. But if these memory errors are at all frequent, the animal has no opportunity even to learn the optimal strategy in the first place. Consequently, its choice pattern will be a moment-by-moment mixture of *extinction* and *sampling*: It chooses some alternative where it has been rewarded, and then, if reward happens not to be available on that trial, persists for a while until its memory of past rewards for that choice grows dim, whereupon it chooses the other. With lapse of time, the first choice grows more attractive (so-called *spontaneous recovery*), and reward may not be available for the second choice – both factors will favor a return to the first choice at some time. Added to these systematic effects is the unsystematic effect of sampling (i.e., unpredictable variation) built in to ensure that the animal does not get trapped (cf. the discussion of behavioral variability in Chapter 3).

If we look just at choices after rewarded trials, animals that follow the optimal sequence look different from animals that do not: Animals that behave optimally always respond to the majority choice after a rewarded trial; whereas animals that do not will divide their choices between the two alternatives – often they divide them approximately in the same proportion as the rewards for the two choices (probability matching), but more usually, the majority choice is more favored. Goldfish tend to match, monkeys to respond exclusively to the majority. Monkeys were thus said to maximize, goldfish to match, but this is misleading. The argument I have just made suggests that both species are maximizing, but because of their different memory capabilities, they perform differently. There is no such thing as maximizing in an absolute sense, the term makes sense only in relation to a set of *constraints*. If the constraints are different, so, usually, will be the behavior.

Whatever the details, it is clear both that the hold procedure will act to

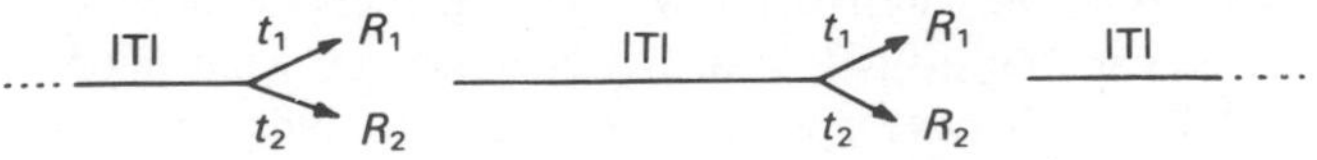

Figure 8.1 Procedure for studying the effects of delay and reward magnitude on choice. After an intertrial interval (ITI), the animal can choose between a small reward, R_1, with delay t_1 or a large reward, R_2, with delay t_2.

maintain nonexclusive choice, even in an animal lacking the capacity to learn the optimal sequence, and that the choice sequence resulting might well lack any simple pattern. Other limitations come from response tendencies that reflect particular apparatus features. For example, rats tend to pick a choice other than the one that was just reinforced (after all, in nature, once you have eaten the food, it is gone and usually will not soon reappear). Inability to overcome this tendency may account for failures to observe optimal sequences in some animals.

The general point is that either momentary maximizing under discriminative stimulus control, or cruder mechanisms involving extinction of unrewarded responses, will tend to produce nonexclusive choice in *hold* procedures. In more complex choice procedures where food is assigned to one or other alternative, but then only made available for a choice with some probability, the optimal pattern hovers around *matching* of choice proportions and payoff proportions. I return to matching in a moment.[4]

Delayed outcomes: "self-control"

In simple probability-learning procedures, reward immediately follows the correct choice. All that varies from condition to condition is the probability of reward for each choice. A modification that sheds considerable light on the nature of choice is to make reward available for either choice on each trial (i.e., abolish the probabilistic aspect), but vary reward amount and add a variable amount of delay to each choice. Such a procedure is illustrated in Figure 8.1. It has three main features; an intertrial interval (ITI), two reward amounts, R_1 and R_2, and two associated delays of reward, t_1 and t_2.

Procedures of this sort are sometimes studied because of their resemblance to human situations involving delayed gratification ("self-control"). If R_1 is larger than R_2, and t_1 is not proportionately larger than t_2, then in some sense it "pays" to choose the alternative with the longer delay, because the larger outcome more than compensates for the longer delay. For example, suppose R_1 is 6-sec access to grain and R_2 is 2-sec access, but t_1 is 4 sec whereas t_2 is 2 sec. Obviously, the expected rate of food delivery at a choice point is 1.5 sec/sec (6/4) for the large reward but only 1 sec/sec (2/2) for the small. In one sense, therefore, the "rational" rat should delay gratification, forego the immediate, small reward, and go for the larger, delayed one – thus satisfying both enlightened self-interest and the Puritan ethic.

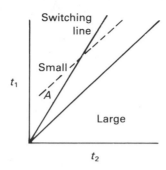

Figure 8.2. Optimal choice for the procedure shown in Figure 8.1. The switching line shows how the animal should weigh delays t_1 and t_2, on the assumption that reward rate over the trial period is maximized. Dashed line shows the effect on predicted choice of adding constant increments to both delays.

I doubt whether this arrangement has anything special to tell us about human delay of gratification, but animals exposed to it do behave more or less as this analysis suggests. For example, experiments with several variants of the procedure[5] have shown that if constant amounts of time are added to both t_1 and t_2, preference can be made to shift: When t_1 and t_2 are both short, the short-delay, small reward is preferred, but when t_1 and t_2 are both long, the longer-delay, larger reward is preferred. This follows because the expected food rate depends upon the proportions, R_1/t_1 versus R_2t_2 , and these change when constants are added to the times.

The formal analysis is as follows: If future rewards are not discounted (i.e., valued less the more delayed they are), the animal should be indifferent between the two alternatives when

$$R_1/t_1 = R_2/t_2, \qquad R_1 > R_2, \qquad (8.1)$$

that is, when the expected food rates are the same for each choice. Equation 8.1 can be rewritten as $t_1 = t_2(R_1/R_2)$, which defines a *switching line*, relating t_1 and t_2, shown in Figure 8.2. The slope of the line is the ratio of reward magnitudes, R_1/R_2, and the line divides the t_1–t_2 space into two regions: Above and to the left of the line, the animal should choose the small reward; below, and to the right, he should choose the large. Imagine that we begin the experiment with t_1 and t_2 chosen to favor the small reward – point A in the figure. Suppose we now progressively (perhaps a step each day) add a constant amount to both t_1 and t_2: This moves the point along the dashed line, parallel to the diagonal, in the figure. It is easy to see that the point must eventually cross the switching line, so that preference should switch from the small to the large reward, as is usually found.

In another experiment, the two delays, t_1 and t_2, were made equal and the procedure was modified slightly to ensure that the animals did not prefer one choice exclusively (i.e., the procedure allowed a graded measure of choice);[6] the effect on preference for the large reward of increasing both delays could then be measured. Increasing both delays, while keeping them equal, is equivalent to movement along the diagonal in Figure 8.2. Because the diagonal diverges from the switching line, such a shift should obviously increase preference for the large reward, and it did.

The strategy I have just described is not as rational as it seems. Equation 8.1 is appropriate only if the animal ignores everything but the trial time. The delay experiment can be interpreted in three ways: (a) As a one-time thing: What should the animal do if it is allowed to make the choice between the two delays only once? This is strictly a thought-experiment, of course, because choices are invariably repeated in these experiments – we cannot instruct the animal about the contingencies in any other way. Obviously, the animal should *always* go for the larger reward when offered a one-time choice – unless the delays are so great that they make up a substantial fraction of its expected life span. We would expect no change in preference over the range of delays typically studied, a few seconds or so. Evidently this is not a realistic view. (b) Granted repeated trials, the animal might ignore everything but the trial time. Something like this is tacitly assumed in many discussions of these experiments, but of course it cannot be correct. Food is necessary for metabolism, and metabolism goes on all the time. An animal that assessed its eating needs only in relation to external circumstances could not long survive. So the third alternative, (c) that the animal takes into account the total rate of food delivery, is most likely to be correct. How should he take account of both trial and intertrial time?

The simplest way is by a rule that takes account of the overall average food rate given that he makes one choice or the other exclusively. If we denote the ITI duration by T, then from Figure 8.1 the average reward rate, R_A, given that he chooses alternative 1 exclusively, is obviously

$$R_A = R_1/(T + t_1).$$

Hence the point of indifference between the two choices is

$$R_1/(T + t_1) = R_2/(T + t_2),$$

which can be rewritten as

$$t_1 = (R_1/R_2)t_2 + (R_1/R_2)T - T.$$

If we rewrite $(R_1/R_2) = w$, for readability, this can be rewritten as

$$t_1 = wt_2 + T(w - 1), \quad w \geq 1 \qquad (8.2)$$

Equation 8.2 is another straight-line switching line, but with an intercept greater than zero on the t_1 axis. We can draw it in the t_1–t_2 space, as shown in Figure 8.3. As before, the region above and to the left of the line denotes exclusive choice of the small reward, R_2, the region below and to the right, exclusive choice of the large reward, R_1. This new switching line has similar

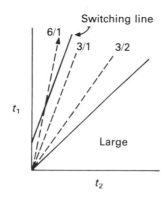

Figure 8.3 Optimal choice for the procedure shown in Figure 8.1, on the assumption that total reward rate is maximized. Reward magnitudes are assumed to be in the ratio $R_2/R_1 = 3$. The dashed lines show the predicted effect on choice of maintaining the delay ratio t_1/t_2 at 6/1, 3/1, or 3/2, while increasing the absolute values of the delays.

properties to the simpler one in Figure 8.2, and explains the effects of increasing t_1 and t_2 in the same way. But it also explains other things.

The three dashed lines in Figure 8.3 denote experimental manipulations of the ratio of delays, t_1/t_2, tried in a pigeon experiment by Green and Snyderman.[7] They held the ratio of delays constant for each line, but progressively increased the absolute value of the delays and measured the effect on preference. The steepest line is for a delay ratio of $t_1/t_2 = 6$, the next for a ratio of 3, and the shallowest for a ratio of 1.5. The ratio of reward durations (seconds access to a food hopper) was held at 3, which is the slope of the switching line.

Green and Snyderman found that at the 6:1 delay ratio, as the absolute values of the delays increased (in the direction of the arrow), preference shifted toward the smaller reward; whereas a similar absolute-delay increase with the 3:2 ratio produced a preference shift in the opposite direction, toward the larger reward. These shifts correspond to the opposite deviations of the 6:1 and 3:2 delay-ratio lines from the switching line.

My simple model predicts that a change in the absolute value of the 3:1 delay ratio should have no effect on preference, because this line is parallel to the switching line. Green and Snyderman in fact found a shift toward the smaller reward, which suggests that the "real" slope of the switching line is somewhat less than 3:1 – just what we would expect if the more-delayed, large reward is somewhat discounted in value.

Pigeons and rats (and no doubt other species whose choice performance has not been as well studied) seem beautifully adapted to these choice procedures, sensitively adjusting their preferences in accordance with the overall food rate to be expected from different patterns of choice. Unless special provision is taken to favor nonexclusive choice, the animals' decisions rule is simply to choose exclusively, according to the switching line. Discounting of delayed outcomes looks like a limitation, but it makes good adaptive sense – the future is always uncertain, and hunger increases unchecked with time, so delayed food is worth less than present food.

We don't really know how the animals determine the switching line, how they estimate the average food rates to be expected under different strategies. Indeed, it is not at all clear that the switching-line analysis has anything much to do with the mechanisms underlying the behavior. Nor do we know what determines the discount function. The answers to both these questions are obviously related to memory in some way, since the animal's only way to predict the future is by remembering what happened in the past. If the same properties of memory account for other things, such as the limitations on spatial learning or the learning of delayed discriminations, then we will be a little closer to our ultimate goal, which is a limited set of principles that explain everything that an animal can do. I return to these questions in later chapters.

I have so far simplified things by assuming exclusive choice. Most choice experiments are designed to ensure nonexclusive choice, however. I show next how the results of these experiments nevertheless follow the same set of principles.

MATCHING AND MAXIMIZING

In nature, food is rarely distributed evenly through the environment; more commonly it occurs in localized regions of high density – patches. Examples are seeds in pine cones, colonies of insects, a bush of favored browse. An individual meal, the zebra the lion has for lunch, for example, is a patch in this sense. The longer the animal spends looking for food in a patch, the more food he is likely to get, although the *rate* at which he gets it is likely to decrease with time as the patch becomes depleted. The relation between amount of time spent and food obtained is just the *feedback function* (see Chapter 5) for the patch.

Suppose that a hungry animal has a fixed amount of time to allocate between two types of depleting patch, "rich" and "lean." This situation is illustrated in Figure 8.4: Cumulative food intake is plotted as a function of time in the patch. Both functions are negatively accelerated because additional food becomes harder and harder to get as food density drops due to depletion; the asymptote (total amount of food in a patch) is higher for the rich-patch type. How should an animal allocate its time between these two kinds of patch? What proportion of time should it spend in each, and how often should it switch from one type to the other? Since the feedback functions in Figure 8.4 are negatively accelerated, we know (see Chapter 7) that optimal behavior will be nonexclusive: The animal should spend some time in both types of patch. To maximize overall rate of payoff, the marginal rate of payoff for each alternative should be the same. It is fairly straightforward to show, in addition, that the marginal rate should also equal the average rate of return from the environment as a whole. This result has been termed the *marginal value theorem* (Charnov, 1976).

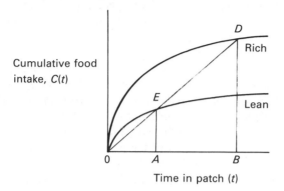

Figure 8.4. Cumulative food intake as a function of time in a patch for "lean" and "rich" depleting patches.

The proportion of time that should be spent in each patch type depends on the form of the depletion (feedback) function. We can get some idea of the proportion by picking a plausible form such as a hyperbola or a power function. Let the cumulative food intake be $C(t)$ for a time t in a patch. Then a flexible form is the power function

$$C(t) = At^s, \quad 0 < s < 1, \tag{8.3}$$

where A and s are constants; if s is less than one, the curve is negatively accelerated. Taking the first derivative to find the marginal value yields

$$dC/dt = Ast^{s-1},$$

which can be rewritten

$$dC/dt = sC(t)/t, \tag{8.4}$$

that is, the derivative of the cumulative food intake function, $C(t)$, can be written as a function of the ratio $C(t)/t$. From the marginal value theorem, dC/dt should be the same for all patches; thus, for two patches 1 and 2, $dC_1/dt_1 = dC_2/dt_2$. If s, the exponent of the depletion function, is the same for both patches, equating the marginal values yields

$$C_1(t_1)/t_1 = C_2(t_2)/t_2,$$

or

$$C_1(t_1)/C_2(t_2) = t_1/t_2, \tag{8.5}$$

that is, the ratio of the times the animal spends in both types of patch is equal to the ratio of the amounts of food obtained; this is termed *matching*. If s is not the same for both patches, then choice proportions should be proportional (not equal) to payoff proportions; this is termed *biased matching*.

This example is an instance of a more general proposition: If animals are acting so as to maximize their rate of payoff, then in situations where the feedback function shows diminishing returns, choice ratios will often match (or be proportional to) payoff ratios.

Concurrent VI VI. Matching is quite a general finding in operant choice experiments, but the account I have just given does not reflect the actual history of the principle. Matching was discovered in a situation that is the logical complement of the example: In patch foraging, the marginal payoff from the patch the animal is in decreases with time, whereas the payoff for switching to another patch stays constant. In the complementary case, the payoff for staying where you are is constant, but the payoff for switching increases with time. This describes concurrent variable-interval variable-interval schedules, where matching was first proposed as a general principle.[8]

Historically, the concurrent VI VI procedure had appeal because on a VI schedule the rate of reinforcement varies little with response rate, providing response rate is high enough. Thus (the argument went), any variation in response rate should be a pure measure of the "strengthening" effects of reinforcement, free of the strong feedback interactions between responding and reinforcement characteristic of, for example, ratio schedules.[9] The no-feedback assumption is not wholly accurate, and the reinforcement-as-strengthening theory that underlay the original use of concurrent VI VI finds fewer adherents these days. As we saw in Chapter 5, there is a fairly well-defined molar feedback relation even on VI schedules; and as I argued in Chapter 7, this relation appears to underlie many, perhaps all, the distinctive molar properties of interval and ratio schedules.

The matching law grew out of the belief that reinforcement contributes to something termed "response strength," for which interval schedules offered a convenient measuring rod. The notion of response strength has not been given up (see Chapter 2) – it is a useful concept in the analysis of behavioral competition in discrimination situations, for example (see Chapter 11) – but its significance has changed. I am inclined to look at response strength as a convenient fiction that provides an intuitive basis for a number of descriptive laws. Here I discuss matching in quite a different way, as something that naturally derives from mechanisms that tend to maximize animals' rate of access to food and other reinforcers.

The simplest variable-interval schedule is one in which the availability of reinforcement is determined by a random process (this is also known as a *random-interval* schedule). The way a random-interval schedule operates is illustrated in Figure 8.5, which shows three "time lines." The top line shows time divided up into brief discrete increments, indicated by the short vertical lines. Imagine that during each discrete interval a biased coin is tossed, yielding "heads" (1) with probability p and "tails" (0) with probability $1 - p$, where 1 indicates that food is available for the next response. As the brief time increment approaches zero, this procedure increasingly approximates a true *random-interval* schedule. The second time line in the figure shows responses, which can occur at any time. The

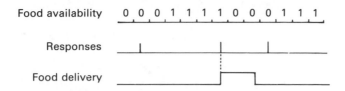

Figure 8.5. Reinforcement contingencies on a random-interval schedule.

third line shows food delivery (reinforcement), which occurs immediately after the first response following a 1 (notice that interval schedules are "hold" procedures, in the terminology used earlier, since food once "set up" remains set up until a response).

Marginal rate of payoff here is equal to the *probability* that a response will yield food. This probability will obviously increase with time since the previous response, because the longer the time, the more likely a "1" will have occurred – and once a 1 has occurred, food remains available until a response occurs. It is easy to work out what this probability should be, since it is equal to 1 minus the probability that *no* 1 occurred during the elapsed time. Let n be the number of time periods since the last response; during each time period, the probability of a 0 (i.e., no 1) is $1 - p$. Each time period is independent, so that the probability that *no* 1 has occurred during period n is equal to $(1 - p)^n$. The probability that a 1 *has* occurred is just equal to 1 minus this probability, hence the probability of food n time period after the previous response is given by

$$P(F|n) = 1 - (1 - p)^n. \qquad (8.5)$$

Since $1 - p < 1$, as n increases, the quantity to be subtracted from 1 gets smaller and smaller; hence the probability a response will get food increases the longer it is since the last response. As the discrete time interval becomes vanishingly small, Equation 8.5 is increasingly well approximated by the exponential function, familiar from earlier chapters, so that

$$P(F|t) = 1 - \exp(-\lambda t), \qquad (8.6)$$

where t is the time since the last response and λ is a parameter equal to the reciprocal of the mean time between VI "setups," that is, 1/the VI value. An example of Equation 8.6 is shown in Figure 8.6.[10]

Suppose an animal is following the hill-climbing rule (momentary maximizing[11]) described earlier; that is, whenever it makes a choice, it picks the alternative with the highest probability of payoff. What form will its behavior take? Clearly the best choice will depend on the relative times that have elapsed since choice A versus choice B. For example, suppose that A and B choices are reinforced according to (independent) VI schedules of the same value (i.e., the same λ). Then momentary maximizing says the animal should always make the response made least-recently – which implies alternation between the two

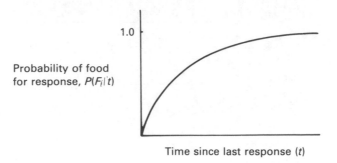

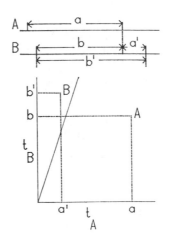

Figure 8.6. Probability of food delivery for a response after time t on a random-interval schedule.

Figure 8.7. Clock-space analysis of the (random) concurrent VI VI schedule. Event lines at the top show a sequence of A and B choices. The graph at the bottom shows the last A and B choices plotted in a space defined by the times since the preceding A and B choices.

alternatives. If the two VI schedules are unequal, the rule is just: choose A if $P(F|t_A) > P(F|t_B)$. From Equation 8.6, after some rearrangement, this leads to the simple *switching rule*, choose A if (if and only if)

$$t_A > t_B(\lambda_B/\lambda_A). \tag{8.7}$$

To see if choices conform to this rule, it is convenient to represent them in a *clock space* whose axes are the times since the last A and B responses, (i.e., t_A and t_B). Equation 8.7 defines the switching line $t_A = t_B(\lambda_B/\lambda_A)$ in such a space. Figure 8.7 shows two responses to each of two alternatives, and illustrates

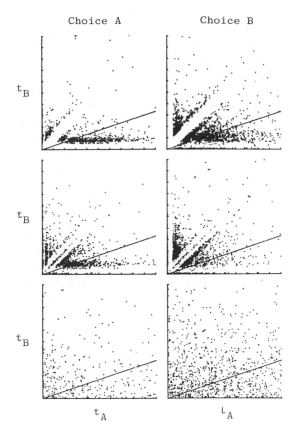

Figure 8.8. Top panels: minority (left panel) and majority (right panel) choice key pecks by a single pigeon during a 1-hr session of a concurrent VI 1 VI 3-min schedule, plotted in the clock space of Figure 8.7. Each dot is a single peck and the axes are the times since the preceding *B* (majority) or *A* (minority) choice. Line through the origin is the momentary-maximizing switching line. During this session, choice and reinforcement proportions matched well. Middle panels: data from the same pigeon during a session when choice and reinforcement proportions did not match well. Bottom panels: simulated choice data, derived from two independent, random processes with response probabilities in 3:1 ratio. Data points are not aligned with respect to the 3:1 switching line (from Hinson & Staddon, 1981).

how they are represented in a clock space. Momentary maximizing requires that all B-responses fall between the switching line and the B axis, all A-responses between the switching line and the A axis. Given an animal well-trained on concurrent VIa VIb (where $\lambda_A = 1/a$ and $\lambda_B = 1/b$) we can plot each response the animal makes in such a space and see how closely its choices conform to this rather stringent rule.

Figure 8.8 shows such data. Each pair of panels represents a single experimental session and each point represents an individual key peck. So that A and B choices can be distinguished, the left panels just show A (minority) choices, the right panels, B (majority) choices. The top two pairs of panels are data from a single, well-trained pigeon responding for an hour or so on a concurrent VI 3 VI 1-min schedule; the switching line is drawn in each panel. The bottom pair of panels show simulated data based on the rule that the animal chooses to respond at random intervals of time (where *random* is defined as in Figure 8.5). There are two differences between the real and simulated data. First, the pigeon shows bands where no responses occur; these depopulated regions correspond to the fixed minimum time it takes for the bird to switch from one response key to another, *switch time*. Second, the pigeon data in the top pair of panels show a strong tendency (confirmed by statistical techniques more refined, if not more compelling, than an eyeball test) for majority choices (B) to cluster above and to the left of the switching line, and minority choices (A) to cluster below and to the right of the line – as the momentary-maximizing hypothesis predicts.

The random data (bottom two panels) fall equally on both sides of the switching line. Although the random probabilities were chosen so that the overall choice proportions would show good matching to payoff proportions, there is no tendency for majority and minority choices to align themselves on either side of the switching line. In other words, obedience to matching does not force conformity to momentary maximizing.

The middle pair of panels show data from the same pigeon as in the top panels, but on a day when he was not matching well (the proportion of minority choices was .30 for the top panel vs. .49 for the bottom; the corresponding payoff proportions were .26 and .29): The alignment of choices with respect to the switching line is poor. It turns out that when pigeons (momentary) maximize well, choice proportions approximately match reinforcement proportions.

The close relation between matching and momentary maximizing is illustrated in Figure 8.9. The figure shows the trajectory traced out in clock space by an imaginary animal that responds every second, and always momentary maximizes. By convention the first response (choice A) occurs at point 1,1. Since the first response is an "A," t_A is instantly reset to zero, as indicated by the leftward dashed arrow, but then both t_A and t_B increase along the 45° line shown (i.e., both times increase at the same rate). After a further second, the representative point is still on the "A" side of the switching line, and another A response occurs; t_A is reset again and the process repeats. By this time, the continued growth of t_B (time since the last B response) has brought the point to the switching line. Our perfect momentary maximizer is now indifferent between A and B, but by convention, we let him make one last A response. The next repeat of the process finds the point on the "B" side of the line, and response B occurs, resetting t_B to zero and leaving the point at 1, 0. Continuation of the process causes the point to track out the triangular trajectory shown in the figure.

Three things are noteworthy about Figure 8.9: First, momentary maximizing obviously entails nonexclusive choice – no matter where the switching line, exclusive responding to one alternative will cause the representative point even-

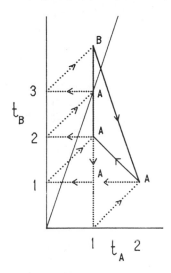

Figure 8.9. Trajectory traced out in clock space when responses are made at 1-sec intervals and choices follow momentary maximizing.

tually to cross the line, forcing a response to the other alternative. Second, the proportion of A's and B's is obviously closely related to the slope of the switching line, hence to the scheduled proportion of payoffs for the two alternatives: the steeper the line, the larger the proportion of A choices entailed by momentary maximizing. Simulations have shown that the commonest result will be *undermatching*, that is, the proportion of responses to an alternative will be somewhat closer to indifference than the proportion of rewards for such responses (e.g., if payoff proportion is 3:1, response proportion will be between 3:1 and 1:1). Third, under most circumstances, the animal should never make more than one response to the minority alternative before switching back to the majority. Choice data conform quite well to these predictions.

Concurrent VI VR

Another procedure that produces nonexclusive choice is concurrent variable-interval variable-*ratio*. This situation can be analyzed in exactly the same way as concurrent VI VI. A variable-ratio schedule is nothing but a one-arm bandit of the Las Vegas variety: Payoff is strictly proportional to number of the lever presses; *probability* of payoff is constant, equal to one over the mean ratio value, M. From Equation 8.6, the condition for switching from the ratio alternative, B, to the interval one is therefore

$$1 - \exp(-\lambda_A t_A) > 1/M,$$

which reduces to

$$t_A > [1n(M/(M - 1))]/\lambda_A, \tag{8.8}$$

so that the switching line is parallel to the t_B (ratio-schedule) axis.

If animals respond (i.e., make choices) at a steady, high rate, two predictions follow from this analysis. First, that most responses will be to the ratio alternative, because it takes a time equal to several interresponse times before the probability of payoff on the interval alternative grows (according to Equation 8.6) to a level equal to the constant ratio payoff probability. Second, there will be perfect matching between choice ratios and reinforcement ratios for the two choices. This is easy to demonstrate: If switching to A occurs when $P(F|t_A)$ is approximately equal to $P(F|t_B)$, then since R_A, the reinforcement rate for A, is equal to $P(F|t_A)$ times x_A, the response rate to A (assuming that A choices occur at fixed intervals), and similarly for B, it follows at once that $R_A/x_A = R_B/x_B$, that is, perfect matching. Given reasonable assumptions about overshoot – an A response occurs only after $P(F|t_A)$ exceeds the fixed $P(F|t_B)$ by some proportion – the expected result is biased matching, favoring the ratio alternative. All these features, a high rate of responding overall, faster responding to the ratio alternative, and biased matching, are characteristic of experimental results with this procedure.

Thus, matching, undermatching (the commonest deviation from simple matching), and biased matching on concurrent VI VR schedules all follow quite simply from momentary maximizing. Active inquiry, and argument, continues, but recent results make it hard to avoid the conclusion that matching on concurrent schedules is derived from a hill-climbing process.

OVERALL MAXIMIZING

The molar feedback relation between response and reinforcement rates on VI schedules depends on both the average rate and the temporal distribution of responding. Response rate is a single dimension, but the temporal distribution of responses can vary in many ways, and we don't know what constraints to put on temporal variation. Without knowing all the constraints, it is not possible to see whether a particular allocation of responding between VI, or VI and VR, alternatives is the one that maximizes total reinforcement rate. If we also include in the objective function the demands for time of other, nonsubstitutable activities (see Chapter 7), the position is even less clear. Nevertheless, we can draw some qualitative and even (given suitable simplifying assumptions) quantitative conclusions.

In Chapter 7, I derived nonexclusive allocation of time to different activities from the fundamental economic axiom of diminishing marginal utility (marginal rate of substitution): The more you have of something, the less valuable each additional increment becomes. The decreasing marginal utility here is a reflection of changes in the individual's internal state, a sort of *satiation*. The situations discussed in this chapter have the same diminishing-returns property, except that here it is the relation between a fixed amount of incremental effort expended and the amount of incremental payoff produced that is decreasing. For example, in patch foraging much more food is obtained during the first second spent looking in a patch than during the tenth second, and more during the tenth than the twentieth. Concurrent VI VI can be thought of in a similar way. Consider the total payoff for an animal responding on a concurrent VI 1 VI 3-min schedule. Suppose that initially the animal spends all its time responding to the VI 1 alternative. Payoff will be about 60/hr if response rate is relatively high. Suppose that the animal now allocates a little time, say 5 min per hour, to the VI 3 alternative. If this time is spread pretty evenly through each hour, the result will be a substantial increase in overall payoff rate, to perhaps 75/hr. A further 5 min spent on the VI 3 will produce a much smaller additional increment, however, and, pretty soon, further increments will produce decreases in overall rate of payoff. A similar argument can be made for concurrent VI VR as the animal allocates more and more time to the VI. At some point in both these situations a single response is likely to produce food with the same probability on each alternative; at this point, marginal values are equal and the animal is maximizing overall payoff. Thus, optimality arguments suggest that the "hold" situations discussed in this chapter, like the allocation of time among nonsubstitutable behaviors discussed in the last, should yield nonexclusive choice – and for the same reason, diminishing marginal returns.

The range of conditions under which this process will also lead to matching is less well understood – although it is relatively straightforward to find conditions where overall maximizing also implies molar matching.

There is an obvious resemblance between hill climbing and the equation of marginals that is a necessary condition for overall maximization. But as we have already seen, momentary maximizing does not always yield the overall maximum. This is partly because responses occur at finite, not infinitesimal, intervals, and partly because the technique can only find local, not global, maxima. The difference between momentary maximizing and a hypothetical optimal strategy is usually, though not invariably, small.[12] In all cases so far examined, the momentary-maximizing pattern yields overall results closer to matching, and thus to empirical results, than the optimal strategy.

Parameter estimation

An animal must settle three issues if it is to follow a momentary-maximizing strategy: (a) It must decide *when* to make a choice; (b) it must assess the relevant variables (such as postresponse times) correctly; and (c) it must in some way assess the relevant schedule parameters (such as scheduled reinforcement rates), so as to correctly estimate the switching line.

The first issue lies outside momentary maximizing itself: The switching-line analysis says nothing about when a choice must be made, only *what* choice should be made, given that a response is about to occur. The second issue can be studied directly by looking at how well animals can discriminate time and re-spond differentially to past events, for example. It turns out that they discrimi-nate time well, but remember individual past events quite poorly – more on this in Chapter 12. The third question is the most difficult. We have no real idea how animals assess the parameters of (for example) variable-interval schedules, al-though all indications are that they do so with considerable precision.

Parameter estimation can make the difference between strategy that sacrifices short-term gains to long-term losses and one that maximizes overall payoff. For example, in the delay experiments discussed earlier, we saw that pigeons seem to take the intertrial interval into account in their choice; they don't just look at reward size and delay, but also take into account the overall rate of payoff associated with different strategies: Their behavior is better predicted by the switching line in Figure 8.3 than the line in Figure 8.2. The animal seems always to pick the best option available, but its estimate of the best usually takes account of more than just the very next event.

This is true even in concurrent VI VI schedules. For example, a common modification in free-operant choice is the *changeover delay* (COD); this is a feature that prevents immediate reinforcement for a switch: If the animal's last response was A, any reward set up for choice B can only be collected for a B response at least t seconds after the first A-B switch, where t is a second or two. The COD means that the probability of reinforcement for a B response after an A (or an A after a B) is zero. An animal that followed momentary maximizing literally would never switch. Imposition of a COD does reduce rate of switching, but not to zero. Hence animals must take account of more than the first postswitch response in estimating the switching line. Most momentary-maximizing errors in concurrent VI VI take the form of overestimating the majority VI value: Animals seem usually to switch late to the minority choice. It turns out that this bias improves overall maximization: Momentary maximizing based only on the next reinforcement (like the delay switching line in Figure 8.2) doesn't maximize overall reward rate (although the difference is small in the concurrent VI VI case). Evidently animals are able to incorporate at least some global information into their switching strategies so as make their performance more efficient than a simple one-step-at-a-time momentary-maximizing strategy. We still know very little about how they do this.

SUMMARY

There are two kinds of choice situation: situations that tend to produce exclu-sive choice, fixation of responding on one alternative or the other; and situations that tend to produce nonexclusive choice, distribution of responses between alternatives. Experiments with the first kind of situation have focused on maxi-mization (picking the best alternative) as the major law of choice; experiments with the second kind of situation have focused on matching (of reward and choice proportions) as the fundamental law of choice.

Situations that assign reward to one alternative or another (single-assignment) and make no new assignment until the reward is collected tend to produce nonexclusive choice, as do *hold* procedures, which keep reward available until the correct response occurs. Situations lacking these features tend to produce exclusive choices. I argue that the same principle – always choose the alternative most likely to pay off (*hill climbing*) – accounts for both exclusive and nonexclusive choice. Hold and single-assignment procedures favor nonexclusive choice because the relative payoff for any alternative increases with time so long as it is not chosen. Such procedures constantly drive the animal away from exclusive choice.

A second theme of this chapter has been the relation between overall (global) maximization and momentary (local) maximization. Animals seem to take each choice as it comes and respond according to a relatively simple decision rule, appropriate to the particular situation. But in every well-studied situation, the decision rule seems sensitively attuned to *global* consequences. Animals do hill climb, in the sense that they evaluate each choice according to a simple rule that embodies what they know of the probable payoff for each choice. But their rules take into account more than the immediate consequences of a choice. The information for this assessment must come from an animal's past reinforcement history, but we have as yet almost no idea of the rules that connect past history to present choice.

NOTES

1. Mathematical learning theory is associated particularly with the names of W. K. Estes (1959; Neimark & Estes, 1967) and Bush and Mosteller (1955) and their students and associates (see reviews in Luce, Bush & Galanter, 1963; and Bower & Hilgard, 1981). The linear learning model briefly described in the text has been ingeniously extended to classical conditioning by Rescorla and Wagner (1972; see notes to Chapter 13). There has been a recent revival of interest in difference- and differential-equation models for molar properties of free-operant behavior (see Myerson & Miezin, 1980; Staddon, 1977a).

2. There is now a substantial theoretical and experimental literature on the matching law. The initial experimental paper is Herrnstein (1961); other influential papers are Catania (1963, 1973) and Herrnstein (1970). See de Villiers (1977) and de Villiers and Herrnstein (1976) for reviews of matching as a general law of reinforcement. More recently, papers deriving matching from various maximizing principles have begun to appear, e.g., Rachlin (1978) and Staddon and Motheral (1978), although these have not gone unchallenged (Herrnstein & Heyman 1979; Heyman, 1979; Staddon & Motheral, 1979). I discuss matching in relation to maximizing later in this chapter, and also in Chapter 11.

3. There is a substantial and rather tangled literature on probability learning and related procedures. In general, most papers are long on experimental results and short on theoretical analysis of the procedures, some of which are quite complicated. For reviews see Sutherland and Mackintosh (1971), Bitterman (1965, 1969), and Mackintosh (1969).

Much of the complexity of this literature derives from the way that animals react to specific physical apparatus. For example, rats in T-mazes are much more inclined to attend to the place where they find food than to other stimulus aspects, such as the pattern of the goal box. Consequently, in experiments where position is irrelevant, the rats often respond to position anyway, either by going back to the side where they were just fed ("reward-following") or, more commonly, *avoiding* that side (*spontaneous alternation*). Another common pattern is to choose the same side repeatedly, even when position is irrelevant (this is termed a *position habit*). These built-in response tendencies obscure patterns in the choice sequence specifically adapted to the choice procedure. The discussion in this chapter is of the ideal case, where these confusing and largely irrelevant patterns have been eliminated. This ideal is approximated by very well-trained animals working in choice situations with highly salient (preferably spatially separated) alternatives. Response patterns such as alternation and reward-following are integral to the acquisition of adaptive strategies, however, and will be discussed in more detail in later chapters.

4. For an extended theoretical discussion of optimal choice on a variety of simple discrete-trial and free-operant choice procedures, see Houston and McNamara (1981) and Staddon, Hinson, and Kram (1981).

5. See, for example, Rachlin (1970), Rachlin and Green (1972), and Herrnstein (1981).

6. Navarick and Fantino (1976). Nonexclusive choice in these procedures is ensured by making both choices continuously available on an interval schedule (rather than presenting them after a fixed ITI). The procedure is as follows: A hungry pigeon is confronted with two response keys, both lit with white light. Independent, equal VI schedules (mean value 2T, so that the average time to reward on both, taken together, is T, the "ITI") operate on both keys. When a VI "sets up," a peck on the key changes the color on that key, and turns off the other key. Food is then delivered after a delay t_i. After the animal eats, the two keys are again lit white, and the cycle begins again. This is known as a *concurrent chain* schedule (in this case, concurrent VI FT, where the VI operates in the first *link* and the fixed-time schedule in the second link). Numerous changes are rung on this basic theme. For reasons discussed later in the chapter, a concurrent interval schedule produces nonexclusive choice. Hence, given two equal first-link VIs, and other things being equal, the animals will tend to more or less alternate between the two alternatives. A shift in preference, because of changes in delays in the second links, shows up as a bias in this preference.

The concurrent-chain procedure is more complicated than the discrete-trial procedure described in the text and more difficult to analyze exactly in optimality terms. The *changes* in preference to be expected from changes in delays can be derived in the simple way I describe, however, even though exact predictions of response proportions probably require something more elaborate. I say a bit more about concurrent chain schedules in the notes to Chapter 14.

7. Green and Snyderman (1980). Their experiment, like the experiment of Navarick and Fantino (1976), used concurrent-chain VI FI schedules.

8. The concurrent VI VI situation can be represented as in Figure 8.4 if we plot on the horizontal axis the proportion of responding (out of a fixed total) devoted to one or other schedule and on the vertical axis the rate of reinforcement associated with the proportion of responses. It is easy to see that for each choice, the rate of reinforcement will rise with the proportion of responses made to that choice, up to an asymptote determined by the VI value. If this molar feedback function (see Chapter 5) has the property that its derivative can be expressed as the ratio of reinforcements obtained to responses made, $R(x)/x$ (using the symbols of Chapter 7), then allocating responses so as to maximize total reinforcement will also lead to matching (see Staddon & Motheral, 1978).

9. Ratio schedules correspond to *non*depleting patches: If "foraging rate" (lever-press rate) is constant, cumulative payoff increases linearly with time. The marginal-value theorem (not to mention common sense) therefore predicts that animals should concentrate exclusively on the richer patch (i.e., the one with the steeper cumulative-gain function). Under most circumstances, they do (e.g., Herrnstein & Loveland, 1975).

10. The curve in Figure 7 has the same negatively accelerated form as the patch-depletion function in Figure 5, but its significance is quite different. *Probability* of reinforcement for a response is equivalent to *marginal* cumulative food intake, that is, to *rate* of food intake. Imagine a response occurring at a fixed interresponse time t with probability q that each response will be reinforced. The expected time between reinforcements, E, is equal to t times q, the probability that a response will be reinforced, plus $1 - q$ times the expected time plus t, that is, a recursive equation of the form, $E = tq + (E + t)(1 - q)$, which reduces to $E = t/q$. Thus, reinforcement rate is equal to q/t: For a fixed response rate, reinforcement rate is proportional to probability of reinforcement for a response.

11. Momentary maximizing was first suggested as an explanation for matching by Charles Shimp (1966, 1969). His initial theoretical analysis of a rather tricky problem was incomplete, and testing the hypothesis appeared complicated, so that his arguments failed to persuade many who were more taken with the empirical simplicities of molar matching. Recently, interest in this view has revived. The simpler theoretical analysis described in the text has appeared and new experiments supporting momentary maximizing as a substantial component, at least, of operant choice have been published (Silberberg, Hamilton, Ziriax, & Casey 1978; Staddon 1980b; Staddon, Hinson, & Kram, 1981; Hinson &

Staddon, 1983) – although dissenting views are not lacking (e.g., Nevin, 1979; de Villiers, 1977).

12. Momentary maximizing may fail to maximize overall payoff even in very simple situations. For example, in a discrete-trial procedure where food is assigned to one or other choice, but is then available for a response only with some probability r, momentary maximizing yields the optimal strategy only when food is not held, that is, when probability r is sampled on every trial. Momentary maximizing is not optimal in concurrent VI VI or its discrete-trial equivalent. In all these cases, the shortfall is small, however (Staddon, Hinson, & Kram, 1981).

The problem with the globally optimal strategies in these situations is that they place a substantial load on memory. For example, in the concurrent VI VI situation with nonrandom VIs, given that the animal has a fixed number of responses to "spend" in a fixed session time, the optimal strategy consists of a *trajectory* through the clock space in Figure 8.7. (Equating marginals would mean placing each choice on the switching line, but the geometry of the situation makes this impossible in most cases.) In order to follow such a trajectory, the animal might need to be guided by several of its past choices and their times of occurrence – a formidable load on memory. The great virtue of hill climbing is that it nearly always approximates the global maximum, while placing minimal demands on memory. In the concurrent VI VI case, the animal need only remember the times that have elapsed since the *last* A and B choices.

9

FORAGING AND BEHAVIORAL ECOLOGY

The last three chapters discussed how regulatory and economic principles apply to experiments on feeding, the allocation of time among various activities, and choice between different ways of getting food. But a critical test of any approach is how well it can deal with the way animals behave in *nature*. It is hard to devise a fair test, of course, because in nature nothing is controlled, and observations will nearly always be incomplete. Nevertheless, applications of these principles to problems of diet selection and foraging in natural or close-to-natural conditions have been increasingly successful. This chapter deals with the application of optimality arguments to problems posed by natural foraging behavior.

DIET SELECTION AND FUNCTIONAL RESPONSE

In recent years, interest in foraging in natural environments has been organized around a few simple quantitative arguments, some discovered independently by several workers.[1] The basic issues are essentially the same as those discussed under the heading of "optimal behavior" in Chapters 7 and 8: how animals choose either between equivalent foods (i.e., perfect substitutes) or between imperfect substitutes. There are also close parallels between many natural situations and some of the reinforcement schedules described earlier. The starting point for theoretical analysis is the *net rate of energy intake*, and the beginning assumption is that animals act so as to maximize it. This assumption has fairly straightforward consequences for choice of diet, allocation of time to patches, search for camouflaged (cryptic) prey, and spatial pattern of search in environments with different food distributions. I will deal with each of these cases in turn.

Functional response

How does the amount a predator takes of a particular prey type depend on the density of that prey type in the environment? The answer to this question is of great interest to ecologists because of its bearing on the stability of predator–prey systems. For example, if individual prey are at greater risk when prey density is high than when it is low (negative feedback), predation will have a stabilizing effect on prey populations; but if risk is less at high densities than low (positive feedback), the stage is set for explosive increases and decreases in prey population size. The question has psychological interest, because the relation between prey density and predator intake is the natural equivalent of a feedback function, with density playing the role of schedule value, and "attack rate" (defined below) as response rate.

There are three simple ways that predation rate relates to prey density. Two of them can be derived from first principles as follows: Consider a single food type, which is uniformly and randomly distributed with spatial density D. An animal foraging at random will encounter prey items at a rate aD, where the constant a (the attack rate) is proportional to the rate at which the animal moves through its environment and is also related to the area it can scan visually – together these define the animal's *search path*. If looking for one prey type does not interfere with looking for others (assuming prey to be randomly distributed in space), then this relation will hold for all available prey types. Thus, for prey type i, the encounter rate will be aD_i.

If a single prey type is encountered at a rate aD, then in a period of time of unit length, the total number of prey items encountered (and eaten) will be

$$R = aDt^*,$$

where t^* is the time when the animal is foraging which, in turn, is total time (i.e., one) less time taken up in handling prey:

$$t^* = 1 - Rh$$

where h is the time to handle one prey item. Eliminating t^* from these two equations yields the result

$$R = aD/(1 + aDh), \tag{9.1}$$

which is known as Holling's *disk equation* (Holling, 1959, 1965). The label "disk equation" derives from Holling's first test of it, in which his blindfolded secretary "foraged" for sandpaper disks scattered on a tabletop. The disk equation says that when a predator finds prey at random, his rate of eating will be a negatively accelerated function of prey density: When prey are thin on the ground, rate of predation is proportional to prey density; but as prey density increases, more and more time is taken up handling each prey item, so that additional increases in prey density have less and less effect. The disk equation obviously corresponds to positive feedback: An individual prey item is protected at very high prey densities because the predator spends less and less time searching and more and more time handling prey – the predator is "swamped" by the prey. This swamping is one functional explanation for the propensity of insects and other small prey organisms to occur in "blooms" of vast numbers of individuals that all emerge at once.

The relation between rate of prey capture, R, and prey density, D, is known as the *functional response* of a predatory–prey system.[2]

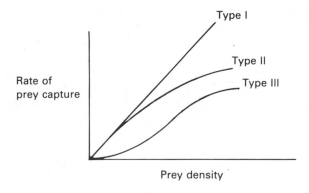

Figure 9.1. Three types of functional response.

It is easy to show that the disk equation can be generalized for n prey types as

$$R = aD_i/(1 + \overset{n}{\sum}aD_ih_i). \tag{9.2}$$

Holling pointed out that there are three simple forms of functional response, illustrated in Figure 9.1: Type I: Linear, with zero intercept – rate of prey capture is simply proportional to prey density. Type II: Negatively accelerated – rate of prey capture increases with prey density but at a negatively accelerated rate, because more and more time is taken up in handling the prey. Type III: Sigmoidal (S-shaped) – rate of prey capture at first increases with prey density at a positively accelerated rate, but then at high densities reverts to the negatively accelerated form.

The type I (linear) functional response is shown only by filter feeders (whales filtering krill, sponges filtering microorganisms, etc.) and other animals where the prey require negligible handling time; it is just a special case of the type II response where $h = 0$. Consequently, the disk equation provides a description of both type I and type II responses. The type II response is the commonest, shown by many predators dealing with a single prey species.[3] The type III response seems to reflect psychological mechanisms in ways I discuss in a moment.

Diet choice

Dietary items such as insects, seeds, nuts, and so forth differ in their energy content and in the time it takes to eat them. For example, a hazel nut will generally take longer to unpack and eat than an ear of grain, although it may also yield more energy. If we ignore for the moment the different *types* of food, it is obviously convenient to summarize the value of each food item by the *ratio* of energy content, E, to *handling time*, h; the ratio E/h is known as the *profitability* of a food item.

Given prey species of equivalent nutritive value, but differing profitabilities, the first question to ask about diet choice is, How many different prey types should the predator take in order to maximize net energy intake? The obvious answer – all – is false, as the following argument shows. Consider just two prey types, with densities D_1 and D_2, energy contents E_1 and E_2, and handling times h_1 and h_2. We need consider just two possibilities: that the predator takes both prey types, or just the most profitable. If the capture rate in the single-prey case is R_1, the net rate of energy intake is just R_1E_1. Thus, at any instant of time, the animal can expect an average rate of energy intake equal to R_1E_1. Suppose it encounters a prey item of a less profitable type. If it attacks the item, its expected rate of energy return is just the profitability of the item, that is, E_2/h_2. Clearly, if E_2/h_2 is less than R_1E_1, the predator does better to pass up the less profitable prey and continue looking for the more profitable one. As this argument shows, the predator's decision is *independent* of the density of the less profitable prey. For a given pair of prey types, there will be a threshold density of the more profitable prey above which it will pay the predator to specialize on the more profitable type.

The same argument obviously holds for more than two prey types. If prey types are ranked according to profitability, as the density of the more profitable types increases, the predator should drop more and more of the low-ranked types from its diet.

The strong prediction of this analysis is that animals should be absolutely rigid in their selection of prey types. This does not always happen. In most tests, animals take a few items outside the "optimal set." For example, in one experiment (Werner & Hall, 1974) a group of ten Bluegill sunfish (*Lepomis macrochirus*) were allowed to hunt for water-fleas (*Daphnia*) of three different size classes. The fish were exposed to three different "mixes" of *Daphnia*, and the proportions caught were estimated by examining the stomach contents of the fish. The proportions of small, medium, and large fleas were the same in all three mixes (equal numbers of each size), but the total numbers of fleas (i.e., absolute prey density) varied. The handling time for all classes of fleas was similar (and small), so that their profitability depended mainly on their size. When absolute density was low, the fish ate fleas in proportion to the frequency with which they were encountered. But as density increased, the fish increasingly favored the more profitable types, although at no density was their preference exclusively for the most profitable type. This study, like a number of others, found that as the density of the more profitable prey type increases, animals become more selective.

There are several reasons why animals might not conform to the strict stereotypy prescribed by simple optimal-diet models: The most obvious is that the assumptions of the simple model – uniform prey distribution, known profitability – are not met. If prey are not uniformly distributed, then prey choices should shift and what is measured will be an average of two or more different choice patterns. Animals must sample to learn about prey profitabilities, and this will produce nonexclusive choice. If many different prey types are encountered, the animal may forget about the profitabilities of some of them (a memory constraint) and may continue sampling to update his information.

An interesting limitation derives from the kind of positively accelerated cost function discussed in Chapter 7. An investor of limited means should spread his investment over several different securities, so that he doesn't lose his shirt if a single stock crashes – the probability that several independent stocks will all crash being much lower than the probability a single stock will fail. For the same reason, an organism is well advised to sample more than one prey type; then a change in the abundance of any given prey type will have less effect on his total intake. It can be shown that this *risk-aversion* corresponds to nonsubstitutability of prey types, and a negatively accelerated utility function (positively accelerated cost function) of the type discussed earlier.[4] This argument implies nonexclusive choice even if different prey types are nutritionally identical, and even if the rate of energy intake is thereby less than maximal.

There is experimental evidence for this tendency to diversify choice, even among identical alternatives. In an ingenious series of experiments, Catania (1980a, b) has shown that pigeons will choose an alternative that gives them several additional choices over an alternative that gives them only one – even if all choices are identical in terms of food delivery. In one experiment, for example, a pigeon was confronted with two rows of response keys: two below and four above. At the beginning of a cycle, only the two lower keys were lit. A peck on the left key occasionally (on a VI schedule) turned it off and turned on the four top keys, three lit green and one red: Each green key delivered food for pecks on the same fixed-interval schedule; pecks on the red key were ineffective (extinction). A peck on the right, lower key occasionally turned it off (according to the same VI schedule) and turned on the four upper keys, as before, but now three were red and one was green. After food delivery, the four keys went off and the two lower ones were reilluminated, to begin the next cycle.

Thus, by pecking left and then pecking any of the three green keys in the second link of this *concurrent-chain* VI FI schedule (see note 6, Chapter 8), the pigeon obtained exactly the same frequency and spacing of food as by pecking right and then pecking the single green key in the second link. The two lower keys were exactly equivalent in terms of food delivery. Nevertheless, pigeons showed a bias in favor of the left key, which yielded a choice among three green keys, over the right key, which permitted no choice. In other experiments, Catania showed that it was the *number* of keys in the second link rather than their size that was important to the pigeons. Clearly this result is consistent with the idea that animals have a built-in tendency to diversify their behavior, when diversification costs them little in terms of overall payoff rate.

Switching and functional response

Nothing I have said so far conflicts with the idea that given two prey types of equal, high profitability, both should be taken whenever encountered. If we

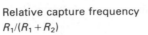

Relative capture frequency
$R_1/(R_1 + R_2)$

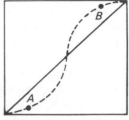

Relative prey density $D_1/(D_1 + D_2)$

Figure 9.2. Predator "switching." The straight line shows the relation between proportion of prey taken and relative prey density when the predator is unselective. The sigmoidal function corresponds to switching, that is, disproportionately more prey taken at high relative densities, disproportionately few at low relative densities.

restrict ourselves to the ideal case in which the two types occur at random in the same area (so that the predator cannot look for one type in one kind of habitat and the other in another), this means that each prey type should be taken in proportion to its frequency of occurrence (D, in Equation 9.1). For a pair of such equivalent prey types, the expected relation between relative density (that is, $D_1/[D_1 + D_2]$) on the x-axis and relative capture rate ($R_1/[R_1 + R_2]$) on the y-axis is shown by the diagonal in Figure 9.2: Relative capture rate should match relative density. (Matching can be derived in one step from Equation 9.2. Note: "Matching" here is not the same as the matching on concurrent schedules discussed in Chapter 8. Matching here is between obtained relative payoff rates and relative payoff (prey) densities; matching on concurrent schedules is between relative *response* rates and obtained relative payoff rates.)

This ideal situation occurs rarely in nature: The number of prey types is usually variable, and they are not usually equal in profitability. Hence it is not surprising that measured relations between prey density and proportion of prey in diet deviate from the matching form. The commonest form of deviation is shown by the dotted line in Figure 9.2: A given prey type is taken disproportionately *less* often when its relative density is low, and disproportionately *more* often when its relative density is high. This has been termed *switching* (Murdoch, 1969) because it corresponds to a switch in preference if the changes in relative density occur over time. For example, suppose that at time t_0 prey types 1 and 2 have densities 10 and 50 per unit area and the animal's diet is made up of 1 and 2 in the proportions 1:10. This corresponds to point A in Figure 9.2. Suppose that after a period of time, the density of prey type 1 increases to 50 and the density of 2 falls to 10. The predator now prefers 1, in the proportion of 10:1, and relative choice corresponds to point B in Figure 9.2.

It is often difficult to be sure of the reasons for the deviation from matching; there are at least four possibilities:

1. Absolute density changes. Figure 9.2 shows relative measures plotted against one another, hence gives no information about changes in absolute density. As

the example shows, relative changes may reflect changes in the absolute levels of either or both prey types. If one prey type is, in fact, slightly more profitable than the other these absolute changes can lead to relative changes in choice of the kind shown by the dotted curve in Figure 9.2. For example, if prey type 1 is, in fact, somewhat more profitable than type 2, then at low densities of 1 it will pay the predator to take both prey types, but if the absolute (and relative) density of 1 rises, it may be better to specialize on 1. This shift in the absolute density of 1 will thus lead to a "switching" pattern of relative choice.

2. Nonrandom spatial distribution. If the spatial distribution of one prey type changes so that it occurs in "patches," rather than being randomly intermixed with the other type, then it may pay the predator to specialize. Thus, if prey type 1 (one of two equivalent prey types) is randomly distributed at low (absolute) density, but distributed in patches at high density, the predator should sample both types when the absolute density of 1 is low, but concentrate on 1 when its absolute density is higher.

Patchy distribution can be further subdivided, into cases where the patches of one prey type are recognizable (i.e., associated with a particular habitat, as when moths of a certain type are associated with a certain type of tree) and cases where the patches are not associated with a reliable signal. When patches are signaled, then obviously the predator would do well to spend most of its time in the signaled area. Even if in the patch it takes prey in proportion to the frequency with which they are encountered, since it encounters disproportionately more of one type, it will take more of that type. If patches are not signaled, they can still be detected by means of *area-restricted search*; this is the vertebrate equivalent of the klinotaxis discussed in Chapter 2: Even without special training, many animals increase their rate of turning when a prey item is encountered; and most species can learn to behave in this way if exposed to patchily distributed prey. If prey are patchily distributed, this pattern of movement obviously increases the animal's chance of encountering additional prey. The result is the same as in the signaled case: Patchily distributed prey are encountered, and taken, disproportionately more often than evenly distributed prey.[5]

3. Changes in profitability with experience. With repeated encounters, predators can often become more efficient in handling prey. If the ease with which they do this depends upon the *absolute frequency* with which prey are encountered, then a change in absolute frequency alone may cause a change in effective profitability, leading to a change in preference of the type discussed under (1), above. There is much evidence that with experience animals become better at handling prey. Oystercatchers (*Haematopus ostralegus*), for example, do not mate until a year or two after attaining sexual maturity, because it takes them that long to learn to open shellfish effectively enough to support a family. If a low-profitability prey can increase in profitability when it is encountered at high density, then presumably the predator should continue to sample it even when it occurs at low density. If profitability depends upon absolute density, then prey choice need not be independent of the absolute density of any prey type.

4. Changes in effective density with experience. I have assumed that the rate at which a predator encounters a prey type is proportional to prey density. This will not be true if animals must learn to discriminate camouflaged prey from their background, or to identify particular objects as potential food. The second kind of learning certainly occurs, and the first kind is very probable. Hence, the assumption that encounter rate is proportional to prey density is not always valid.

The possibility that encounter rate can change with experience is interesting because it is a striking instance of an intimate relation between the processes of learning and memory and a behavior pattern − relative predation − critically important to the distribution and abundance of species. I have not yet discussed learning and memory in any detail, but rather general knowledge about memory provides some obvious pointers to features that are likely to be important. For example, a predator will be more easily able to remember features of a prey item if items of the same type are encountered in "runs," without intervening items of a different type; in this way interference from the other type is minimized. In nature, runs of the same prey type are much more likely if prey are patchily distributed, so that learning to detect a prey type, as a cause of "switching," is often likely to be confounded with nonrandom distribution.

There are at least two kinds of learning that might lead to changes in encounter rate: learning that a particular object is in fact a potential prey, and "learning to see" a cryptic prey type, that is, formation of what has been termed a *specific searching image*.[6] The first type, classification learning, is relatively straightforward; the second, perceptual learning, is less so.

Chicks provide an example of classification learning as they learn to select nutritive over nonnutritive objects (see Chapter 14). All omnivores sample new foods and learn either to avoid them or incorporate them into their diet. These cases involve learning the significance of readily identifiable objects, not learning to discriminate the object from its background. An experiment with captive jays (*Garrulus garrulus*) and chaffinches (*Fringilla coelebs*) is a less clear case: de Ruiter (1952) found that the birds initially failed to treat stick caterpillars as food, but after they found one accidentally (perhaps by treading on it and causing it to move) they quickly found others. At first they treated the caterpillars like twigs, but afterward they treated them as prey. The birds in this experiment learned that a familiar object is, in fact, prey, and to this extent these results require no perceptual interpretation. But they may also have been learning to discriminate stick caterpillars from sticks, which does imply perceptual learning.

Search image. The idea of a search image derives initially from human subjective experience. von Uexküll, the great German ethologist, describes his experience as follows:

During an extended stay at the home of a friend, an earthen water jug was always set before my place at lunchtime. One day, however, the servant broke the jug and in its place put a glass carafe. At the next meal, when I looked for the jug, I did not see the carafe. Only after my host assured me that the water stood in its accustomed place did the various glittering reflections off knives and plates suddenly fly together through the air to form the glass carafe. . . The Searching Image obliterates the Receptor Image. (von Uexküll, 1934, p. 83 – trans., Croze, 1970)

Many people have had similar experiences, and they make plausible the idea that animals and people have to "learn to see" new objects, especially camouflaged objects. Nevertheless, when subjective experience is ruled out, as it must be when we are dealing with nonverbal organisms, these two types of learning – to identify a new prey type, and search–image formation – are difficult to distinguish in practice.

Consider, for example, Marian Dawkins' ingenious experiments (1971a, b), in which domestic chicks were allowed to search for colored rice grains on various backgrounds. Three results are of special interest: (a) Confronted with green and orange grain on a green background, the chicks consumed the orange grains first. The green (cryptic) grain was only consumed after a delay; moreover, the cryptic grains were consumed at an increasing rate, once they had been detected. (b) The lag in consuming cryptic grains was not due either to color preference (green and orange grain and backgrounds were counterbalanced) or simply to classifying them as nonprey. For example, chicks that had just been eating conspicuous orange grains (i.e., on a green background) were still deficient in taking (cryptic) orange grains on an orange background. (c) The ability to take cryptic grain was abolished both by the passage of time and interpolated experience with another type of grain: The chicks forgot within 24 hr, and the peck latency to a cryptic grain type was inversely related to time since the chick pecked the preceding grain of that type. Apparently, the chicks were learning something specific about green or orange grain *in its cryptic context.*

Compelling as these results are, they still do not force us to the conclusion that the chicks were undergoing a perceptual change of the same kind as von Uexküll's with the water carafe. Key facts are the transience and susceptibility to interference of the learning in Dawkins' experiments. One is inclined to attribute the changes in performance to perceptual modification, because chicks probably do not as readily forget to attack noncryptic prey (although this has not been rigorously established, as far as I know – indeed, it might be quite hard to prove, given that chicks have a predisposition to peck at certain kinds of objects, quite apart from particular experience with them). It is also clear that learning to detect cryptic prey is more *difficult* than learning to attack conspicuous prey, however, although we don't really know why. Difficulty in and of itself provides a basis for rapid forgetting, because the more memory resources devoted to a given task, the more likely that time and interpolated tasks will interfere with its retention (see Chapter 12). Hence, learning to recognize cryptic prey may just be more difficult than other kinds of learning, not necessarily different in kind.

ROC analysis. One *objective* concomitant of the search–image hypothesis is that animals should improve their ability to *detect* cryptic prey, in the detection-theory sense touched on in Chapter 5. The meaning of "detect" is illustrated by

Attack?

		No	Yes		
Prey	Present	30	70	100	$p(Y \mid P) = .7$
	Absent	50	50	100	$p(Y \mid {}^\sim P) = .5$

Table 9.1.

example in the accompanying contingency table (Table 9.1), which shows 200 hypothetical "encounters" between a predator and a cryptic prey. The prey (e.g., caterpillar) was only present on half the occasions; on the remainder, the animal was looking at something else (twig). When there was no prey, the animal nevertheless attacked 50% of the time: $p(Y \mid {}^\sim P) = .5$; but when prey was present, the animal attacked on 70% of occasions: $p(Y \mid P) = .7$. The fact that $p(Y \mid P) > p(Y \mid {}^\sim P)$ shows that the animal was able to detect the prey, albeit imperfectly.[7]

These two values of $p(Y \mid {}^\sim P)$ and $p(Y \mid P)$ are plotted at the filled dot labeled "A" in the contingency space of Figure 9.3. The curve through A illustrates the limitations on the animal's ability to detect this hypothetical cryptic prey type. The curve is traced out by considering what we would expect if the animal were more or less cautious in his attacks. For example, suppose he becomes less cautious, attacking when he merely suspects that the twig is a caterpillar: Obviously the number of "hits" $(p(Y \mid P)$ will go up; but since he cannot detect prey any better, so also will the number of "false alarms." This rasher policy corresponds to point B in Figure 9.3. Conversely, if he really hates to peck at twigs rather than caterpillars, he can be more conservative, and not peck unless he is really sure he sees a caterpillar and not a twig. This strategy, point "C" in the figure, reduces $p(Y \mid {}^\sim P)$, but also drops $p(Y \mid P)$, so that he misses more prey items.

The curve through points *A, B,* and *C* is called an *isosensitivity curve*; another name is *ROC curve* (the initials stand for "receiver operating characteristic" – this kind of analysis was first used for evaluating the properties of telecommunications equipment). The difference between the ROC curve and the diagonal in Figure 9.3 is a measure of *detectability* (sometimes abbreviated as d', which is one measure of detectability). The diagonal corresponds to complete inability to detect the prey, and obviously the more the ROC curve pushes up into the upper left-hand corner, the better the animal is able to detect the prey: Curve ABC corresponds to a more cryptic prey (or a dumber animal) than the curve above and to the left, to a less cryptic animal than the curve below and to the right. The shape of the ROC curve is determined by the distribution of prey items

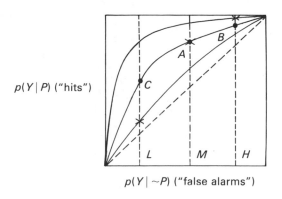

$p(Y\,|\,P)$ ("hits")

$p(Y\,|\,{\sim}P)$ ("false alarms")

Figure 9.3. Search image as detectability change. Three ROC curves, corresponding to high, medium, and low detectability. Filled circles indicate the hypothesized relation between hits and false alarms on the assumption that different densities of cryptic prey affect only response bias (all three points are on the same ROC curve). The x's correspond to the search-image view: Detectability is low at low prey densities, but high at high densities.

(caterpillars: the *signal*) and distractors (twigs: the *noise*) along all the stimulus dimensions that are important for detectability.

The idealized one-dimensional case is illustrated in Figure 9.4, which shows overlapping "signal" and "noise" distributions. Because the two distributions overlap, it is impossible for the animal to get everything right: Wherever he sets his *criterion* (the solid vertical line in the figure: potential prey items falling to the right he attacks, items to the left he doesn't), he will make some false alarms (the vertically crosshatched area of the left distribution) and miss some positive cases (the horizontally crosshatched area of the right distribution). As the animal shifts his criterion (sometimes also called *bias*) from left to right, the areas under the distributions corresponding to false alarms and hits (the two areas to the right of the criterion) shift in the way shown by the ROC curve. Points on the ROC curve corresponding to Low, Medium, and High criterion settings are indicated in Figure 9.3 and by the dashed lines in Figure 9.4. The separation of the two distributions in Figure 9.4 defines the detectability of the prey: the less the overlap, the better the detectability. The three ROC curves in Figure 9.3 correspond to different separations between the signal and noise distributions.

The position the animal sets his criterion should depend on the relative costs of the two kinds of errors: misses and false alarms. Two things affect this ratio: costs of each type of error and their relative frequency. The cost of a false alarm will depend on the time lost in responding mistakenly, the danger of attacking something dangerous, and the physical effort involved. The cost of a miss will depend on the animal's state of hunger and the profitability of the prey. The relative frequency of the two types of error depends on prey density. If prey

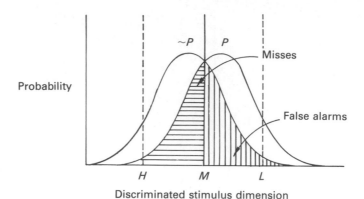

Figure 9.4. One-dimensional signal-detection analysis: The distributions represent the relative frequency with which either prey (right distribution) or nonprey (distractors: left distribution) occur at different values of the discriminated stimulus dimension. The vertical, solid line is the response criterion: Items to the right are attacked, to the left, not attacked. The crosshatched areas represent the probability of the two kinds of error: misses ($p(N|P)$ and false alarms ($p(Y|P^\sim)$.

density is high, then in the nature of things, the animal will be more likely to encounter prey than if prey density is low – the total in the upper row of Table 9.1 will be higher. Consequently, the animal can afford to be more lenient in his criterion (move it more to the left) when prey density is high than when it is low. It turns out that the optimal criterion level is directly related to the ratio of prey and distractor densities: The higher the density of the prey in relation to the density of things confused with the prey, the more lenient the animal should be, and the farther to the right in Figure 9.3 should be the point representing his behavior.[8]

Two things follow from this argument: (a) The risk to an individual cryptic prey item should increase as prey density increases; thus optimal adjustment of predator bias alone, with no increase in detectability, should have a stabilizing effect on the prey population (more on this in a moment). (b) Because the effects of a change in bias are different from the effects of a change in prey detectability, we should be able to see experimentally whether change in prey density has one effect, both effects, or neither. Figure 9.3 shows three ROC curves, corresponding to Low, Medium, and High levels of detectability. The filled circles show the expected effect of changing the absolute density of prey, from *Low*, through *Medium*, to *High*, on the hypothesis that this affects bias only: As the absolute frequency of prey increases, $p(Y|^\sim P)$, the probability of attacking a distractor (a "false alarm"), and $p(Y|P)$, the probability of attacking P (a hit), increase along the same ROC curve. The x's show the effect of increasing prey density, on the hypothesis that detectability increases: As the absolute frequency of A increases, $p(Y|^\sim P)$ and $p(Y|P)$ increase along successively higher ROC curves.

It is simple in principle to test these two hypotheses. In practice, uncertainty about the proper form for the ROC curves makes testing less sure. One possibility is to compare the ROC curves generated by varying prey frequency with those obtained by varying choice payoffs.[9] Since there is no reason to expect that variations in payoff will affect detectability, we might accept the form of the ROC curve obtained under these conditions as the norm. If a different curve is obtained when frequency is varied we might want to conclude that frequency variation affects detectability, as the search-image hypothesis requires. As far as I know, this test has not been carried out.

Ecological implications. Experimental results agree in showing that animals take some time to learn to attack cryptic prey, and that this hard-won ability is easily lost either after lapse of time, or after the interpolation of some other learning. This susceptibility to interference is sometimes referred to as the "incompatibility of search images," since it implies that animals will find it more difficult to look for two different cryptic prey types than for one. These two factors – difficulty in learning to detect cryptic prey, and enhanced difficulty of detecting more than one cryptic type – have implications for the composition and size of populations. For example, if predator efficiency increases as absolute prey density increases, this will constitute a negative feedback tending to limit the size of prey populations.

I just showed that a stabilizing effect of crypticity on prey population can be derived without assuming perceptual learning. Bias in favor of attack, which we can measure by $p(Y|P)$, should increase with prey density. But $p(Y|P)$ is just the risk incurred by an individual prey item. Hence, the risk to each individual cryptic prey item should increase with absolute prey density. Thus, crypticity in and of itself should tend to exert a stabilizing effect on population growth since it provides less protection at high population densities than low.

Moreover, this process will tend to shift a type-II functional response (Figure 9.1) in the direction of the type-III response, because it will cause a given cryptic prey to be taken disproportionately less at low densities. In terms of the disk equation (Equation 9.1), this change in bias means that the attack rate parameter, a, instead of being constant, increases with prey density, D. The simplest possible quantitative assumption is that a is proportional to D, that is, $a = kD$, which yields the modified equation:

$$R = kD^2/(1 + hkD^2). \qquad (9.3)$$

It is easy to see that Equation 9.3 has the required properties: When density is low (D close to zero), the equation reduces to $R = kD^2$, a curve of positive, rather than negative, acceleration; but when D is large, acceleration is negative as the function approaches its asymptote of $1/h$.

The obtained form of functional response for vertebrate predators foraging for cryptic prey quite often corresponds to the sigmoidal form of Equation 9.3 (Holling, 1965; Murdock & Oaten, 1975). Available data are not sufficient to decide whether the type III response is uniquely characteristic of cryptic prey, as this analysis implies. There are some indications that attack rate increases with prey density – the "feeding frenzy" of sharks and many other predators in a patch of abundant prey is a well-known example. But again, we have no evidence that the effect is stronger in the case of cryptic prey and competition among predators provides another functional explanation for feeding frenzy.

There is evidence that prey distributions in nature are in accordance with what one might expect from this analysis. For example, cryptic prey are usually dispersed, keeping density as low as possible, whereas aposematic (bad-tasting, poisonous) potential prey are usually clumped (Benson, 1971; Ford, 1945; Tinbergen, Impekoven, & Franck, 1967).

Equation 9.3 can readily be generalized to N prey types; for two prey types it takes the form:[10]

$$R_1 = aD_1^2/(1 + ah_1D_1^2 + ah^2 D_2^2),\tag{9.4}$$

where D_1 and D_2 are the densities of the two prey types, h_1 and h_2 their handling times, and a is the attack-rate parameter.

It is easy to derive switching from Equation 9.4. For two prey types, the relation between the ratio taken and the ratio of densities is

$$R_1/R_2 = (D_1/D_2)^2,\tag{9.5}$$

or, in terms of proportions,

$$R_1/(R_1 + R_2) = [D_1^2/(D_1^2 + D_2^2)],\tag{9.6}$$

which is a sigmoidal function.

Thus, both switching and the type-III functional response can be derived from the assumption that animals foraging for cryptic prey adjust their response criterion in an efficient way.[11] These effects need not depend on the predator being able to improve its ability to detect more frequent prey, nor even on interference between two prey types: It is not necessary that a predator be able to forage more efficiently for one cryptic prey than for two – although that is usually taken as an essential implication of the search-image idea. It is important to make this point, because all these effects have at one time or another been tied to the concept of search image – which assumes more than is necessary to account for them.

Theory aside, learning to detect cryptic prey is susceptible to interference; consequently it is, in fact, more efficient for animals to forage for one prey type than two. For example, in an experimental test of the search-image notion, Pietrewicz and Kamil (1979) found that captive bluejays (*Cyanocitta cristata*) are better able to identify color slides of cryptic *Catocala relicta* moths if they are presented alone, than if they are intermixed with slides of *C. retecta*, another cryptic species. It is easy to show that an advantage of this sort should always lead to specialization on one or other prey type, that is, to switching of the most extreme kind (Staddon, 1980b).

This experiment is perhaps the best evidence that learning about cryptic prey does involve an improvement in detectability, not just a shift in response bias: In the Pietrewicz and Kamil study, net percentage of correct responses (that is, correct positives plus correct negatives, a measure of detectability) improved within runs of the same prey type, but not across a series where the two types were intermixed. Hence this study provides real evidence for the search-image idea. Few other studies do so. Some can be dismissed because of inadequate experimental design, but many cannot. For example, there are numerous experiments on attention in human subjects, and a few with animals,[12] that could provide the necessary evidence, but for the most part do not. In these experiments, cryptic prey are represented by a small set of symbol types (e.g., x's and z's) that must be spotted by a subject in briefly presented video displays. These target symbols are always embedded in a larger number of distractor symbols, so they are hard to pick out. The number of types of target symbols is termed the *memory-set size*, and the idea that search images are incompatible would lead one to expect that the accuracy with which members of the memory set are detected should be inversely related to memory-set size: the more different types the subject is looking for, the worse his performance on any given type. The surprise is how little difference memory-set size makes: Most experiments show that subjects can search for two, three, or even four types almost as easily as for one. Of course, these subjects are very well trained, so once again one is led to the idea that the changes in detectability go along with learning, and may not represent an intrinsic limitation of animals' perceptual systems.

The possibility that search images interfere with one another – that animals can forage more efficiently for one cryptic prey type than for two – implies that cryptic prey populations should tend to be *polymorphic*: If low density affords relative protection then rare morphs should be favored. This steady selection pressure for variation in crypsis might be expected to lead to polymorphism in the population as a whole. Poulton (1888) pointed out many years ago that polymorphism probably reduces the risk to individual prey organisms, and several natural populations (*Cepea* snails, some butterflies) have been looked at from this point of view. There are limitations on the possibilities, however, because the different cryptic morphs must satisfy two conflicting conditions: to be cryptic with respect to the common background, but not to be confusable with each other. The more heterogeneous the background, the more possibilities for polymorphism, but there will be some limit in every case.

There is no doubt that animals learn something about cryptic prey, that such learning can be difficult and is unusually subject to decay and interference, and that these limitations mean that it is often advantageous for predators to specialize on an abundant cryptic prey type. There is also some evidence that such learning involves an improvement in detectability, as opposed merely to a change in response tendency. Nevertheless, it is also true that phenomena such as switching and type-III functional response that are sometimes attributed to these properties of search image may simply reflect an adaptive adjustment of the attack threshold for cryptic prey. The set of data for which search image is a necessary explanation is, in fact, rather small and to some degree inconsistent with laboratory results.

Nutrient constraints and sampling

The analysis so far is simplified in two important respects: It ignores qualitative differences among foods, and it assumes an unchanging environment. Both these factors are important to real (as opposed to model) animals foraging in natural environments, although neither is easy to tackle theroretically or experimentally.

The need for specific nutrients such as vitamins and minerals means that foods cannot always be compared just in terms of their energy content. There are two possibilities to consider: (a) when the animal must choose among a set of "pure" foods, that is, foods each containing just one nutritional ingredient; and (b) when the animal must choose among a set of foods each of which contains a mix of essential ingredients in approximately constant proportions. The second case obviously corresponds more closely to reality. In both cases, the situation confronting the animal can be described most compactly using the economic techniques introduced in Chapter 7.

The way an animal should treat each of these situations obviously depends upon its metabolic needs. Since nutritional ingredients such as amino acids are always utilized to form combinations, a sensible animal's preference for one ingredient should depend upon the amount it already has of the other ingredients that enter into the compounds. Hence preferences will rarely be independent of one another.

The problem can be illustrated by considering a simplified creature that requires only two essential nutrients. Its needs can be represented as shown in Figure 9.5.[13] The horizontal axis shows the animal's nutritional state with respect to nutrient A, the vertical axis its nutritional state with respect to nutrient B. Thus, the animal's total nutritional state at any time can be represented by a point in this space (P in Figure 9.5). P will generally be some distance from point Q, the "ideal" nutritional state. Surrounding Q, we assume, will be a set of closed, concentric indifference contours representing the animal's ordered preferences for nutritional states different from Q (see Figure 7.12). Consider now a particular food item of nutrient B. If ingested, the item will (the animal's individual and evolutionary history tells it) move the animal's nutritional state from P in Figure 9.5 to point P_1. For convenience I will term the line joining P and P_1 the *improvement vector* for the food item in question. An alternative food item contains both A and B; if ingested it will move the animal's state from P to P_2. Clearly, the animal's preference in this case will be for the second item, since it shifts the total state closer to Q, the "ideal" state. In this way, the animal can choose from any set of potential food items. This analysis of the problem can obviously be extended to any number of essential nutrients.

This is strictly a regulatory analysis. It does not incorporate environmental features, such as taste and circadian factors, dealt with by the linear model of

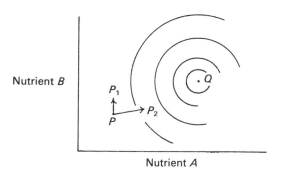

Figure 9.5. State–space representation of diet choice, given nutrients of different types. Point Q is the optimal mix of the two nutrients, point P is the animal's actual nutritional state. Vectors P_1 and P_2 represent the improvements in nutritional state to be expected from two food items: an item containing nutrient B only, and an item containing both A and B.

Chapter 6. However, it does, of necessity, incorporate information about the environment, in the form of the animal's estimate of the improvement vector.

This scheme converges on the one discussed at length in Chapter 7 if we make three transformations. First, a space of potential actions (behaviors) is set up, corresponding to the space in Figure 7.2 and several subsequent figures. Second, the axes of the space in Figure 9.5 are rescaled so that the lengths of the improvement vectors are proportional to the "real" improvement in physiological state associated with a given action.[14] Third, a set of indifference contours corresponding to the real improvement associated with any action is drawn in the behavior space.

The general conclusion is that the problem of a balanced diet is just one aspect of the broader problem of allocating time and energy to particular activities. Although it is sometimes convenient for us to distinguish dietary problems from others that the animal faces, the animal itself cannot really do so: Since all action draws from the same common stock of time and energy, the animal must balance all against all.

In practice, it is highly unlikely that animals make choices on the basis of a scheme like this. Rather, they have evolved sets of relatively simple rules that can handle the vagaries of their natural environments and ensure a plentiful and balanced diet. For example, rats and many other animals when confronted with a number of novel foods will sample them one at a time, waiting a sufficient time between new choices to see if each new food has bad effects. Similarly, a rat suffering from experimentally induced vitamin deficiency will widen its choice of foods, and, by a slow process of learning add foods to its diet that can reduce the deficiency (Rodgers & Rozin, 1966; see Rozin & Kalat, 1971, for a review). Contrary to the implications of the economic model, animals are rarely able by taste or sight to evaluate foods in terms of their specific nutritional constituents

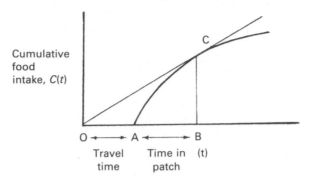

Figure 9.6. The marginal-value theorem. Optimal time in a patch (AB) is defined as the point where the instantaneous rate of prey capture (slope of cumulative curve at point C) is equal to the overall mean capture rate, that is, cumulative food intake divided by travel time plus time in patch ([OA + AB]/OB).

(specific hungers for taste for sugar, salt, and specific aversions to some poisonous substances are exceptions, but there is no special sensitivity to the great majority of essential vitamins and minerals). As the analysis in Chapter 6 suggested, environmental factors are exceedingly important in motivating the animal to eat versus doing something else; they are much less important in guiding it in a selective way to particular foods. Omnivores become "bored" with a monotonous diet or, to say the same thing a bit differently, a given food type becomes more attractive the longer the time since it was last sampled. Specific satiation–deprivation mechanisms like this function to promote dietary diversity and thus ensure dietary balance in general while not guiding the animal toward particular nutrients. These mechanisms also ensure that the animal will not become too fixed in its behavior (although there are some striking exceptions to this, to be discussed later) and will continue to sample alternatives that may be useful in the future as the environment changes.

NATURAL FEEDBACK FUNCTIONS

Prey in nature are distributed in various often nonrandom ways, are subject to depletion by predators, and may also recover after lapse of time. These three characteristics – nonrandom, particularly patchy, spatial distribution, depletion and repletion – interact with the behavior of the predator to define a *feedback function*, precisely equivalent to the feedback function defined for operant reinforcement schedules that was discussed extensively in earlier chapters.

For example, if prey are patchily distributed, the activity of a predator foraging in a patch is subject to diminishing returns, as shown schematically in Figure 9.6 (a slightly modified version of Figure 8.4). At first, the predator can take prey at a high rate (indicated by the steep initial slope of the cumulative curve); but as the patch is depleted, the density of prey drops and the rate of return decreases. This situation is equivalent to an operant-conditioning experiment in which the rate of return for responding to one alternative decreases with time, while the rate of return to be expected from another alternative (i.e., other patches, in the natural case) remains constant. An example might be a two-choice procedure: On one response key a ratio schedule where the size of the ratio increases with each food delivery and on the other a constant ratio schedule, the decreasing ratio would be reset to its initial value after each switch to the constant ratio (Hodos & Trumbule, 1967; Lea, 1976). Obviously, the optimal course is for the predator to leave a patch when the "instantaneous" rate of return drops below the rate of return from the habitat as a whole. If all patches are of equal richness, this optimal strategy is illustrated by the diagonal line in Figure 9.6. The average rate of return is just BC/OB, that is, the mean cumulative food intake per patch divided by the mean travel time between patches (OA) plus the mean time in a patch (AB); point C is where the instanteous rate of return within a patch just equals this average, hence AB represents the optimal time in a patch (Charnov, 1976; Parker & Stuart, 1976).

One way to test this analysis is to let animals forage in an artificial environment where the richness of patches and the travel time between them can be controlled. Cowie (1977) did this with great tits (*Parus major*) in a large artificial aviary. The birds were looking for mealworms hidden in sawdust-filled plastic cups (the patches) on the branches of artificial trees. Travel time was manipulated by putting loose or tight lids on the plastic cups – a tight lid was assumed to have the same effect as a longer travel time. The cumulative food-intake curve was measured directly as the bird consumed the mealworms in each identical cup. Cowie looked to see if the average time spent on a branch varied with the travel time as predicted from the optimal-foraging analysis. The birds did indeed spend longer in each patch when the travel time was longer, but on the average they seemed to spend even longer than predicted by the analysis. When the energetic, as well as time, costs of traveling were taken into account, the fit between theory and data was quite good. Other studies, with bumblebees, water-skaters, and other animals have also provided support for the marginal-value approach, although it is not clear that it works in every situation (Whitham, 1977; Cook & Cockrell, 1978; Davies, 1977; see Krebs, 1978, for a review).

There is no standard reinforcement schedule precisely equivalent to foraging in depleting patches. But there are common natural situations analogous to both interval and ratio schedules. Within a nondepleting patch – a large cloud of insects or *Daphnia* prey, for example – return is directly proportional to attack rate (if handling time is excluded). This is the natural equivalent of a ratio schedule. Correspondingly predators in such situations usually respond at a high rate.[15]

When a food source replenishes itself after lapse of time, the predator faces a situation analogous to an interval schedule. For example, a predator that has exclusive access to a self-renewing foraging area should follow a foraging path that allows each region the optimal time for replenishment. Studies of wagtails

(*Motacillidae*) patrolling their territory along a river bank at which insect prey arrive more or less randomly show that the feedback function here corresponds to a random-interval schedule. The birds seem to adjust their search path appropriately.[16] Hawaiian honeycreepers (*Loxops virens*) defend territories of nectar-producing flowers; here the pattern of repletion is more predictable and corresponds to a fixed-interval schedule. This study and others have shown that predators under these conditions adjust their foraging patterns so that depleted flowers are not revisited until their nectar stores have substantially recovered (Kamil, 1978; Gill & Wolf, 1978).

SUMMARY

The study of natural foraging is difficult because of uncertainties about the feedback function relating the animal's behavior to its consequences. Nevertheless, much has been learned about diet choice and foraging patterns by looking at foraging in terms of energetic return and by making simple assumptions about foraging patterns. For example, logic suggests that animals should specialize more when food is abundant than when it is scarce, and they do, although not as strictly as theory implies. Exceptions seem to reflect either the animal's lack of knowledge about its environment, or an evolutionary bias that favors nonexclusive choice (risk aversion).

The natural equivalent to a feedback function is the predator's *functional response*: the relation between prey density and predation rate. Two types of functional response can be derived from simple arguments about random search and prey-handling time. The third type, often associated with foraging for cryptic prey, seems to involve psychological mechanisms that take the form either of changes in the animal's criterion for attack, or in the actual detectability of prey (*search image*), as a function of experience. Simple type I and type II functional responses correspond quite closely to ratio feedback functions. Situations where the prey replenish with time (foraging by nectar feeders, for example) resemble interval schedules. Animals seem to behave as these characterizations imply, responding rapidly under ratiolike conditions, but allowing time to elapse between visits when the probability of finding prey increases with time away.

NOTES

1. Reviews of this work appear in Schoener (1971), Pyke, Pulliam, and Charnov (1977), Kamil and Sargent (1981), and Krebs (1973, 1978). The principle (discussed later in the chapter) that an optimal diet from a set of nutritionally equivalent prey differing in profitability is just the N most profitable types (where N depends on the relative profitabilities and abundances of the most profitable types) was independently proposed by at least nine different people. Some other principles were borrowed from economic theory.

2. See Murdoch and Oaten (1975) and Hassell (1976) for reviews of predation and population stability.

The disk equation is a feedback function, in the sense that it says how the animal's obtained rate of reward depends upon its "attack rate," *a*. The hyperbolic form is reminiscent of the result I derived earlier for a VI schedule, but this similarity is misleading. The natural foraging situation corresponds to a ratio schedule, in the sense that encounter rate is strictly proportional to attack rate during the time the animal is searching. The negative acceleration derives from the increasing time taken up with handling prey, not from the animal's inability to predict when reinforcement is set up.

3. See, for example, the review in Hassell (1978); an introductory summary of much of the same material is Hassell (1976).

4. This argument has been made by Real (e.g., 1980).

5. The general point is that any departure from randomness introduces *redundancy*, in the information-theory sense, into the spatial distribution of prey. Providing the predator has some way of taking advantage of the extra information provided by nonrandomness, experienced prey densities will always be higher than average levels. Consequently, predictions about optimal diet based on average levels will usually be wrong.

6. This concept is originally due to von Uexküll (1934). It was first applied by L. Tinbergen (1960) as an explanation for variations in the diet of bird predators in pine forests. Tinbergen found that the proportion of different cryptic insect prey species in the birds' diets bore no simple relation to the relative abundance of the insects in the birds' environment: An increase in the proportion of a prey type was generally followed by an increase in its proportion in the birds' diet only after some delay. Tinbergen interpreted this lag as the time taken for the birds to "learn to see" the new type.

7. Dawkins' (1971a) data show that, in fact, birds make very few "misses," rarely pecking at nonprey – perhaps because this entails some risk. One way to increase prey detectability is to adjust foraging speed, searching more slowly for cryptic prey. Gendron (1982; Gendron & Staddon, 1983) has shown that quail adjust their search speed as this suggestion implies, searching more slowly for more cryptic prey. This choice to trade off speed for accuracy makes no difference to the optimality analysis in the text. It poses a problem for the animals when foraging simultaneously for prey of differing crypticities, however, since the search speed appropriate for the more cryptic type is unnecessarily slow for the less cryptic, the speed for the less cryptic too fast for the more cryptic. Hence this is another way in which foraging for two prey types is less efficient than foraging for one.

8. *The effect of prey density on prey risk.* The optimal adjustment to changes in the density of cryptic prey can be derived quite simply from the ROC analysis. First, we need to estimate the costs and benefits for the four outcomes in Table 9.1. These are shown in the *payoff matrix* in Table N9.1. Thus, the cost of a false

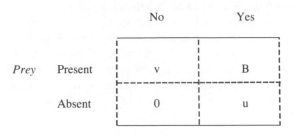

Table N9.1.

alarm is u, of a miss, v, and the benefit of a hit is B; I assume that a correct failure to attack carries neither cost nor benefit. The net benefit, H, to an animal of a particular strategy (criterion) is proportional to these costs and benefits, multiplied by their probabilities of occurrence and weighted by the densities of prey and distractors (nonprey):

$$H = D_P[p(Y|P)B - (1 - p(Y|P)v] -- D_N p(Y|{}^\sim P)u,$$

where D_P and D_N are the densities of prey and nonprey, respectively, and costs u and v are expressed as nonnegative numbers. For readability we can replace $p(Y|P)$ with p and $p(Y|{}^\sim P)$ with q; simplifying then yields

$$H = D_P[p(B + v) - v] - D_N qu. \tag{N9.1}$$

In this equation p and q are not independent; they are related by prey detectability, which defines a particular ROC curve (see Figure 9.3). A simple approximation to standard ROC curves is the power function

$$p = q^s, \qquad 0 < s \leqslant 1, \tag{N9.2}$$

where the exponent, s, is a measure of crypticity: The larger the value of s, the more closely the ROC curve approaches the diagonal $p = q$, hence the more cryptic the prey. Substituting Equation N9.2 in Equation N9.1 and simplifying yields

$$H = pD_P(B + v) - vD_P - p^{1/s} \cdot D_N u. \tag{N9.3}$$

In these equations p represents the animal's criterion, since he can set p anywhere he wants by being more or less stringent. Consequently, finding the best criterion is equivalent to finding the value of p in Equation N9.3 that maximizes H. Differentiating Equation N9.3 with respect to H, setting the result to zero to find the maximum, and simplifying yields

$$\hat{p} = [s(D_P/D_N)((B + v)/u)]^r, \qquad r > 0, \tag{N9.4}$$

where $r = s/(1 - s)$.

It is easy to see that Equation N9.4 has the properties one might expect: Prey risk, $\hat{p}$, is directly related to prey density, D_P, and inversely related to density of nonprey, D_N; $\hat{p}$ is directly related to the sum of the benefits of a hit, B, and the cost of a miss, v, and inversely related to the cost of a false alarm, u. In addition, the steepness of the function depends on the crypticity, s: The more cryptic the prey (the higher the value of s), the more sensitive $\hat{p}$ should be to prey density. (See Staddon & Gendron, 1983, for a fuller account.)

9. This would have to be done using a modified version of the operant conditioning procedure used by Pietrewicz and Kamil (1979), discussed later in the chapter. One way to proceed is as follows. Pigeon or bluejay subjects could be confronted with three response keys. A color slide of either a potential cryptic prey item, against a background ("signal + noise": S) or a background without prey ("noise": N) would be projected on the center key at the beginning of each trial. A few pecks on this "sample" stimulus would then turn on the two outer keys. A peck on the left key would be correct if the sample were "S"; a peck on the right key would be correct if the sample were "N." Both types of error could be "punished" by timeouts (i.e., periods in the dark when no reward is possible). Correct responses would be reinforced with food. After reward or timeout, a new sample would be presented and the cycle would continue as before.

Prey frequency would be varied by varying the proportion of S and N trials; payoff could be varied either by rewarding correct responses on an intermittent basis, by varying amount of reward, or by varying the magnitude of punishment. ROC curves could easily be traced out by any of these methods. If the curve derived by varying prey frequency shows at high prey frequencies significantly higher values for d' than the other curves, one would have strong evidence for the search-image idea.

10. *Switching and functional response.* The general form of Equation 9.4 is

$$R_i = aD_i^2/(1 + a\Sigma h_i D_i^2), \tag{N9.5}$$

for N prey types. It is also not necessary to assume that attack rate, a in equation 9.1, is strictly proportional to prey density, D. Almost any positive monotonic relation can be well fitted by the power function

$$a = kD^m, \qquad m \geq 0. \tag{N9.6}$$

Substituted in Equation N9.5 this yields a general formula for functional response:

$$R_i = aD_i^{m+1}/(1 + a\,h_i D_i^{m+1}). \tag{N9.7}$$

Equation N9.7 is similar to the general form suggested by Real (1977). When $m = 0$ the equation corresponds to the type I ($h_i = 0$, $\forall i$)) or type II ($h_i > 0$) response; when $m > 0$, it corresponds to the type III response. When $m = 1$, and $N = 2$, Equation N9.6 is reduced to Equation 9.4. The relative choice function derived from Equation N9.7 is obviously

$$R_1/R_2 = (D_1/D_2)^{m + 1}, \qquad\qquad (N9.8)$$

which corresponds to switching when $m > 0$.

11. This analysis implies that "switching," in Murdoch's (1969) sense, does not require a type III functional response for each prey type considered separately (i.e., with the densities of all others held constant). However, if each prey type shows the type III response in isolation, then the predator should show switching when confronted with both, assuming that it treats each type independently: The type III response implies switching, but not conversely.

12. Blough (1979) presents elegant experimental data on visual search in pigeons. Experimental results with humans are described in Schneider and Shiffrin (1977), Shiffrin and Schneider (1977), Green and Swets (1966), and Rabbitt (1978).

13. McFarland and his associates have pioneered the application of models of this sort to the analysis of motivational systems. See, for example, McFarland and Houston (1981) and McCleery (1978). An analysis of the nutrient-constraint problem is provided by Pulliam (1975).

14. *Fitness and utility – again.* The definition of "real improvement" is not a trivial matter. Early approaches to animal motivation (e.g., Hull, 1943), identified physiological state as the key factor. Hull, for example, supposed something termed "tissue need" to be the driving force behind action. I argued at length in Chapter 6 that this cannot be correct: It is very poor design indeed to arrange that the animal becomes hungry only when it is beginning to suffer real resource depletion. Hull (1952) eventually recognized this deficiency and added *incentive* as a determinant of behavior: Thus food-related behavior was presumed to be proportional to the *product* of "drive" ("tissue need") and incentive (roughly, the expected rate of energy acquisition) (see also McCleery, 1977).

The final extension was made by thoroughgoing selectionists (e.g., Sibly & McFarland, 1976) who identified Darwinian *fitness* as the key variable. Thus "improvement" in Figure 9.5 must be "improvement in fitness." This has become the working assumption for evolutionary biologists (see Oster & Wilson, 1978; Maynard Smith, 1978) and is the contemporary version of the adaptationist manifesto that essentially every phenotypic feature serves some adaptive function. Although theoretically more justifiable than the much simpler assumption that some measurable aspect of food intake (say) is maximized, it suffers from two practical problems: All motivational systems (for food, water, sex, exploration, etc.) are now lumped together, since all must deal in the common currency of Darwinian fitness. And fitness is essentially impossible to measure directly. It is after all not even *current* fitness that is required, but fitness in the ancestral populations in their "selection environments." In addition, the mechanisms of embryonic and postembryonic development, plus the availability of appropriate variation (in the form of gene combinations and mutations), constrain the set of possible phenotypes in ways that are impossible to pin down in detail. As a practical matter, fitness is impossible to measure directly, and the existence of unknown constraints means that some phenotypic features are likely not to be optimal.

The practical resolution of all this is to deal with measurable aspects of behavior, such as energetic efficiency, but to recognize that these are plausible surrogates for fitness. Despite the obvious philosophical flaws of this position, no real alternative exists. The approach is an essential part of any attempt to understand adaptive behavior and, warts and all, has been exceedingly successful.

15. The alert reader will notice that according to my argument, herbivores are also on a ratio schedule, yet are not noted for the tempo of their foraging. One difference lies in the relation between the cost of foraging activity and its energetic return: For herbivores the relative return on foraging is modest, because of the low energy content of their food. Thus, their ratio schedule is a high-valued one. But as we saw in Chapter 7, on very high ratios, animals will respond slowly; it is only at intermediate values, yielding payoff rates close to the peak of the bitonic response function, that ratio schedules generate high response rates. Thus the leisurely foraging of herbivores does not constitute a paradox.

As a practical matter, herbivores may also be limited by the time it takes them to process their food: Since processing takes so long, and holding capacity is limited, there is also a limit to the useful rate of foraging.

16. *Foraging in a repleting food source.* Davies and Houston (1981) assumed that insects and other prey items arrive in a random way at the wagtail's riverbank territory. Hence, the number of food items arriving within a unit length of territory since the wagtail's last visit is given by

$$x(t) = K(1 - \exp(-wt)), \tag{N9.9}$$

where t is the time since the last visit (the *return time*) and K and w are constants (for a given observation period). Equation N9.9 is of the same form as Equation 8.6, which describes reinforcement probability as a function of interresponse time on (random) variable-interval schedules where K is the magnitude of reinforcement and w is the average rate of reinforcement.

Davies and Houston develop the analogy more fully, as follows. If the wagtail's territory is of length L, then the total number of items obtained is

$$N(t) = Lx(t) = LK(1 - \exp(-wt)), \tag{N9.10}$$

assuming that the same return time holds for all points on the territory, and that there are no invasions by other wagtails (a strong assumption, sometimes!). The return time is given by

$$t = L/v,$$

where v is the speed at which the wagtail walks. If the animal's rate of energy consumption is proportional to v, then the net rate of energy gain is given by

$$N(t)/t - kv, \tag{N9.11}$$
$$= [LK(1 - \exp(-wt)) - kL]/t,$$

where k is a constant.

Now consider an animal working on a VI schedule with scheduled reinforcement rate, w. If the animal adopts a constant interresponse time, t, then the expected reinforcement rate, R, is given by Equation N9.10, where K is the reinforcement magnitude. If each response has a constant energetic cost, k, then, obviously, the net energy gain is given by Equation N9.11.

It seems likely animals adapt so well to standard reinforcement schedules because the schedules resemble common natural situations.

10

STIMULUS CONTROL AND COGNITION

To behave adaptively is to behave differently in one situation than in another. As we move up from paramecia to pigeons and people, the number of different modes of possible behavior increases enormously, and with it the number of different situations for which a unique behavior is appropriate. How do animals organize this knowledge? And how are situations recognized? The two questions are not really separate, since some kinds of organization make recognition quick and accurate, whereas others make it slow and unreliable.

The ability to recognize when particular adaptive behaviors are appropriate has been taken for granted in previous chapters. In this chapter I look at one aspect of recognition and the organization of individual knowledge.

What does it mean to *recognize* something? In a formal sense the answer is simple: It means to be in a unique state (as defined in Chapter 4), so that in the presence of object w, the animal is always in state W, and state W never occurs at any other time. This is also a necessary condition for the animal to *discriminate* w from things that are not w. But the formal answer conceals a great deal. For example, it isn't much help in constructing machines that can recognize, for which much more specific information is required: We need to know how to process particular visual (or auditory, touch, or whatever) inputs; how to direct the visual apparatus on the basis of past information – where should the machine look next? How to distinguish objects from their backgrounds, how to identify the same object from different points of view; and how to encode all this information so that it provides a useful basis for action.

Most of these questions are about *perception*, and I cannot do justice to them here. In order to get on with the study of learning and motivation we must take for granted the processes that translate a particular physical environment into some internal representation that allows the animal to recognize the environment on subsequent occasions (or, more cautiously phrased, the processes that allow the animal to behave in the same way – or ways that are the same in essential aspects – every time he is in the same environment). Perceptual processes are not trivial; in many respects they are much more complicated than the things I do deal with. We pass them by not because they are negligible, but because they are little understood – and because our main interests are elsewhere. This chapter is about the last step in the process: the encoding of information in ways useful for action.

I am concerned not with how the animal "sees" a Skinner box or a colored light, but with how these things resemble or differ from other things in his world.

It is not at all clear that the best way to answer this question is to begin with the concept of *stimulus*, where a stimulus is some very specific, physically defined event. Nevertheless, because the study of learning in animals grew up under the earnest ministrations of behaviorists, the relevant terms, experimental methods, and concepts have all evolved from "stimulus-response" psychology. Hence, "stimulus" is the natural place to start. The term's behaviorist ancestry at least means that we will know what we are talking about, even if we are uncertain about how useful it will prove to be.

I first discuss the concept of stimulus and stimulus element, then describe how control by simple stimuli is measured in transfer tests. The results of transfer tests can often be summarized by rules describing the effects of stimulus compounds in terms of the effects of their constituent elements. Tests are also useful for assessing the effects of reinforcement on stimulus control. The chapter ends with a discussion of similarity and the invariant relations among simple and complex stimuli that may underlie performance in stimulus-control experiments, and in animals' natural environments.

THE DEFINITION OF STIMULUS

Discriminative and eliciting stimuli

The etymology of "stimulus" implies the existence of a response: A stimulus is a stimulus *for* something. As I noted in Chapter 4, this definition is too narrow: A stimulus may change the organism's internal state without immediately eliciting a response. But some stimuli are clearly more important for the actions they produce, others for the change of state. Some stimuli are goads to action; others function more as signals. The stimulus for an "ideal" reflex is a goad: The response follows the stimulus invariably and at once. A "pure" eliciting stimulus is one that produces a reaction but has no other effect on the animal. Obviously, few, perhaps no, stimuli fit this description exactly, but stimuli for simple protective reflexes of decerebrate organisms (light for pupillary contraction, air puff for blinking, touch for the scratch reflex) come close. In reality, of course, the phenomena of temporal and spatial summation, refractory period, and so on (see Chapter 2) show that even eliciting stimuli have effects on the animal that go beyond the response they produce.[1] And normal, not decerebrate, animals can remember past stimuli, which can, therefore, affect future behavior after long delays.

The distinctive property of *discriminative* or *controlling* stimuli (I use these terms synonymously) is that they define a certain *state* of the organism, defined as a set of stimulus-response and stimulus-state-change relations different from those associated with other states. Discriminative stimuli serve as signals defining situations in which a particular course of action is appropriate. For example, young bluejays will attack and eat most butterflies. Monarch butterflies are mildly poisonous and make birds sick. After some unfortunate experiences with monarchs the birds learn to avoid them. The distinctive red and black pattern of the monarch signals (*controls* in the conventional terminology) a pattern of behavior different from that elicited normally by butterflies. Thus, discriminative stimuli often also elicit a response; but their special property is that in the presence of a discriminative stimulus, the animal behaves according to a set of rules different from those applying in the presence of other discriminative stimuli.

The signal defining the situation need not be the thing attacked or avoided; the signal stimulus may be different from the stimulus responded to. For example, on rainy days birds may look for worms brought to the surface by waterlogging of their burrows, whereas on dry days they may look for other prey: The weather is the signal, but the worm is attacked. A hungry dog may rush to its food bowl at the sight of its master if this usually means a feeding. The number of caterpillars a female digger wasp (*Ammophila campestris*) brings back to each of her separate larva is determined during a daily checkout visit: The food store for each larva is the signal controlling subsequent foraging. Even primitive animals can show contextual reactions of this sort; we saw in Chapter 3 that the protozoan *Paramecium* under suitable conditions is geopositive in the light and geonegative in darkness: I termed this kind of flexibility *systematic variation*. For mammals, birds, and a few other animals, systematic variation – sensitivity to context, control by discriminative stimuli – is often acquired during the life of the individual: This is called *learned* systematic variation.

This view of stimulus effects is not too different from the commonsense idea that animals perceive the world as a set of situations, each with its own regularities.

This is an approach, not a finished theory. I present it at the outset because the experimental methods I discuss in a moment, and the history of this field, foster the deceptively simple alternative that operant behavior can be understood solely in terms of stimulus-response relations, where both stimuli and responses are defined as physical events. Skinner[2] enlarged the definition of stimulus and response to embrace classes of physical events linked by a common consequence: A stimulus was the class of all physical stimuli that signaled an operant contingency; a response was the class of all acts (physical topographies) that satisfied the contingency. But he was silent on the relations among these classes and their structural properties. The stimulus-response view, in either its simple or enlarged form, is experimentally convenient and epistemologically hygienic, but provides few clues to understanding how animals work. Because it never goes beyond single stimuli, or stimulus classes, to the relations among them, it is little help in understanding how animals get about in natural environments, why some complex stimuli appear similar whereas others look different, how pigeons differ from people, or why stimuli that are psychologically simple to identify (faces and other natural objects, for example) are often physically complex.

There are two main approaches to these questions. One is perceptual and physiological: to identify the transformations imposed on the physical stimulus by the animal's nervous system that allow it to detect complex invariances, such as the fixed shape of a three-dimensional object perceived from different angles or the constant size of the same thing seen at different distances. Perception is difficult to study with animals, and as I suggested earlier, our knowledge even of human perceptual processes falls far short of answering these questions.   The second approach is functional: The evolutionary function of knowledge must be as a guide to action; hence, the animal's task is always to organize the welter of physical stimulation in ways that enable it to deploy its behavioral resources with maximum efficiency. Physical stimuli that signal the same set of regularities (in the life of the individual, or in the lives of his ancestors) should be treated as equivalent, and come to control the same pattern of adaptive behavior: The class of such physical stimuli then constitutes a stimulus in the functional sense. The relations of similarity and difference between stimuli so defined constitute the animal's knowledge about his world. The functional approach to stimulus control (which turns out to be also a cognitive approach) therefore begins with the study of similarities and differences and hopes to end with some representation of knowledge.

Stimulus equivalence and data structure. The experimental study of stimulus effects therefore boils down to two questions – about stimulus equivalence, and about what may be termed (borrowing from computer science) data structure: (a) *Stimulus equivalence* – What stimuli are equivalent in their effects to a given stimulus? (b) *Data structure* – How many different equivalence *classes* are there and how are they related to one another? Stimulus equivalence is usually studied under the rubric of *stimulus generalization*. The question of data structure has rarely been studied directly, although it is related to the traditional problem of *discrimination.*[3] I return to data structure in the section on similarity.

Species differ greatly in complexity of stimulus classes and data structures. Simple invertebrates are often guided by signals that can be identified with relatively simple physical properties, so they classify only crudely. For example, ticks locate their prey by seeking a certain height above the ground. There the tick waits, until an animal passes close enough to provide the necessary chemical stimulus – whereupon the beast releases its hold and drops on its unwitting host. For a tick, evidently, the world is divided into different heights – most bad, a narrow range good – and the presence versus absence of butyric acid. No doubt the animal can also identify acceptable ranges of temperature, illumination, and humidity, but other features, patent to us, are ignored. The music of Mozart, the beauty of a summer evening, even the difference between one grass and another, all pass it by. The tick asks rather few questions of the world and is content with simple answers. The more intelligent the animal, the less will this be true. Mammals and birds can identify dozens or hundreds of different situations and obviously react to complex properties of their environments. Occasional demonstrations that birds and mammals can be tricked by simple *sign stimuli* (such as the male robin's foolish attacking of a red piece of fluff, or the stickleback's attack on primitive models) attract experimental attention in part because they are easily studied exceptions to the prevailing complexity.

Although everyone knows that mammals and birds must have rich world-models, in practice we are usually reduced to studying one or two simple stimuli and a comparable number of responses. Nevertheless, it is worth remembering that from the animal's point of view each stimulus class has meaning only in relation to others – just as each state of the finite-state systems described in Chapter 4 makes sense only in relation to the whole set of states. A state is a *relation*, defined by difference from other states, not a thing. Consequently, to really understand the effects of any particular stimulus, we need to compare its effects with the effects of a wide range of others.

The needs of a tick are simple, its responses few, and its information-processing capacity limited. It functions rather like a guided missile, using an easily measured cue sufficient to identify its intended host. Butyric acid is evidently an adequate signal for a meal for a tick, as heat suffices to guide the sidewinder missile to a plane or a tank. More complex animals can do and recognize more things, and their niches demand more complex relations between what they can recognize and what they can do. But both simple and complex animals need differentiate among states of nature only to the degree that the states make a difference – both in terms of consequences of value, and in terms of the animal's ability to respond appropriately.

"Consequences of value" are, of course, simply the events termed "reinforcers." States of nature that make no difference, now or in the lives of ancestral animals, will not usually be differentiated – will not produce different states of the animal (see Chapter 4). I describe how stimulus control depends on the function of stimuli as reward signals in a moment.

"Ability to respond appropriately" is harder to define. Animals are limited in three ways: by the physical properties to which they are sensitive; by the responses they can make; and by their ability to integrate stimulus information and use it as a guide to action.

For example, bees are sensitive to near-ultra violet light, but humans are not; hence bees can respond to an aspect of the world that we cannot. Sensitivity to the appropriate physical property is a necessary but not a sufficient condition for adaptive response. Similarly, birds can fly, pigs cannot; some responses are physically possible for an animal, others are not.

But the most interesting constraints are information-processing ones: Octopi are reported to have difficulty in discriminating between a figure and its vertical reflection (between \ and /, for instance). The necessary information is evidently available at the retinal level (the octopus can discriminate between \ and |, for example) but processing limitations prevent the animals from making appropriate use of it. The digger wasp just mentioned is subject to a particularly interesting limitation. A female usually maintains several burrows, each containing a larva. Hence she must keep track of the provisioning requirements at several locations – a considerable feat of memory. Yet she updates her information on each burrow only on the first visit each morning. Experiments show that the information is not updated on subsequent provisioning visits, even if circumstances have changed: If, in between visits, some food is removed, the animal does not make up the loss that day.[4] Again, the information is available, but the animal cannot process it appropriately. The functional explanation for this limitation, of course, is the absence of such interventions in the life of wasp ancestors: A predator that removes the larva's food is likely also to remove the larva, so there can have been no direct selection of rechecking after the first daily visit. (We can confidently predict that the wasp will notice the absence of her larva on later visits, however.)

The most widespread information-processing limitations derive from the imperfections of memory. Most animals are very poor at learning sequences. For example, rats cannot learn to make a sequence of choices such as ABAABBAAABBB to get food. Instead they will make A and B choices in an irregular way so that occasionally by accident the correct sequence occurs and food is delivered. For the rat, the key property – the property it can detect as a predictor of food – may be not the sequential arrangement of As and Bs, but their relative frequency. Perfect responding here requires either learning a *rule* – or exact determination of the nth choice by up to eleven preceding choices, that is, accurate recall of these events in order. People are partly immune from this limitation because symbolic representation – language, numbers, and so on – provides a *digital* method of representing past events. In contrast, "lower" animals appear to encode past events in *analog* form, so that accuracy decreases as sequence length increases. Thus, a rat finds it very difficult to learn sequences of the form NAB (AB, AAB, AAAB, etc.) where N is greater than two or three; a person can count to N and has no difficulty.[5] Nevertheless, in situations where symbolic coding is difficult or impossible, the temporal resolution even of human memory is far from impressive. For example, when similar events occur in temporal alternation, as in changes of serve during a tennis match, after several alternations it may be hard to remember which event occurred most recently.

The "situations" into which animals organize the world are, therefore, limited by their ability to detect certain kinds of relations and not others. Animals are usually good at detecting the times at which food will occur and at identifying situations where it occurs with highest frequency; they are not good at picking up complex stimulus sequences.

Measurement of stimulus control

Stimulus equivalence can be studied with simple stimuli by training an animal with one stimulus and seeing to what extent the trained response will occur to others (see Chapter 4). This approach raises two general questions: (a) How is control established? (b) How do the physical properties of stimuli affect stimulus equivalence? The answer to the first question was foreshadowed in Chapter 5: A stimulus comes to control behavior when it predicts something about positive or negative reinforcement. For example, suppose we take a hungry pigeon and place it in a Skinner box in which every minute or so a red key light comes on for 8 sec. Now imagine two possible experiments with this basic procedure. In the first, the animal is given brief access to food 60 times per hour, with the occurrence of food and onset of the red light determined by independent, random processes. In the second experiment, food again occurs 60 times per hour, but now always at the end of the 8-sec light. In the first experiment, the pigeon will probably look at the light the first few times it comes on, but will soon ignore it. The pigeon will spend no more time near the light than near any other feature of the apparatus an equal distance from the feeder (no doubt he will spend more than a chance amount of time near the feeder).

The result of the second experiment is quite different. The bird will attend more and more to the light instead of less and less, and within 30 or 40 food deliveries is likely to peck at the red key. Once it has developed, this *autoshaped* pecking will be maintained indefinitely. Autoshaping is quite a general result.[6] The particular response to be directed at the stimulus will depend on details of the apparatus and the species of animal. Rats, for example, are notoriously unwilling to peck a key, but they may press or chew an illuminated lever. Most species will learn to approach the signal stimulus when it comes on, and many will also learn something about its fixed duration, approaching the food hopper as the end of the 8 sec approaches. Thus, the rule for the development of stimulus control is that there be a stimulus-reinforcer *contingency*, in the sense described in Chapter 5.

The only way to find out which physical properties of a stimulus are important to the behavior it controls is to vary the stimulus and look at the effect. This is termed *transfer testing*, since the idea is to see how much of the original behavior transfers to the control of the stimulus variants during the test. Transfer testing involves two steps. The first is to identify a physical stimulus that controls an identifiable aspect of behavior. Sometimes the control has been established through explicit training, but it may also be a natural relation. The second step is to vary aspects of the stimulus under conditions where any associated change in behavior can be attributed solely to these stimulus variations.

Vocal communication of the American brown-headed cowbird (*Molothrus ater*) provides a neat example of natural stimulus control. The cowbird is a brood parasite: Like the European cuckoo, it lays its eggs in other bird's nests. This unlovely habit has made it of great interest to biologists from Darwin onward, but for present purposes the important thing is the vocal repertoire of the male. The male cowbird produces a song consisting of a series of whistlelike sounds that elicits a distinctive "copulatory posture" from a receptive female. This response is rapid and easily recognizable; hence it provides an excellent way to measure the effectiveness of song variants. A picture of a typical song is shown as Figure 10.1. In one experiment (West, King, Eastzer, & Staddon, 1979), preliminary tests with tape-recorded songs showed that the song can be divided up into three significant units: phrase 1 (P1), phrase 2 (P2), and the interphrase unit

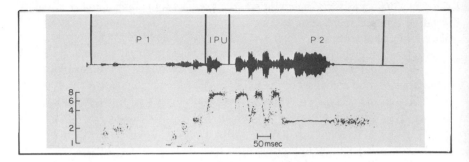

Figure 10.1. Typical song of male cowbirds reared in isolation. The upper display shows the changes in amplitude (intensity) through the song. The lower display shows the associated changes in instantaneous frequency (frequency in kiloHertz on the vertical axis). (From West, King, Eastzer, & Staddon, 1979, Figure 1.)

(IPU). In subsequent tests, songs in which one or two of the three segments had been deleted were played to females.

Table 10.1 shows the results, represented as follows: S = complete song; P1, P2 = phrases 1 and 2; IPU = interphrase unit; S – IPU = complete song with IPU deleted (the same as P1 + P2); P1 + IPU = phrase 1 followed by IPU; P2 + IPU = IPU followed by phrase 2. (Data are not shown on the IPU presented in isolation, since it then had no effect.) Each of five receptive females heard about 200 songs, equally divided among these variants. The table shows the percentage of positive responses over the whole group for each song variant.

The results can be summarized by two statements: (a) P1 and P2 contribute to song effectiveness in an independent additive way: If p_1 and p_2 are the probabilities of response to P1 and P2 presented alone, then the probability of response when both are presented (assuming independence) is just

$$1 - (1 - p_1)(1 - p_2) = p_1 + p_2 - p_1 p_2,$$

that is, one minus the probability that the animals respond to neither P1 nor P2. Plugging in the values of p_1 and p_2 from Table 10.1 yields a predicted joint probability of .41, which compares well with the obtained value (for S–IPU = P1 + P2) of .38. (b) Addition of the IPU approximately doubles the effectiveness of a song variant (compare S – IPU with S, P1 + IPU with P1, and P2 + IPU with P2). Because probabilities cannot exceed unity, the proportional increase is somewhat less as the base value increases (e.g., 75/38 < 46/19).

This example illustrates one plan of attack when attempting to measure stimulus control. First, you need some idea of the general features of the physical stimulus that are likely to be important. In the cowbird case, for example, preliminary work showed that the amplitude of the signal was less important than changes in its frequency over time (frequency modulation). Then these critical features are varied, either by selective omission, as in this example, or by graded variation, as in *generalization testing*.

Table 10.1.

Song variant	Percentage Response
S	75
S − IPU	38
P1 + IPU	62
P1	27
P2 + IPU	46
P2	19

In either case, the spacing and frequency of tests must be chosen with care, so that the response does not change for reasons unrelated to the stimulus. These confounding effects are principally *habituation*, for naturally occurring stimulus-response relations (as in the cowbird example), and *reconditioning* or *extinction* for relations established through differential reinforcement. Habituation is the eventual weakening of a response following repeated elicitation (see Chapter 4). Female cowbirds habituate during repeated song playbacks unless the playbacks are relatively infrequent, as they were in this experiment. Reconditioning, that is the establishment of control by a test stimulus that would otherwise be ineffective, can occur if reinforcement continues to occur during the generalization test. Conversely, the response may extinguish if reinforcements are omitted during the test. I explain the experimental solution to these conflicting requirements in a moment.

If the critical stimulus features have been correctly identified, the results of the tests should lend themselves to a simple description of the sort just offered for the cowbird song. If no simple pattern can be discerned, then it may be either that no simple pattern exists – or we have failed to define the essential stimulus features.

STIMULUS GENERALIZATION

Cowbirds need no training either to make or respond to calls of the type shown in Figure 10.1. (Indeed, one result of this experiment was to show that isolated male cowbirds have more effective songs than males reared normally in the company of their fellows.) Every animal species shows examples of stereotyped, innate relations of this sort.[7] For many, these reactions, together with primitive orienting mechanisms of the type described in Chapters 2 and 3, constitute the animal's entire behavioral repertoire. In mammals and birds, and a few other species, however, much behavior consists of reactions acquired to relatively arbitrary stimuli.

A simple procedure for studying learned stimulus control of this type is as follows. A hungry pigeon is first trained to peck a key for food reinforcement,

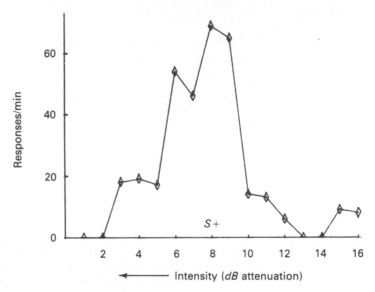

Figure 10.2. Key-peck rate of an individual pigeon as a function of the intensity of a green keylight during a single 60-min generalization test (Staddon, unpublished data).

delivered on a variable-interval schedule of intermediate value (e.g., VI 60 sec). The response key is illuminated with the stimulus of interest (termed $S+$ or S^D, D for discriminative). In early studies $S+$ was usually chosen to be physically simple – light of a single wavelength, a single vertical line, a tone. More recently, stimuli complex in a physical, but not necessarily psychological, sense – pictures of scenes or animals, for example – have begun to receive more attention.[8]

After the pigeon has learned to peck $S+$ for food, variants on the test stimulus are presented for relatively brief periods; for example, if five variants are tried, each may be presented for 1 min in an irregular order, for a total of perhaps 60 presentations. Thus, each variant is presented repeatedly, so that any slow changes in the tendency to respond are shared equally by all. No reinforcement occurs during a *generalization test* of this sort, so as to avoid reconditioning. Extinction is prevented, or at least mitigated, by the VI training. On the VI, long periods without food are common; hence, the animals do not soon cease to respond when food is omitted entirely. The *extinction curve* is quite gradual, and over a 60-min period, responding generally shows little decline. Thus, the average rate of response in the presence of each test stimulus is likely to be an accurate measure of its tendency to facilitate or suppress responding relative to $S+$.

Figure 10.2 shows the results of a typical generalization test. The horizontal axis shows the physical values (here intensities of a monochromatic green light) of eight test stimuli. The vertical axis shows the average number of pecks per minute made by an individual pigeon to each of these values during a single

60-min test session. The resulting symmetrical curve is typical: Responding is highest to $S+$ and decreases more or less smoothly as the physical value of the test stimulus departs from the $S+$ value.[9] Because it is highest at $S+$ and declines on either side, the gradient in Figure 10.2 is termed an *excitatory* or *decremental* generalization gradient.

A great many physical stimulus properties – wavelength of light, line tilt, roughness, spatial frequency, and others – have been tested like this. The almost universal result, with pigeons, rats, monkeys, goldfish and people, is the kind of gradient shown in Figure 10.2: Responding is maximal at (or near) $S+$, and falls off systematically with the physical stimulus difference between $S+$, and the test stimulus.

In some respects this result is unsurprising: Why shouldn't behavior bear an orderly relation to the properties of the physical world? Often the physiological basis for the relation seems obvious. Tonal frequencies are spatially represented on the basilar membrane of the cochlea, for example; many neural transducers, such as those for pressure and light, fire at a rate directly related to physical stimulus intensity.[10] But in other cases, the result is puzzling. Color perception, for instance, depends upon central integration of information from three or four types of photoreceptor, each with a different wavelength-absorption peak. Wavelength, as a continuum, finds no simple representation in such a system. Yet wavelength generalization gradients are among the most orderly, and show no sudden slope changes, even at color boundaries. In a moment I discuss evidence from human experiments suggesting that generalization gradients probably represent something like stimulus *similarity*, a cognitive rather than a purely sensory property.

Compounding of elements

I have now described two kinds of stimulus control: by a stimulus *element* (the cowbird example), and by a stimulus *dimension*. These two kinds of control reflect the different test operations: When a stimulus dimension is varied, but the stimulus element is always present, then we are measuring dimensional control; when the element is sometimes removed entirely, we are measuring control by the element.

Elements and dimensions can be defined by simple physical properties, such as wavelength, or in some other way. For example, rather than splitting up the cowbird song into elements, we could have considered "proportion of total song" as a stimulus dimension: This is a perfectly objective property, but it would it not have been very useful because, as we have seen, the elements (P1, P2, etc.) vary greatly in their effectiveness as elicitors of the response, and two of them combine additively, whereas the other one seems to act multiplicatively. In other words, the things we choose to vary in a generalization test cannot be arbitrary. We must judge the correctness of our choice by the comprehensibility of the results. The justification for labeling P1, and P2, and IPU as elements in the cowbird song is that they *do* behave in sensible ways when compounded.

Subjective experience suggests two ways that stimulus elements can combine, and these seem to correspond to different algebraic rules. For example, a visual stimulus element, such as a triangle, must have some color: Neither form nor color can exist in isolation. People and animals tend to perceive colored objects as wholes; they don't normally attend to form and color separately. Dimensions treated this way are described as *integral*, and they roughly follow a multiplicative rule – a value of zero on either dimension and the stimulus has no effect: A form with no color cannot be seen. On the other hand, it is easy to imagine visual displays whose elements are not so intimately related: A pigeon might attend to the stars or to the stripes in a picture of Old Glory, for example. Stimulus elements of this sort are termed *separable* and follow an additive rule.[11] In the cowbird experiment, the two phrases P1 and P2 appear to be perceived by the birds as separable elements, since their effects were independent. But the interphrase unit, IPU, looks more like an integral element, since it had little effect on its own but greatly enhanced the effect of other elements with which it was compounded.

Animals must behave with respect to objects or states of the world, not stimulus dimensions or elements. One use for the elegant technique of generalization testing, therefore, is to shed some light on the way that the animals classify things as a guide to behavior. Since objects differ not in one dimension but in many, the interactions among dimensions have first claim on our attention. Unfortunately, rather little is known about multidimensional stimulus generalization. One reason is technical: It is no simple matter to create and manipulate at will multidimensional stimuli. A second reason is that once invented, techniques take on a life of their own. We know that rewarding an animal for pecking a key illuminated with monochromatic light will cause him to attend to wavelength. Why not look at the effects of reinforcing two or more wavelengths, of alternating reinforced and unreinforced stimuli of different wavelengths, of successive versus simultaneous comparison, and so on? All these things have been done, usually with orderly results not devoid of interest – the next chapter is largely concerned with them. But the relation between these neat manipulations and the animal's knowledge about its world, its *umwelt*, in von Uexküll's phrase (1934), is not always clear. The technique of generalization testing has generally been used to study the effects of reinforcement rather than either stimulus equivalence or data structure.

Stimulus control and reinforcement: attention

Although unidimensional gradients leave us rather far from understanding cognition, they are useful tools for the study of reinforcement mechanisms. The steepness of the gradient is a measure of the degree to which the animal's natural tendency to vary – to respond indiscriminately – is restrained by the schedule of

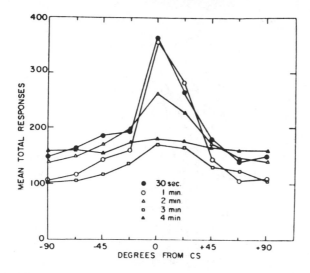

Figure 10.3. The total number of key pecks made by five groups of pigeons, each trained with a different VI schedule (ranging from VI 30-sec to VI 4-min), to a line-tilt stimulus during a generalization test. *S +* was a vertical line (0°). (From Hearst, Koresko, & Poppen, 1964.)

reinforcement. Gradient steepness is also affected by the availability of other sources of control.

For example, Hearst, Koresko, and Poppen (1964) trained pigeons to peck a key on which a vertical line is projected and looked at the effect of overall rate of reinforcement on the steepness of the gradient for line tilt. Different groups of birds were trained with different VI schedules. Hearst et al. found that the higher the rate of reinforcement, the steeper the gradient. These results are shown in Figure 10.3. Others have found similar results. Evidently, the better the correlation between a stimulus and food, the tighter the control of a given stimulus dimension over operant behavior – the greater the restraint of the animal's natural tendency to vary. The critical factor is the predictive nature of the stimulus. In general, the stimulus, or stimulus aspect, that best predicts reinforcement will be the one with sharpest control.

A related effect is that control by a highly predictive stimulus aspect tends to weaken control by a less predictive aspect. Thus, in another condition of the Hearst et al. experiment, pigeons were trained to peck a vertical line, but the schedule was spaced-responding (food was delivered only for pecks separated by more than 6 sec) rather than variable-interval. In a spaced-responding schedule, *postresponse time* is the best predictor of food delivery. The stimulus on the response key guides the actual peck, but has no other significance. On the VI schedule, on the other hand, the stimulus on the key is the best predictor of food.

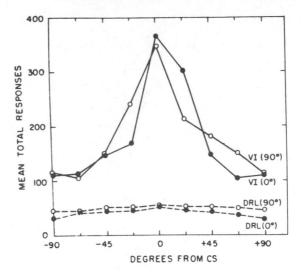

Figure 10.4. Generalization gradients of line tilt following VI or spaced-responding training. There were two groups in each condition, one trained with a vertical line as $S+$, one with a horizontal line as $S+$. (From Hearst et al., 1964.)

As this difference would lead one to expect, a generalization best showed that control by the line-tilt dimension was much worse for spaced-responding animals than for VI animals. This result is shown in Figure 10.4.

These results lead naturally to the concept of *attention*, and to the generalization that animals, sensibly enough, attend preferentially to stimuli, and stimulus properties, that best predict the available goodies (or baddies). The steepness of the generalization gradient provides that the animal is attending to a given dimension; a flat gradient that he is not (more on attention in Chapter 13).[12]

Attention to different dimensions

Animals may attend differentially to stimulus dimensions, or stimulus elements. But comparing the control exerted by different dimensions is not always as obvious as in the preceding example, because gradients on different dimensions are not directly comparable. For example, suppose we train a pigeon to respond to a monochromatic light (green, of 550 nm, say) of certain luminance (say 20 dB above threshold[13]). In a two-dimensional generalization test we present stimuli differing both in luminance and wavelength from $S+$; five values of each yield a total of 25 test stimuli.[14] Suppose that as test wavelengths vary from 400 to 550 nm (at a constant 20 dB luminance), response rates vary from 20 to 60 per minute. As luminances vary from 0 to 20 dB (at a fixed 550 nm wavelength), response rates vary from 30 to 60. Are we justified in saying that since rates vary over a wider range when wavelength is varied and intensity held constant than in

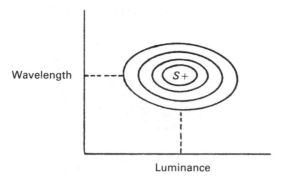

Figure 10.5. A hypothetical two-dimensional generalization gradient. Contours are lines of equal response rate.

the converse case, the animal attends more to wavelength than intensity? Not really, because we could have chosen a different range of intensities – from –10 to 20 dB, for example, over which the range response variation might have been greater.

There is, of course, a natural scale imposed by the range over which the animal is sensitive to each dimension. If these ranges are known, then the range of experimental variation on the two dimensions can be made comparable in terms of them. For example, if the animal is sensitive to a range of sound intensities of 1 to 10^{10} (i.e., 10 log units – 100 dB), but only to a range of 1 to 10^2 tonal frequencies (2 log units), then a change of 10 to 1 in intensity (1 log unit) is roughly equivalent to a change of only 1.58 to 1 (.2 log units) in frequency.[15]

The results of a hypothetical two-dimensional generalization experiment are illustrated in Figure 10.5, which shows the two independent stimulus dimensions of luminance and wavelength and the response rates to each luminance–wavelength combination, represented as contours of equal response rate. The two-dimensional gradient is a "hill" with its peak at $S+$. If the two dimensions are appropriately scaled, "equal attention" to both dimensions is represented by circular contours – a symmetrical hill. Of course, we cannot always be sure that axes are, in fact, scaled appropriately, but in any case *changes* in the contours of Figure 10.5 can always be assessed.

Figure 10.5 represents how two physical dimensions, such as luminance and wavelength, may combine to affect response rate when one particular wave-length–luminance pair is associated with reinforcement. Yet this representation is probably not an accurate model of how the animal sees things, for two reasons: First, for people, at least, the similarity relations among colored objects are not consistent with a space in which luminance and wavelength are at right angles (I describe the evidence for this in a moment). Second, Figure 10.5 does not separate how things look (i.e., how they are represented internally) from how the animal should act with respect to them. I just suggested that "attention," defined

Table 10.2.

	Test stimulus	Bird no.		
		65	66	71
	C	.39	.07	.24
	F	0	.22	.10
(S+)	CF	.51	.24	.62
	c	0	.01	0
	f	0	.21	0
(S−)	cf	0	0	0
	cF	0	.01	0
	Cf	.10	.23	.05

as the effect of reinforcement on stimulus control, corresponds to the shape of the contours in Figure 10.5, but there is no reason to suppose that the way colors look, and how similar they are to one another, is much changed by the association of some colors with food (although the idea of search image, discussed in the last chapter, is in partial contradiction to this – but search images are probably restricted to stimuli that are intrinsically hard to discriminate, so they may represent a special case). The notion that animals can separate the physical properties of events from their hedonic consequences seems to be a prudent working hypothesis.

Attention is usually guided by reinforcement predictiveness (contingency) but when several stimulus aspects are equally predictive, individuals may differ in their attention to different features. The contours in Figure 10.5 are likely to differ somewhat from animal to animal, and even at different stages of training. Individual differences in attention can be particularly striking when different elements are involved. For example, in one experiment,[16] pigeons were trained on a *successive discrimination*: Two color–form compound stimuli alternated at 60-sec intervals. In the presence of one compound stimulus (S+, e.g., a white triangle on a red background) food reinforcement occurred on a VI 60-sec schedule; food was never delivered in the presence of the other color–form compound (S−, e.g., white cross on a green background). This arrangement, in which successive stimuli signal different conditions of reinforcement, is termed a *multiple schedule*; this particular version is a multiple VI EXT. After the birds had learned to peck at S+ but not at S−, they were given a generalization test with the four individual stimulus elements (two colors, and two forms projected on a black background), and the four possible color–form compounds.

Typical results from three pigeons are shown as Table 10.2. C and F denote the color and form elements of S+, c and f the elements of S−; CF, cf, and so on are the test compounds. The entries in the table are the proportions of responses during the entire generalization test made to each of the eight test stimuli. The different results from the three pigeons are typical: One bird, no. 65, apparently attended almost entirely to color; the bird responded not at all to individual elements other than C. A second bird, no. 66, attended primarily to form, responding negligibly to C. A third animal, no. 71, clearly attended to both color and form, allocating substantial proportions of responding to both C and F.

More careful scrutiny of Table 10.2 shows that these simple characterizations are approximate at best. For example, bird no. 65 responded only to stimuli containing C, the $S+$ color, but it also responded much more to CF, $S+$ itself, than to C alone. Clearly, F, the $S+$ form, has an effect in the compound, even if it elicits no responding by itself. Bird no. 71 shows a similar superadditivity: The proportion of responding to CF is considerably greater than the sum of response proportions to C and F. Bird no. 66 responded much more to form than color, but it also responded to f, the $S-$ form. The response rule for this animal looks like, "respond to any form, except when the background is the $S-$ color."

These results illustrate three general principles. One is familiar: Animals will learn to respond for food in the presence of a stimulus[17] that signals responding will be effective and will learn not to respond in the presence of a stimulus signaling that responding will be ineffective. That is, they can learn a successive discrimination. In any discrimination task, $S+$ and $S-$ are likely to differ in several ways. Hence, discrimination *performance* – responding to $S+$ but not to $S-$ – is consistent with several different patterns of *control* by the elements and dimensions of $S+$ and $S-$. The second principle is that different animals are likely to pick different patterns of control: In learning, what is not explicitly constrained (by the contingencies of reinforcement) is very likely to vary. Third, test results show that stimulus control involves both *excitatory* and *inhibitory* effects. Excitatory stimulus control is familiar, but as the results for bird no. 66 show, a positive result can also be achieved by a mixture of excitatory and inhibitory effects. In fact the evidence reviewed in the next chapter shows that all stimulus control may be excitatory, but inhibitory effects are produced by excitatory control of activities antagonistic (in the sense that they compete for the available time) to the measured response.

Since different animals can reach the same end – respond to $S+$, but not to $S-$ – in different ways, tests are required to pinpoint the particular rule followed by a given animal. The rule postulated for bird no. 66 in Table 10.2 – respond to any form, unless the background color is c – implies two things not explicitly tested: that bird no. 66 would respond even to a novel form, and that c, the $S-$ color, is an inhibitory stimulus. To test the first implication, a new test stimulus, say a circle, would have to be presented; the prediction is that the bird would continue to respond. To test the implication that c is inhibitory, c must be compounded with a new form known to produce responding when projected on a "neutral" background; if c is indeed inhibitory, the result should be a decrease in the level of response. This is the standard test for inhibitory control by a stimulus element.

Inhibitory control by a stimulus *dimension* is also associated with the production of an inhibitory or incremental generalization gradient. All these effects reflect interactions among competing activities, each controlled in an excitatory way by different stimulus aspects; I return to these topics in the next chapter.

We are obviously much farther along in understanding the ways in which particular physical stimulus elements control the overt behavior of animals than in understanding how they put these things together to represent the world. Yet intuition suggests that underlying the malleability of behavior in response to shifting contingencies of reinforcement must be some invariant structure corresponding to the unchanging aspects of the physical world. Time, three-dimensional space, the properties of solid objects in relation to one another, all are independent of the animal. Although by judicious manipulation of rewards and punishments we can cause animals to do different things with respect to different physical features, still the features themselves are not changed. In a properly designed organism, one feels, there should be some representation of these invariances, an *umwelt*, that remains even as behavior shifts in response to the exigencies of reinforcement. Some promising beginnings have been made toward measuring how people represent objects, and some of these ideas have implications for animal studies. I conclude this chapter with a discussion of this work.

SIMILARITY

People can readily group things on the basis of similarity. Asked to classify birds into three or four groups, most would place *hawk* and *eagle* in one class and *robin* and *blackbird* together in another. Experiment has shown that people can give a number to the *dis*similarity between pairs of things: *hawk* and *robin* would get a high number, *hawk* and *eagle* a low one. These numbers can be used to define a *similarity space*, in which the distances between objects correspond to their dissimilarity: *Hawk* and *eagle* would be close together in such a space, *hawk* and *robin* would be far apart. Other experiments have shown that distances in similarity space predict the time people take to switch attention from one object to another: If the two objects are similar (close together in the space), the time is short; if they are dissimilar (far apart in the space), it is long.[18]

There are two kinds of similarity space: The simpler kind takes physically defined objects and transforms their physical dimensions so that the transformed values reflect similarity. The second kind is derived by multidimensional scaling techniques from empirical data; the dimensions of such a space need not be simply related to physical dimensions. I give some examples of the first approach now. Data on similarity relations among colors, and among Morse-code characters, discussed later, provide examples of the second type.

The essence of the first kind of space can be illustrated by a simple example. The left panel of Figure 10.6 shows the results of a hypothetical experiment in which generalization gradients were successively obtained with three different

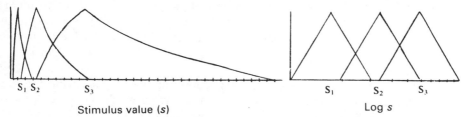

Figure 10.6. Left panel: three hypothetical generalization gradients in which range is proportional to $S+$ value. Right panel: gradients as they appear when the stimulus axis is logarithmic.

$S+s$: S_1, S_2, and S_3. The figure shows that the gradient spread is directly related to the value of $S+$: The same stimulus difference produces a large decrease in response rate in the test with S_1 as $S+$, a much smaller decrease when S_3 is $S+$. This difference implies that the psychological effect of the same physical difference is substantial at the low end of the scale, but less impressive at the high end. This dependence of differential sensitivity on base value, the Weber–Fechner relation, is characteristic of many physical dimensions, notably sound and light intensity, and time: For example, a second is a lot when one is judging intervals on the order of 3 or 4 sec; it is not so much when judging intervals on the order of minutes.

The right panel of Figure 10.6 shows the transformation of the stimulus axis that reduces these three gradients to the same form, namely, $s' = \log s$, where s' is psychological (transformed) stimulus value, and s the physical value. Thus, in this hypothetical case (which is not too different from many actual cases that obey the Weber–Fechner relation), equal psychological *differences* are represented by equal physical *ratios*. Note that the transformation here applied to the physical stimulus axis is unrelated to the actual form of the generalization gradient. The essential property of this transformation is just that it renders equal deviations from $S+$ equally effective, in terms of the measured change in responding.

Figure 10.6 illustrates a simple transformation that relates the unidimensional world of physical intensities to its psychological representation. The same strategy can be applied to more complex aspects, such as color. For example, numerous experiments have obtained judgments from people of the subjective similarity of color samples. In an experiment by Ekman (1954), subjects were asked to rate the similarity of all possible pairwise combinations of 14 spectral (single-wavelength) colors. The results can be plotted as a series of 14 generalization gradients: For each color (wavelength), the height of the gradient is just inversely related to the judged similarity between it and any of the other 13 wavelengths. Shepard (1965) found that these gradients could not be made similar by any transformation of the wavelength axis that preserves the straight-line form with red and blue at opposite ends. He showed that the gradients can be made similar by a transformation that allows the wavelength axis to curl around so that the shortest wavelengths are adjacent to the longest. The transformed gradients are shown at the top of Figure 10.7; the middle panel shows the 14 transformed gradients superimposed on each other, and the bottom panel shows the transformation used: It is just the familiar *color circle*, in which the blue end joins up with the red end, with purple in between.

Two things about the color circle are worth noting. First, although there seems to be only one physical dimension involved, wavelength, the space is two-dimensional. Second, although distances (dissimilarities) are measured directly from one wavelength to another, the region in the center of the circle does not contain wavelengths. In fact, of course, regions inside the circle correspond to *desaturated* colors (single wavelengths diluted with white light). Thus, the two-dimensional similarity space does correspond to two psychological dimensions: color, around the rim of the circle, and saturation, varying radically, from white in the center to highly saturated at the rim.

The color space is rather unusual as spaces go. It is two-dimensional and Euclidean (in the sense that distances conform to the Pythagorean theorem), but the physical dimensions of the stimuli in it don't follow a simple pattern; "north–south" does not correspond to one physical dimension and "east–west" to another. Physical dimensions need not follow any particular pattern in a similarity space. Nor is it necessary that psychological dimensions such as color or saturation be located in any particular position within it. Its essential property is that it accurately represent the invariances in a set of behavioral data.

In both these examples the "objects" dealt with have been simple physical quantities. This reflects the usual bias of the experimentalist, but it is not necessary – and it may not even be the best way to understand how stimuli are represented by animals. The power of the similarity-space approach is that one can begin with essentially *any* object, even (perhaps especially) "natural" objects, such as color pictures of actual senses.

Practical difficulties have meant that, with one exception, rather little work of this sort has been done with animals. The exception is of course work on *orientation*, how animals find their way about their natural environments. There is now ample proof that rats, for example, do so by means of a map that represents the distances and directions of objects in a more or less Euclidean way. The function of particular stimuli is not so much to directly elicit approach or avoidance as to tell the animal where he is in relation to his map, that is, to function as landmarks. We saw a primitive example of this in the light-compass reaction discussed in Chapter 3. Many animals use this reaction (modulated by an internal clock) to define north–south by the position of the sun; given a single landmark or a cue for latitude, they can then locate themselves perfectly.

In the laboratory, experiments with mazes show that visual cues outside the maze act as landmarks rather than simple push–pull stimuli. For example, in a recent series of experiments, rats have been presented with an eight-arm radial maze with food boxes at the end of each arm. Rats very quickly learn to pick up each piece of food without revisiting an arm. To do this, they need to know

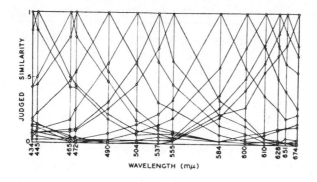

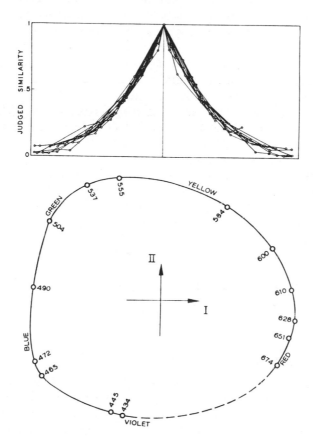

Figure 10.7. Top panel: generalization gradients derived from similarity judgments of color samples, with the stimulus axis transformed as shown in the bottom panel. Middle panel: transformed gradients from the top panel superimposed to show similar form. Bottom panel: stimulus transformation derived by multidimensional scaling techniques – a close approximation to the color circle. (From Shepard, 1965.)

where each arm is in relation to their current position, and they seem to do so by forming a map of the maze with extramaze cues as landmarks. Thus, if, after picking up four pieces of food (i.e., visiting arms), a rat is covered up in the center of the maze and the arms are rotated, it will revisit arms. However, these revisited arms will be in the spatial position (in relation to the rest of the room) of arms the animal had not visited before the maze was rotated. Well-trained rats do not explore the maze in any systematic way, but rather know the spatial location of each arm, and can remember the locations they have not visited.

In one ingenious experiment, a radial maze was covered with a black cloth dome onto which stimuli could be fixed. In a series of tests, this stimulus array was altered in various ways to see if individual stimulus elements were guiding the animals. The results showed that the animals were using the whole array to orient their map with respect to the maze. Small changes in the array produced no effect; large changes caused the animals to behave as if in a new maze. No individual element had a dominant effect.[19]

So far as we know, cognitive maps are pretty accurate Euclidean representations. When dimensions other than spatial are involved, however, physical and psychological dimensions are rarely identical. For example, in an early study by Rothkopf[20] a large group of human subjects was asked to identify Morse-code signals, a set of 36 different patterns. Rothkopf obtained a 36×36 *confusion matrix*, in which the entry in cell (i, j) (row i, column j) indicated the percentage of trials that i was identified as j – an obvious measure of the similarity of signals i and j. Shepard applied multidimensional scaling (a technique that automatically finds the space with the smallest number of dimensions necessary to accommodate a given set of similarities[21]) to these data and came up with the two-dimensional Euclidean space shown in Figure 10.8. As in the color example, simple physical stimulus properties are regularly (but not orthogonally) arranged in the space. There is a gradient corresponding to proportion of dots to dashes going from left to right and another gradient corresponding to number of dots or dashes going from top to bottom.

Spatial representation as a data structure

Spatial representation is a very general idea – we saw it applied with some success to motivational questions in Chapter 7, for example. The world of simple animals, such as the tick referred to earlier, lends itself easily to a spatial description. The animal is sensitive to relatively simple physical properties such as temperature, illumination, altitude above ground (although this may be computed indirectly, from things like illumination, temperature gradient, and crawling time), humidity, time of day, and various chemical stimuli. These together define a multidimensional space, and any environment a point in such a space. The representation itself immediately solves the problem of recognition: Since the space is defined by just those physical dimensions that are important to the animal, the location of the representation point constitutes "recognition" of a

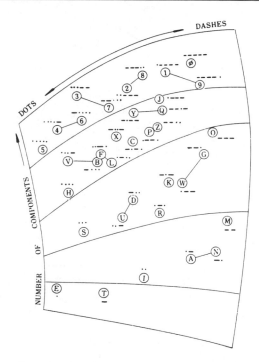

Figure 10.8. Two-dimensional similarity space for the Morse code, derived by nonmetric multidimensional scaling by Shepard (1963) from the complete confusion matrix giving percentage of "same" judgments, obtained by Rothkopf (1957).

situation. The beast comes equipped by its heredity with a set of preferences: indifference contours in this "world space." By moving around, the animal changes its location not only in the real world, but also in its world space. Simple orientation mechanisms of the type discussed in earlier chapters can then be used to hill climb up the value gradient. In this way the tick finds the right height, temperature, humidity, and other things his genes tell him will promote its survival.

Spatial representation is not just a theoretical trick, it is an efficient data structure. It represents environmental features vital to ticks in a form that allows the animal to compare any set of dimensional values and make a decision about where to go next. In essence, it provides a *relational data-base-management system* (to borrow another computer-science term); that is, a system for representing data that allows open-ended questions to be asked of it: No matter what environment the tick finds itself in, it is never at a loss to evaluate it or decide what to do about it. Spatial representations are efficient, but that doesn't mean evolution has built organisms to use them. Nevertheless, something like this is probably our best current guess about how animals represent their worlds.

The similarity-space idea can be adapted to represent the findings of the

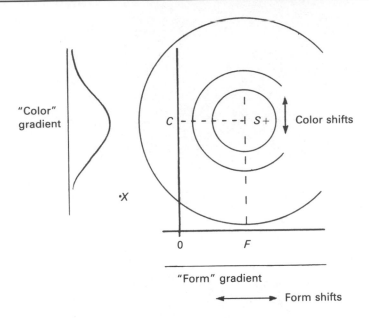

Figure 10.9. Location of the representative point in a hypothetical two-dimensional color–form similarity space sufficient to represent attention to color (a peaked gradient) but not to form (flat gradient). Point X is the converse case.

two-dimensional attention experiment described earlier (Table 10.2). Figure 10.9 shows a set of isosimilarity contours in the vicinity of $S+$ in a two-dimensional color–form space. I have made the simplest possible assumption about the form of these contours, namely, that "color" and "form" are equally weighted, so that the contours are circular. The vertical axis corresponds to "similarity to $S+$." As in the Morse-code example, we can expect that colors and forms will correspond to gradients in this space. For simplicity I assume these gradients are at right angles, but they need not be.

Now recall the experimental result (for two pigeons) that in single-element tests, only one element of $S+$ (color, C, for one bird, form, F, for the other) sustained responding. Nevertheless, in the two-element tests, responding to $S+$ was always greater than responding to the single element: CF > C, for the color animal, CF > F for the form animal. Suppose we represent the color-dominant bird by point 0 in Figure 10.9. Varying color then corresponds to movement along the vertical line through 0. Since this line cuts several isosimilarity contours, similarity to $S+$ varies as color varies and this animal will show control by color: Response rate at point C ($S+$ color) will be higher than response rate at point 0 (no color, or at least a color different from the $S+$ color). This control by color is indicated by the gradient on the left. But this same animal will show no control by form, which corresponds to movement along the horizontal line through 0 – because this line cuts no isosimilarity contours.

The other bird showed opposite results: control by form, not by color. Nevertheless, he can be represented in the same similarity space – by point X. As before, change in color corresponds to vertical movement, but since a vertical line through X cuts no isosimilarity contours, the form-dominant animal will show no control by color. Conversely, horizontal movement from X does cut isosimilarity contours, and this animal will, therefore, show control by form. Both animals will respond more to $S+$ (CF) than to either C or F alone, however.

Each animal in Table 10.2 corresponds to a particular location of the representative point in similarity space. Thus, the location of the point corresponds to "attention": When the point is closer to $S+$ on the "form" axis than the "color" axis, the animal appears to attend more to color (in the sense that color variations cause larger response-rate changes than form variations); when it is closer to $S+$ on the "color" axis, it attends more to form. Thus, the representation in Figure 10.9 separates attention (i.e., response-rate changes) from stimulus similarity; it provides what we are looking for: an invariant representation of the world that underlies diversity of *performance*.

The representation in Figure 10.9 can also describe results showing that generalization gradients are steeper when reinforcement rate is higher, and shallower when other things predict food better than the stimulus being varied. If the "hill" in Figure 10.9 has the bell shape indicated on the left, then the farther away from $S+$ the representative point is, the shallower will be the gradient when any physical dimension is varied. If "attention" to a stimulus corresponds to how close the representative point is to the stimulus in similarity space; if alternative sources of reinforcement tend to move the point toward the stimuli that predict each source; and given that movement toward one stimulus will generally mean movement away from another – then the point will tend to be relatively far away from stimuli that are relatively poor predictors of reinforcement, such as stimuli associated with low rates of reinforcement. Consequently, control by such stimuli will be poor.

Figure 10.9 is perhaps best thought of a small section of similarity space. In reality, other stimuli, such as $S-$ and the context in which $S+$ and $S-$ occur, must also be represented. Unfortunately, in the absence of a detailed, experimentally based map, Figure 10.9 is just a suggestion of the possible similarity relations that may underlie stimulus control. The figure contains many arbitrary features: the orthogonal axes, the circular isosimilarity contours, the assumed proportionality between response rate and similarity to $S+$, and, most arbitrary of all, the assumption that the space is two-dimensional. Nevertheless, such a scheme is sufficient to bring together a number of experimental results, and is open to verification – although the experimental effort required to map out similarity spaces is obviously substantial. Practical limitations on the number of stimuli that can be presented in a generalization test, and limits to the reliability of response-rate measures, may mean that better methods are needed before we can assess animals' cognitive representations of simple objects as well as we can those of people.

SUMMARY

This chapter began with the problem of how animals recognize situations. I took recognition for granted in earlier chapters, where we simply looked at how animals adapt to various procedures without asking how they know when a particular strategy is appropriate. Traditionally this problem was handled by the concept of *stimulus control*. Particular physical stimuli, or classes of physical stimuli, were said to control particular responses, or classes of responses. This worked well, as far as it went, but it cannot explain relations among complex stimuli. To answer these and other questions, some notion of an *internal representation* seems to be required. A simple and powerful one is to represent things in a multidimensional space. The problem of recognition is solved at once, because any given environment defines a particular point in such a space. If, either innately or as a consequence of past training, different regions of the space have different values, then this representation provides a guide for action, in the sense that it lets the animal know whether his actions have made things better or worse. Thus some kind of spatial representation is our best current guess for the way that animals represent much of their world.

But it is only a guess: Traditional experimental methods of transfer testing with varied elements or physical stimulus dimensions are directed at very different questions and, so far, tell as rather little about the problem of representation. They do tell us something about the rules describing how physical stimulus elements combine, and about the effects of reinforcement on stimulus control. Stimulus elements can combine additively, multiplicatively, and according to other quantitative principles, all of which can be represented spatially. Stimulus control, measured by the steepness of generalization gradients, is directly related to how well a stimulus feature predicts a valued outcome.

Human experiments on the judgment of similarity provide useful clues to the way that nonverbal objects are represented internally. Some of the results of generalization experiments can also be represented in this way. It is possible that extensions of this approach to animals can tell us more about what they know.

This chapter has been a most unsatisfactory one to write, because the techniques that have been developed to study stimulus effects seem so inadequate to the real task. Generalization testing with single wavelengths was fine so long as we all believed that everything we needed to know about animals could be built up in a routine fashion from atomic stimulus-response units. Research in cognition and machine intelligence has shown, I believe conclusively, that behaving as intelligently as even simple animals behave requires much, much more. To recognize hundreds or even thousands of different situations and to respond rapidly, precisely, and adaptively to them demands a sophisticated world model, as well as a substantial repertoire of stimulus-response routines. We have made much progress in understanding what animals do, but we are still groping to understand what they know.

NOTES

1. The effects of eliciting stimuli such as those for simple reflexes are usually thought of as invariable: A patellar tap always yields a knee jerk. This may well be true of many protective reflexes, but there is no reason to restrict the term *elicitation* just to these. The normal human reaction to stimuli such as a fire alarm or a red stoplight is no less elicited because it is also dependent on contextual cues.

2. The term *discriminative stimulus* was coined by B. F. Skinner (1935, 1938) to refer to the special relationship between an "arbitrary" stimulus and a response reinforced in its presence. Such stimuli were said to "set the occasion" for the response, even though in many cases the relation is at least as immediate and powerful as the relation between the reflex (eliciting) stimulus and its response (see Staddon, 1967, 1973). The definition I am proposing here does not force one to draw theoretical distinctions between stimulus-response relations that are empirically indistinguishable. It nevertheless accommodates actual usage of the term discriminative stimulus. (See also Chapter 4.)

3. *Discrimination and psychophysics.* I am here passing over in silence several issues important in the history of stimulus control. One such is the relatively unhelpful influence of *psychophysics*, the psychology of sensory processes. In the study of human vision, hearing, and touch, it has often been useful to look at the limits of performance of particular sensory systems, attempting to answer such questions as, What is the dimmest light that can be seen, or the quietest sound that can be heard? This is the problem of *absolute threshold*, briefly mentioned in Chapter 2. A related question is the discriminability of differences, What is the smallest change in light or sound intensity that can be detected? This is the problem of *difference threshold*. It is natural to suppose that generalization, the problem of equivalence, might have something to do with difference thresholds. For example, over regions of the wavelength spectrum where the difference threshold is small, a given physical stimulus difference will encompass many *just noticeable differences* (JNDs), as the difference threshold is termed; the same physical difference over a region where discriminability is poor will span fewer JNDs. Hence, the same physical difference should appear larger in the first case than in the second.

This argument led early students to the hypothesis that the steepness of the generalization gradient at different parts of the wavelength spectrum should be related to the size of the JND. This turned out to be false under most circumstances. Unless special steps are taken to push sensory systems to their limits this result is typical: Generalization effects are not simply related to psychophysical measures. What is left (I will argue shortly) is the notion of *similarity*, a cognitive rather than psychophysical idea. Ultimately stimulus equivalence is determined by the demands the environment makes on the animal. Situations that make similar demands – either now, or in the history of the species – will be treated as, and will appear, similar.

The study of stimulus effects on operant behavior has traditionally been divided into the two areas of stimulus generalization and stimulus discrimination. Stimulus generalization is not too different from the more general question of stimulus equivalence, although both practical difficulties and the influence of psychophysics have tended to concentrate experimental effort on simple physical dimensions such as sound intensity and wavelength of light. But stimulus discrimination has been almost exclusively studied as a psychophysical problem, hence one of more relevance to sensory physiology than to the behavior of the whole organism. For behavior-as-a-whole, the properties of the set of stimulus-equivalence classes – the set of all those things perceived as different by an animal – are of much more relevance than the limits of sensory transduction.

4. The work on the digger wasp was largely done by G. P. Baerends and his associates in Holland. Excellent accounts in English are by Tinbergen (1951, 1958).

5. The limitations on analog coding derive from the psychophysical fact that error is often proportional to the absolute value of the coded quantity (the Weber-Fechner relation). Thus, in a counting schedule, error is proportional to N. In digital coding, the intrinsic error rate is lower and not proportional to N. Of course, we have no real idea how things are coded in the brain, and use of the terms *analog* and *digital* here is to some degree metaphorical – although the digital nature of symbolic communication, at least, is self-evident.

Despite this evidence for analog coding of times and stimulus values in animals, there is no reason to suppose that the fundamental operations of brains are anything but digital, although the evidence is indirect and not conclusive: No man-made analog machine can perform with anything like the intelligence of a suitably programmed digital computer. There are human "lightning calculators" who can perform extraordinary computational feats, apparently often in a largely unconscious fashion – yet accurate numerical computation demands digital processing. It is hard to see how any complex analog process can avoid the cumulation of error. Every highly developed information process that we know of, from the genetic code to the integrated circuit, is digital.

6. Brown and Jenkins (1968). See Schwartz and Gamzu (1977) for a review. I discuss autoshaping again in Chapter 14.

7. In the older ethological literature, unlearned reactions of this sort to well-defined, and often simple, stimuli termed *releasers*, are referred to as *innate releasing mechanisms* (IRMs); the stereotyped reactions are termed *fixed action patterns* (FAPs). As so often happens, additional research has blurred this simple picture; these reactions are neither so stereotyped nor so independent of experience as was at one time believed. See Hinde (1970) for a review. I return to this topic in Chapter 13.

8. See, for example, the experiments by Herrnstein and his students on "concept formation" by animals, in which pigeons easily learn to recognize color slides of natural scenes exemplifying categories such as "people," "trees," and "water" (Herrnstein & Loveland, 1964; Herrnstein, Loveland, & Cable, 1976; Cerella, 1979).

9. This technique was first used by Guttman and Kalish (1956) to study wavelength generalization. See Rilling (1977) and Honig and Urcuioli (1981) for reviews of the extensive work since then.

10. These physiological facts do not really explain generalization, of course, because the relation between transducer coding and behavior of the whole organism is not understood. The fact that neural firing is proportional to sound intensity doesn't force the animal to *behave* in any particular way with respect to intensity differences.

11. Integrality and separability correspond to different kinds of rule defining the transformation necessary to derive similarity relations among compound stimuli from their physical properties. The effect of integral stimuli can be represented by a multiplicative rule: $E = S_1 \times S_2$, whereas the effect of two separable elements can be represented by an additive rule: $E = S_1 + S_2$. There are obviously many possible combination rules so that the terms *integral* and *separable* do not define all the possibilities.

12. *Attention.* The term *attention* is used rather differently by animal and human experimental psychologists. For both it refers to control of behavior by one stimulus aspect rather than another. But in studies of human attention, individuals typically shift back and forth between attending to this or that several times within a single experimental session, usually in response to verbal instructions. A comparable animal experiment might be to require a pigeon to attend to the color difference between two stimuli in the presence of a tone, but to their size difference in the absence of the tone. This would be termed a *conditional discrimination* experiment in animal discrimination learning. Pigeons don't find such discriminations easy, and attention in animals is studied in other ways. When steady-state behavior is the main interest (as in the study by Reynolds, 1961a, referred to later) the term *attention* is used to refer to selective stimulus control resulting from a history of differential reinforcement. When attention is studied in the context of stochastic learning models, attention usually refers to particular parameters of the models (see Chapter 13). (Mackintosh, 1975, is a good review of some of this work).

13. The decibel is a logarithmic unit that expresses the *ratio* of two physical energies. Thus 20 dB means that $10 \log_{10} (L/L_0) = 20$, where L_0 is the threshold luminance and L is the luminance of $S+$. A 3-dB increment represents an approximate doubling of energy. Decibels are more commonly used to represent sound than light intensities. For sound, $dB = 20 \log_{10}(P/P_0)$, where P_0 is the threshold sound-pressure reference level.

14. This example illustrates another practical difficulty in studying multidimensional generalization gradients: the large number of test stimuli required. For N stimulus values on M dimensions, N^M test stimuli are needed. The practical exigencies of the transfer-test procedure (steady-state generalization gradients, discussed later, are partially exempt from these limitations) limit the number of different test stimuli to a dozen or two, effectively precluding serious study of more than two stimulus dimensions at a time.

15. The hypothesis that the range of sensitivity predicts the psychological effect of a given physical change has been applied with some success to human psychophysical data (see E.C. Poulton, 1968; Teghtsoonian, 1971).

16. These results are from an unpublished experiment in my laboratory which was a modest extension of an earlier study by Reynolds (1961a).

17. This description introduces yet another meaning for the term *stimulus*, namely a physically defined *object* presented to an animal by an experimenter. This is different from defining a stimulus in terms of its *effect* on the organism, or as a physically defined event, which are other common usages. It is usually clear from the context which meaning is intended, but use of the same term for so many different concepts sometimes leads to confusion.

18. The idea that stimuli can be represented in a multidimensional Euclidean space seems to have been first proposed by Richardson (1938). Subsequent developments are due to Torgenson (1952), Attneave (1950), Shepard (e.g., 1964, 1980), and Lockhead (1970, 1972).

19. For reviews of the radial-maze work, see Olton (1978, Olton & Samuelson, 1976; Roitblat, 1982; and Dale & Staddon, unpublished). The dome experiment is reported in Suzuki, Augerinos and Black (1980). In a most ingenious series of human experiments, Shepard and Metzler (1971; see also Shepard, 1975). (See Cooper, 1982, and Shepard & Cooper, 1982, for a review) have measured the rotation of mental images. Subjects were presented with an asymmetrical target object (a set of cubes arranged to form a figure, but a picture of a left hand would have done as well). They were then shown pictures either of the same object rotated through a variable number of degrees (i.e., the hand turned through some angle), or of its mirror image (a right hand). The subjects had to respond one way if the object was the same, another way if it was different. Shepard and Metzler found that the *time* taken to respond to the "same" object was proportional to the angular difference in orientation between it and the reference object. Evidently, the subjects had to rotate (at a constant speed!) their mental image of the reference object to bring it into registry with the projected figure before they could decide whether or not the two were the same. In this case, the images were of three-dimensional objects projected on a screen, but the alignment of cognitive maps by means of landmarks is a closely similar process. Indeed, it is something many people become aware of in the course of normal life: For example, many will have noticed the perceptible time it takes to reorient oneself after emerging from an unfamiliar subway entrance, or from a novel department-store exit.

20. Rothkopf (1957). The application of multidimensional scaling techniques to Rothkopf's data is discussed in Shepard (1980) and earlier papers therein referred to.

21. *Multidimensional scaling*. Multidimensional scaling (MDS) refers to a whole library of techniques representing extensions and elaborations of linear regression. Linear regression finds the straight line that best fits a set of points. MDS finds the plane or higher-dimensional surface that best fits the distances (dissimilarities) between a set of entities.

For example, suppose we have just three items, A, B, and C, and the dissimilarity relations are A-C: 2 (A and C are quite dissimilar), A-B: 1 (quite similar), and B-C: 1 (quite similar – note it is assumed that the dissimilarity of A vs. B is the same as the dissimilarity of B to A this is not a trivial assumption, but it makes the math easier). The simplest space that will accommodate these dissimilarities is obviously one-dimensional: A, B, and C just lie on a line. But if the dissimilarity of A-C had been 1 rather than 2, the three would have had to be arranged as points of a triangle in a two-dimensional space.

The similarity relations among N objects can always be accommodated by a space of $N-1$ dimensions. The space need not be Euclidean, however. For example, suppose we had the following similarity relations:

$$A-B: 1,$$
$$B-C: 2,$$
$$A-C: 4.$$

There is no way these can be represented in a Euclidean space, but they can be accommodated in a two-dimensional non-Euclidean one.

Obviously, multidimensional scaling becomes useful only when the similarity space is Euclidean or has some other relatively simple form, and the number of dimensions is significantly fewer than the number of entities.

11

STIMULUS CONTROL AND PERFORMANCE

The last chapter concluded that animals develop an internal representation of their world that guides action. We are uncertain both about the properties of such an internal representation, and the effects of reward and punishment on it. The most parsimonious assumption is that the representation of simple objects is independent of reward and punishment. Search images (*concept formation* is the psychological term for a very similar notion) may be an exception to this: Representations of very complex objects may perhaps be acquired only through a history of explicit reinforcement. Medieval teachers believed that Latin is learned only through the birch, and this general view of the motivation required for complex learning was almost universal until recently. Still, for recognition of simple stimuli, no special training seems to be required. The effect of reward and punishment is to give value to certain objects or places, as represented, rather than to create or modify the representations themselves.

Is performance then determined solely by the animal's external environment, as internally represented? In this chapter I argue that there is at least one other factor that must be taken into account: competition among activities for available time. These two factors, competition and external stimuli, taken together account for numerous experimental results on generalization and discrimination. The rest of the chapter explains how competition and stimulus control contribute to discrimination, behavioral contrast, generalization, and peak shift.

INHIBITORY AND EXCITATORY CONTROL

Animals need to know both what to do and what not to do; hence stimuli can have both inhibitory and excitatory effects. But, as we saw in Chapter 7, when an animal is not doing one thing it is probably doing something else. Moreover, animals are highly "aroused" under the conditions typical of operant conditioning experiments – hunger, combined with frequent access to small amounts of food (see Chapter 14). Behavioral competition is then especially intense: The animals have a lot to do and limited time in which to do it. A stimulus that signals the absence of food ($S-$) not only lets the animal know that he need not act in ways related to food, it also tells him that other activities are free to occur.

Since different activities compete for the available time, it is difficult to decide

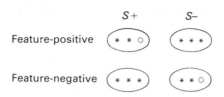

Figure 11.1. Examples of feature-positive and feature-negative stimulus displays.

whether a stimulus that suppresses activity A acts directly on A, or indirectly by facilitating some other, antagonistic activity. Perhaps the question is not even a useful one (although it has exercised heavy thinkers in this area for at least two decades). It *is* clear that a stimulus that acts to suppress activity A changes the balance of behavioral competition in favor of a complementary class of activities, ~A. As we shall see, it is possible to demonstrate direct, excitatory controls by $S-$ of activities that compete with the food-related behavior controlled by $S+$.

Feature effects

There are also striking experimental results showing how difficult it is to establish inhibitory stimulus control when competing activities are weak and stimulus generalization tends to facilitate the response to $S-$. For example, in a well-known series of studies, Jenkins and Sainsbury (1970; see also Hearst, 1978; and Hearst & Jenkins, 1974) trained pigeons on a simple discrete-trials discrimination procedure of the following sort. At random intervals averaging 60 sec a response key is illuminated with one of two stimuli ($S+$ or $S-$). Four pecks on $S+$ produce food; four pecks on $S-$ turn off the stimulus but have no other effect. In either case, if the animal fails to make four pecks, the stimulus goes off after 7 sec.

The birds quickly learn not to peck during the dark-key intertrial interval. Do they also learn not to peck $S-$ (which amounts to inhibitory control by $S-$)? Not always, and the conditions when they do and don't suggest the importance of behavioral competition.

Figure 11.1 shows two kinds of $S+$, $S-$ stimulus pair that Jenkins used: The upper pair may be termed *feature-positive*, in the sense that the feature distinguishing $S+$ from $S-$ appears on $S+$. The lower pair are *feature-negative*, since the distinctive feature appears on $S-$. Pigeons easily learn the feature-positive discrimination, pecking on $S+$ and not on $S-$. But they *fail* to learn the feature-negative discrimination, pecking indiscriminately on $S+$ and $S-$ (this is termed the *feature-negative* effect). Why?

There seem to be three reasons for the feature-negative effect. One is perceptual: The displays in Figure 11.1 consist of separate elements; their dimensions are separable, not integral (see Chapter 10). Moreover, $S+$ and $S-$ share elements in addition to those in the display: They are usually on for about the same time, and they both appear on a white-lighted key (as opposed to the dark key during the intertrial interval). Separability means that the elements are responded to separately; $S+$ and $S-$ are not treated by the animal as discrete entities, but rather in terms of the elements they contain. Since $S-$ contains two of the three separable elements in $S+$, any tendency to respond to $S+$ must generalize powerfully to $S-$.

The second reason for the feature-negative effect is that pigeons naturally peck at features that signal food (*sign tracking*); this is the basis for the phenomenon of autoshaping, briefly mentioned in the previous chapter. Sign tracking is, of course, just a special case of hill climbing: going for the thing that best predicts reward. Thus, in the feature-positive case, the animals learn to peck at the distinctive element in the $S+$ display (the o in Figure 11.1) even before they show much decline in pecking $S-$. In the feature-negative case, however, the stimulus element that signals food is present in both the $S+$ and $S-$ displays. Restriction of pecking to this feature is therefore incompatible with refraining from pecking $S-$. Pigeons can master a feature-negative discrimination if the display elements are very close together, so that $S+$ and $S-$ are perceived as a unit, that is, they become integral rather than separable (Sainsbury, 1971). Pecking then is directed at the whole $S+$ complex, not to individual elements within it.

A third reason for the feature-negative effect may be the weakness of competing activities in $S+$ and $S-$. The discrimination procedure used in these experiments allows much time during the intertrial interval for activities other than key pecking: The ITI averages 60 sec, whereas $S+$ and $S-$ are at most 7 sec in duration. The birds soon learn not to peck during the ITI, so that the time is free for other activities. Hence (for the reason elaborated in Chapter 7 and later in this chapter, diminishing marginal value) the animal's tendency to engage in non–key-pecking activities during the relatively brief $S+$ and $S-$ periods must be low. Thus, any contribution to discrimination performance made by the facilitation of antagonistic activities in $S-$ will be small. In support of this idea are reports (see note 13 and Figure 11.10) that the feature-negative effect is not obtained in more conventional successive discrimination procedures, where $S+$ and $S-$ simply alternate at perhaps 60-sec intervals, with no intervening ITI.

The competition argument is diagrammed in Figure 11.2. The upper diagram shows the stimulus-control factors acting in the feature-positive case. $S+$ is made up of two kinds of element, E_1, the distinctive feature (o in Figure 11.1), and E_2, the element common to both $S+$ and $S-$. $S-$ is made up of just E_2, the common element. E_1 controls T, the *terminal response*, of pecking; E_2 controls the *interim response*, the collective term for activities other than the food-related terminal response (more on terminal and interim activities in a moment). The horizontal lines symbolize the reciprocal inhibition (competition) between T and I activities. Obviously, in the feature-positive case there is nothing to facilitate key pecking (T) in the presence of $S-$. It is otherwise in the feature-negative case. Pecking must be controlled by E_2, the common element, so that suppression of pecking in $S-$ must depend on reciprocal inhibition from interim activities. If

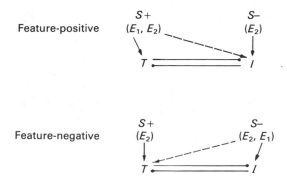

Figure 11.2. Excitatory control of terminal (T) and interim (I) activities by stimulus elements in feature-positive and feature-negative discriminations.

these are weak, there is every reason to suppose that pecking will occur in $S-$ as well as $S+$, as it does.

The general point of this argument is that discrimination performance is determined by two factors: (a) The nature of the stimuli: How discriminable are they? Are they integral or separable? (b) The availability of competing (interim) activities: How strong are they? What aspects of $S-$ are likely to control them?

Several predictions follow from the idea of behavioral competition as an essential component of discrimination. First, discrimination performance should be better when a substantial competing activity is available than when it is not. Second, the operant response rate in one component of a successive discrimination (multiple schedule) should generally increase when reinforcement rate in the other component is reduced. This is the much-studied phenomenon of *behavioral contrast*. Third, inhibitory generalization gradients should generally be less steep than their excitatory counterparts. Fourth, discrimination gradients obtained following training where $S+$ and $S-$ are on the same dimension should differ in predictable ways from gradients obtained after training with $S+$ and $S-$ on separate dimensions. I will discuss each of these issues in turn.

BEHAVIORAL CONTRAST AND DISCRIMINATION PERFORMANCE

About 25 years ago, when the operant-conditioning movement was enjoying its first flush of success, stimulus control was thought of in a simple way: A stimulus came to control an operant response when the response was reinforced in its presence. Each stimulus was thought of as an independent entity, maintaining behavior strictly according to the conditions of reinforcement associated with it. Many data appeared to support this view. For example, if the schedule associated with one stimulus was variable-interval, and with an alternating stimulus, fixed-interval, then the behavior in each stimulus soon became appropriate to the schedule in force. During the fixed-interval stimulus (*component*) the

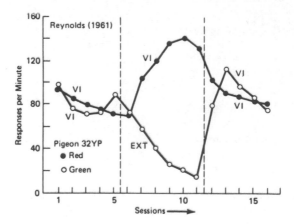

Figure 11.3. Positive behavioral contrast. Filled circles: response rate in the unchanged (VI → VI) component. Open circles: response rate in the changed (VI → EXT) component. Right third of the figure shows that the rate changes produced by the shift to extinction in the changed component are reversible. (From Reynolds, 1961b.)

animal would show "scallops" of accelerating responding between food deliveries; during the variable-interval component, a steady response rate would prevail.

It was, therefore, quite a surprise when George Reynolds in 1961[1] published a simple experiment that violated the rule of stimulus independence. His experiment had two conditions. In the first (the *prediscrimination phase*), hungry pigeons were trained to peck a key for food reinforcement delivered according to a moderate (e.g., 60-sec) VI schedule. The stimulus on the key was either a red or a green light; the stimuli alternated every 60 sec. In the second condition (the *discrimination phase*), the schedule in one stimulus (green, say) was changed from VI to extinction (*multiple VI, EXT*). Conditions in the other stimulus remained unchanged. The result was a reliable and substantial increase in response rate in the unchanged (VI) component (*positive behavioral contrast*). This result is illustrated in Figure 11.3, which shows response rate in the two components before and after the shift from VI-VI (VI reinforcement in both components) to VI−EXT (VI reinforcement in one component, extinction in the other). Before the shift, response rate is roughly the same in both components; afterwards, as rate decreases in the extinction component, it increases in the (unchanged) VI component.

Reynolds also demonstrated the converse effect, negative contrast, which occurs when conditions of reinforcement in one component are improved, rather than degraded, for example, when a multiple VI 60 VI 60 sec schedule is changed to mult VI 60 VI 20 sec: the usual result is a decrease in response rate in the (unchanged) VI 60 component.

Behavioral contrast is a widespread but not universal effect. For example, it is much larger and easier to obtain with pigeons pecking keys than rats pressing levers – or pigeons pressing a treadle. If reinforcement rate in the VI-VI condition is very high (e.g., VI 15 sec), contrast effects are not obtained.

There are two obvious alternative explanations for contrast: The decrease in response rate in the component shifted to extinction, and the decrease in reinforcement rate in that component. The response-change account rests on the unstated hypothesis that the animal has only so many responses to "spend," so that if he spends fewer in one component (because responses are no longer reinforced there), he will have more to spend in the still-reinforced component. The reinforcement-rate-change account rests on the general notion that response rate is guided by relative, rather than absolute, reinforcement rate.[2]

Reynolds and others attempted to discriminate between these two hypotheses by experimentally separating the response-rate and reinforcement-rate drops in the changed component. For example, in one experiment, pigeons were reinforced for *not* responding for 6 sec – all periods of 6 sec without a key peck ended with the delivery of food. This very effectively abolished pecking, but it failed to produce an increase in responding in the unchanged component – no contrast. In other experiments, food in the changed component was delivered independently of pecking or its availability was signaled; both maneuvers reduce or abolish pecking, but neither reliably produces contrast.

The general conclusion is that a change in reinforcement rate is usually sufficient to produce contrast, but a change in response rate unaccompanied by a change in reinforcement rate is generally ineffective.

Terminal and interim activities: schedule-induced behavior

These various effects can all be brought together by the idea of response competition, provided one further thing is conceded: that time is taken up by food-related activities, even when no explicit response is required to procure food. There is good evidence for this in general, although not much in some of the specific situations used to study behavioral contrast.

The clearest data come from the simplest situation. When hungry pigeons are daily exposed for an hour or so to a periodic-food (*fixed-time*, FT) schedule, they spend a good portion of their time near the feeder. If food delivery is frequent and they are sufficiently hungry, they will peck the wall in front of the feeder, even though the pecks are both unnecessary and ineffective. This activity is nicely synchronized with the delivery of food (although this aspect is not critical to the present argument).

Some typical results are shown in Figure 11.4. The figure shows the behavior of a single, well-trained pigeon averaged across three 30-min daily sessions of a FT 12-sec schedule (from Staddon & Simmelhag, 1971). Each curve refers to a separate activity of the animal, assessed by visual observation. Activity R7 is "pecking the feeder wall," R8 a pacing movement from side to side in front of the wall, and so on. The curves show the probability of occurrences of each activity

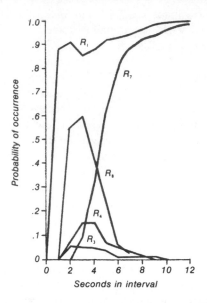

Figure 11.4. Average behavior of a single hungry pigeon receiving food every 12 sec. Each curve shows the probability that the numbered activity will occur in each of the 12 sec in between food deliveries. R1 is "being in the vicinity of the feeder," R7 is "pecking the feeder wall"; other activities are described in the text. (From Staddon & Simmelhag, 1971.)

in each of the 12 sec between food deliveries. Thus, R7 (pecking) never occurred during the first 2 sec after food, but thereafter it occurred with increasing frequency, reaching a probability of almost one by the end of the interval.

All the animals showed the same pattern, namely, two classes of behavior within each interfood interval, a single activity that increases in frequency up to the time that food is delivered – this is called the *terminal response*; and other activities that occur during earlier parts of the interval – these are termed *interim activities*. Thus, R7 (pecking) is the terminal response; R8 (pacing), and two other movement patterns are interim activities in Figure 11.4. This division into terminal and interim activities has subsequently been demonstrated with rats and hamsters as well as pigeons (Staddon & Ayres, 1975; Anderson & Shettleworth, 1977; see also Staddon, 1977a). The particular activities that fall into the terminal and interim classes differ from species to species; all activities become less vigorous at longer interfood intervals; and the terminal response need not always be the same as the response normally elicited by food, but the general pattern is well established.

Terminal responses are obviously food-related, and occur at times when food delivery is likely. Interim responses are not generally food related, and occur at times when the food delivery is unlikely. This is the basis for earlier labeling activities controlled by $S+$ as "terminal" and those controlled by $S-$ as "interim" in Figure 11.2.[3]

When food delivery depends upon a response (a fixed-interval rather than fixed-time schedule), this response follows the same course as the terminal response in Figure 11.4. The *time taken up* by food-related activity is little affected by whether or not a response is necessary to get food – although, of course, the *type* of response shown is likely to depend upon the response contingency. When food delivery is aperiodic, as in a variable-time schedule, rather than periodic, the same division into terminal and interim activities can often be seen, especially if there are definable periods when the probability of food delivery is zero. For example, if the shortest interval between food deliveries is 5 sec, then interim activities will typically occur during the first 2 or 3 sec after food delivery. When food deliveries are random in time, no definable postfood period is available for interim activities and they take on a vestigial form, occurring in alternation with food-related activities. Thus, on a VI schedule, key pecks often alternate with brief turning-away movements.

Whether food delivery is periodic or aperiodic, whether a response is required or not, animals spend much time in food-related activities. The rest of their time is spent in activities that tend to occur at times, or in the presence of stimuli, that signal the absence of food. And the competition between these two classes of activity seems to be especially intense in conditioning situations.

Intertemporal effects

The existence of interim as well as terminal responses, with the properties just described, together with the property of diminishing marginal competitiveness (the higher the rate of an activity, the weaker its tendency to displace other activities), sets the stage for contrast effects. The argument is as follows: In the prediscrimination condition, with the same VI schedule operative in both components, interim activities must occur in both. Because of competition for available time, key-peck rate must, therefore, be at an intermediate level. When one component is changed to extinction, the tendency to peck is weakened; interim activities are consequently free to increase their time allocation in that component. The more total time allocated to interim activities, the less competitive they become (because of diminishing marginal utility). Since the factors tending to produce pecking in the unchanged component have not altered, but its competitor has become less effective, pecking must increase in the unchanged component – which is positive contrast.[4]

A similar argument accounts for negative contrast. When the reinforcement rate in one component is increased, key pecking takes up more time. Hence, interim activities have less time and (because of diminishing marginal competitiveness) consequently become more competitive in the unchanged component, suppressing pecking there – which is negative contrast.

There are some straightforward limitations on this argument. For example, when the VI-reinforcement rate is high, very little time may be taken up by interim activities in the prediscrimination phase. Consequently there is little room for further increases in the terminal response, should the interim activities become less competitive. Thus, contrast should be less at high reinforcement rates. There is a similar limitation when the rate of the terminal response is very low, rather than very high. In this case, the animal may be spending as much time as it needs in the interim activities, even in the prediscrimination phase, so that the opportunity to spend even more time in the discrimination phase actually makes no difference. Thus, contrast effects might well be reduced whenever the rate of the terminal response in the prediscrimination phase is low.

The competition explanation for negative contrast depends upon displacement of competing activities from the changed to the unchanged component caused by an increase in the time allocated to terminal responding in the changed component. The VI response functions in Figure 7.15 show that such an increase is not to be expected over the whole range of reinforcement rates. If reinforcement rate in the prediscrimination phase is already high, a further increase (in the changed component) might reduce, rather than increase, the proportion of time allocated to the terminal response in that component. Procedural details also enter in ways that are rarely made explicit. For example, an increase in reinforcement rate in the changed component means that unless the component duration explicitly excludes eating time, the time available for responses other than eating is reduced, whether or not the level of terminal responding also increases. This factor effectively shortens the changed component and will always tend to promote negative contrast by displacing interim activities to the unchanged component.

Finally, contrast should depend upon the factors tending to strengthen or weaken interim activities. For example, rats are relatively inactive in a Skinner box when they are not actually pressing the lever; without environmental support (a running wheel, for example), rats show little evidence of interim activities. Pigeons, on the other hand, spend much time in the kinds of pacing and turning movements described earlier (Figure 11.4).[5]

These limitations account for the limitations on contrast described earlier: its absence at high reinforcement rates, with treadle pressing by pigeons (a very low-rate activity), and (usually) in rats. The fact that time is taken up by terminal responding, even if no instrumental response is required, accounts for failures to find positive contrast when food is signaled or presented independently of responding in the changed component: These manipulations abolish the measured response, but may not free up time for additional interim activities.

Figure 11.5 shows the results of a simple experiment that demonstrates directly the role of competition in contrast and discrimination. Rats were run in the standard two-condition contrast paradigm: first trained with VI 60 sec in both components of a multiple schedule, then shifted to VI EXT in one component. The experiment was done in two ways, either with or without a running wheel available to the rats. The top panels in Figure 11.5 show the levels of lever

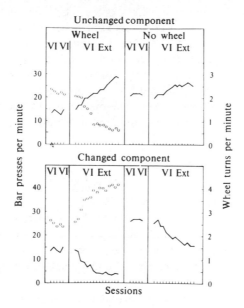

Figure 11.5. Reallocation of competing behavior during behavioral contrast. Mean daily rates of bar pressing (solid line) and wheel turning (open circles) for four rats in changed and unchanged components, with and without a running wheel available. (From Hinson & Staddon, 1978.)

pressing and wheel running in the unchanged component. With no running wheel available (right quarter), lever-press rate is relatively high and the contrast effect (proportional increase in response rate in the unchanged component) small. With a wheel (left quarter), lever-press rate (solid line) is lower, because of competition from running, but the contrast effect following the shift to VI EXT is substantial. As response rate increases in the second panel, rate of wheel running (open circles) decreases. The lower panels show the concomitant changes in the changed component. With no wheel (right quarter), lever-press rate decreases little in the extinction component – discrimination performance is poor. With a wheel (left quarter), wheel running increases and lever pressing decreases substantially in the extinction component – discrimination performance is good (Hinson & Staddon, 1978). Thus, the availability of a strong interim activity can both improve discrimination performance and facilitate behavioral contrast.

The competition account of contrast is mechanistic not purposive. It refers only to the relative competitiveness of each activity, where the competitiveness of the terminal response in each component is assumed to be directly related to the reinforcement obtained, and the competitiveness of the interim activity a negatively accelerated function of its overall rate. In a moment, I will make these assumptions explicit with the aid of a simple model. It is worth noting, however, that the competition hypothesis does a pretty good job of *intertemporal utility maximization*. This is economese for a process that allocates resources among different temporal periods so as to maximize benefit to the individual. For example, the efficient executive will read reports or dictate memos on his commuter train, when he cannot engage in other species-typical activities such as telephoning, chairing committees, or eating at the Four Seasons. In like fashion, the rat trapped in $S-$ might as well "do" his interim activities so he can put in more time pressing the lever in $S+$. The mechanism of behavioral competition solves this problem in a way that requires no foresight and puts no load on memory.

COMPETITION AND MATCHING

In Chapter 7, I showed how a particular model for diminishing marginal utility could account for response functions obtained on numerous simple schedules. I now apply a simplified version of this approach to multiple schedules. It leads both to the qualitative predictions of behavioral contrast just described and to quantitative predictions of matching and undermatching on concurrent and multiple schedules.

This section involves some simple mathematics. Here, as elsewhere, the mathematics has three functions. The most important is to ensure that our explanations are clear: If an explanation cannot be framed in a formal way, then it is not clear – the idea of "competition" is not an obvious one; it is as well to be precise about what we mean by it. Second, even if clear, verbal explanations do not lend themselves to the derivation of complex predictions. When there are many steps in the argument, or its outcome has a quantitative aspect (prediction of the shape of a gradient, for example), verbal argument is unsatisfactory. (I vividly remember being present some years ago at an argument between an eminent behavior theorist and a student about what the theorist's heavily published but nonquantitative theory predicted in a certain case. Someone suggested that they should settle the argument by implementing the theory as a computer program. The theorist bridled at this suggestion, but I have never understood why, since a theory whose predictions are a matter of opinion is worse than useless.) Third (and probably least important for biology and psychology), mathematical models make possible quantitative, rather than merely qualitative, predictions.

There is, of course, a cost in making formal models – even beyond its chilling effect on mass book sales. At an early stage of knowledge, theorizing is necessarily imprecise. To insist on quantitative rigor may be just to exchange an exact irrelevance for a useful hunch. I believe that the experimental facts on discrimination performance are sufficiently solid and coherent to justify something more than verbal description.

The task is to apply the optimality analysis of Chapter 7 separately to each successive component of a multiple schedule, where the components may differ in the value of VI schedule associated with them. It will be simplest to begin with a concurrent schedule, which can be thought of as a multiple schedule in which the component durations are brief and, most importantly, under the control of the animal. The components share interim activities (for simplicity I will assume a single interim activity), since these can occur in both. I assume that in each component, the animal adjusts its level of responding so as to minimize cost (maximize value), in the sense defined in Chapter 7. Given the single further simplifying assumption that the cost of the instrumental response is negligible, minimum-distance-type models reduce to simple reinforcement maximization. The condition for optimal behavior is then that the marginal change in reinforcement rate be equal to the marginal change in value of the interim activity, formally

$$dR(x)/dx = dV(z)/dz, \qquad (11.1)$$

where $V(z)$ is the value of interim activity as a function of its level, z, and x is the level of responding in one component, and $R(x)$ its rate of reinforcement. If we assume that the marginal value of interim activity is *constant* across both components of the schedule, the optimality condition is that

$$dR(x)/dx = dR(y)/dy, \qquad (11.2)$$

where y is the response rate in the other component.

The expressions in Equation 11.2 can be evaluated if we know the feedback functions, $R(x)$ and $R(y)$, for interval schedules, but it is not necessary to know these functions with exactness. In Chapter 7, I pointed out that many negatively accelerated functions (including two, the power and hyperbolic forms, that have been proposed as interval schedule functions) have the property that $dF(x)/dx = G(F(x)/x)$, that is, the marginal of the function can be expressed as a function of the ratio of the function and its argument. Thus, for the hyperbolic function, $R(x) = ax/(a + x)$, $dR(x)/dx = a^2/(a + x)^2$, which can be rewritten as $[R(x)/x]^2$. Since $R(x)$ and $R(y)$ are interval schedule functions, we can rewrite Equation 11.2 as

$$G(R(x)/x) = G(R(y)/y),$$

which is obviously equivalent to

$$R(x)/x = R(y)/y, \qquad (11.3)$$

that is, matching of response and reinforcement ratios – a result discussed in Chapter 8 in connection with concurrent schedules.

To go further and derive contrast from this result, we need to incorporate two other things into the analysis: (a) a time-allocation constraint; and (b) a model for the diminishing-marginal-utility property of $V(z)$, the function relating the value of z (the interim activity) to its level.[6]

As in Chapter 7, I assume that activities x, $R(x)$, y, and z are scaled in time units, so that the time-allocation constraint is just Equation 7.8 generalized to four activities, namely

$$x + R(x) + y + z = 1. \qquad (11.4)$$

The simplest way to incorporate diminishing marginal utility into $V(z)$ is to assume the same type of negatively accelerated relation between z and $V(z)$ as between x and $R(x)$ or y and $R(y)$. The argument for *some* kind of negatively accelerated function is that under free conditions, animals do more than one thing within the time period typical of an experimental session. As we saw in Chapter 8, nonexclusive choice implies diminishing marginal utility and a negatively accelerated value function. The defense for picking a function the same as the VI feedback function is simplicity – the conclusions I draw would not be much affected by choosing other negatively accelerated functions, however.

We are left, then, with two relations: Equation 11.4, and (if the cost-of-deviation of the instrumental response, X, is also assumed negligible) Equation 11.3 generalized to three activities,

$$R(x)/x = R(y)/y = V(z)/z. \tag{11.5}$$

Consider now three cases of increasing complexity: (a) The single-response case (i.e., no response y); (b) the two-response concurrent schedule; and (c) the two-response multiple schedule.

1. Simple VI (single-response case). In this case we have

$$R(x)/x = V(z)/z \tag{11.6}$$

and $x + z + R(x) = 1$; hence $z = 1 - x - R(x)$. Substituting in Equation 11.6 yields

$$R(x)/x = V(z)/[1 - x - R(x)],$$

which reduces to

$$x = \{R(x) - [R(x)]^2\}/[V(z) + R(x)], \tag{11.7}$$

which is the response function for VI given the simplification that the cost of the instrumental response is negligible.

If the time taken up by $R(x)$ is small, the squared term can be neglected and Equation 11.7 reduces to

$$x = R(x)/[V(z) + R(x)], \tag{11.8}$$

which is a well-known response function for simple VI schedules proposed by Herrnstein.[7]

2. Concurrent VI VI (two-response case). We have already dealt with this case in the derivation from Equation 11.2; the result is the prediction of matching, which here requires only that the CoDs of the two instrumental activities be negligible and assumes nothing about the feedback function for the interim activities beyond its constancy. It is also pretty obvious that matching can be derived directly from a generalization of the single-response case (Equation 11.8), namely,

$$x = R(x)/\Sigma R \text{ and } y = R(y)/\Sigma R, \tag{11.9}$$

where ΣR is the sum of reinforcement rates for all activities ($R(x) + V(z)$ in Equation 11.8, $R(x) + R(y) + V(z)$ for two instrumental responses), which reduces to

$$x/y = R(x)/R(y),$$

that is, matching of response and reinforcement ratios.

Contrast is derivable from this analysis just by comparing response rate, x_{pre}, in the unchanged component in the prediscrimination condition (VI in both components) with rate, x_{post}, in the postdiscrimination condition (VI in one component and EXT in the other). From Equation 11.9, $x_{pre} = R(x)/[R(x) + R(y) + V(z)]$, and $x_{post} = R(x)[R(x) + V(z)]$, hence the *ratio* of pre and post (a measure of contrast) is just

$$x_{post}/x_{pre} = C_0 = [R(x) + R(y) + V(z)]/[R(x) + V(z)]. \tag{11.10}$$

Equation 11.10 has some of the properties described earlier for the informal competition model of contrast. For example, if $R(x)$ (and $R(y)$, since they are equal) is small relative to $V(z)$, contrast effects will be small, since the ratio C_0 will be dominated by $V(z)$; thus, the analysis predicts small contrast effects when the absolute reinforcement rate for the instrumental responses is small (i.e., reinforcement is infrequent or the animal is only weakly motivated).

If $R(x)$ and $R(y)$ are large relative to $V(z)$ Equation 11.10 predicts large contrast effects, and as we have seen this is contrary to fact. But Equation 11.10 does not take into account the time constraints that affect multiple but not concurrent schedules, that is, the fact that responding in each component cannot take up more than the proportion of total time devoted to that component: In multiple schedules, component duration, hence the maximum disparity between numbers of responses in each component, is set by the experimenter; in concurrent schedules, it is set by the animal.

3. Multiple VI VI. Responses x and y in these equations are expressed as proportions of the total time. This raises no difficulties in either the single-response or concurrent cases, because responses x, y, and z can occur at any time. But on multiple schedules, the time available for x and y is limited to the durations of their respective components, that is (in the equal-component case) to 50% of the total time each. This additional constraint introduces a discontinuity into the matching function, since neither x nor y can exceed .5 no matter what the reinforcement proportions. The result is shown in Figure 11.6, which is a plot of reinforcement proportions versus response proportions. Perfect matching corresponds to the diagonal, but the constraint owing to "saturation" of responding in a component causes the function to deviate toward *undermatching* (i.e., too much responding in the minority component) when $R(x)$ and $R(y)$ are very different.[8] Each S-shaped function is for a different value of $V(z)$: the larger $V(z)$

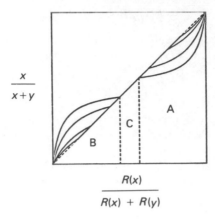

Figure 11.6. Matching relations predicted by competition model. Region C indicates range of $|R(x) - R(y)|$ for perfect matching for the outermost curve; as size of $V(z)$ relative to $R(x)$ and $R(y)$ increases, region of perfect matching also increases, as shown by the other curves.

relative to $R(x)$ and $R(y)$, the longer the linear part of this function – in accord with experimental results showing that matching on multiple schedules of food reinforcement *improves* when body weight is allowed to approach normal levels, that is, as the animals become *less* motivated.[9]

Figure 11.7 shows experimental results from an early experiment on matching in multiple schedules.[10] The S-shaped form of the functions is clear, but they lack the discontinuities shown in Figure 11.6. There are several possible reasons for this: The hypothesized feedback relation $V(z)$ may not be identical to the VI feedback functions for x and y; the marginal of the VI feedback function may not be expressible as the ratio of the function and its argument; the assumption that instrumental responses X and Y are costless is only an approximation; any "noise" in the data would blur discontinuities. These deviations from the simple mathematical model (should they turn out to be real) are not surprising. It is more important to attend to the fact that the general form of the deviations from matching in Figure 11.7 is just what we would expect from competition, and saturation.

Saturation also means that contrast is reduced at high as well as low values of $R(x)$, as shown in Figure 11.8, which shows C_0 as a function of $R(x)$ for a fixed value of $V(z)$: Contrast is a maximum when $R(x) = V(z)$ (given the assumptions about $V(z)$ just described), and declines asymptotically to 1 (no contrast) as $R(x)$ increases beyond that value.

This analysis of contrast and matching shows that behavioral competition provides a simple way for animals to allocate their time efficiently: One arrives at very similar conclusions by looking at behavior in competition terms, or in terms of marginal utility. In addition to accounting for behavioral contrast and its

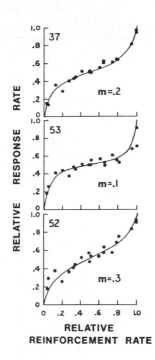

Figure 11.7. Relative response rate $(x/[x + y])$ vs. relative reinforcement rate $(R(x)/[R(x) + R(y)])$ from a multiple-schedule experiment by Reynolds (1963a) in which relative reinforcement rate was varied in three ways: One component held at VI 3 min, the other VI varied; the other component held to 1.58 min, the other varied; both varied. (From Herrnstein, 1970.)

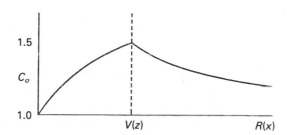

Figure 11.8. Contrast ratio, C_0 (Equation 11.10), as a function of VI reinforcement rate, $R(x)$, predicted from competition and diminishing marginal competitiveness. Contrast in multiple schedules is at a maximum at an intermediate value of reinforcement rate.

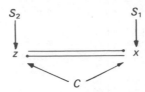

Figure 11.9. Causal factors in disjoint stimulus control: each activity is facilitated by a disjoint stimulus element (S_1 or S_2) and common factors (C), and inhibited by the other activity.

limitations, the approach also provides a rationale for Herrnstein's equation for the VI response function, for matching and for the general form of deviations from matching on multiple schedules. The theoretical basis for these phenomena is far from settled, however, and there are other accounts for positive contrast and some aspects of matching[11] – although none which attempts to relate all these effects to one another.

Inhibitory generalization gradients

Performance on reinforcement schedules is determined by the joint effects of external stimuli and competition. The message of the preceding section is that on variable-interval schedules the competitiveness of an activity is directly proportional to the reinforcement for it, and inversely proportional to its overall rate. Can we find a comparably simple way to describe the effects of external stimuli on behavior?

In the steady-state concurrent and multiple-schedule situations I have been discussing, discrimination is perfect or near perfect: The stimuli are easy to tell apart, and changes in response rate reflect not failures of discrimination, but efficient allocation of behavior. Even when a rat continues to respond in $S-$ (as in Figure 11.5, bottom right, for example), there is little doubt that this is not because he cannot tell the difference between $S+$ and $S-$, but rather because the cost of lever pressing is very low and he has nothing more appealing to do. In a generalization test, response rate falls off gradually rather than abruptly; yet given appropriate training, extremely sharp gradients can be produced. Presumably the psychophysical discriminability of the stimulus continuum (i.e., the cognitive and perceptual properties discussed in Chapter 10) has not altered, so that this change reflects a change in behavioral competition. What form does this change take?

The static analysis of molar behavioral allocation developed in this chapter and in Chapter 7 provides a relatively simple way to deal with these interactions among stimuli, competing activities, and differential reinforcement for those activities. Figure 11.9, a simplified version of Figure 11.2 (and similar in form to Figure 2.11 which was used to model reciprocal inhibition), provides a starting point. The figure shows the factors affecting terminal (x) and interim (z) activities in a successive discrimination of the sort already discussed. S_1 and S_2 represent *disjoint* (i.e., perfectly discriminable – another term is *orthogonal*) elements that differentiate $S+$ and $S-$: For example, if $S+$ is a dark pecking key with a white star in the center, and $S-$ a green key, then we can assume both that there is zero perceptual overlap between these two stimulus elements (green and star) and, since they are the only aspect of the key that is different in the positive and negative schedule components, that they must be the basis for any discrimination. Thus x is controlled (*facilitated* might be a more accurate term[12]) just by S_1, and z is facilitated just by S_2; in addition, each activity is inhibited (because of competition for time) by the other. C in the figure refers to all those effective stimulus factors that have effects on both x and z (i.e., common factors). Let's consider how the causal diagram in Figure 11.9 applies to an experiment with a simple, successive discrimination procedure.

In this experiment[13] two groups of pigeons were exposed to two alternating stimuli, one associated with VI reinforcement, the other with extinction (a multiple VI EXT schedule, of the type by now familiar). For one group of animals, $S+$ was a vertical line, $S-$ a blank key; for the other group, $S-$ was a vertical line, $S+$ a blank key. Since the vertical line is the distinctive feature here, the first group is *feature-positive*, and the second *feature-negative*, in the sense these terms were used earlier.

After training sufficient to produce good discrimination between $S+$ and $S-$ (i.e., few $S-$ responses), both groups of animals were given generalization tests in which peck rate was measured in the presence of different line tilts. The results are shown in Figure 11.10. The result for the feature-positive group is the familiar excitatory gradient, with a peak at $S+$. The feature-negative group shows an *inhibitory* or *incremental* gradient, however, in which response rate increases as the test stimulus departs from $S-$; moreover (and this is typical of many similar subsequent studies), the inhibitory gradient is somewhat shallower than the excitatory one, even though the two experiments used the same procedure.

The diagram in Figure 11.9 implies that in the feature-positive group, the vertical line will facilitate x, the instrumental response, whereas z, the interim activities, will be facilitated by the blank key. Variation in line tilt during a test should therefore weaken x much more than z, yielding the usual decremental gradient for x. In the feature-negative group, however, the vertical line is $S-$, hence must facilitate z much more than x. Consequently, variation in line tilt must weaken z much more than x, allowing x to *increase* (since its facilitating factors – the blank key background to the line – are still present), producing the incremental, *inhibitory* gradient shown in Figure 11.10.

The shallower slope of the inhibitory than the excitatory gradient is also easily explained. The excitatory gradient directly measures the effect of stimulus variation on the activity it facilitates, x, (pecking). However, the inhibitory gradient is an indirect measure of the effect of $S-$ variation on z (an interim activity), the activity it facilitates. If we were to measure this interim activity directly (which

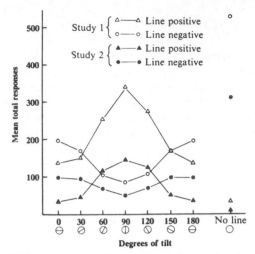

Figure 11.10. Inhibitory generalization gradients. Open triangles: excitatory line-tilt generalization gradient for a group of pigeons trained on multiple VI EXT, with a vertical line as $S+$, blank key as $S-$ (feature-positive group). Open circles: inhibitory line-tilt generalization gradient from a group trained with vertical line as $S-$, blank key as $S+$ (feature-negative group). Closed symbols are from a repeat experiment. (From Honig, Boneau, Burstein, & Pennypacker, 1963.)

has unfortunately not been done in these experiments) – as in the contrast experiment shown in Figure 11.5 – we would expect to see an excitatory gradient as the $S-$ feature is varied. The effect on the measured instrumental response, however, depends upon what proportion of the interim activities are directly facilitated (controlled by) the $S-$ feature varied in the test. If the activity under the control of the $S-$ feature comprises essentially all the interim activities, and time-allocation conservation holds (i.e., $x + z = 1$), then any decrease in z implies an equal increase in x. But if, as seems more likely, some interim activities are not controlled by $S-$, then any decrease in z, caused by variations in its controlling dimension, must be shared between x and this other activity. Consequently, a given decrease in z will generally produce a smaller increase in x (see the discussion of substitution relations in Chapter 7). Under any but the most restricted conditions the existence of a third class of activities, not controlled by either $S+$ or $S-$, is highly likely. Hence, we would expect inhibitory gradients to be generally shallower than excitatory ones, as they are.

Conjoint stimulus control and peak shift

This is the last of the three topics promised at the beginning of the chapter. The cases discussed so far, behavioral contrast and inhibitory generalization gradients, have been based on successive-discrimination training with disjoint stimuli: In all these experiments, $S+$ and $S-$ are so easy to tell apart it is unlikely that

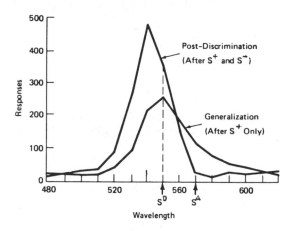

Figure 11.11. Wavelength generalization gradients after exposure to $S+$ only and after exposure to successive discrimination involving $S+$ and $S-$; difference between the two gradients illustrates positive contrast and peak shift. (After Hanson, 1959.)

the animals ever get confused as to which is which. When this is not true, when $S+$ and $S-$ differ slightly on only a single dimension, the resulting generalization gradient is different in informative ways.

Figure 11.11 shows the result of a classic experiment on generalization after *conjoint* discrimination (i.e., discrimination with similar stimuli). Pigeons were trained on the familiar multiple VI EXT procedure; $S+$ was a monochromatic green light of 550 nm, $S-$ was a greenish-yellow light of a slightly longer wavelength, 570 nm. In the control condition of the experiment, a group of pigeons was trained with $S+$ alone (simple VI), then given a generalization test. The result is the lower gradient, peaked at $S+$, in Figure 11.11. The other group was trained on the multiple VI EXT procedure, and then tested. The result is the higher gradient with its peak at about 540 nm, that is, shifted from $S+$ away from $S-$.

The increased responding in $S+$ in the second, postdiscrimination, gradient is just positive behavioral contrast, which we have already analyzed. The *peak shift* is a related phenomenon. It represents an effect of conjoint stimulus control. Figure 11.12 shows the causal model for conjoint control. The figure is the same as Figure 11.9, except for the added diagonal dotted lines, which represent the partial confusion caused by the similarity of $S+$ (S_1) and $S-$ (S_2): Because the two stimuli cannot be discriminated perfectly by the animal, S_1 has some tendency to facilitate z, the interim activity, as well as a stronger tendency to facilitate x, the terminal response; and S_2, similarly, has some tendency to facilitate x, together with a stronger tendency to facilitate z.

To show that this partial confusion between S_1 and S_2 can cause peak shift, it is necessary to give the pictures in Figures 11.9 and 11.12 some quantitative

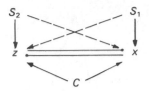

Figure 11.12. Causal factors in conjoint stimulus control. The stimulus element controlling the instrumental response, x, also has some facilitatory effect on z, and vice versa.

properties. A simple way to do this is to let the effects depicted in Figure 11.12 be linear and additive.[14] This means writing two equations, one to describe the factors acting on activity x, another for those acting on z, thus

$$x = S_1 - k_{zx}z + C_x, \tag{11.11A}$$

and

$$z = S_2 - k_{xz}x + C_z, \tag{11.11B}$$

where S represents the additive facilitating effect of the two controlling stimuli, k represents the subtractive effect of z on x and vice versa, and C is the additive effects of stimulus factors not varied during a generalization test.

Notice that if by convention we take x to be the measured response, then Equation 11.11A represents the relation between x and z in the feature-negative case (S_2 varied) and Equation 11.11B represents the relation in the feature-positive case (S_1 varied).

Because these subtractive effects are due to competition for available time, the gain in all other activities associated with unit decrement in activity z must be equal to one – there cannot be a total gain. But when three or more activities are involved, k_{zi} can even be negative for some activities – if y strongly inhibits x, and z strongly inhibits y, for example, then an increase in z might well lead to an *increase* in x, because of reduced inhibition of x by y; this is termed *disinhibition*.

To illustrate the kinds of interaction I'm thinking of, imagine that there are three activities involved, x, y, and z, measured in time proportions so that $x + y + z = 1$. Suppose that the free levels of these three activities are .2, .3, and .5, say, but that for experimental purposes, z is progressively restricted to values of .4, .3, .2, and .1. The linear inhibition assumption predicts that the levels of x and y will increase in such a way that plots of x against y or z, or of y against z, will all be linear. Corresponding to the given values of z, the values of x might be .25, .30, .35, and .40 and of y, .35, .40, .45, and .50 (for convenience, these are all listed in Table 11.1). In this instance, the decrements in z are shared equally between x and y, but this need not always be the case.

In this example, $x = .45 - .5z$, and $y = .55 - .5z$ so that $k_{zx} = k_{zy} = .5$, and $S_1 = .45$. To find k_{xz} and S_2 experimentally, it is necessary to hold x to various levels and measure the level of z.

Table 11.1.

	Free	z Constrained			
x	.2	.25	.30	.35	.40
y	.3	.35	.40	.45	.50
z	.5	.40	.30	.20	.10
Total time	1.0	1.0	1.0	1.0	1.0

In the disjoint case, where variation in a stimulus element or dimension affects only one activity directly, Equations 11.11A and B imply that there should be a linear relation, with slope generally between −1 and 0, between the activity whose controlling stimulus is being varied and any other activity. Since we can expect the controlled activity to show an excitatory gradient, this means that measurement of an activity not directly controlled by the varied dimension should show an inhibitory gradient of generally shallower slope, as I argued in the preceding section.

There are few direct tests of the linear prediction. However, Figure 11.13 shows three sets of data that provide one direct and two indirect tests (see also Figure 7.1). The figure shows plots of one response, controlled by the stimulus dimension being varied (abscissa), versus a competing response whose controlling stimulus remains constant (ordinate). The experiment most directly related to the present analysis is by Catania, Silverman, and Stubbs (1974).[15] They trained hungry pigeons to peck at two keys for food delivered on independent VI schedules. The stimulus on the right key was a vertical line, on the left key a color. Line tilt was varied in the generalization test, and the result was an excitatory generalization gradient for pecking on the right (line-tilt) key and an inhibitory gradient for pecking on the left (color) key. The x's in Figure 11.13 represent average response rate on the left plotted against average response rate on the right at each line-tilt value: The points are tolerably well fitted by a straight line, as predicted by Equation 11.11. The other two lines are from two similar experiments by Honig et al., already shown in Figure 11.11. These data are a less direct test, just because they represent a between-groups, rather than within-group or within-animal, comparison. Nevertheless, when response rate for the feature-positive group (abscissa) is plotted against rate for the feature-negative group (ordinate) at the same line tilt, the points again are approximately collinear. The slope for the Catania et al. data is close to −1, suggesting that pecking on the two keys occupied most of the available time; the slope for both the Honig et al. experiments is between −1 and 0.

So far we have established that both terminal responses and interim activities have controlling stimuli; that variation in these stimuli yields excitatory generalization gradients; and that because of competition, an excitatory gradient in terms of one response will usually be associated with a (shallower) inhibitory gradient in terms of other responses. To show in addition that these processes can lead to

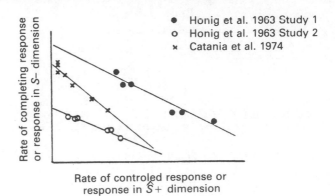

Figure 11.13. Linear relations between levels of interim and terminal responses in successive discrimination. Response rates in inhibitory and excitatory gradients from the studies shown are plotted against each other. (Data from Honig et al., 1963; and Catania, Silverman, & Stubbs, 1974.)

peak shift we need to make some assumption about the way in which variation in a controlling stimulus dimension affects the stimulus contribution in Equation 11.11.

The terms S_1, S_2 in Equation 11.11 represent the excitatory contributions of controlling stimuli to the level of each activity. If the properties of the stimulus are varied from the $S+$ value, then this contribution will decrease – presumably in proportion to the *similarity change* associated with a given change in the physical properties of the stimulus. If we make some definite assumption about the relation between physical change and similarity change, it becomes possible to predict generalization effects. I first give the equation for the disjoint case (interdimensional discrimination); then derive from them the necessary equation for the conjoint case (intradimensional discrimination), from which peak shift can easily be derived.

The form of the competition equations in the *disjoint* case (Figure 11.9) is

$$x = S_1 \cdot S(u_0, u) - k_{zx}z + C_x, \qquad (11.12\text{A})$$

and

$$z = S_2 \cdot S(v_0, v) - k_{xz}x + C_z, \qquad (11.12\text{B})$$

where u and v are physical dimensions controlling x and z independently, u_0 is the value associated with $S+$ and v_0 is the value associated with $S-$, and S is a function representing the similarity between u and u_0 and v and v_0. The properties of S are straightforward. For example, when $u = u_0$, $S = 1$ (i.e., $S = 1$ means identity, the highest value of similarity), so that Equation 11.12A becomes equal to Equation 11.11A; when $u > u_0$ or $u < u_0$, $S < 1$. In addition, we might expect that the changes in S as u deviates increasingly from u_0 will be gradual rather than abrupt and that as the deviations become very large similarity will approach zero.

Many functions satisfying these rather minimal conditions are sufficient to predict the properties of inhibitory gradients and peak shift. For example, a simple one is

$$S(u_0, u) = 1/[1 + (u_0 - u)^2],$$

and another is the familiar bell-shaped Gaussian curve,

$$S(u_0, u) = \exp[-(u - u_0)^2/D], \tag{11.13}$$

with mean u_0 and standard deviation D, which is the one I shall use.

Conjoint (in this case *intradimensional*) control incorporates both feature-negative and feature-positive cases, since variation in the same stimulus dimension affects both x and z directly. Hence Equation 11.12B must be rewritten as

$$z = S_2 . S(u_1, u) - k_{xz}x + C_z, \tag{11.14}$$

where u_1 is the value of $S-$ on the varied dimension. Eliminating z from Equations 11.12A and 11.14 and rearranging yields response rate x as a function of the stimulus value, u:

$$x(u) = [S_1 . S(u_0, u) - k_{xz}S_2 . S(u_1, u) - k_{zx}C_z]/(1 - k_{zx}k_{xz}), \tag{11.15}$$

which is just the weighted difference of the two similarity functions: $S(u_0, u)$, which is centered on the $S+$ value, and $S(u_1, u)$, which is centered on the $S-$ value, that is

$$x(u) = A_1 S(u_0, u) - A_2 S(u_1, u) + A_3, \tag{11.16}$$

where A_1, A_2, and A_3 are lumped constants made up of the various constant terms in Equation 11.15. Substitution of Equation 11.13 for the similarity functions in Equation 11.16 then allows prediction of the postdiscrimination generalization gradient.[16]

Figure 11.14 shows the kind of prediction that results: The two identical bell-shaped curves are the Gaussian similarity functions for $S-$ and $S+$; the higher curve is the postdiscrimination gradient derived from Equation 11.16. There are three things to note about the predicted postdiscrimination gradient: (a) It is steeper than the underlying similarity functions; this points to the same conclusion we came to earlier, that behavioral competition aids discrimination performance. The hypothetical animal in Figure 11.14 discriminates perfectly between $S+$ and $S-$ (in the sense that no responses are made to $S-$) even though the similarity functions for the two overlap considerably. (b) The postdiscrimination gradient is higher than the similarity functions; this is just behavioral contrast. (c) The peak of the postdiscrimination gradient is shifted away from $S+$ in a direction opposite to $S-$; this is *positive peak shift*.

The term C_x (the contribution to x made by factors common to $S+$ and $S-$) in Equation 11.14 acts like an additive constant; when C_x is high (early in training, for example, before the discrimination is fully developed), the horizontal axis in Figure 11.14 may be displaced downwards. Response x then occurs during $S-$ as

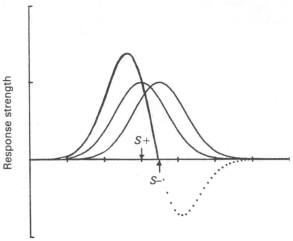

Figure 11.14. Positive and negative peak shift predicted from linear reciprocal inhibition and Gaussian similarity gradients. Two light curves are the similarity gradients; heavy curve is the predicted postdiscrimination gradient, showing positive peak shift; dotted portion shows negative peak shift.

well as $S+$, so that a *negative peak shift* (Guttman, 1965) – the dotted curve – can sometimes be observed.

Thus, the basic properties of discrimination performance, inhibitory control and generalization gradient peak shift, seem all to reflect behavioral competition and direct excitatory control of behavior by stimuli.

Several other effects also fit into this general picture. For example, some years ago Terrace did an experiment in which pigeons were first trained to peck a distinctive $S+$ (e.g., a key illuminated with white vertical line) for food presented on a VI schedule. Once this pecking was established, Terrace occasionally presented a brief $S-$ with a very different appearance (e.g., a dark key). Pecks on this stimulus did not lead to food, but in fact most animals never pecked it. Terrace then progressively increased the duration of $S-$ and faded in a color. In this way he was able to establish discriminations without the animals making any "errors" ($S-$ responses). The animals turned away from the key as soon as $S-$ went off, so that presumably no activity came under the explicit control of $S-$ Perhaps for this reason, these animals failed to show inhibitory generalization gradients (responding was close to zero at all values of the varied $S-$ dimensions). They also failed to show behavioral contrast. Other work has confirmed that contrast and inhibitory gradients seem usually to go together. The present analysis suggests they should, because both depend upon stimulus control by $S-$ of activities antagonistic to the terminal response. Contrast and peak shift also tend to go together, although there are some dissociations: Contrast can occur without a peak shift, and vice versa, for example. These dissociations are not surprising, because although peak shift and contrast both depend on the strength of inhibitory interactions (the k values in Equation 11.11) as well as the similarity difference between $S+$ and $S-$, the quantitative form of the dependence is quite different.[17]

The static view of free-operant discrimination performance I have just presented is undoubtedly much too simple: It assumes linear competition, additive stimulus effects, and invariant similarity relations – and all of these things are probably true approximately at best. Nevertheless, this view brings together a range of experimental facts not easily related in any other way. It is likely to require modification in at least three respects: to accommodate nonlinear effects; to incorporate acquisition processes – *how* discrimination is learned as well as *what* is learned; to accommodate dynamic effects. I shall have little to say about nonlinear models, and acquisition is dealt with later. I end this chapter with a brief account of dynamic effects.

Dynamic effects

Most of this chapter has been concerned with the effects of a very simple successive-discrimination procedure in which two stimuli, $S-$ and $S+$, are presented in alternation. I have said little about the significance of the duration of each component, although in fact it makes a great deal of difference to the outcome. For example, suppose we compare the results of two contrast experiments, Reynolds' original, in which 60-sec components are presented in alternation, and another in which $S+$ is presented for 60 min each day for several days, followed by $S-$ for a comparable time, followed by a return to $S+$. What differences might we expect between these two experiments?

The two studies will be similar in one respect: Animals will in both learn to respond when $S+$ is present and not when $S-$ is present. But the similarities end there. In the second study, response rate in $S+$ on its second presentation is unlikely to be higher than during its first – no behavioral contrast. If anything, rate in $S+$ the second time around is likely to be *lower* than at first (this is sometimes termed *induction*[18]). In a generalization test, there will be no peak shift nor will it be possible to demonstrate inhibitory generalization gradients around $S-$.

There is an obvious lack in the static scheme that accounts for these differences. Recall that the basic explanation for contrast is that z, the competing, interim activity, is reallocated to the extinction component (or the component with lowered reinforcement rate if reinforcement rate is merely reduced in $S-$); with more total time spent doing z, z becomes less competitive in $S+$ allowing the rate of the terminal response to rise (positive contrast). The fine print here is the word *total*: Over *what period* is this total to be measured? In explaining the feature-negative effect, I assumed that the length of the 1-min ITI relative to the 7-sec $S-$ was sufficient to reduce the competitiveness of interim activities in $S-$. We saw in Chapter 6 that animals seek to regulate the rate of activities not just over days, but even over periods of a few minutes. Yet the static model assumes that the competitiveness of z (which is proportional to $V(z)$ in the model) is constant and depends upon its average across a whole experimental session, or at least across one $S+-S-$ cycle. This is obviously wrong when component duration becomes longer than a few seconds. For example, if component duration is 30 min, say, who can doubt that the competitiveness of interim activities at the end of a "rich" component, when there has been little time to engage in anything but the terminal response, is likely to be considerably higher than at the end of the alternating "lean" component, when there has been much free time. Thus the assumption of constant $V(z)$ will increasingly be violated at long component durations. What effects might this have?

To answer this question we need to know something about the second-by-second processes that determine the strength[19] of an activity. Our knowledge here is unsatisfactory because it is practically impossible to study activities in isolation. Anything we can measure is likely to be multiply determined (see note 12), making identification of individual causal factors a formidable task. Nevertheless, we can make some reasonable guesses based on the analysis of choice in Chapter 8. Recall that nonexclusive choice depended at the level of individual responses on negative feedback. In patch foraging, for example, the longer the animal works one patch, the smaller the incremental payoff. The result is that he switches from patch to patch, rather than remaining fixed on one. Similarly, on concurrent VI VI schedules, the longer the animal continues to respond to one alternative, the better the payoff for responding to the other; hence he spends some time on both. Similar negative feedbacks must underlie the animal's allocation of time to different activities under free conditions. The only difference is that the feedback must then be internal, rather than via external contingencies of reinforcement. The effects in both cases are the same: An activity becomes less attractive the longer it is engaged in. I will term this the *principle of satiation* by analogy (but only analogy, no other similarity is intended) to satiation for food and water.[20]

There is, of course, a converse principle, *deprivation*: the longer the time since the last occurrence of an activity, the stronger its tendency to occur. Both principles are combined in the generalization that an activity becomes *relatively* less attractive as more time is spent on it.

These principles do not apply to every activity at all times. Under restricted conditions, even opposite effects can be demonstrated, such as *sensitization*, when the occurrence of an activity for a while *increases* its strength. Conversely, some activities may become less likely if they are prevented from occurring. Still others apparently occur at random, with no dependence on time at all. Nevertheless, satiation–deprivation has the same kind of generality as the principle of diminishing marginal utility: Without it, there could be no preferred mix of activities and no regulatory response to contingencies that force a deviation from the preferred mix.

In multiple schedules, there are three activities to consider: the terminal responses in each component, x and y, and the interim activities, z. Since animals

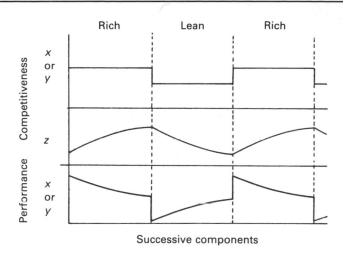

Figure 11.15. Model for dynamic changes in interim activities. Top panel shows step changes in competitiveness of the terminal response, associated with stimuli in each multiple-schedule component. Middle panel shows satiation–deprivation effects on interim activity – becoming less competitive in lean component, more competitive in rich component. Bottom panel shows local contrast effects predictable from the difference between top and middle panels.

on these schedules are generally very hungry, we might expect relatively little weakening within a session of the tendency to make the food-related response (i.e., *x* or *y*). The interim activities are presumably less urgent than eating, however, so should be more subject to satiation. If the rates of deprivation and satiation are relatively rapid, we might then expect to see changes in the competitiveness (strength) of *z* from component to component of the sort depicted in Figure 11.15. The center panel shows the changes in the strength of *z*, on the assumption that it occurs only at a low rate in $S+$ (the rich component) so that deprivation causes an increase in strength; z can presumably occur at a higher rate in $S-$ (the lean component), so that satiation then causes a decrease in strength. In $S+$ the strength of the terminal response is high, in $S-$ it is low, as shown in the top panel. Since $z + x$ or *y* is constant, because of time allocation, the actual values of *x* and *y* will be the complements of *z*, as shown in the bottom panel. We have seen this general pattern before; it is the *exponential lead* discussed in Chapter 3, and represents a direct relation between the controlled variable (here response rate, *x* or *y*) and the *rate of change* of the controlling variable (here rate of reinforcement).[21]

Figure 11.16 shows experimental data confirming that *local contrast* effects of this sort actually occur. The figure shows response rate for two individual pigeons during successive 30-sec periods within 3-min components of a multiple VI 6 VI 2-min schedule. As in the theoretical figure, response rate is highest at the beginning of $S+$ components and lowest at the beginning of $S-$ components,

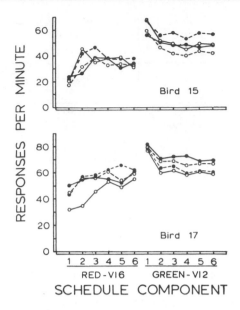

Figure 11.16. Local contrast effects in a multiple VI 6 VI 2-min schedule; each curve is an average of a single experimental session. (From Nevin & Shettleworth, 1966.)

subsequently converging toward an intermediate level. Other experiments have shown that the cause of these effects is indeed the relative richness of the two schedule components: There is negative local contrast in a lean component preceded by a rich one, and positive local contrast in a rich preceded by a lean.

Relative response and reinforcement rates in multiple schedules conform better to matching when component duration is short – a few seconds – than when it is long – a few minutes (Shimp & Wheatley, 1974; Todorov, 1972). Satiation-deprivation dynamics suggest why: the longer the components, the less valid the assumption (necessary to the derivation of matching) that $V(z)$ is constant throughout. $V(z)$ will tend to increase during a long, rich component, and to decrease during a long, lean one. Hence, z will be more competitive during the rich component than during the lean, so that (terminal) response rate will be lower during the rich component than required by matching; the result will be undermatching, which is the usual deviation from matching on multiple schedules.

An intriguing fact about these local contrast effects is that they are usually *transient*, occurring during early exposure to the multiple VI VI schedule, but dissipating after a few days. On multiple FI FI schedules, however, they persist indefinitely; and they can be reinstated on multiple VI VI by manipulations that make the stimuli harder to discriminate or remember.[22] Under other conditions they may fail to occur at all, or occur in a different way – being correlated with the succeeding rather than the preceding component, for example.[23]

The transience of local-contrast effects has not yet been satisfactorily explained, but one possibility is as follows. The magnitude of local contrast obviously depends upon the relative importance as causal factors of competition and facilitating stimuli: If the effect of stimuli is large ($S_1 >> k_{zx}z$ in Equation 11.11A, for example), then changes in the strength of competing activities will obviously have little effect; conversely, if the effect of stimuli is relatively weak, changes in competition will have large effects – of course if the effects of stimuli are zero, there can be no effects at all, since changes in response rates will not be correlated with stimulus changes. It turns out that the things that promote local contrast are also things that should weaken the animal's ability to identify $S-$ and $S+$ or weaken control by $S+$ and $S-$: making $S+$ and $S-$ more similar; introducing a competing source of control (multiple FI FI, where responding is under both temporal control and control by the discriminative stimuli, compared to multiple VI VI, where only the stimuli are predictive); and intermixing many different stimuli, so that memory limitations make it difficult to recognize individual stimuli. Thus the transience of these effects is consistent with some kind of additive model in which response rate is jointly determined by stimuli and competition from other activities.

There is no obvious explanation for local effects in which rate changes seem to reflect the following, rather than the preceding, component. The procedures in which these effects are found are often complex – with more than two components, for example – but this is not a sufficient explanation in itself. The effects may be *path dependent*, that is, they could depend not just on the conditions of reinforcement *now*, but also on prior history; that is, the sequence of conditions leading up to the present arrangement. Metastable (see Chapter 5) patterns are relatively common on multiple schedules, and frequent switching between conditions is necessary if one is to be sure of the stability of any given pattern.[24]

Given the kind of complex individual dynamics I have described, it is far from obvious what we should expect when there are several schedule components and their durations are close to the time constants of the hypothesized satiation-deprivation processes. It is not too difficult to provide a formal model for the simple case in which only the single competing activity z is subject to satiation-deprivation effects. If there is more than one such activity, and the time constants for each are different, or if the instrumental responses are subject to satiation-deprivation, it is not trivial to provide an exact account. But without an exact account it is impossible to be sure what form local-contrast effects should take. Hence the jury is out on the significance of effects other than standard local-contrast effects.

Stimulus effects. Local-contrast effects are interactions along the dimension of time, true dynamic effects. Very similar interactions take place along stimulus dimensions in *maintained generalization gradients*. A maintained gradient is obtained when instead of two very different stimuli being used as $S+$ and $S-$, many similar stimuli are used, one or a few being associated with reinforcement,

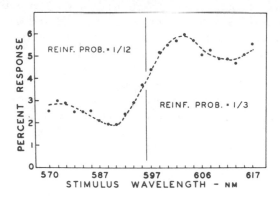

Figure 11.17. Edge effects in a maintained generalization gradient. Each point is the average response probability to the indicated wavelength; stimuli on the right signaled reinforcement, those on the left, its absence. (From Blough, 1975.)

the rest being unreinforced. For example, Figure 11.17 shows elegant data from an experiment by Blough (1975; see also Catania & Gill, 1964; Malone & Staddon, 1973) in which pigeons were reinforced for pecking at wavelength stimuli from 597 to 617 nm and not for pecking at shorter wavelengths, from 570 to 595.5 nm. After many days training on this procedure, the average percentage of responses to each wavelength followed the smooth curve shown in the figure. Of special interest are the positive and negative "shoulders" on either side of the boundary between $S-$ and $S+$ stimuli. The shape of the curve bears an obvious resemblance to the perceptual "edge effects" discussed in Chapter 3. Like those effects, these also hint at the existence of inhibitory interactions along the stimulus dimension.

Blough explains these and other similar data by means of a linear learning model of the type briefly discussed in Chapter 8 (see also Chapter 13) together with a generalization assumption analogous to the similarity gradients discussed in the previous section. Like most local-contrast effects, these stimulus effects are transient and disappear after more or less extended training. Blough's dynamic model handles both the general form of the inhibitory "shoulders" and their eventual disappearance, but has not been applied to behavioral contrast, peak shift, matching, and so on, nor to the effects of stimulus- and memory-related manipulations on the transience of local contrast.

SUMMARY

In this chapter I have argued that the static properties of performance on multiple (successive discrimination) and concurrent (simultaneous discrimination) schedules can be derived from four assumptions. The first two are: (a) Inhibitory stimulus control – the suppression of an ongoing activity by presenting a stimulus

– is associated with excitatory control of antagonistic activities, that is, inhibitory control of activity A is excitatory control of complementary activity ~A. The two complementary classes correspond to the *terminal* and *interim* activities observed in periodic-food experiments. (b) Competition between activities is governed by diminishing marginal utility (see Chapter 7). Given certain simplifications, these assumptions permit two conclusions: that the competitiveness of an activity is inversely proportional to its rate of occurrence and directly proportional to its rate of reinforcement; and that activities are in equilibrium when all are equally competitive. Matching, on concurrent VI VI schedules, deviations from matching, on multiple schedules, and behavioral contrast, together with the effects on contrast of species, absolute reinforcement rate, and response type, can all be deduced from these assumptions.

Two other assumptions allow prediction of generalization-gradient peak shift and the relations between contrast and peak shift: (c) In the steady state, response rate is additively determined by stimulus factors and competition from other activities. (d) Stimulus factors may either be independent (disjoint) so that the stimulus element facilitating one activity has no effect on its antagonist, or overlapping (conjoint), so that the same stimulus element has effects (of different magnitudes) on both activities.

The static approach assumes constant competitiveness, which is obviously unrealistic. As an activity continues to occur at a high rate, its competitiveness must in reality decline, and as time goes by without the activity, its competitiveness must increase. A moment-by-moment satiation–deprivation process of this sort is the dynamic counterpart of static diminishing marginal utility. Satiation–deprivation processes can account for local-contrast effects, and the dependence of stimulus effects on memory can account for the revival of these effects under conditions where stimulus identification or control is weakened and their disappearance when stimulus identification is easy.

Stimuli are additive terms in these equations, but of course no discriminative stimulus acts directly on behavior. To be effective, a stimulus must be recognized, and recognition depends on the properties of memory in ways that are the topic of the next chapter.

NOTES

1. Reynolds (1961b). An effect similar to behavioral contrast had been shown some time earlier by Crespi (1942), with rats running in runways. But the *Crespi effect* (as it is sometimes called) depends on a single change in amount of reward in the same situation, rather than different frequencies of reward in the presence of different, alternating, stimuli. Behavioral contrast is, in fact, hard to show when only amount of reward is varied and it is a steady-state effect, rather than an effect of a one-time manipulation. These differences in species and apparatus, and the different theoretical tradition within which Crespi's work was done, meant that it did little to diminish the impact of Reynolds' experiment.

Research on behavioral contrast since Reynolds' first experiment is reviewed in Schwartz and Gamzu (1977), and there is a good brief review in the textbook by Fantino and Logan (1979).

2. The response-change and reinforcement-change accounts of contrast are conceptually rather different. The response-change account is perfectly mechanistic, and just depends upon conservation of key pecks: What is lost in one component is gained by the other. But the reinforcement-change account demands a bit more in the way of computation by the animal, since it assumes control of behavior by *relative* rather than absolute reinforcement rate.

The competition view is similar to the response-change account in one respect, that it emphasizes activities rather than reinforcement. But it differs in three important ways: It does not require that pecks lost in $S-$ exactly equal pecks gained in $S+$ – which is fortunate, since the two are not generally equal; it considers activities other than the measured instrumental response; and it acknowledges that animals spend time in food-related activities even when food is delivered free.

3. I have argued that there are at least two different types of interim activities: *facultative* and *schedule-induced* (Staddon, 1977a). Facultative activities are things like wheel running (for a rat) or pacing (for a pigeon): they occur on food schedules but are not facilitated by them. Indeed, the usual effect of a food schedule is to reduce the proportion of time devoted to facultative activities. Schedule-induced interim activities, on the other hand, are facilitated by a schedule. The most striking example is drinking by rats; other examples are schedule-induced attack, by pigeons, and perhaps wood chewing by rats. Most interim activities appear to be of the facultative variety (see Roper, 1981). The arguments I will make apply to facultative rather than schedule-induced interim activities.

4. In support of the competition view see Estes (1950), Henton and Iversen (1978), and Hinson and Staddon (1978). Bouzas and Baum (1976) and White (1978) have shown that contrast involves time reallocation. The main arguments in this section are taken from a theoretical chapter by Staddon (1982).

5. It is true that "lying around" is as much an activity as "pacing" and similar vigorous behaviors, but it is also likely to be much more easily displaced by other activities (i.e., have a lower cost-of-deviation, see Chapter 7) – which is the important point for the competition argument.

6. *Matching as a Consequence of Maximizing.* Matching, and Herrnstein's equation for response functions on a single VI, can be derived from minimum-distance-type models in the following way: Recall that the objective function from the minimum-distance model (Equation 7.9) is

$$C[x, R(x), z] = a(x_0 - x)^2 + b(R_0 - R(x))^2 + c(z_0 - z)^2, \quad \text{(N11.1)}$$

where x is the rate of instrumental responding, z is the level of interim activity, and $R(x)$ is the obtained rate of reinforcement. The constraints are due to time allocation:

$$x + R(x) + z = 1,$$ (N11.2)

and the VI feedback function:

$$R(x) = Ax/(A + x),$$ (N11.3)

where A is the programmed (i.e., maximum) VI reinforcement rate. Equations N11.1, N11.2, and N11.3 are then combined to form the Lagrangian:

$$L[x,R(x),z,\lambda_1,\lambda_2] = a(x_0 - x)^2 + b(R_0 - R(x))^2 + c(z_0 - z)^2$$
$$-\lambda_1(1 - x - R(x) - z) - \lambda_2(R(x) - Ax/(A + x)).$$ (N11.4)

If we assume that the cost of x is negligible, then $a = 0$ and the first term in Equation N11.4 vanishes. Taking partial derivatives to find the minimum then yields:

$$\partial L/\partial x = \lambda_1 + \lambda_2 A^2/(A + x)^2,$$

which can be rewritten as

$$\partial L/\partial x = \lambda_1 + \lambda_2[R(x)/x]^2,$$ (N11.5)

plus four other expressions. A similar exercise for the other instrumental response, y, yields an expression exactly parallel to Equation N11.5. By the hypothesis, these two marginals must be equal to the fixed marginal for interim response, z; hence they can be equated, which yields $R(x)/x = R(y)/y$, the matching relation. Adding the time constraint yields Herrnstein's hyperbolic equation, as shown in the text.

7. Herrnstein (1970). Herrnstein's equation is (in his symbols) $P = kR/(R + R_0)$, where P is response rate, R is reinforcement rate, R_0 reinforcement rate for "other" behavior, and k a constant. If response rate is proportional to time spent, this equation is equivalent to Equation 11.8 in the text, with k being the number of responses per unit time. When eating time is excluded from the rate measures, experimental results over a range of VI values that includes all but very short interfood intervals fit this relation well.

8. *Matching and contrast: formal derivation and a mechanical model*. Figure 11.18 shows a very simple mechanical analog for the equations in the text. The figure shows a cylinder containing two free pistons that divide the unit volume into three compartments of volume x, y, and z, respectively. Clearly, $x + y + z = 1$, as in Equation 11.4 ($R(x)$ can be neglected). In equilibrium, the pressure of gas in each of the three compartments must be the same, or else the pistons would move. Hence we can assume that $P(x) = P(y) = P(z)$. From Boyle's law we know that for each compartment, $P(i) \cdot i = NT$, where i is the volume, N is the number of molecules per unit volume, and T is the absolute temperature.

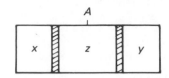

Figure 11.18. Pressure model for matching and contrast effects in multiple and concurrent schedules. Figure represents a cylinder containing two free pistons dividing unit total volume into three compartments, of volumes x, y, and z. For equal-component multiple schedules there is a detent at A preventing x or y from exceeding half the total volume.

Suppose we identify N with the reinforcement rate for activity i, so that we can write $P(i) \cdot i = R(i) \cdot T$, or $P(i) = TR(i)/i$ for all three activities. Equating the values of P leads at once to Equation 11.5 in the text: $R(x)/x = R(y)/y = V(z)/z$.

If the pistons in Figure 11.18 are completely free to move anywhere in the cylinder, the model is appropriate for concurrent VI VI schedules, since the component duration is there completely under the control of the animal. Multiple schedules require the addition of a *detent* in the middle (for equal-duration components) of the cylinder that prevents x or y from taking up more than half the total volume. It is obvious from this analog that perfect matching can hold in the multiple case only if $P(z)$ is sufficiently large that neither x nor y expands to more than half the total volume.

These conclusions can be derived formally in the following way. Behavioral contrast is defined by the ratio of response rates in component X of a multiple schedule under two conditions: a *prediscrimination* phase, in which responses X and Y receive equal VI reinforcement; and a *postdiscrimination* phase in which reinforcement for X continues as before, but reinforcement for Y is abolished. The limitation on the maximum values of x and y relative to z, the level of the interim activity, means that the analysis must consider separately the two cases where $V(z)$ is less than or equal to, or greater than, $R(x)$:

1. $V(z) \leq R(x)$: In the prediscrimination phase, $R(x) = R(y)$; hence it is obvious that z takes up equal volumes in compartments x and y. If $V(z) > 0$, neither x nor y can "saturate," so that the simple matching law applies. Thus,

$$x_{pre} = R(x)/[2R(x) + V(z)], \qquad (N11.6)$$

in the prediscrimination phase. When reinforcement for Y is abolished in the postdiscrimination phase, however,

$$x_{post} = R(x)/[R(x) + V(z)], \qquad (N11.7)$$

but since $V(z) \leq R(x)$, this expression is $\geq .5$, the maximum possible value for x. Consequently, $x_{post} = .5$ (i.e., x has saturated) and the magnitude of contrast is therefore given by

$$\begin{aligned} x_{post}/x_{pre} = C_o &= [R(x) + .5V(z)]/R(x) \qquad (N11.8) \\ &= 1 + V(z)/2R(x), \qquad V(z) \leq R(x), \end{aligned}$$

which is a positive function in $V(z)$, and negative in $R(x)$. Contrast is at a maximum, at $C_o = 1.5$, when $R(x) = V(z)$.

2. $V(z) > R(x)$: As before, $x_{pre} = R(x)/[2R(x) + V(z)]$. However, the matching relation yields a value for x_{post} that is less than the maximum possible, because $R(x) < V(z)$; consequently, x_{post} is given by Equation N11.7 and contrast is therefore given by

$$C_o = [2R(x) + V(z)]/[R(x) + V(z)], \qquad V(z) > R(x), \qquad \text{(N11.9)}$$

which is a negative, hyperbolic function of $V(z)$.

Equations N11.8 and N11.9 show contrast to be a nonmonotonic function of both $V(z)$, the reinforcement for ("strength" of) other behavior and $R(x)$, reinforcement rate in the unchanged component. Both functions are discontinuous and bitonic, with maxima at $R(x) = V(z)$.

The predictions for matching in multiple schedules must be derived in a similar piecemeal way, except that there are now three cases to consider:

(a) If $R(x) \geq R(y) + V(z)$, then $x = .5$ (X saturates) and   (N11.10)
$$y = .5R(y)/[R(y) + V(z)],$$

i.e., matching between Y and Z in a compartment of duration .5.

(b) If $R(y) \geq R(x) + V(z)$, then $y = .5$ (Y saturates) and   (N11.11)
$$x = .5R(x)/[R(x) + V(z)].$$

(c) If neither (a) nor (b) is true, then $V(z) > |R(x) - R(y)|$; neither X nor Y saturates, so that overall matching applies: $x/y = R(x)/R(y)$.

If a constraint such as $R(x) + R(y) = K$ is placed on $R(x)$ and $R(y)$, then the function relating $x/(x + y)$ to $R(x)/[R(x) + R(y)]$ can be derived. Examples are shown in Figure 11.6. Each function has three regions corresponding to cases (a), (b), and (c). Simple matching holds in the central region defined by $|R(x) - R(y)| < V(z)$ (i.e., $V(z)$ greater than the absolute value of the difference between $R(x)$ and $R(y)$). As the value of $V(z)$ increases relative to $|R(x) - R(y)|$, region (C) increases relative to regions (A) and (B), the curvature of the function in regions (A) and (B) decreases, and the function as a whole approximates more and more closely to matching.

9. Herrnstein and Loveland (1974); see Reynolds (1963b) for data showing that contrast effects are small when reinforcement rates are high.

10. These data are from Reynolds (1963a); the fitted curves are from an equation proposed by Herrnstein (1970), based on his analysis of matching (see note 12, above). In my symbols, his equation is

$$x = R(x)/[R(x) + KR(y) + V(z)],$$

where K is a constant. It is obviously similar to Equation 11.10 in the text, and for similar reasons is forced to predict that contrast must always be reduced by increases in $V(z)$ – which is false. Parameter K indicates the degree of interaction between components, being high for short components, low for long. The equation must, therefore, predict that absolute response rate in both components should decrease as component duration decreases, which is also inconsistent with the data (Edmon, 1978). Herrnstein's multiple-schedule equation is evidently an inappropriate extension of his successful rule for simple VI response functions.

11. *Theories of contrast.* There are two other theoretical accounts of contrast that have achieved wide currency: additivity theory and the matching law. I have shown that those features of the matching-law account that fit the data also follow formally from the competition view. Additivity theory (see Schwartz & Gamzu, 1977, for the most extensive account) argues that contrast experiments are also autoshaping situations, in the sense that reduction of reinforcement rate in the changed component improves the signal properties of the stimulus in the unchanged component. Since pigeons will often peck at stimuli that signal food (the theory continues), these "autopecks" add to the pecks already maintained by the response contingency signaled by $S+$ to produce contrast. Ingenious experiments have shown that if the signal for $S+$ is presented on a key separate from the pecking key, additional (and useless) pecks are indeed delivered to it when the changed component is shifted to extinction – and little or no contrast is then observed on the pecking key. Nevertheless, the theory has run into serious trouble because it cannot easily handle negative contrast or contrast in situations that do not produce autoshaping, such as rats pressing levers or pigeons pressing treadles. Attempts to show that autopecks are quantitatively different from "operant" pecks (shorter duration, less vigorous) are also open to the alternative interpretation that these two classes just differ in strength (see note 19 to this chapter and the discussion of reflex strength in Chapter 2). (See also Ziriax & Silberberg, 1978; Schwartz & Gamzu, 1977; Farley, 1980; Whipple & Fantino, 1980).

A possible interpretation of autoshaping in terms of the competition view is that pecking and similar food-elicited activities will occur in situations that signal food where competing tendencies are also at a minimum. For example, the account of dynamic factors later in the chapter suggests that the most effective way to weaken competing activities is to lengthen $S-$ relative to $S+$; but this is also the best way to get pigeons to peck at $S+$ (Gibbon, Locurto, & Terrace, 1975).

Rachlin (1973) has proposed a nonquantitative theory of contrast similar to the dynamic model sketched out at the end of this chapter.

12. Use of the term *controlled* to refer to the effect of a stimulus on a response is conventional, but it is not quite accurate. Behavioral competition means that the level of an activity is jointly determined by its stimulus and by competing activities (hence by their stimuli); discriminative stimuli exert less than total control over the level of their responses. Terms such as *facilitated* or *excited* are more accurate in the present theoretical context.

13. Honig, Boneau, Burstein, and Pennypacker (1963). The argument in the next few paragraphs about time allocation and the slope of inhibitory gradients was first made by Jenkins (1965).

Note that in this situation, with relatively long $S+$ and $S-$ components and no timeout periods in between, pigeons have no difficulty handling a feature-negative discrimination.

14. In earlier discussion, response rates x and y were involved in two other sets of equations: time-allocation conservation (Equation 11.4) and reinforcement (Equation 11.7). It is natural to ask about the relation between these two and Equations 11.11A and B, the competition equations. This is a gray area, but a possible answer is as follows: The first two equations (certainly the reinforcement equation) refer to conditions at equilibrium. They say nothing about the moment-by-moment determinants of activity. But the competition equations refer to the results of generalization tests, which are brief probes that give information on the current "forces" acting on behavior. There is no reason, therefore, why the two sets of equations should be simply related.

Animals seem to adapt to these steady-state procedures not by learning specific stimulus-response "connections" but by constructing a *routine*, analogous in some ways to a computer program, that has some stimulus "inputs" to be sure, but also involves other processes (see Staddon, 1981b). If the world changes, the effect on the animal depends on how large and how persistent the change is. If it is relatively brief, and not too large – as in a generalization test – then the routine built up by the training procedure continues to function, and the results of the test can tell us something about it. But if the change is large, other processes come into play, the old routine is partially or completely abandoned, and the animal goes about constructing a new routine to cope with the changed circumstances. The temporary nature of the structure probed by the generalization-test procedure is emphasized by the changes that take place during a test (the gradient becomes steeper, for example) and the transience of phenomena like the peak shift: With repeated testing, the shift disappears and the peak of the gradient moves back to $S+$.

The competition equations are an attempt to make a model of the routine set up by successive-discrimination procedures. The parameters of these equations must be such that the balance of behavior satisfies the earlier time-conservation and reinforcement equations, but the way the system responds to stimulus changes is not itself predictable from these equations.

15. The other study in Figure 11.13 is by Honig et al. (1963).

16. *Spence's theory of transposition.* Equation 11.16 has a family resemblance to the most famous model offered for an effect like peak shift, that due to Spence (1937). In my symbols, Spence's model amounts to

$$x(u) \propto S(u_0,u) - S(u_1,u),$$

i.e., responding is proportional to the simple difference between excitatory and inhibitory gradients. There are procedural and formal differences between the two models. Spence was concerned with data from simultaneous (rather than successive) discriminations, in which rats were trained to respond to the larger (say) of two squares. After training, the animals were then confronted with two new squares, the smaller of which was equal in size to the larger of the previous pair. Would the animals respond to the same-size square (absolute responding, congenial to stimulus-response theory) or to the still-larger square (*transposition*: relational responding congenial to gestalt theory)?

Of course, the clever animals usually picked the larger square, giving aid and comfort to the gestaltists and, until Spence, spreading gloom and despondency among S-R theorists. Spence's insight was that this kind of result can be derived from S-R assumptions: If reinforcing responses to $S+$ causes the development of an excitatory gradient (something widely accepted at the time), then *non*reinforcement of responses to $S-$ should produce an inhibitory one. If these two gradients are smooth, and the slope of the inhibitory one is greater than the slope of the excitatory one (at least in the vicinity of $S+$), then the difference between the two gradients will have its peak not at $S+$, but away from $S+$ on the side opposite to $S-$. Given $S+$ and a still larger stimulus, therefore, Spence's model predicts that response strength associated with the larger stimulus might well be greater than response strength associated with $S+$. Moreover (and this was the *coup de grace*), Spence's model also predicted that if the new stimulus were *too* large, the animal would not transpose but would show *transposition reversal*, preferring the original $S+$. Animals indeed do this, so that Spence's model received solid support.

More than twenty years later, after the invention of the technique of free-operant generalization testing and the discovery of peak shift following intradimensional discrimination, Spence's theory was applied to the peak shift. But here it has several flaws: The least important is that it was devised for simultaneous situations, but peak shift does not occur in operant concurrent experiments. More critical is its requirement that the inhibitory gradient be steeper than the excitatory one, and that responding to $S+$ after the formation of a discrimination be at a lower level than before. As we have seen, measured inhibitory gradients are generally shallower than excitatory ones, and postdiscrimination $S+$ responding is generally at a higher level than before (positive contrast).

S-R theorists took Spence's success as further proof that animals do not respond to relations, but just to simple physical properties. Later work shows that they won the battle but lost the war. Experiments in which animals can learn to recognize complex stimulus categories, such as "people" or "trees" (see note 17, Chapter 10), studies in which they learn to pick the "odd" stimulus, or one of intermediate value in a set of three, show that animals do indeed possess the complex perceptual abilities favored by the gestaltists. Spence's contribution was to show that these talents are probably not involved in some tasks that might seem to demand them.

Spence was also correct in rejecting "relational responding" as a satisfactory explanation for anything. It's fine to show that pigeons can solve tricky perceptual problems; but to term this "relational responding" and leave things there is just to label the mystery, not explain it. A proper explanation must, like Spence's, eventually get down to the physical properties of the stimulus and their transformation by the animal. We are very far from being able to do this, even for things like object recognition. Animals are able to perform feats of recognition that still cannot be duplicated by even the most sophisticated pattern-recognition programs.

17. See Staddon (1977b) for a quantitative account of the different predictions for contrast versus peak shift. For reviews of experimental results see Terrace (1966), Mackintosh (1974), and Rilling (1977).

18. This *induction* is quite different from the *induction* of Pavlov and Sherrington discussed in Chapter 2: Pavlovian induction is much closer to local contrast (shortly to be discussed) than to Skinner's induction. Why Skinner, who was certainly aware of the prior usage, chose to use the term in this very different way is a puzzle.

19. *Response strength.* It seems impossible to talk about choice and the allocation of individual behavior without using terms like "tendency," "causal factors," and "strength" that refer to the likelihood one activity will win out over others. The term *response strength* has a long history in experimental psychology, with its meaning changing according to the currents of theoretical fashion (see discussion of reflex strength in Chapter 2). Applied to instrumental behavior the term has been consistently tied to the supposed strengthening effects of reinforcement. When the influence of operationism was at its height, earnest attempts were made to define the term operationally. These met with only limited success, and some alarm was caused by experiments showing that supposedly equivalent measures of strength such as *latency, vigor, probability*, and *resistance to extinction* of a response did not reliably rise and fall together: A short latency need not always go along with high vigor, probability or resistance to extinction (see Osgood, 1953, for a review of this controversy).

Skinner (e.g., 1950) added his nail to the response-strength coffin by pointing out that response properties like latency and vigor can be shaped by suitable contingencies of reinforcement. A measure of reinforcement that can itself be altered by reinforcement is about as much use as a rubber ruler, argued Skinner. He went on to advocate *response probability* (by which he meant *rate* of response) as the only true measure of response strength. Unfortunately, response rate is just as amenable to shaping by means of reinforcement as response vigor, so that the special status of response probability could be reserved only by bringing in the idea of time as a discriminative stimulus. Low-rate behavior might then be deemed strong at particular points in time. For example, on spaced-responding schedules, probability of a response is low just after a response, and rises to high values at times approaching the spaced-responding value. Time clearly can act like more conventional stimuli, but response rate under these conditions ceases to be a useful quantity. At this point researchers who wished to be thought methodologically sophisticated spoke little of response strength. The idea was widely held, but little discussed.

Recently, response strength has staged something of a comeback. One approach has been via empirical laws relating response rate to relative rate of reinforcement: The equations of the matching law have sometimes been proposed as an appropriate measure – the strength of a response being just its relative rate of reinforcement. J. A. Nevin (e.g., 1974) has taken a different approach, suggesting that response strength be defined as *resistance to change*. The idea is that a strong response is one whose rate is changed only slightly by operations usually effective in weakening or abolishing operant behavior, such as extinction and the presentation of free food. Nevin's idea is obviously close to the concept of competitiveness, nor is it remote from the matching relations (see also Staddon, 1978, for a theoretical discussion of this view).

These recent developments follow a current and I believe correct philosophical trend, namely, to let the definition of a term like *strength* grow out of valid theory. Operationism put the cart before the horse. P. W. Bridgman (its physicist founder) noticed that physical terms like force and energy could be reduced to sets of measuring operations. He erroneously drew the conclusion that requiring operational definitions for fledgling terms in young sciences would be helpful to their growth – forgetting that even in physics, the theory came first, operations after. Infants grow taller as they mature, but stretching babies is no recipe for speeding development.

20. This is not a new principle. Hull's (1943) *reactive inhibition* has similar properties; sensory adaptation and reflex habituation (see Chapter 2) are also similar in some respects. More recently the theoretical consequences of self-inhibition have been most extensively explored by Atkinson and Birch (1970). See also numerous papers by Grossberg (e.g., 1981, 1982) for related mathematical discussions of inhibition and competition.

21. One form of the satiation-deprivation assumption applied to activity z is

$$dV(z)/dz = K_1 - K_2 z, \tag{N11.12}$$

where K_1 and K_2 are constants: When z (the rate of the interim activity) equals zero, the rate of growth of $V(z)$, the "value" of z, is maximal; as z increases, the rate of growth of $V(z)$ decreases, eventually becoming negative when $z > K_1/K_2$. The solution to Equation N11.12 is just the exponential lag function derived in Chapter 3; the complement of this function (i.e., the time-course of the terminal response) is the exponential lead.

22. Staddon (1969); the effects may also be persistent on multiple VI VI when the components are very long; for example, Rachlin (1973) has reported large and apparently persistent local effects with 8-min components. Conversely, local effects may be weak when components are short (Hamilton & Silberberg, 1978). The dynamic analysis suggests that local effects should be stronger at longer component durations. Effects of number of components have been shown by Malone and Staddon (1973), of stimulus discriminability by Catania and Gill (1964).

23. See Buck, Rothstein, and Williams (1975) for some of these unusual effects.

24. Under some conditions, activities that happen to occur in a highly predictive stimulus may persist long after the changes that caused them have passed. For example, in an experiment by Kello, Innis, and Staddon (1975) pigeons were trained on a multiple FI 1 FI 3-min schedule, in which 12 FI 1 intervals occurred in one component (green key) and 4 FI 3 intervals occurred in the other (red key). After some training, all animals showed typical local-contrast effects: elevated response rate in the first FI-1 interval, depressed responding in the first FI-3 interval. Then the green key in the FI 1 component was changed to blue for just one interfood interval. All the animals at once responded at a high rate – this is a relatively familiar finding on FI: Temporal control (see Chapter 12) by the food is context-dependent, so that a novel stimulus abolishes it, leading to a high response rate. The novelty of this result was that the effect persisted indefinitely, so long as the blue stimulus occurred only during one FI-1 interval out of 12 in each component. This result shows that under certain not-very-well-specified conditions, unusual effects can be maintained for long periods on multiple schedules.

12

MEMORY AND TEMPORAL CONTROL

Memory is probably the most protean term in psychology: It has as many technical as nontechnical meanings, and the numbers of both are large. Psychologists have, at various times, written of long- and short-term memory, of working and reference memory, of episodic and semantic memory, of primary and secondary memory, and others. In common speech the term memory refers to a purely private event: "I remember...," something not directly accessible to observation. Little wonder that there is still no consensus on what we mean by memory, or on its relations to learning.

In Chapter 4, I defined memory simply as a change of state caused by a stimulus: Memory is involved if how the animal behaves at time t_2 depends on whether event A or event B occurred at previous time t_1. Breaking a leg is a change of state in this sense, of course, so we need to restrict the definition to effects that are *specific* and to some extent *reversible*: The difference in behavior at t_2 should bear some sensible, informational relation to the difference between prior events A and B; and we should be able to change the effect by additional experience. Nevertheless, the advantage of the formal definition is that it commits us to no particular theoretical position – and it draws attention to the memorylike properties of habituation, dishabituation, spontaneous recovery, and, particularly, temporal control – phenomena not traditionally considered memorial.

Much is known about temporal control. The first part of this chapter reviews the properties of temporal control and derives some general principles about the discrimination of recency. In the middle part of the chapter, I show that these principles also apply to more traditional situations used to study memory in animals, such as successive discrimination reversal and delayed matching to sample. The last part of the chapter brings together the idea of internal representation, described in Chapter 10, and the principles of memory described in this chapter, to explain behavior in the radial maze and related spatial situations.

TEMPORAL CONTROL

As we saw in earlier chapters, animals readily detect periodicities: If a pigeon is rewarded with food for the first key-peck T sec after eating (i.e., a fixed-interval T-sec schedule), he will usually not begin to peck until perhaps two-thirds of the time has elapsed, that is, his *postreinforcement pause* will stabilize at close to .67T.[1] The animal is able to do this by using food delivery as a *time marker*. Control of behavior by a past event is termed *temporal* control, to distinguish it from control of behavior by a present stimulus, which might be termed *synchronous* control (see Chapter 4). Temporal control is the instrumental equivalent of trace conditioning.

A simple, if not totally accurate, way to represent temporal control by food is to say that food "resets" the animal's "internal clock," and pecking is initiated when the clock reaches a value which is an approximately constant proportion of T.

Many features of temporal control are consistent with the clock idea. For example, as in a real clock, the *error* in timing is proportional to the interval to be timed: A clock that is 1 min fast after an hour will be 6 min fast after 6 hr; if instead of a constant error, the clock is simply variable from day to day, then the variation in its error over an actual time T will be proportional to T. This is termed the *scalar timing* property.[2] The *reset* property of the time marker can be demonstrated by omitting it or replacing it with something that is not treated as a time marker. For example, Staddon and Innis (1969) trained pigeons and rats on a fixed-interval 2-min schedule and then shifted to a procedure in which food was delivered at the end of only 50% of intervals. The change this produces is shown in Figure 12.1. Intervals ending with "no food" (a brief blackout of the same duration as food delivery) are indicated by "N" in the figure. The animals show the usual pause after food, but if food is *omitted* at the end of a fixed interval, then the animals continue to respond until the next food delivery: In the absence of a "reset" (food), responding continues. With continued experience of reinforcement omission, some animals learn to pause a bit after the nonfood stimulus, but for most, the effect persists indefinitely in the form shown in the figure. This absence of pausing after a nonfood stimulus is known as the *reinforcement-omission* effect.

Synchronous discriminative stimuli seem to tell the animal what to expect at a certain time, rather than affecting his estimate of time directly. For example, Church[3] trained rats on a procedure in which 30-sec and 60-sec fixed intervals were intermixed, each with its own distinctive (synchronous) discriminative stimulus (i.e., a multiple FI FI schedule). The animals soon developed pauses appropriate to the signaled interval. Then, in test sessions, the stimulus in the 30-sec interval was changed abruptly to the 60-sec stimulus. The change could occur 6, 12, 16, 24, or 30 sec after food. The rats behaved as if they had a single postfood clock and used the stimulus just to scale their rate of responding to the clock setting. Thus, an animal's rate of responding t sec after food in the presence of the 60-sec stimulus was the same, whether the stimulus appeared just after the food (i.e., simple FI 60) or at some later time. The rats seem always to know what time it is; the synchronous stimulus just tells them whether to expect food at that time or not.

The clock idea is a convenient simplification, but I show later that the reset

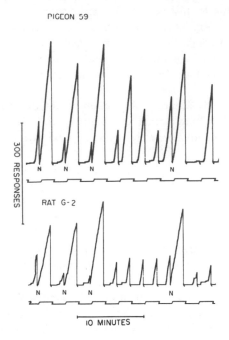

Figure 12.1. Sample cumulative records of stable performance in an experiment in which animals were trained on a fixed-interval 2-min schedule. At the end of intervals marked "N," food delivery was omitted and a brief 3-sec stimulus was presented in its stead. Record at the top is from a pigeon, at the bottom from a rat. The recorder pen was reset at the end of every interval. (From Staddon & Innis, 1969, Figure 3.)

property is far from absolute: Under many conditions, earlier events, preceding the resetting stimulus, can affect behavior. These interference effects, and the conditions under which they occur, show that temporal control reflects the same process studied in more conventional memory experiments: Temporal control and memory are different aspects of the same thing.

Excitatory and inhibitory temporal control

In the preceding chapter we saw that synchronous stimuli can either enhance or suppress an instrumental response. Temporal control can also be excitatory or inhibitory, depending on circumstances. All the examples discussed so far are inhibitory, because the instrumental response is suppressed immediately after the time marker. As with inhibitory synchronous control, suppression of the instrumental response is usually associated with facilitation of competing responses. These are the *interim* activities discussed in the preceding chapter: The postfood "pause" is a pause only in instrumental responding; other activities, such as

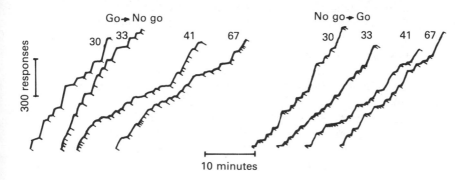

Figure 12.2. Cumulative records of four pigeons trained either on a procedure that reinforces a high rate after food (left, go → no-go schedule) or a low rate after food (right, no-go → schedule). Diagonal "blips" indicate food deliveries. See text for details. (From Staddon, 1972a, Figure 2.)

pacing in pigeons, drinking or wheel running in rats, occur at their highest rates during the pause.[4]

Animals can learn to respond rapidly after a time marker just as easily as they can learn to pause: Excitatory temporal control is as easy as inhibitory control. For example, in one experiment (Staddon, 1970, 1972a) pigeons were trained on a VI 60-sec schedule where the response required for food was either pecking, or *refraining* from pecking, for at least 10 sec. The signal telling the animals which response was required was postfood time: At postfood times less than 60 sec, pecking was required to obtained food; at longer postfood times, not-pecking was required. Cumulative records on the left in Figure 12.2 show stable performance on a "temporal go → no-go" schedule of this sort. The animals show a sustained period of high-rate responding for about 60 sec after each food delivery (indicated by the blips on the record) followed by no responding until the next food delivery. The records on the right show the "FI-like" behavior produced by the reverse, "no-go → go," schedule in which not-pecking is required at postfood times less than 60 sec, pecking at longer times.

Like synchronous control, temporal control shows generalization decrement: Variation in an inhibitory temporal stimulus produces increases in subsequent responding; variation, in an excitatory stimulus, decreases. Because the effect of a temporal stimulus is delayed, discrimination is not as fine as in the synchronous case: Larger changes in the stimulus must be made to produce comparable percentage changes in response rate.

It is easy to show effects of variation in the stimulus complex associated with food delivery on fixed-interval schedules. For example, in one study (Kello, 1974) with pigeons, fixed intervals ended unpredictably with one of three events: food paired with 3-sec extinction of the key and house lights and 3-sec illumination of the feeder light (F), all these events without feeder operation (NF), or extinction of the lights alone (N). The pigeons paused longest after F, least after N, and an intermediate amount after NF, and response rates over the interval following each kind of event were in the reverse order. Comparable experiments with the excitatory procedure (Figure 12.2, left) have shown the opposite result: slowest responding after N, highest after F. In both cases, the effect of the test time marker is of the same sort as the training time marker, and directly related to their similarity.

These experiments are examples of control by stimulus elements (see Chapter 10). It is trickier to demonstrate temporal control by a stimulus dimension, simply because it is harder to establish temporal control by "neutral" stimuli, such as colored lights and line tilts (I explain why in a moment). Nevertheless, when good control is established, gradients of the standard sort are obtained. If the time marker is inhibitory, then responding following it increases as the test stimulus varies from $S+$; if the time marker is excitatory, then responding following it decreases as the test stimulus varies from $S+$.[5]

Conditions for temporal control

Under what conditions will a stimulus such as food come to serve as a time marker? The general answer is the same for both temporal and synchronous stimuli: when it predicts something. A temporal stimulus acquires control when it reliably signals a period free of food delivery – such as the early part of each interval on a fixed-interval schedule; or a period when the conditions of reinforcement are different from earlier or later periods – as in the go ⁺ no-go schedule just described.

The signaled period need not be immediately after the time marker. Consider a modified fixed-interval T-sec schedule in which food is sometimes available (in perhaps 50% of intervals, on a random basis) at postfood time $T/6$, as well as always available after time T – this is a simple (three-interval) variable-interval schedule. Such a schedule is illustrated in Figure 12.3. The top panel shows the probability of food delivery at different postfood times: It shows just two vertical lines, one, of height .5 at postfood time $T/6$, the other, of height 1.0 at time T. The bottom panel shows the average response rate as a function of postfood time for a pigeon that has adapted to this procedure. It shows a high rate just after food (roughly corresponding to the first probability "spike"), followed by a period of low response rate, ending finally with a high response rate towards the end of the T-sec interval.[6] Thus a time marker can initiate a sequence of alternating "respond–not respond" periods.

If the availability of food is random in time then at any instant the probability that food will become available for a response is constant – food is no more likely at one postfood time than another. This is a *random-interval* schedule of the type already encountered in Chapters 5, 7, and 8. We would not expect, nor do we find, much patterning of responding as a function of postfood time: Average response rate is approximately constant. As we saw in Chapters 7 and 8, how-

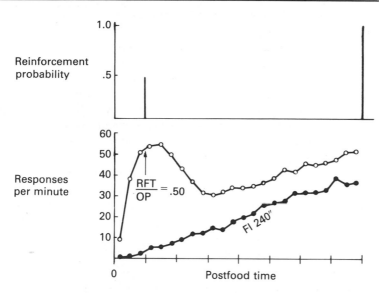

Figure 12.3. Top panel: food reinforcement probabilities as a function of postreinforcement time on a modified fixed-interval schedule. Reinforcement occurs with probability .5 30 sec after food and with probability 1.0 240 sec after food. Bottom panel: average rate of key pecking as a function of time since food for a pigeon trained on this schedule. (From Catania & Reynolds, 1968, Figure 19.)

ever, probability of food does increase as a function of post*response* time on these schedules, since the longer the animal waits, the more likely that the random food-availability programmer has "set up." We might expect, therefore, that probability of response will be low immediately after a response. This is true to some extent, although the effect is somewhat masked by the tendency of pigeons to respond in "bursts" of two or three pecks at a time. Animals on concurrent variable-interval, variable-interval schedules show in their patterns of choice that they are quite sensitive to this property of variable-interval schedules, as we saw in Chapter 8. *Spaced-responding* schedules make the temporal requirement explicit, only reinforcing responses longer than time *T*. If *T* is fairly short (less than a minute or so), pigeons and rats adapt by spacing their responses appropriately.

What determines exactly when an animal will begin to respond after a time marker? The functional answer suggested by the discussion of optimal behavior in Chapter 7 is that it will depend on what other activities are available to the animal and the constraints to which he is subject. An important but hard-to-estimate constraint is set by the psychophysical limits to the animal's ability to tell time. For example, suppose that in addition to pecking the key for food, the animal has at least one other thing he likes to do. On a fixed-interval schedule, the amount of time available for the "other" activity is critically determined by the accuracy with which the animal can estimate the interfood interval. If he is very accurate, then he can defer key pecking until just before the end of the interval (i.e., to a late setting of his internal clock), with the assurance that he will only rarely respond late, and so receive food after a longer time than necessary. On the other hand, if the animal is very inaccurate, he cannot safely defer pecking until almost the whole interval has elapsed, because by so doing he will often wait much too long. This problem was discussed earlier in Chapter 5 (Figure 5.18). The animal's choice of pause "setting" is necessarily a compromise: If he sets it long, so as to waste few responses and have a maximum amount of time for interim activities, he gets food less frequently than he might if he set it shorter – but then he would have less interim time and waste more terminal responses. Moreover, as the timer is set longer, the variance of the pause distribution increases, according to the scalar-timing property, which worsens the terms of the tradeoff.

The costs and benefits associated with any particular setting for mean pause can be derived from the minimum-distance model discussed in Chapter 7. This could be done analytically, but a qualitative analysis is quite sufficient to show that typical behavior on fixed-interval schedules is just about what we would expect. We already know from the earlier discussion that the *cost* of deviations from the preferred rate of eating is typically high relative to the cost of deviations from the preferred rate of key pecking or lever pressing. Hence, we can expect that animals will set their pause distributions so as to keep the area to the right of the line in Figure 5.18 quite small (so that eating rate is as close as possible to the maximum allowed by the interval schedule). On the other hand, we might expect that if offered a relatively attractive interim activity, the pause setting might shift to longer values. This usually happens: Rats on a fixed-interval schedule in the usual bare Skinner box will show shorter pauses than rats also offered the opportunity to run in a wheel. Pigeons trained in a small box, or restrained in a body cuff, show shorter pauses than animals responding in large enclosures (Frank & Staddon, 1974).[7]

Characteristics of the time marker. Food is not the only stimulus that is effective as a time marker, but it is more effective than neutral stimuli, such as tones and lights – or even the animal's own response (recall that pigeons and rats can only learn to space their responses if the delay times are quite short). For example, consider again the procedure illustrated in Figure 12.1. Food was omitted at the end of half the fixed intervals in that experiment, but something happened even at the end of no-food intervals: The light on the response key went out (for the pigeon) and the "house" lights went out (for both rat and pigeon) for about 3 sec – a period equal to the duration of access to food at the end of food intervals. This brief timeout period tells the animal exactly as much about the time until the next opportunity to eat as does food; in both cases, the next food opportunity is after 2 min. Yet both rat and pigeon paused after food but not after the timeout. Why?

There are obviously two possibilities: Either the original hypothesis – that the pause is determined by the predictive properties of the time marker – is wrong, or there is something special about food (and electric shock and other hedonic

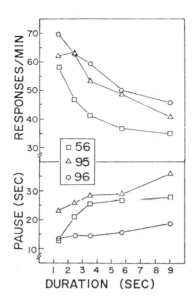

Figure 12.4. Top panel: Response rate in intervals following the access-to-food durations shown on the abscissa for three pigeons trained on a fixed-interval 60-sec schedule in which intervals terminated unpredictably with one of five different food durations. Bottom panel: time-to-first-response (postfood pause) following different food durations. (From Staddon, 1972b, Figure 1.)

stimuli) that makes it more effective than a neutral stimulus. There is too much other support for the predictiveness idea to give it up, and there is much evidence that there is indeed something special about hedonic stimuli such as food.

Consider some other experimental results. In the first experiment (Staddon, 1970b; see also Lowe, Davey, & Harzem, 1974), pigeons were trained on a fixed-interval 1-min schedule in which each interval ended with food reinforcement, with the special proviso that the *duration* of access to food varied unpredictably from interval to interval. There were five different feeder durations, ranging from 1.3 to 9.0 secs. In this experiment the pigeons had food available as a time marker in every interval; there is no obvious reason why they should not have paused in much the same way as on the usual FI schedule in which all food deliveries are of the same duration. But they did not. The results, for three pigeons, are shown in Figure 12.4. The bottom panel shows the average pause after each of the five food durations; all the pigeons show a positive relation: the longer the food duration, the longer the pause. The upper panel shows similar data on rate of responding over the whole interfood interval. Response rate decreases as food duration increases.

There is an uninteresting explanation for this result. Perhaps the pigeons simply take longer to swallow their food after eating for 9 sec as compared to eating for 1 or 2 sec? This plausible explanation is wrong for at least two reasons.

First, it implies that pigeons should pause after food on any schedule, not just fixed-interval, but as we know, they do not – they show minimal pauses on variable-interval schedules, for example. As we have seen with the go → no-go schedule it is also relatively easy to train animals to respond *especially fast* after food, rather than pausing. Second, and more directly relevant, other experiments[8] have shown that the differential-pause effect depends upon the animals' experiencing different food durations *within the same experimental session*, or at least in an intermixed fashion. If, instead of daily experiencing five different food durations, the animals are given several days at one duration, then several more days at another and so on, then pausing after the short durations increases from one session to the next (or decreases, if the food duration is long) so that soon all differences disappear and the animals pause about the same amount after any food duration. The differential-pause effect depends on *intercalation* of different food durations.

The effectiveness of a stimulus as a time marker depends on its freedom from interference from other remembered events. The long feeder durations were evidently less susceptible to interference than the short, when long and short were intercalated, so that postfood pause was longest after the long. When only short intervals occurred, however, they did not interfere with each other, so pause lengthened.

The destructive effects of interference between intercalated stimuli can be shown directly. In the following experiment (Staddon, 1975a), pigeons' ability to use a brief stimulus as a time marker was impaired by intercalating it with another stimulus with no predictive significance. The birds were first trained to respond for food on a variable-interval 1-min schedule. After a little experience, the birds showed characteristic steady responding, with no postfood pausing. In the second phase, every 2 min a brief (3-sec) stimulus (three vertical lines) was projected on the response key. This stimulus signaled that the *next* reinforcement would be programmed on a *fixed*-interval 2-min schedule. Thus, after food, or at any other time, the animal could expect food after some unpredictable time averaging 1 min; but after the 3-sec vertical-line stimulus, the animal knew that food would be available only after exactly 2 min.

The pigeons adapted to this new time marker by developing a poststimulus pause appropriate to the 2-min fixed-interval duration. This pattern is illustrated for one animal by the cumulative record on the left in Figure 12.5. The record reset at the end of each 4-min cycle (i.e., after each FI 2-min food delivery), and the recording pen was depressed during the fixed interval. The pause after the brief stimulus is clear in four of the five intervals in the figure, a typical proportion. This result shows that when there are no interfering events, pigeons can learn to use a brief, neutral stimulus as a temporal cue.

The right-hand record in Figure 12.5 shows the effect of an apparently trivial modification of this procedure. Instead of scheduling the 2-min fixed interval exactly once during each 4-min cycle, it was scheduled on only half the cycles. By itself, it is not likely that this change would have had any significant effect,

All vertical   Horizontal and Vertical

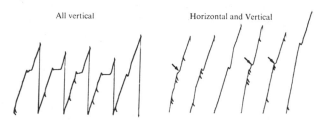

Figure 12.5. Left record: cumulative records of stable performance on a variable-interval 1-min schedule in which a brief vertical-line stimulus (indicated by depression of the response record) occurred every 4 min; the stimulus signaled a 2-min fixed-interval schedule. The record reset at the end of each FI 2 reinforcement. Right record: performance on this schedule when some 4-min periods were initiated by a brief horizontal-line stimulus (arrows) signaling "no change" (i.e., the VI 1 schedule continued in effect). (From Staddon, 1975, Figure 3.)

providing the brief vertical-line stimulus continued to provide a reliable temporal cue. However, during those cycles when no fixed interval was scheduled (i.e., when the VI 1-min schedule remained in effect), a brief *horizontal-line* stimulus was projected on the response key. The first effect of this change was that the pigeons paused indiscriminately after both horizontal and vertical stimuli. This, of course, led to their sometimes waiting longer than necessary after the horizontal stimulus. Eventually, the animals ceased to pause after either stimulus, as shown in the right-hand record in Figure 12.5.

Pigeons have no difficulty in telling vertical from horizontal lines when they are required to do so in standard simultaneous or successive discrimination procedures. So their failure to pause differentially after the two stimuli in this experiment doesn't reflect some kind of perceptual limitation: They can recognize a vertical-line stimulus when they see it, and learn to respond differently in the presence of vertical and horizontal lines. The problem seems to be that in this experiment the animals were not required to respond *in the presence* of the stimuli. Instead they had to behave differently *after* the stimuli had come and gone – pausing after the vertical lines, not after the horizontal lines. In the intercalated (horizontal and vertical) condition, the vertical lines produced only brief pauses: The effect of the stimulus was restricted to a relatively brief poststimulus period. Why?

A commonsense explanation is just that in the intercalated condition, the animals *could not remember* for more than a few seconds which stimulus had just occurred. A more descriptive way to put the same thing is to say that the effect of the informative vertical-line stimulus was abolished by prior presentation of the (uninformative) horizontal-line stimulus, an effect termed *proactive interference* – impairment of recall of a stimulus by occurrence of a prior stimulus.

There is a complementary effect, well known in human memory studies – *retroactive* interference: impairment of control by an earlier stimulus by the interpolation of a later one. Is there a parallel effect in the study of temporal control? The evidence is less clear than for proactive interference, but there is a common phenomenon that may be related: If a novel stimulus is presented during the pause on a fixed-interval schedule, animals will often begin responding at once. This is a sort of *disinhibition*, first seen by Pavlov in delayed-conditioning experiments. This effect may be attributable to retroactive memory interference, the later event (the novel stimulus) impairing recall of the earlier one (food, the time marker). The converse effect, insulation from proactive interference by means of differential stimuli during the fixed interval, has been demonstrated, as I show in a moment.

If the interference demonstrated in these experiments has something to do with memory, we can make some additional predictions. All theories of memory agree that under normal circumstances an event is better remembered after short times than after long. If, therefore, the effects of reinforcement omission reflect failure of memory they should be reduced when the time intervals involved are short. This seems to be the case. For example, if we repeat the fixed-interval reinforcement-omission experiment with 15- or 30-sec intervals rather than 60- or 120-sec intervals, the pigeons pause as long after brief stimuli presented in lieu of food as they do after food (Neuringer & Chung, 1967; Starr & Staddon, 1974).

Interference between to-be-remembered events can also be reduced if they are associated with different situations or contexts. For example, if a person is required to learn two lists of similar items, words or nonsense syllables, one after the other in the same room, and then asked to say whether a particular item comes from a particular list, he will often make mistakes, identifying an item from list B as coming from A and vice versa. If the same experiment is done with the two lists either separated in time, or learned in different rooms or from different experimenters, confusions of this sort are much reduced.

A similar experiment has been done with temporal control.[9] Two pigeons were trained on a fixed-interval 2-min schedule in which intervals ended either with food (F) or a brief blackout (N) with probabilities 1/3 and 2/3. The response key was either red or green during each fixed interval. Both stimuli gave the same information about the *outcome* of an interval: In either case, the probability the interval would end with food was 1/3. But when the stimulus during the interval was green, that interval had *begun* with N (i.e., blackout) – the green stimulus was a consistent context for remembering N (green *retrodicted* N). The red stimulus was an ambiguous context, because red intervals began indiscriminately with N or F.

The critical question, obviously, is whether the animals were better able to use the neutral time marker beginning green intervals than the same time marker when it began red intervals: Would they pause longer following N in green than following N in red? The answer is "yes"; these two pigeons, and two others similarly trained with a shorter fixed interval, all paused almost as long after N as after F in green, but paused much less after N in red. Evidently, the distinctive context was able to mitigate the usual interference between N and F in fixed-interval reinforcement-omission procedures where intervals beginning with N and F are not otherwise distinguished.

Some "neutral" stimuli are more memorable than others. For example, the extraordinary human memory for faces has often been noted. The reason why people are able to identify hundreds, or even thousands of faces, but only a few (for example) telephone numbers is still the object of active research, but a popular suggestion is that it has something to do with the *multidimensional* property of "natural" stimuli like faces and scenes (see Chapter 10). There is some evidence that animals' ability to use a stimulus as a time marker in fixed-interval schedules is similarly affected by stimulus complexity. For example, in an unpublished experiment in my own laboratory we have found that if instead of the usual simple color or blackout stimulus we present a color slide of a pigeon as the neutral omission stimulus, the birds show essentially normal fixed-interval pauses.

Events of longer duration are usually easier to remember than events of shorter duration. Suppose, in the standard reinforcement-omission paradigm, we present neutral events of variable duration at the end of half the fixed intervals – being sure to time the intervals from the end of the event in every case. Will animals pause longer after longer stimuli, as they did when the stimulus was food, and as the memory argument implies? The answer is "yes" for pigeons, but (except for a transient effect) "no" for rats (Staddon & Innis, 1969; see also Roberts & Grant, 1974).

Conclusion: the discrimination of recency. The reinforcement-omission effect – shorter pausing after a neutral stimulus presented in lieu of food than after food on fixed-interval schedules – seems to reflect a competition for control of the animal's behavior between two past events: food, which is the earlier event, and the neutral stimulus. The animal must attend to the most recent event and ignore the earlier one. Both events have the same temporal significance, but food is more valued. Evidently, a few seconds after the neutral stimulus the animal attends to food rather than the neutral stimulus. Since the last food delivery is relatively remote in time, the animal responds (long postfood times signaling further food), resulting in a too-short pause after the neutral stimulus (the reinforcement-omission effect). The same process accounts for diminished pausing after short FI feeder durations when long and short are intercalated.

The general conclusion is that trace (temporal) stimulus control is vulnerable to the kinds of proactive and retroactive interference studied in memory experiments. Things that give a stimulus value, such as reinforcing properties and stimulus complexity or "meaningfulness," facilitate temporal control. Separation, in time, or by context, minimizes interference between events. Conversely, the occurrence of similar interfering events (the horizontal–vertical experiment) or more memorable events with similar significance (the reinforcement-omission effect) impairs temporal control. When the interfering event is similar in properties, but different in temporal significance, to the event of interest, the resulting

Table 12.1. *Payoff matrix for signal detection study of time estimation*

		Response	
		"Short"	"Long"
Time interval	$>T$	C_1	V_2
	$<T$	V_1	C_2

impairment of temporal control may be termed a *recency confusion* effect, since the animal is evidently uncertain about which stimulus just occurred. When the interfering event is highly salient or valued, the resulting impairment is better termed a *recency overshadowing* effect, since the more salient, older, event exerts control at the expense of the more recent, less salient, event.[10]

Other methods for measuring temporal control and memory. Fixed-interval schedules might be termed a *production* method for studying temporal discrimination in animals, in the sense that the animal determines how long he waits. Animals and people can also be asked to *estimate* time intervals. For example, in one popular discrete-trial procedure the animal is provided with two response alternatives (e.g., two pecking keys, for a pigeon), one signifying "too long," the other "too short." Each cycle of the procedure has two parts: In the first part the keys are dark and ineffective; after a variable period of time, t, the keylights come on and the animal must respond. If time t is less than some target time, T, a response on the left key (say) is reinforced with food and a response on the right key is either unreinforced or mildly punished by a timeout. If $t > T$, a response on the right-hand key is reinforced, and one on the left punished. The advantage of this procedure is that the costs and benefits to the animal associated with different kinds of errors and correct responses can be explicitly manipulated, rather than being an accidental consequence of the interim activities that happen to be available (as in the fixed-interval situation). In this way, we can get some idea of the limitations on the timing process itself, apart from biases to respond or not respond associated with competition from activities other than the measured response.

   This experimental arrangement is obviously well suited to ROC analysis (see Chapter 9). Bias (the animal's criterion) can be manipulated either by varying payoffs, or by varying the relative frequency of too-long versus too-short time intervals. The payoff matrix is illustrated in Table 12.1: V_1 and V_2 represent the benefits (reinforcement probabilities or magnitudes) associated with the two kinds of correct responses (correct identifications of "long" and "short" intervals); C_1 and C_2 represent the costs (timeout probabilities or durations) associated with the two different kinds of errors (long → short confusions and short → long confusions).

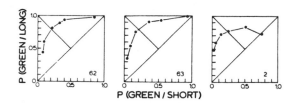

Figure 12.6. ROC plots for three pigeons trained on a procedure in which the duration of a time interval was judged; responses on a green key signaled "too long," to a red key, "too short." Abscissas show the probability of a green-key response given a sample interval shorter than the target duration; ordinates show the probability of a green-key response given a sample interval longer than the target duration. (From Stubbs, 1976, Figure 2.)

Figure 12.6 shows results from a pigeon experiment of this sort (Stubbs, 1976; see also the chapters by Stubbs & Platt in the volume edited by Zeiler & Harzem, 1979) in which pecking a green key was the correct response for a "long" stimulus and pecking a red key the correct response for a "short" stimulus. All three pigeons show ROC curves of the expected type, indicating that there is a stable limit on temporal discrimination that can be separated from the bias induced by payoffs – although other experiments already discussed suggest that this situation is not as pure a measure of the limits of time perception as it might appear, because the intercalated long and short test intervals undoubtedly interfere somewhat with one another.

A very similar procedure, *delayed matching to sample* (DMTS), can also be used to study memory interference. In this procedure, a pigeon (for example) is confronted by three response keys. At the beginning of each cycle, only the center key is illuminated, with one of two stimuli, S_1 and S_2. One or a few pecks on this *sample* key turns it off. After a delay of a few seconds, the two side keys come on, one showing S_1 the other S_2. The animal's task is to peck the key showing the sample stimulus. A correct response yields food, an incorrect a timeout. After either event, the cycle resumes. The location of S_1 and S_2 in the choice phase varies unpredictably from trial to trial, so that the animal must recall the most recently presented sample to make a correct choice.

This procedure offers the same possibilities for confusion as the temporal-control experiments just described. For example, on each choice trial the animal must be able to discriminate between the most recent sample and earlier samples. This suggests that there should be fewer errors if the delay value is short or if sample stimulus duration is long, and both effects are generally found. One also suspects that performance is likely to be better if there is a substantial interval between trials, because each sample is then more widely separated in time from preceding samples, which should reduce interference from them. This intertrial effect has not been shown reliably with pigeons with nontemporal sample stimuli, but experiments with rats have shown it (Riley & Roitblat, 1978; Lett, 1975). On the other hand, no experiment appears to have been done in which the intertrial interval varies from trial to trial. The earlier results with variable food and timeout duration on fixed-interval schedules suggest that intertrial-interval variation within each session may be necessary to get a reliable effect. Choice performance is more accurate the longer the preceding intertrial interval when the stimulus is itself a time interval, however (Riley & Roitblat, 1978).

An older version of the DMTS procedure nicely demonstrates that it is the memory for the *sample* that is important, not the subsequent choice arrangement. In an experiment by Harrison and Nissen[11] with chimpanzees the procedure was as follows: The animals were presented with two buckets, one covering food the other not. The animal saw the trainer baiting one or other bucket. After a delay, out of sight of the buckets, the animal was released and allowed to select one. In between baiting and choice, the buckets were moved closer together or farther apart. Nissen concluded that it is the separation of the buckets *at the time of baiting* that is important to accurate choice, not their separation at the time of choice. Evidently it is the way the sample is represented, or *coded*, by the animal that determines how well it can be responded to after a delay. If the baited and unbaited buckets are coded with very different spatial coordinates, they are not confused, even after substantial delays. But if their spatial coordinates (or other properties) are similar, then with lapse of time the animal cannot reliably distinguish between them.

As might be expected from the importance to recall of stimulus value, performance in this task depends upon the magnitude of the bait. If large and small baits are shown on different trials, accuracy is better on trials with the large baits even if the animal is always rewarded with bait of the same, small size.

Proaction and retroaction

The laws of memory define the limits on control of present action by past events. These limits are of two kinds: *proaction* effects, in which an earlier event (S_1) interferes with control of behavior by a later event (S_2), and *retroaction* effects, in which a later event interferes with control of behavior by an earlier event. The degree to which one event interferes with control by another depends on two properties: the *similarity* of the two events (including similar time coordinates), and the *difference* between the behavior controlled by each. For example, suppose that S_1 normally elicits R_1 and S_2 normally elicits R_2. In the DMTS situation S_1 might be a red sample key and S_2 a green key, R_1 would then be pecking the red-choice key and R_2 pecking the green-choice key. In DMTS R_i is always the next response required after S_i, so that only proaction effects are possible. Since the responses required by S_1 and S_2 are very different (S_1 and S_2 are not confused when both are present), interference in this situation depends upon the similarity of the stimuli: The more similar are S_1 and S_2, the worse the performance. If instead of red and green we used red and pink as sample stimuli, we could expect

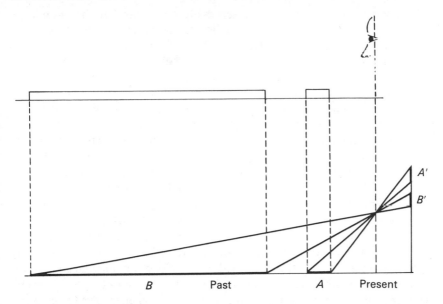

Figure 12.7. Perspective metaphor for the temporal resolution of memory. Top panel: the "mind's eye" viewing past events *B* (long duration, remote past) and *A* (short duration, recent past). Bottom panel: Perspective projections of *A* and *B*, *A'* and *B'*, from the viewpoint of the present.

choice accuracy to decrease: As we just saw, animals make more accurate choices when the two baited buckets are far apart than when they are close together. Response similarity makes much less difference than stimulus similarity, because the responses are usually available all the time and are not subject to interference or decay: The animal always knows what the response alternatives are on each trial; he is less sure about which alternative is correct (indeed, this asymmetry is a logical necessity, since the animal cannot even learn the task unless he can perfectly discriminate the choices). Knowing the choices requires only that the animal have learned a discrimination; knowing which choice to make requires in addition that he be able to remember which of the discriminated events *occurred last*.

In temporal-control experiments, however, the responses controlled by the interfering events can be either the same or different: in the FI reinforcement-omission experiments, F and N control the same behavior; in the first condition of the horizontal–vertical experiment (Figure 12.5, left), F and N (vertical lines) controlled different patterns. Interference was less when the two events controlled different patterns. Under some conditions, therefore, interference seems to depend on response as well as stimulus aspects of the task. The ways in which stimulus and response factors interact are yet to be fully worked out.

Proaction and retroaction effects in these experiments measure the *temporal resolution* of memory.[12] A useful analogy here is shown in Figure 12.7. This model assumes that past events are separately represented by the animal and implies that the (temporal) similarity relations among events change continuously with the passage of time. The upper part of the figure shows the "mind's eye" looking back over the record of past events, arranged in a time line, where the distance of the event from the eye (the present) is proportional to how long ago the event occurred. The bottom part of the figure shows how the events seen by the eye will appear on the "retina." Let us suppose that this eye (unlike a normal visual system) perceives the size of events solely in terms of their retinal size. Then the "salience" of the various events is given by their projected size, as shown by the vertical line on the right in the bottom half of the figure. Thus, a long-lasting but remote event may appear only as large as a shorter but more recent event: B' is the same size as A', even though B is an event of longer duration than A. Moreover, the relative sizes (saliences) of events will change with lapse of time, that is, as the vantage point moves to the right.

The various effects I have described are generally consistent with this metaphor. For example, events of longer duration are obviously easier to "see" than events of short duration. This fits in with the results of reinforcement-omission experiments in which longer events produce longer postevent pausing. Events widely separated in time are easier to tell apart (i.e., interfere less) than events close together in time; this is consistent with the effect of long intertrial intervals in improving DMTS performance. Moreover, the interference (proximity in the projected "memory image") between adjacent events should increase with time. No matter how brief an event, at short delays it will appear more salient than any earlier event; but as time elapses, longer, long-past events will gain relative to shorter, more recent events – just as a mountain will loom over a house when both are viewed from a distance, but the house will blot out the mountain when the viewpoint is close to the house. As time passes, A', the representation of event A, will therefore lose in size relative to B', the representation of earlier, but longer, event B. This is a venerable principle in the study of memory, *Jost's law*: Given two associations of the same strength, but of different ages, the older falls off less rapidly in a given length of time.[13] This principle accounts for the ability of even a not-very-salient event to control behavior over brief time intervals.

Figure 12.7 shows that any limit on the animal's "visual acuity" means that in a given context, only a limited number of past events can be distinguished: This model therefore implies a limit to the capacity of *event memory*.

Hedonic value, as an important factor in memorability, can be represented in this analogy by the dimension of *height*: Just as a distant high structure looms over a lower, closer one as the viewpoint recedes, so a preceding reinforcement seems to overshadow a neutral stimulus in the fixed interval reinforcement-omission paradigm. Thus both confusion and overshadowing effects fit easily into the analogy.

I show in a moment that this visual analogy corresponds to a decaying-trace view of memory.

Discrimination reversal and learning set

I end this section on methods of studying animal memory with an account of two procedures much used to study species differences in intelligence. These two procedures exemplify the memory processes already discussed and also illustrate the fallacy of comparing species in terms of their performance on some task, rather than in terms of the processes that underlie performance differences – a variation on the theme first played in the discussion of probability learning in Chapter 8.

Memory limitations enter in an interesting way into two tasks originally devised to study "higher mental processes" in animals: *discrimination reversal* and *learning set*. Both tasks were intended to assess animals' flexibility by requiring them frequently to learn a new task, either a discrimination opposite to the one already learned (discrimination reversal) or a completely new discrimination (learning set). There are several versions of each procedure. One that has been used with pigeons is as follows.[14] The animals are trained on a multiple schedule, familiar from Chapter 11. Two 1-min components occur in strict alternation. In one component, key pecks produce food according to a VI 60-sec schedule; in the other, pecks are ineffective (extinction). The extinction stimulus ($S-$) changes to the VI stimulus ($S+$) after 60 sec only if no peck has occurred in the preceding 30 sec; thus by pecking on $S-$ the animal can prolong its duration indefinitely. This "correction" procedure imposes a cost for responding to $S-$ – and also provides an additional cue to the identity of $S+$ each day (if a stimulus changes within 30 sec of a peck, it must be $S+$). The stimuli are red- and green-key lights. After an initial period during which the animals learn a specific discrimination (e.g., GREEN: VI, RED: EXT), the significance of the two stimuli is changed daily, GREEN signifying VI reinforcement on odd-numbered days, EXT on even-numbered days.

The first question is: Do the pigeons improve in their reversal performance from one discrimination reversal to another? Figure 12.8 shows the percentage of "correct" (i.e., VI-stimulus) responses on the first day of each reversal for animals reversed every day, or less frequently. The results for all six pigeons are similar: a steady improvement in performance, settling down to perhaps 90% correct responses after several reversals. What does this result tell us about the flexibility of these animals' learning processes?

Two other results from this experiment – the effects of a shift to a new pair of stimuli and of days off – shed some light on this question.

After good performance had been achieved on the red–green reversal problem, the two stimuli were changed to blue and yellow. The pigeons were given a total of 11 daily reversals with this new pair of stimuli. Then the animals were not run for a period of four days, then run for a single day, then not run for a further eight days. The effect on discrimination performance is shown in Figure 12.9: The animals performed quite well on the first day with the new stimuli, but discrimi-

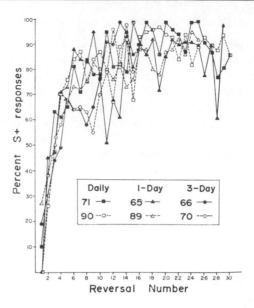

Figure 12.8. Performance of six individual pigeons exposed to daily reversal of a red–green successive discrimination, where $S+$ was reinforced according to a VI 60-sec schedule, no reinforcement occurred in $S-$, and the change from $S-$ to $S+$ occurred only if no $S-$ response preceded the change by less than 30 sec. The animals had learned the simple red–green discrimination perfectly before being exposed to reversal training; this experience, plus the correction procedure, accounts for the close to 0% $S+$ responses on the first reversal. (From Staddon & Frank, 1974, Figure 1.)

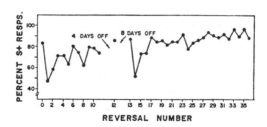

Figure 12.9. Average % $S+$ responses for six pigeons following a shift to blue–yellow reversal after extended training on red–green. (From Staddon & Frank, 1974, Figure 8.)

nation was very poor the next day, that is, on the *first reversal* after the change, and took several further reversals to recover almost to its previous level. In a similar way, the pigeons performed well after the four days off and after the next eight days off, but on the first *reversal* after the eight days off performance was poor and remained so for several subsequent reversals.

To interpret these results, consider the two things that are necessary for good performance on this task: (a) To *ignore* on day N the significances established for the stimuli on day N−1.[15] (b) To attend each day to the cues to the correct stimulus, that is, the delivery of food reinforcement in the presence of $S+$ and the delay contingency for pecking on $S−$. Presumably good performance depends upon the balance of these two factors. For example, if the animal is good at detecting the cues for $S+$ and $S−$ but poor at disregarding the significance established for red and green on the previous day, then performance each day will not be very good: The animal will always begin by responding a lot to $S−$ and not much to $S+$, thus ensuring a mediocre discrimination score. Conversely, if the animal treats the stimuli afresh each day but is poor at detecting $S+$ and $S−$, performance will also be inferior. The properties of memory are involved in the first prerequisite – control of behavior on day N by the significances established on day N−1.

We can get an idea of the relative importance of these two factors, memory and speed of learning, from the way that performance changes within and across experimental sessions, and from the results in Figure 12.9. For example, consider the animal that treats the stimuli afresh each day (this would be a beast capable of only local memory, in the sense of Chapter 4). If such an animal learns fast, then the absolute level of performance will be quite good, but there will be no improvement across successive reversals. In the case of this animal neither the days-off manipulation, nor the shift to a new problem (i.e., new pair of stimuli), should have any effect. Certainly, there is no reason at all to expect any special effect on the *first reversal*.

How about the animal that learns fast, but remembers the stimulus significances established on the previous day? The critical thing here, obviously, is *how well* the animal remembers. The previous discussion of the effects of stimulus intercalation makes some suggestions on this point. Early on, one might expect animals to remember quite well. In particular, on the second day with any new pair of stimuli there will be no sources of interference and animals should remember well the significances established on the first day. But since the correct response is now the opposite one, we might expect to see especially *poor* performance on the second day, the *first reversal*, of any new problem – exactly what is shown in Figure 12.9. In short, anything that *improves* the animal's memory for what happened in the previous experimental session should *impair* discrimination-reversal performance.

Days off is such a factor: We might reasonably expect that the longer the time elapsed since discrimination N−1 – especially if discriminations N−2, N−3, and so forth precede N−1 and provide sources of proactive interference – the smaller the effect the significances established then should have at the outset of discrimination N. So it proved in the experiment shown in Figure 12.9. After a days-off period, performance is slightly better than before the days-off period (reversals 12 and 13 in the figure): If the pigeons are run on Saturday (say), but then not run again until the following Thursday, performance on Thursday is good. But the very factors that minimize interference on Thursday from what was learned on Saturday act to *maximize* interference on the next day, Friday, from the significances established on Thursday. Thursday is isolated temporally from the discrimination sessions preceding it, so that its effect on Friday is unimpaired Saturday. The result is poor performance on Friday.

This same line of argument leads us to expect that performance at the beginning of each experimental session should change systematically with successive reversals. Early on, the animal should respond incorrectly at the beginning of each session, responding most to $S-$ (i.e., the previous day's $S+$). But with continued training, recall of the previous day's $S+$ and $S-$ should be progressively impaired, so that at the beginning of each experimental session the animal should respond more or less equally, at an intermediate level, to both stimuli. This is more or less what happens with pigeons: At first, errors are high chiefly because the animal consistently picks the wrong stimulus at the beginning of each session. With continued training, this initial bias disappears and the animal appears more hesitant, responding at a slower rate, but more or less equally to both stimuli (presumably the hesitancy reflects the ambiguous status of both stimuli: The animal cannot, at this stage, recall which stimulus was $S+$ yesterday, but it has no difficulty recalling that both stimuli have served as both $S+$ and $S-$).

The discrimination-reversal task is not ideal as a test of "intelligence" in animals, because good performance can be achieved in several ways, not all of which correspond to superior ability. For example, poor temporal resolution of memory, that is, a relative inability to distinguish yesterday's $S+$ from $S+$ the day before that, can aid performance on the task. It is possible to imagine three types of performance on the task, depending on the temporal resolution of memory: (a) At the lowest level, temporal resolution is exceedingly poor (this amounts to just local memory). Hence each day is treated as a separate experience, and discrimination-reversal performance is little different from simple discrimination performance. There should be no improvement across successive reversals. (b) At an intermediate level, temporal resolution is intermediate, hence discrimination-reversal performance is initially poor, but improves as proactive interference accumulates and weakens the effect of day $N-1$ training on day N performance. (c) At the highest level, temporal resolution is sufficiently good that the animal can show *spontaneous reversal*, using the $S+$ on day $N-1$ as a cue to $S+$ on day N. Spontaneous reversal is not possible at the two earlier stages, because late in training the day-$N-1$ $S+$ cannot be recalled, so cannot be used as a cue, on day N. (Spontaneous reversal may fail to occur even if memory permits because of information-processing constraints: The animal may, in some sense, know that today's $S+$ is opposite to yesterday's but be constrained to respond to the most recent $S+$ – yesterday's – anyway, as in negative automaintenance. All three of these cases permit good steady-state reversal performance. They differ in the means used to achieve it – but these differences can be revealed only by appropriate tests.

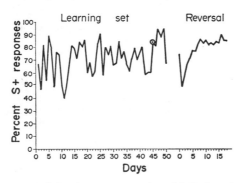

Fig. 12.10. Performance of six pigeons on a series of daily learning-set problems after extended reversal training. Circled problem (no. 45) was also used for the reversal series on the right-hand side of the figure. (From Staddon & Frank, 1974, Figure 10.)

Available results suggest that goldfish correspond more or less to case (a), pigeons and rats to (b), and primates to (c).

The learning-set task, in which a new pair of stimuli must be discriminated each day, seems like a better test of "learning ability," whatever that might be, because the role of temporal resolution is minimized. Performance in this task does depend on something that might be termed "cognitive resolution," however: the ability to keep separate (not confuse) a number of different pairs of stimuli, since if new stimuli are confused with old ones, then on some days the animals will pick the wrong one of the two new stimuli as $S+$ and this will retard acquisition of the discrimination. Thus, learning set is subject to the same dual process as discrimination reversal: Any improvement across problems (i.e., pairs of stimuli) can reflect improvement in attending to the procedural features – differential reinforcement, correction – that signal $S+$ and $S-$. But improvement may also reflect increasing confusion among past stimuli with concomitant reduction in their ability to affect preferences for new stimuli.

If the pairs of stimuli used each day are very different (i.e., the animals have good "cognitive resolution"), then animals will treat each day as a fresh discrimination. If the animals are pretrained on discrimination reversal, they should transfer perfectly to such a task, having already learned how to identify $S+$ each day. Figure 12.10 shows learning-set performance of a group of six pigeons shifted to learning set after extensive experience with discrimination reversal. The performance of the group changed little across a series of 50 daily problems. Moreover, resumption of discrimination reversal again showed the first-reversal performance decrement and slow improvement required by the memory analysis.[16] Final performance on the reversal problem was at about the same level seen when that problem was one of the learning-set series. All these characteristics – little or no improvement in learning-set performance after reversal training, first-reversal decrement after learning-set training, and similar performance within

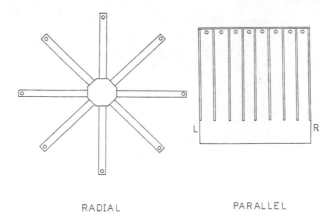

RADIAL   PARALLEL

Figure 12.11. Left: eight-arm radial maze. Right: a maze with eight parallel arms used to assess the role of spatial encoding in radial-maze performance.

a learning-set series and at asymptote in a reversal series – are what we would expect from the properties of event memory just discussed.

MEMORY AND SPATIAL LEARNING

This discussion of reversal learning and learning set emphasizes a major difficulty in studying learning and memory: We see and can measure *performance*, but the same performance can usually come about in several ways, and even simple tasks call on more than one ability. Performance can never be taken at face value; we must always ask about the component abilities that make it up. An analysis into components can never be satisfactorily proved by a single "crucial" experiment. The best we can do is to take our hypothesized basic abilities or processes and show how by putting them together in various combinations we can bring together a wide range of facts. The smaller the number of processes, and the larger the number of facts explained, the more reasonable the set of processes will appear.

   In this last section, I take the principles of memory derived earlier and put them together with the notion of spatial representation discussed in Chapter 10. I use the resulting set of temporal and spatial principles to explain a wide range of experimental results with the radial-arm maze and other spatial apparatus.

The radial-arm maze[17]

The multiarm radial maze, first discussed in Chapter 10, has turned out to be an ideal arena for demonstrating the properties of memory in rats and other small animals. The standard radial-maze experiment involves an open eight-arm maze (Figure 12.11, left) with a baited food cup at the end of each arm. Hungry rats are allowed to choose freely among these arms, until either all eight have been visited or a fixed time (amply sufficient for eight choices) has elapsed. One or a few trials of this type are given each day until performance ceases to improve further. The usual finding is that rats soon learn to visit each arm almost without repetition, so that all eight arms are visited within the first eight or nine choices of a trial.

Several experiments have shown that under normal conditions this efficient performance depends primarily upon the animal's memory for arms it has previously visited, rather than upon any kind of response strategy such as "choose the next clockwise arm."

I first describe the factors involved in radial-maze performance, then develop a model based on representation of events in the maze by a two-part code. The two parts are an identification code for places visited, and a temporal code reset by each visit. The identification code is usually spatial (see Chapter 10); the temporal code has the properties described in the first part of this chapter. I then show how the major experimental findings with the radial maze and related T-maze tasks fit into this model. The last part of the chapter deals with apparent exceptions posed by findings from delay experiments with fewer than eight choices and more than a few trials per day. These exceptions can be accommodated by the assumption that a limited number of events are distinguishable in event memory, something implied by the kind of model illustrated earlier in Figure 12.7.

An overview of radial-maze performance. Performance in the radial maze seems to be determined by three factors: the coded values of past choices plus a *response rule*, response strategies, and "error," i.e., unaccounted-for factors.

These three factors represent three ways of solving the maze problem. The first method is the most satisfactory: Given some way to identify each arm, choices can be made in such a way as to avoid previously entered arms (appropriate for the radial-maze task), or select a particular arm (appropriate in most traditional maze experiments). Arms might be identified either by cues outside the maze (extramaze cues, such as room features) or by cues within the maze (intramaze cues, such as arm color and texture, etc.). Intramaze arm identification is more difficult, since the arms are usually similar – and may also be less useful, since a code based solely on intramaze cues lacks spatial information. A code based on extramaze cues can be maplike, with information about adjacency and other spatial properties.

The second method, response strategies, is less efficient than either of the arm-identification methods, for two reasons: A given arm can be reached only after entering all the preceding arms in the sequence; and if the sequence is interrupted for some reason, succeeding arm choices will be inappropriate unless the sequence is resumed exactly where it was left off – which requires accurate memory. Since the main advantage of the response-sequence solution is that it makes minimal demands on memory, rats should depend on response strategies only when memory is impaired or arms cannot be accurately identified.

The third method, unsystematic choice, is not a solution, but serves to guide choice when other factors provide no guidance – early in training, for example.

The balance among these three methods is determined by factors such as stage of training, availability of extramaze cues, maze configuration, and the animal's information-processing capabilities.

Under usual conditions, behavior is guided by extramaze rather than intramaze cues. This permits animals to develop a maplike representation of the maze, rather than responding in push–pull fashion to specific stimuli (Suzuki, Augerinos, & Black, 1980).[18] Absence of extramaze cues favors a more primitive representation, tied to intramaze cues – or a response pattern. Rats seem to behave as this reasoning suggests, adjusting their relative reliance on extramaze, intramaze and response-pattern factors so as to do as well as possible: Rats trained with few extramaze cues are more likely to show response patterning than rats trained under normal conditions, for example. The use of extra- or intramaze cues depends upon memory: No matter how each arm is encoded, the animal must be able to distinguish entered from unentered arms. Hence animals with impaired memories should show more reliance on response patterning, the only strategy left open to them. Young rats have poorer memories than adults and also show more response patterning.

Even when extramaze cues are available, and spatial coding is therefore possible, rats often enter arms in a systematic sequence. Experiments in which choices early in a trial are determined by the experimenter (forced-choice trials) show that these patterns are not necessary for them to learn the maze, however, and many successful animals show no obvious response pattern as they learn. Well-trained rats will abandon their response patterns, without loss of accuracy, if patterning conflicts with correct choice.

The radial-maze task constrains the order in which animals can learn different things. Since animals do not know the requirements of the task at the start of training, they must learn that only the first visit to each location (arm) is rewarded. To do so without relying on a response strategy, they must be able to identify each arm. They must also know whether an arm has been previously visited or not, which implies some form of temporal code. Only then can the animal apply the appropriate response rule based on this arm-identification and temporal knowledge. The response rule can neither be learned nor used effectively until the identification and temporal codes are relatively unambiguous.

A two-part code

Several studies show that rats can reliably determine which arms of the radial maze they have just visited, and that arms visited within the last few minutes are not confused with those arms visited on previous days. These findings suggest that animals encode two properties of each maze arm visited: the identity of the arm, and the time at which the arm was last visited (temporal location). The temporal code corresponds roughly to what Olton (1978), Honig (1978) and others have termed *working* memory. The spatial code is one aspect of what Honig and Olton have termed *reference* memory.

Under normal conditions (ample extramaze cues, spatially separated arms) arms seem to be identified by a maplike spatial code. Temporal location can be represented by a decaying trace, or *tag* (a temporal code). As each arm is visited, a temporal tag is attached to the spatial code for that arm; the tags for all arms decay at the same rate so that the *recency* with which any arm has been visited is given by the current value of its tag.

This scheme implies that rats represent the maze as a *list* of pairs, of the form P_iT_i, where P_i corresponds to the spatial ("place") coordinate for the ith arm, and T_i to the temporal tag for the ith arm. The reinforcement contingencies (e.g., all arms baited or only one arm baited) then determine how the rat responds with respect to his internal representation, his *response rule*. Let's look at the properties of the spatial and temporal codes and the response rule.

Spatial code. Classic studies of human memory show the power of mnemonics prescribing that the to-be-remembered items be "stored" in spatially separated locations.[19] The more widely separated the locations the better: Items stored in different bureau drawers are more likely to be confused than items stored in different rooms in one's house, for example – recall the sample-separation experiment by Harrison and Nissen, discussed earlier. These results strongly suggest that the spatial code incorporates information about spatial proximity: Neighboring locations should have similar codes and be confused more easily than disparate locations.

The spatial coordinate may be bivariate, to reflect the two-dimensional structure of the maze, so that the animal's experience with the maze is represented as a set of triples. It is likely that the form of the code depends both on the form of the apparatus – an hierarchical apparatus lends itself to an hierarchical code, for example – and on memory constraints such as the rodent equivalent of Miller's[20] "magical number seven": Humans cannot remember more than about seven unrelated items; no doubt rats are similarly limited.

Temporal code. The temporal code carries information about *when* an arm has been visited. Performance in the radial maze and related situations can be accommodated by two assumptions: (a) When the rat visits an arm, a single trace variable is reset to a maximum value; and (b) that the trace decays with negative acceleration thereafter. The second assumption is not controversial: Trace decay is old hat in theories of memory. The assumption that trace values are completely reset after each arm visit is less easily accepted because it implies that rats cannot learn to behave differentially depending on whether they have visited a place once or more than once: Yet under appropriate conditions, rats can discriminate perfectly well the number of occurrences of a repeated stimulus. I point out later

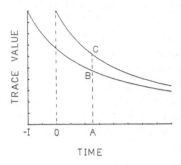

Figure 12.12. Decay functions for the temporal codes for two arms, entered at times −I and 0. The absolute recency (AB) and differential recency (BC) change at different rates.

that, in fact, each arm visit is probably represented separately – but other memory limitations, built in to the trace model, mean that we can treat the eight-arm radial maze as if memory traces are reset (i.e., as if each arm visit is represented only once).

The form of the temporal code (trace) is constrained by two times: the time over which animals can distinguish entered from unentered (i.e., never-entered, or last-entered-a-long-time-ago) arms, and the time over which they can distinguish the least-recently entered arm from the next-to-least recently entered arm – *absolute* and *differential* recency, respectively. Experiments show that the time of absolute recency is on the order of hours, of differential recency, minutes.

In one experiment, for example, delays ranging from 5 sec to 24 hr were imposed between rats' first and last four radial-maze choices. Choice accuracy over the last four was very high with delays as long as 4 hr, and above chance even after 12 hr. Evidently absolute-recency discrimination is very good. Another experiment ran three successive trials each day, with only a minute or so between trials. Performance on the second and third trials was worse than on the first, showing that differential-recency discrimination is relatively poor (Roberts & Dale, 1981; Beatty & Shavalia, 1980).

Absolute recency is determined by the animal's ability to discriminate whether the trace associated with a given arm is different from zero; differential recency by its ability to discriminate which of two traces is higher or lower in value. The relation between absolute and differential recency is illustrated in Figure 12.12, which shows the decaying traces associated with two arms, entered at times $−I$ and O. The levels of these traces at time t (point A) are denoted by heights AB and AC. Thus, height AB represents the animal's ability to discriminate an arm that has been entered from one that has never been entered (absolute recency) – height AC is the absolute-recency value for the newer trace. Height BC ($= AC − AB$) represents the animal's ability to tell which of two arms has been entered most (or least) recently (differential recency).

The simplest model for a memory trace is exponential decay, but this cannot account for Jost's law, one of the best-established memory principles. A simple function that is satisfactory (it corresponds closely to the perspective metaphor in Figure 12.7) is the hyperbolic function,

$$T(t) = 1/(a + bt), \tag{12.1}$$

where $T(t)$ is the trace value at time t after the event and a and b are constants.

Using the hyperbolic function, the absolute value of one trace and the differential recency value for two traces can be derived from Equation 12.1 (see Figure 12.12) as follows:

Absolute recency: $\quad AB = 1/(a + b(t + I)), \tag{12.2}$

Differential recency: $\quad BC = AC - AB = BI/(a + b(t + I))(a + bt)$
$$= bI/(b^2t^2 + (2ab + Ib^2)t + a^2 + abI)$$
$$= bI/(At^2 + Bt + C), \tag{12.3}$$

where I is the time between arm choices, t the time since the most recent choice (OA in Figure 12.12) and A, B, and C are lumped constants. An important property of the hyperbolic function is that differential recency decreases faster than absolute recency since the first is inversely related to t, the second to t^2: In other words, the difference between two traces declines faster than the absolute value of a trace, as Figure 12.12 shows.

Response rule. The task in the radial maze is to find a particular place (spatial code) or find the place visited least- or most-recently (temporal code), depending on the reinforcement contingency. In terms of the processing the animal must carry out on his representation of the maze, these possibilities correspond to different *sorts* of the list of time–place pairs that represent the maze: The animal can sort by either spatial or temporal code, depending on which is appropriate.

An animal's ability to learn any discrimination task involving delay will be directly related to the *discriminability* of the traces for the stimuli at the time when a choice is made. In delayed matching to sample, for example, the animal has to identify the stimulus seen *most recently*. We can be no more certain of the details of trace discrimination than of discrimination among sets of more familiar unidimensional objects such as wavelengths or loudnesses (see Chapter 10); but whatever the details, discriminability is obviously related to differential recency as defined by Equation 12.3.

In the radial maze, the probability an arm will be chosen is directly related to the discriminability of its trace value from the set of trace values for already-chosen arms. Under usual conditions (widely spaced trials) this means that the subject chooses the arm with trace value closest to zero (i.e., the *least-recently visited* arm). In terms of the list representation, the animal is always selecting the time-place pair with the lowest value of T. Consequently, pairs will always be confused because of the similarity of their temporal, rather than their spatial, codes.

Experimental results

This model has three parts: spatial and temporal codes, and a response rule. Let's look at some experiments relating to each:

Response rule. Rats require little training to learn not to revisit arms in the eight-arm maze, but if the four arms entered in the first half of a trial are also the ones baited on the second half, they can learn (albeit more slowly) to repeat their first four choices rather than choosing the other four, unbaited arms. It's conventional to assume that these different performances reflect a difference in response rules, rather than a difference in the process by which arms are encoded, but it's important to realize that this is an assumption, not something that can be taken for granted. What's the evidence?

Under usual conditions (all eight arms baited) the response rule is that the spatial code with the oldest trace is selected over others (least-recent choice). There are three types of evidence for this rule: (a) Evidence that in spatial situations rats behave spontaneously in accordance with the least-recent rule. (b) Results showing that tasks for which this rule is appropriate are learned rapidly, whereas tasks for which the rule is inappropriate are learned more slowly. (c) Results showing that even if the rule is appropriate, learning is rapid only if choices are spatially encoded, that is, if extramaze cues are available and goal directions are varied.

Rats and many other animals have a spontaneous tendency to avoid places recently visited. This tendency was first remarked as *spontaneous alternation* in T-mazes, but the same tendency is exhibited as "patrolling" in residential mazes, in mazes with more than two alternative goal-routes – and in the radial maze. This least-recent tendency makes adaptive sense from two points of view: The least-recently visited place is the one where things are most likely to have changed. Consequently, if there is value in keeping up to date about the state of the world, the least-recent rule is the one to follow. For an opportunistic forager like the rat, many food sources will correspond to a random-interval schedule: depleted by each visit, and replenishing unpredictably with time. The least-recent strategy is optimal for exploiting such sources (see Chapter 9, note 16).

Rapid learning in the radial maze is consistent with the least-recent rule: Spatial encoding is ensured because the mazes are typically large and open, with ample extramaze visual cues, because the arms of the radial maze differ in two ways, direction and location, and because the goal boxes are widely separated. The usual reward contingency (bait in every goal) makes the least-recent response rule appropriate. This rule is also appropriate for the parallel maze (Figure 12.11, right), but here the arms differ only in location, and goal boxes are adjacent, making spatial encoding difficult – and rats find this maze much harder than the radial maze. Spontaneous alternation does not occur in a maze with two parallel arms – further support for the idea that rats have a natural tendency to select the least-recently visited *place*.

Spatial effects. If information about spatial proximity is included in the spatial code, then anything that reduces the distance between maze arms should impair learning. Conversely, the learning even of very similar mazes should not be subject to interference, providing the mazes are in separate locations. There are experiments of both types.

Distance between maze arms can be varied either by making them adjacent, as in the parallel maze (Figure 12.11, right) or by reducing the size of the maze. Learning the parallel maze is much more difficult for rats than learning a radial maze. It is not yet clear whether the difficulty of the parallel maze reflects the physical proximity of the arms, their similar direction, or both.

Rats can learn radial mazes in different places without interference, up to a limit set by memory capacity. For example, one experiment showed that exposure to a second radial maze halfway through a trial on a maze produces subsequent performance decrements on the first maze only when the second maze is in *exactly* the same position as the first (not next to it or even 2 feet above it), and when there are several intervening trials (on identical mazes in similar rooms) before the interrupted trial is completed. The first result is expected from the spatial-coding hypothesis: Mazes in different places do not interfere because their arms have different spatial codes.

The interfering effect of a large number of intervening mazes may reflect memory-capacity limitations: Each room provides a different context, and as we have already seen, separating potentially interfering events by means of different contexts can reduce interference. But when the number of contexts becomes very large, memory limitations are a factor. For example, if an animal is trained with the same maze in three separate rooms, he might encode each arm in the form $R_i A_j$, $i = 1-3$, $j = 1-8$, where R denotes the room code and A the arm code. Since the number of rooms is small, we may expect few confusions among the R-codes, hence no interference among mazes. If the number of rooms is large, however, the number of room codes may exceed the rat's memory span for unrelated items and confusion will occur. These results suggest both a context-specific code, in which each arm is related to a particular maze (whose center may be coded in terms of absolute spatial location), and that the rats can restrict their trace sort to one context.

A puzzling feature of radial-maze experiments is the complete failure to find evidence for spatial, as opposed to temporal, generalization. For example, several experiments have found that when rats make mistakes – that is, reenter already-chosen arms – they do not selectively repeat arms near to correct (i.e., unchosen) arms. If locations are identified according to a spatial code, why don't spatial confusions occur?

The present model gives a straightforward answer to this question: If the maze is represented by a bivariate code, and arm choices are made by sorting place–time pairs on the basis of the *time* coordinate, errors *should* have nothing to do with the place coordinate. The only way that spatial generalization can occur is in the mapping of the animal's spatial representation onto the physical maze. For example, if the animal's identification of "true north" deviates by more than 22.5° from the actual north, then in selecting the coordinate for the north arm he may actually enter the NE arm. Granted the multiplicity of extramaze cues typically available, it is unlikely that rats make errors as large as this in relating their spatial representation to the actual maze. Consequently, spatial generalization in the radial maze is not to be expected.

But what of the parallel maze? We know that rats find this much more difficult to learn than the radial maze, so we might expect measurable errors in their mapping of their representation onto the actual maze. Yet here also there is no evidence for spatial generalization. One answer is that behavior in the parallel maze is not guided by a spatial representation. The rats seem to be guided by intramaze cues and a nonspatial representation in preference to a spatial-map method made unreliable by the spatial proximity and identical direction of the arms.

Evidence for nonspatial representation in the parallel maze comes from experiments on spontaneous alternation and with blind and sighted rats. I have already discussed evidence suggesting that spontaneous alternation is conditional on the formation of a spatial representation, and noted that rats do not show spontaneous alternation in a two-arm parallel maze. Sighted rats learn the radial maze much faster than blind rats, but show no such superiority in the parallel maze. Under normal circumstances (i.e., when they have not been trained before being blinded) blind rats do not seem to use a spatial representation in either type of maze. The similar performance of blind and sighted rats on the parallel maze implies that even sighted rats do not encode the maze spatially. If the parallel-maze arms are not encoded spatially, spatial generalization is not to be expected even in sighted animals.

But it is not even necessary to invoke differences in the way that radial and parallel mazes are represented. The lack of spatial-generalization errors follows from the task constraint that gives priority to arm identification: The animal cannot apply the temporal tag and relevant response rule *until he can recognize an arm without error*. Consequently, if the maze problem can be solved at all, and response strategies are precluded, arm identification *must* be close to 100% accurate. If extramaze cues are degraded so that a spatial code becomes inaccurate enough to show spatial-generalization errors, performance based upon such a code cannot approach typical levels of accuracy. Hence, accurate performance under these conditions implies a *non*spatial code. In either case, if the animal can learn the task at all, he will not show spatial-generalization errors.

Temporal effects. The model assumes that the temporal tag associated with each maze arm is initially zero, is reset to a maximum on each visit, and then decays

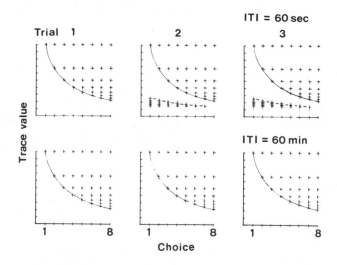

Figure 12.13. Trace values for each of the eight arms of a radial maze immediately after each of eight choices on three successive trials. The trials are either 60 sec apart (upper) or 60 min apart (lower). Choices within a trial were always 10 sec apart. The solid line indicates the lowest trace value among the arms in the "previously chosen" set; the dashed line shows the highest trace value among those arms yet to be chosen on a trial.

hyperbolically with time thereafter (Figure 12.12, Equation 12.1). The pattern of trace values to be expected on the first and two subsequent trials within a day is illustrated in Figure 12.13. The curves in the figure are based on the assumptions that the rat chooses an arm every 10 sec during a trial and that the intertrial interval is either 60 sec (top three panels) or 60 min (bottom three panels). Parameters a and b in Equation 12.1 were determined by Equations 12.2 and 12.3 so that absolute recency and differential recency are in approximately the right ratio (roughly 60:1). These numbers are not critical; the figure shows a pattern to be expected under a wide range of conditions.

The +s and curves in Figure 12.13 were derived as follows. Look at the upper-left panel, which is typical of all. The leftmost upper + represents the maximal (reset) trace value associated with the first arm entered on this first choice of the first trial of a day. The descending curve beginning from that point shows the hyperbolically decaying trace value associated with that arm; this line is that value of the *oldest* trace among all the arms visited on that trial. The second + from the left in the upper-left panel represents the reset value for the trace of the second arm entered; this trace value also decays with time, as indicated by the +s just above the solid line showing the oldest trace. The other +s similarly represent the trace values for the other arms entered.

The dashed line below the solid line in upper panels 2 and 3 represents the decaying trace of the arm entered last on the preceding trial; the +s below it represent traces for arms entered earlier. The difference between the solid and dashed lines provides a measure of the recency discriminability of chosen versus unchosen arms: the closer the two lines, the harder it should be for the animal to distinguish arms chosen on this trial from unchosen arms (i.e., arms chosen on the previous trial).

Discriminability (the separation of the solid and dashed envelopes) decreases during each trial, and is much lower on the second and third trials when trials are 60 sec apart, but not when they are 60 min apart. The first prediction is consistent with the universal experimental finding that the probability of repeating an arm increases during a trial, even when the increased opportunities for repetition are corrected for. The second prediction is consistent with the finding that choice accuracy decreases between Trials 1 and 2 when the trials occur 1–2 min apart, but not when they are separated by an hour. Figure 12.13 also predicts that choice accuracy will not decrease any further after trial 2, since the trace values are the same at the start of all trials after the first. This surprising prediction is also confirmed by data.

Figure 12.13 shows that discriminability is always high after the first choice of a trial, even if the trial follows quite closely on a preceding trial. Figure 12.13 also suggests, and the data confirm, that at the start of a trial rats should prefer arms chosen early in the preceding trial. Both these results argue against a special trial-end "resetting" process, sometimes proposed to account for high choice accuracy at the beginning of each trial.

Figure 12.13 also shows that if trials are well spaced, discriminability depends only on choices already made (i.e., nonzero trace values) not on choices yet to be made, which will all have zero trace values. Hence, rats should show the same accuracy on the first N choices of any radial maze, no matter how many arms it may have (and so long as it can be learned at all). In confirmation, experimental results show that accuracy over the first eight choices on a 17-arm radial maze is the same as choice accuracy over all eight choices on the 8-arm maze.

After eight correct choices have been made, the set of trace values appears as shown in the rightmost column of $+$s in each panel in Figure 12.13. The least-recent response rule implies that the animal should have difficulty deciding on his ninth choice because the oldest traces are all close together in value: The set of traces does not divide easily into a set of chosen and a set of unchosen arms, so that discriminability is low. This suggests that animals should hesitate after visiting all eight arms in a trial.

After the ninth choice has been made, however (first column of $+$s in each panel), the traces once again divide up into two relatively discriminable sets because the trace for the arm just chosen is at a high value and the other seven are relatively low; hence, choice time should decrease. Experimental results fit in with this picture. Rats typically spend only a second or so in the center between arm choices, until all arms have been chosen; then they wait in the center of the maze 30 sec or more before making the ninth choice. But after the ninth choice, arms are again chosen at a high rate.

Other spatial situations

The notion of an identification code, a resetting, hyperbolic trace, and the least-recent response rule accounts for most radial-maze experiments, and experiments on spontaneous alternation, that have used one or just a few trials per day. These assumptions also account for many results from delayed-alternation experiments and, if the response rule is suitably changed, for results from delayed-reaction and delayed-matching-to-sample experiments as well. The theory runs into difficulties when the number of choices is small, as in the T-maze, and when there are many trials per day, so that the ratio of intertrial interval to retention interval is relatively small. The critical problem here is the resetting assumption. I first show how the theory works, when it does; then show where it fails to work – then reconcile the two.

Animals have difficulty learning a discrimination when reward is long delayed after the correct choice. Lett (1975; reviewed in Revusky, 1977) has done a series of experiments in which rats learned to either return to, or avoid, one arm of a T-maze, even though they were rewarded only after delays of from 1 min up to an hour or more. For example, in one experiment, Lett gave rats one trial per day on a simple T-maze discrimination, removing the animals to their home cages as soon as they entered the correct arm, and then rewarding them in the start box after a 60-min delay. Over a period of ten days or so, the rats learned to perform correctly.

Two features of the experiment seem important to its success: removal of the rat from the goal box as soon as it entered, and the long intertrial interval in relation to the delay interval.

Removal from the goal box seems to be important to minimize context-specific retroactive interference from activities that may occur in the goal box after it is entered. The long intertrial interval is necessary for the animal to be able to discriminate the most recent goal entry from previous goal entries: The task in this experiment is to discriminate between the trace of the correct goal box, visited 60 min ago (the retention interval), and the trace of the wrong box, visited perhaps as recently as the previous trial (i.e., with a time separation equal to the intertrial interval plus twice the retention interval): The animal must discriminate between two traces originating t and $2t + I$ sec ago. Equation 12.3 shows that discriminability of one trace from another is a positive function of intertrial interval (I), with asymptote inversely related to the retention interval (t). Hence, the model suggests that animals should have no difficulty in delayed-reward experiments so long as the intertrial interval is large in relation to the retention interval, as it was in the Lett experiments.

In delayed-alternation experiments, animals are forced to visit one arm of a T-maze then, after a delay, are rewarded for visiting the other arm. The required rule is exactly the same as in the radial maze, but the number of arms is smaller. Despite the similarity of the tasks, delayed-alternation performance typically falls to chance after delays much shorter than the several hours required to produce impairment of radial-maze performance. An important variable here seems to be the intertrial interval, which is typically less than the 24 hr usual in radial-maze studies. The shorter time would lead one to expect more rapid impairment in the delayed-alternation task. Yet this cannot be the whole story, because (as we have already seen) radial-maze performance is quite good even when trials are only 60 sec apart. Why should rats be able to discriminate arms visited 30 sec ago from arms visited 90 sec ago in the radial maze, but not the T-maze?

There is a straightforward explanation for this apparent paradox. The perspective metaphor in Figure 12.7 implies that each visit to a maze arm is separately represented by the rat, but that (for a given context) there will be some limit to the number of visits that can be separately represented – simply because as time passes, past events are bunched closer and closer, so that at some point they must cease to be discriminable. This limitation is part and parcel of the trace model: Equation 12.3 shows that the differential recency of any pair of events decreases very rapidly with the passage of time. We might also expect that the larger the number of different traces (events), the more difficult it will be to discriminate one from the others. Thus, both *time* and *number of intervening events* should affect animals' ability to identify a particular trace. The data seem to agree: Both time and number of intervening choices have been shown to affect radial-maze performance, for example.

Suppose that this limit on event memory means that after perhaps six to eight visits to the same arm or different arms, information begins to be lost. In the eight-arm radial maze, this will occur at about the time when every arm has been visited. On a succeeding trial, therefore, as arms are revisited, information about prior visits will be lost, precluding any interfering effects from such visits. The effect of this capacity limitation on event memory is that only the most recent visit to an arm will be represented, so that traces will appear to be completely reset after every tour of the radial maze.

Not so in T-maze delayed alternation: Here several visits to each arm will be separately represented, so that the animal cannot just sort the set of place–time pairs by the time coordinate. For example, suppose we denote the two arms by L and R (T-maze) or $A-H$ (radial maze), and the trace values by numbers. Then the radial-maze task, after seven choices in a day, is to sort a list of pairs A99, B95, C90, D87, E83, F79, G76, H2, say, by their trace values alone. In this example, 99 (close to 100%) is the value of the most-recent trace, and 2 the value of the least-recent (the last arm chosen on the previous day): 2 would be selected, and H would be the correct choice. In T-maze delayed alternation, after the same number of choices (i.e., three and a half trials, each a forced choice followed by a free choice, and assuming no errors) the set of pairs might be L99, L95, R90, L87, R83, L79, R76, L2 (L2 is again the last arm chosen on the previous day). The animal cannot solve this problem merely by picking the oldest trace and choosing the associated arm. He must identify the most-recently visited arm, and then choose the other – a much more difficult tasks, for two reasons: (a) Rats find the least-recent-choice rule much easier than most-recent choice; and (b) trace-value difference between the most-recently visited arm and the arm visited next-most-recently (4 in the illustration) is necessarily much smaller than the difference between the oldest trace and the next oldest (74).

Thus, the poorer performance in T-maze tasks versus radial-maze tasks may reflect response-rule and recency-discrimination limitations that combine to favor the radial maze. The animal's limited event memory, together with the large number of arms in the radial maze and the typically long ITI, protects performance from interference from remote trials; whereas the smaller number of arms in the T-maze, and the typically short ITI, ensure that proactive interference will occur.

Delays on the order of seconds (not minutes or hours) between sample and choice stimuli severely impair the performance of pigeons on delayed matching to sample. DMTS experiments can be explained in the same way as delayed-response, with two added difficulties: Spatial stimuli, such as two maze arms, are almost certainly easier to remember than the colored lights and patterns typically employed (with pigeon subjects) in DMTS studies; and the animals have no built-in response rule appropriate for the reinforcement contingencies, as rats do in delayed-alternation or the radial maze. Taken together, these factors seem ample to account for the poor performance of most animals in the DMTS procedure with even quite moderate delays.

SUMMARY

Memory in the most general sense is implied by any difference in behavior that can be traced to a difference in an organism's past history. Control of behavior by a time marker, that is, temporal control, is perhaps the simplest example of a memory effect. I showed in the first part of the chapter that control by a time marker can be impaired by both prior and subsequent events – interactions corresponding to the proactive and retroactive interference much studied in human and animal memory experiments. These interactions limit animals' ability to discriminate the absolute and differential recency of events. This memory for events is sometimes known as *working* memory.

Recency discrimination depends on the particular properties of time markers. Hedonic stimuli, such as food and electric shock, are particularly effective as time markers. Hedonic events preempt the animal's attention so that a remote hedonic event may control behavior even though a more recent, but less salient, event is a better predictor.

When interference is minimized, neutral events can serve as time markers, and they then behave in the same way as synchronous stimuli: Excitatory temporal stimuli show decremental generalization gradients, inhibitory stimuli incremental gradients. This aspect of memory, encoding of the properties of particular stimuli, is sometimes known as *reference* memory.

Event recency can be represented by a trace variable that decays in negatively accelerated fashion with elapsed time. Such a model implies a context-specific limit on the number of events that can be separately represented in event memory. Animals learning tasks such as successive discrimination reversal and delayed matching to sample seem to behave as this model suggests: Reversal performance improves with practice, is unaffected by a lapse of several days or by shifting to a new pair of stimuli, but is usually impaired on the first reversal after the shift.

The latter part of the chapter showed how trace discrimination, together with spatial encoding of maze arms (a bivariate spatiotemporal code) can explain a variety of experimental results with the radial maze and related apparatus. When every arm is baited (in the radial maze) or when alternation is rewarded (in T-maze delayed-alternation experiments) animals appear to "sort" the bivariate codes representing arm visits according to the trace value, selecting the spatial coordinate associated with the oldest trace.

This model explains why the learning of mazes in different spatial locations does not interfere and why there is no spatial generalization in either radial- or parallel-maze experiments. It also accounts for the dependence of maze performance on the time and number of intervening choices, for the similarity of performance on successive trials after the first within a day, and for pauses within a trial after each arm has been visited once.

Rats in the eight-arm radial maze behave as if each arm visited is represented only once. This "reset" property of the trace conflicts with other memory results showing that repetition affects performance in delayed-reaction and DMTS situations. A resetting-trace model also cannot explain why delayed-alternation performance in a T-maze is much more sensitive to retention-time delays than performance in the radial maze, since the two tasks are identical save for the number of choices involved. This contradiction is resolved by the limit on event memory implied by a trace model. The number of arm visits within a trial in the radial maze exceeds this event limit so that a trial interferes only with the following trial and not with later trials – this is why the resetting-trace model works for the radial maze. The smaller number of different-arm choices in the T-maze means that unless the retention interval is short relative to the intertrial interval, multiple visits to the same arm are represented, complicating the animal's task and impairing choice accuracy. This analysis accounts for the effects of sample and nonsample repetition and for the effects of different times between repetitions, in delayed-alternation experiments – as well as for the sensitivity to delay of delayed-response and DMTS performance.

The discussion of memory in this chapter begins to get at the problem of learning: change in performance with experience. From this perspective, learning can be thought of as the change in an animal's performance brought about by the change in its representation of past events and their significance – where the change in representation depends both on the animal's behavior (acting through a feedback function to produce changes in the environment) and the passage of time. The next two chapters deal directly with learning.

NOTES

1. Schneider (1969). The book *Time in Animal Behaviour*, by Richelle and Lejeune (1980), provides a comprehensive review of work on temporal control as well as some discussion of circadian rhythms in animals.

2. The data support the assertion that error is proportional to the mean interval being timed, but this follows from the clock model only under certain conditions. For example, if the clock rate is constant from reset to reset, but varies randomly around the true mean from one reset to the next, then error will be proportional to the timed interval. On the other hand, if clock rate varies randomly from moment to moment, mean error will not bear such a simple relation to mean interval timed. This process can be analyzed as a *random walk*, for which the expected deviation will be proportional to the square root of the timed interval. Scalar timing, the evidence for it and its theoretical properties, is discussed at length by Gibbon (1977).

3. This experiment, and other work related to the clock hypothesis, is described in Church (1978; see also 1980).

4. In Chapter 11 I noted that behavioral competition can improve discrimination. Effects that seem to reflect competition, such as local contrast, increase when stimuli are made more difficult to discriminate or stimulus control is weakened in other ways. It is possible, therefore, that the animal uses interim activities as a way of sharpening discrimination performance. Since time is a continuous dimension, recognizing the time-of-arrival of periodic events poses special problems. This difficulty may perhaps account for the special strength of interim activities (especially schedule-induced interim activities) on periodic schedules. Schedule-induced interim activities, such as drinking by rats, are *not* seen in the $S-$ component of multiple schedules, for example; they seem to occur only in the temporally defined $S-$s on periodic schedules.

5. The difficulties encountered in obtaining intradimensional generalization gradients of temporal control, and some examples of such gradients, are described in Staddon (1975).

6. The Catania and Reynolds (1968) monograph from which this figure is taken describes a mass of data on the effects of postfood reinforcement probability on the pattern of postfood responding by pigeons.

7. *The postfood pause on fixed-ratio schedules.* It is tempting to apply this temporal-discrimination account to fixed-ratio schedules also: The postfood pause on FR has usually been explained by the roughly fixed time between food deliveries that emerges as a secondary consequence of the approximately constant rate of responding animals show on ratio schedules. Once the time between food deliveries stabilizes, the argument goes, a fixed-ratio schedule looks like a fixed interval, and the animals therefore begin to pause after food. The problem with this argument is that if the schedule looks like FI, the animals should also slow down (because typical response rates on FI are lower than rates on comparable FR, see Chapter 7). But if they slow down, the interfood interval should increase, so the pause should increase, and the result should be a very unstable pattern. This explanation also assumes that animals are in some sense unable to tell the difference between interval and ratio schedules, even though features of their behavior other than the pause are quite different on the two schedules.

The satiation–deprivation process discussed at the end of the last chapter provides a better explanation for the postfood pause on ratio schedules. Satiation–deprivation implies changes in the competitiveness of interim activities early and late in the postfood interval. The instrumental (terminal) response necessarily occurs just before food. Hence, the tendency to engage in interim activities will always be highest *after* food, i.e., after a period when they have not been occurring.

Animals pause much less on variable-ratio schedules, and this has usually been attributed to aperiodicity of food deliveries. But animals also respond more slowly on VR, so there is time for interim activities after each response or group of responses, diminishing the tendency for them to occur after food (the same argument accounts for lack of postfood pausing on VI). The small pauses sometimes seen on VR may reflect the time taken up by eating (when interim activities cannot occur). Postfood pausing on VI is negligible because the tendency to make the instrumental response is weak just after a response (reflecting the lowered probability of food at that time), allowing time there for the occurrence of interim activities and correspondingly weakening their tendency to occur after food.

One prediction of this analysis is that the postfood pause on fixed-ratio schedules should be reduced by postfood timeouts longer than the typical pause, and this seems to be the case (Richards & Blackman, 1981), unless the timeouts are very long (which may produce countervailing increases in food motivation).

8. This point was recently confirmed experimentally by Hatten and Shull (1983).

9. Staddon (1974). Context effects in conventional animal-memory experiments have been shown by Roberts (1972) and Grant (1980).

10. *Overshadowing.* The term *overshadowing* is originally due to Pavlov (1927), who used it to describe the effect of an intense, or salient, stimulus element in gaining control at the expense of a less salient element. For example, if an animal is trained to discriminate between a positive stimulus consisting of a loud sound and a dim light and a negative stimulus consisting of darkness and silence, then in a test it is likely that the sound will be the most important controlling element: The animal will *attend* (in the sense the term was used in Chapter 10) primarily to the sound dimension. Later classical-conditioning experiments have shown that a similar effect can be produced by pretraining: If an animal is first trained to discriminate between a tone and silence, then a light is added to the tone, little or no control will be acquired by the light. This is termed *blocking* and I discuss it further in Chapter 13.

Overshadowing by hedonic time markers obviously conforms to Pavlov's usage, because it is the salience of these stimuli that leads to their prepotency, rather than any kind of pretraining. The temporal usage is different in that it refers to a phenomenon that can be repeatedly demonstrated in "steady-state" responding in individual animals, rather than to differences in the ease with which some behavior can be *acquired*. There may be no difference at the level of fundamental processes, however, since the ease with which a discrimination can be learned obviously depends upon the memorability of the stimuli to be discriminated.

11. See Fletcher (1965) for a review of this and related experiments on animal memory. Roberts (1972) and Medin (1969) have reported similar findings.

12. D'Amato (1973) seems to have been the first to propose that many memory effects can be explained in terms of the discrimination of recency.

13. This is actually Jost's second law; the first is "Given two associations of the same strength, but of different ages, the older one has greater value on a new repetition." Both refer to the relative gain in strength of old vs. new associations. This version is from Hovland (1951, p. 649), which also contains the original Jost (1897) reference. Jost's law implies that any decay component in memory cannot be simply exponential, at least if the same decay rate is assumed for all memories (Simon, 1966).

14. Staddon and Frank (1974); see Mackintosh (1974) for a review of the numerous studies on discrimination reversal. The learning-set task was devised by Harlow (1949) as a test of intelligence in subhuman primates; it became a standard procedure in work with primates, but is now little used. Miles (1965) has reviewed much of this work.

15. An even better possibility is to *remember* yesterday's $S+$, and then respond to the other stimulus, so-called *spontaneous reversal*. This requires excellent memory on day N for the stimulus significances established on the previous day, and seems to occur in some higher primates. The best that other animals can do is to disregard prior stimulus significances. I discuss these differences later.

16. The first-reversal decrement shown in Figure 12.11 proves that the pigeons were able to keep the various stimulus pairs distinct, and that lack of interference across days was due to this rather than to complete confusion about past stimuli. If the animals had been completely confused each day about the identity of $S+$ on the previous day, then there would have been *no* decrement on the first reversal because the animals would not have been able to recall the identity of the previous-day's $S+$ (because of proactive interference from earlier, presumably similar, $S+$s). They did show a first-reversal decrement, hence must have been able to recall on reversal 1 $S+$ on day 0, even though that $S+$ was preceded by 50 days with novel pairs of stimuli each day.

17. The material in this section is based on an unpublished theoretical paper by Dale and Staddon. The original experiments on the eight-arm radial maze were done by Olton (1978; Olton & Samuelson, 1976).

18. The original idea that animals get around with the aid of *cognitive maps* is owed to the Berkeley psychologist E. C. Tolman (1886–1959). Tolman proposed what would now be called an information-processing theory of animal learning that made much use of concepts such as *expectancy*, *means-end-readiness*, and the like, whose descendants (conscious or otherwise) are to be found in the writings of human psychologists such as Simon (e.g., 1979). Tolman's best known works are *Purposive Behavior in Animals and Men* (1932) and the contribution to the collection edited by Koch (1959).

19. These mnemonics are entertainingly reviewed in Crovitz (1970) and Yates (1966).

20. Miller (1956) wrote a paper, now a classic, entitled "The magical number seven, plus or minus two: Some limits on our capacity for processing information" in which he reviewed a range of experimental results pointing to this limitation on *immediate-memory span*.

13

LEARNING, I: THE ACQUISITION
OF KNOWLEDGE

Most animals are small and do not live long; flies, fleas, bugs, nematodes, and similar modest creatures comprise most of the fauna of the planet. A small, brief animal has little reason to evolve much learning ability. Because it is small, it can have little of the complex neural apparatus needed; because it is short-lived, it has little time to *exploit* what it learns. Life is a tradeoff between spending time and energy learning new things, and exploiting things already known. The longer an animal's life span, and the more varied its niche, the more worthwhile it is to spend time learning.[1]

It is no surprise, therefore, that learning plays a rather small part in the lives of most animals. Learning is not central to the study of behavior, as was at one time believed. Learning is interesting for other reasons: It is involved in most behavior we would call intelligent, and it is central to the behavior of people.

"Learning," like its complement, "memory," is a concept with no generally agreed meaning. Learning cannot be directly observed because it represents a change in an animal's *potential*. We say that an animal has learned something when it behaves differently now because of some earlier experience. Yet all learning is *specific*: The animal behaves differently in some situations, but not in others. Hence a conclusion about whether or not an animal has learned is true only for a range of situations; a negative conclusion may always be invalidated by a change observed in some new situation not previously tried. And as we saw in Chapter 4 even this test is one that cannot be perfectly applied in practice because we never have exact replicas of the animals we study. Lacking replicas, we cannot really compare the same animal with different histories to see which differences in history make a difference to the animal's behavior.

In defining a particular change as due to "learning" some kinds of experience are usually excluded. A change in behavior is necessary but not sufficient: For example, a change brought about by physical injury or restraint, by fatigue or by illness doesn't count. Short-term changes, such as those termed *habituation, adaptation,* or *sensitization* (see Chapters 2 and 4) are also excluded – the change wrought must be relatively permanent. *Forgetting* has an ambiguous status: The change is usually permanent and does not fall into any of the forbidden categories, yet it is paradoxical to call forgetting an example of learning. Evidently it is not just *any* quasi-permanent change that qualifies. *Learning* is a category defined largely by exclusion.

There is obviously no point in attempting a neat definition for learning. What we need is some understanding of the ways in which an animal's environment can produce long-lasting changes in its behavior. If we can understand how these effects come about and, especially, if we can find the general principles that underlie them, then learning will be one of the phenomena explained. An adequate theory of learning must be just one aspect of an adequate theory of behavior in general.

The first step in understanding how the environment produces persistent changes in behavior is obviously to classify, but to classify *what*? Not types of change, per se; we would not expect the particular behavior affected – pecking, lever pressing, singing, or whatever – to be especially revealing. Rather it is the *way in which* the animal's past experience guides its future behavior that is important – the set of *equivalent histories*, in the terminology of Chapter 4. If we can also come up with a compact description of the *changes produced* (the problem of "what is learned?") we will have a solid basis from which to make generalizations.

For convenience, I shall distinguish two main types of learning, *template learning* and *reinforced learning*. The first type tells us much about the kinds of change produced by learning; the second tells us much about how these changes occur. The discussion in this chapter and the next also shows that these two aspects of learning are not independent.

Imprinting and the song learning of birds are discussed under the first heading. Classical conditioning and related forms of learning about causal relations that do not involve the animal's own behavior are discussed under the second.

TEMPLATE LEARNING

Imprinting

The young of many precocial[2] birds (such as ducks and chickens) learn about their parents during the first day or so after hatching. This learning, known as *imprinting*, takes place spontaneously, without any explicit "training" by the parents. The function of filial imprinting (i.e., imprinting on the parent) is to bring the young under the parent's control, so that they will attend to warnings of danger and follow the parent when necessary. In a few species, filial imprinting leads also to sexual imprinting, which enables sexually mature animals to seek out for mating individuals of the correct sex and species.

Imprinting is not restricted either to birds or to visual stimuli: Goats and shrews imprint to olfactory signals; ducklings while still in the egg form preferences for the frequency of the maternal call; Pacific salmon imprint to the odor of the stream where they were hatched; touch, temperature, and texture are important in the attachment of a young monkey to its mother.[3] But we know most about the imprinting of precocial birds.

Chicks imprint in two steps. First, the chick tends to follow moving objects. If the only object it sees is a box with flashing lights, it will follow that. After a little experience of this sort (a few minutes can be sufficient), the young bird learns enough about the characteristics of the followed object to pick it out from others and to follow it on future occasions. With further experience, the young animal begins to treat the imprinted object as a member of its own species. In a few species, filial imprinting may persist into later life and lead to aberrant mate choice. In the laboratory the imprinting of chicks can appear strikingly unintelligent, in the sense that the young bird can become imprinted to a variety of strange objects – a human experimenter, a hand, a box with flashing lights. So described, the process seems quite mechanical and mindless.

Yet on closer examination the mechanisms that underlie imprinting are more selective than may at first appear. For example, if provided with a *choice* of moving objects during the first posthatch day, the young bird is much more likely to imprint on the object resembling its natural parent.[4] Even after imprinting on some artificial object, the initial imprinting may wear off when the animal is exposed to the natural object later – and this is more likely the more unnatural the original imprinting object.

Imprinting also involves more than just a stimulus-response connection. For example, in one experiment mallard ducklings were raised with adults of another species. When the mallads were released as adults onto a lake containing dozens of species of geese and ducks, the males attempted to mate almost exclusively with females of the species that raised them. Obviously the birds learned more than just to hang around with their foster parents. This kind of generality is characteristic of most learning by birds and mammals.[5]

What kinds of experience are necessary and sufficient for imprinting? And precisely what changes do they produce?

There seem to be two features essential to imprinting: an object of a certain type, and a *critical period* when the bird is sensitive to such objects. The young bird must be exposed to a moving (or flickering[6]) object within the first one or two posthatch days. There has been much argument about the importance of the critical period. There are two problems: Imprinting when it occurs is hard to reverse, so that the animal's *first* experience with an imprintable object blocks any effect of later objects. And after the very early period, young birds become *neophobic* – afraid of new things – which interferes with potential imprinting. The growth of neophobia leaves a brief time "window" when the young bird can imprint to a novel, moving object; afterward, such objects elicit only fear. Of course, neophobia implies prior learning, since the unfamiliar is defined only in contrast to the familiar, and familiarity is learned. Until it has learned about the familiar, the animal cannot even recognize what is unfamiliar, much less be afraid of it. The critical feature of imprinting may, therefore, be its irreversibility, plus a built-in tendency for young birds to fear the unfamiliar.

In nature, the female duck interacts with her chicks in ways that are not possible for an artificial, imprinted object. Maternal calling is an important part of these interactions. It is likely that they serve to confirm imprinting in nature and make it less easily altered than some imprinting in the laboratory.

The essential change produced by imprinting seems to be one of *classification*. Young precocial birds seem to be innately equipped with an empty mental slot labeled "parent" (I define what I mean by "slot" more precisely later). The imprinting experience provides them with a set of natural objects to put into that slot. The animals have a ready-made repertoire of filial and, later, sexual behavior and need to know only where to direct it. Experiments have shown that the imprinting object can function as a reinforcer, in the sense that the young animal will learn to make some more or less arbitrary response to bring it closer. Chicks are clearly "comforted," in some sense, by proximity to the imprinted object: They cease making "distress calls" and begin to make "comfort calls," they move around more, and so on. Yet the concept of reinforcement has little to do with their learning: The imprinting object functions as a reinforcer because it has been classified in a certain way; it is not so classified because it is reinforcing. The real driving force seems to be the young animals' need to fill the empty "maternal" slot.

The slot may initially be empty, but it is not shapeless. Clearly some objects will fit it better than others, since the young birds imprint much more effectively to some objects than to others. This is the basis for placing imprinting under the category of *template learning* – a term originally coined to describe song learning in birds (about which more in a moment).

Although the effects of sexual imprinting are delayed, the major effects of filial imprinting are immediately apparent. Moreover, we do not know if the learning involves *practice* in any obvious sense. The young chick will follow the imprinting object, and imprinting is clearly better if it is allowed to do so; but following is not absolutely essential. In some learning by passerine birds, the effects are nearly always delayed and practice is essential.

Song learning

Songbirds differ in the extent to which song development depends upon experience. Song sparrows (*Melospiza melodia*), with one of the most intricate and beautiful songs, need relatively little experience. In one experiment[7] song sparrows were foster-reared by canaries in a soundproof room. The songs of the adults were indistinguishable from the songs of wild-reared birds: Song development was not impaired by foster rearing, and the sparrows showed no tendency to imitate the canary song. Other experiments have shown a small effect of depriving song sparrows of early experience, but clearly much of the information for the song-sparrow song is innate. At the other extreme, white-crowned sparrows (*Zonotrichia leucophrys*) develop severely abnormal songs if deprived of the opportunity to hear species' song during the first six months of life. The upper sonogram in Figure 13.1 shows 2 sec of the song of a wild sparrow; the middle sonogram shows the much simpler song of a deprived animal.

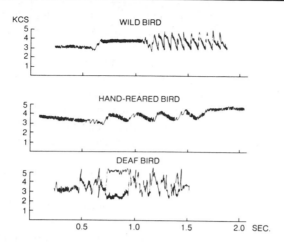

Figure 13.1. Sonograms of songs of three white-crowned sparrows raised in different auditory environments: a wild bird, a hand-reared individual, and a bird deafened before the onset of singing. The sonograms show frequency as a function of time. Since most passerine song is close to being a pure tone (so-called whistle song) the sonograms mostly show just a single line of varying frequency. (From Konishi & Nottebohm, 1969.)

The Oregon junco (*Junco oreganus*) is an intermediate case. Individually isolated nestlings when adult produce wild-type songs but with a simpler syllabic structure. But juncos raised in isolation in groups, or in general laboratory rooms where they could hear other species, produce more complex syllables and larger song repertoires, including sound patterns present neither in wild junco populations nor copied from other species. This study (Marler, Kreith, & Tamura, 1962) and work with song sparrows show that the ability to develop normal song in isolation does not preclude the ability to copy and to improvise new sounds.

Song learning need not be delayed as it is in the white-crowns: Mockingbirds, starlings, mynah birds and many parrot species imitate other sounds, and do so soon after hearing them. And the learning in this case is usually reversible: Starlings seem to come up with a new repertoire every season, and most other imitative birds continue to add and delete items from their repertoire throughout life.

The song learning of male white-crowned sparrows and many other songbirds proceeds in two phases. For white-crowns the first phase occurs during a critical period from 10 to 50 days after hatching, when the immature males must have the opportunity to hear adult male song. Birds kept in isolation, or deafened, during this critical period never sing normally. But if the animals are prevented from hearing song either before or after this period, the later development of adult song is unimpaired.

At about 8 months of posthatch age, the birds become sexually mature and begin to sing; this is the second phase. The onset of singing is not immediate, but is preceded by a period of *subsong*, a twittering unorganized pattern quite different from adult song; adult song emerges gradually from subsong over a period of several weeks.

Experiments[8] have shown that the first of these phases is essential for the bird to store a model, or *template*, of adult song. As in imprinting, the animal is selective in what it will store. For example, white-crowns exposed during the critical period only to songs of the song sparrow incorporate none of this into their model. Song development is just as impaired as if the birds had been reared in complete acoustic isolation. On the other hand, in the wild, the birds' early receptivity permits them to learn the local *dialect* of white-crown song (indeed, this may be the evolutionary function of song learning in this species). Figure 13.2 shows sonograms of dialect variation in the San Francisco Bay area: Although the dialects are different from one another, there is obviously a family resemblance. Evidently the birds' mechanism for template formation restricts the class of possible templates. It is as if the bird were a musician programmed to learn a single piece of Baroque music: Such an individual would be immune to the charms of Brahms or the Beatles, but would fixate instantly on anything by Bach or Vivaldi.

Even after the template has formed, song learning is not complete. The period of *practice*, when the bird is showing subsong and incomplete versions of adult song, is also essential. The final song is evidently built up by a circular process in which the bird sings, hears the results of its own singing, and slowly modifies what it sings until its production matches the stored template.

This loop can be interrupted in two places: either by disrupting the motor apparatus, to prevent practice, or by deafening the bird, to prevent its hearing its own production. Experimentally deafened birds show normal subsong, but it never evolves into normal adult song. An example of the grossly abnormal song of an early-deafened bird is shown in the bottom panel of Figure 13.1. Moreover, this is not owing to some disruptive effect of the deafening operation: Birds reared in a noisy environment so that they cannot clearly hear their own song also develop abnormal songs. These results prove the importance of auditory feedback. It is obviously difficult to impair motor function in a reversible way so as to prevent practice but permit singing in adulthood. As far as I know, this has not been attempted. It is conceivable that a bird denied the opportunity to practice would nevertheless sing normally as an adult when motor function is restored, but it seems unlikely. Almost certainly, both halves of the loop – motor and perceptual – are essential for normal song development in white-crowns, chaffinches, and many other songbirds.

The requirement for practice is not limited to birds whose template is formed by learning. Song sparrows require little experience with conspecific song to develop normal singing, but they do need to hear their own early singing efforts. Song sparrows deafened at an early age produce extremely abnormal song. On the other hand, several nonpasserine[9] birds, such as domestic chickens and ring

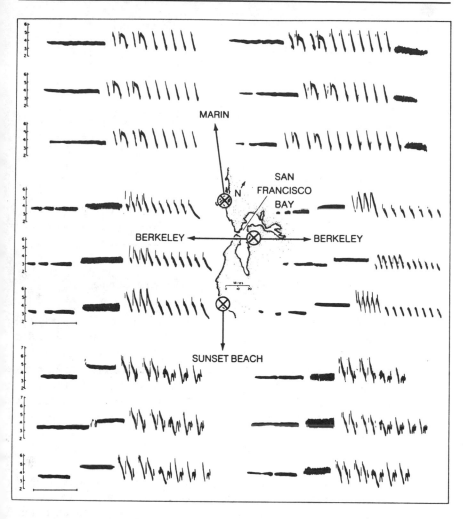

Figure 13.2. Sonograms of the songs of 18 male white-crowned sparrows from three different localities within the San Francisco Bay area. Songs from a given area are similar to each other, and different from the songs of other areas. (After Marler & Tamura, 1964.)

doves (*Streptopelia risoria*), need no auditory experience and develop their normal rather rudimentary song even if deafened soon after hatching.

Once song sparrows, white-crowns, and other songbirds have developed their adult song, it is largely immune to further interference. For example, white-crowns deafened as adults continue to sing essentially normal songs for many months.

These diverse processes of song development presumably have an ecological basis, although this is not always obvious. For example, the sensitivity of young

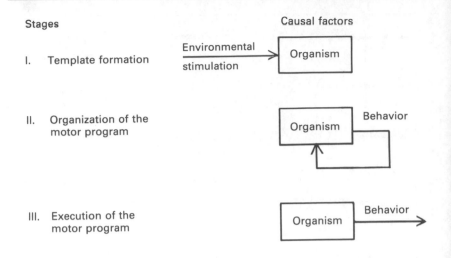

Figure 13.3. The three stages of passerine song learning.

white-crowns allows them to learn the dialect of their location: What advantage is conferred on a male that learns a dialect, rather than being innately provided with an adequate song? White-crowns occupy a wider range of habitats than some other similar species, suggesting that genetic subpopulations may be especially well adapted to particular habitats. Perhaps kin selection leads males to be more hostile to males singing a different dialect, thus favoring males that can learn dialects of the place they were born. There are probably other possibilities. The imitative abilities of mockingbirds may be related to the evolutionary benefits of a large song repertoire, something which is itself not well understood.[10]

Conclusion. For all their apparent diversity, imprinting and the song learning of passerines share two central features. In both cases, behavior is based upon a stored model or template and the perceptual learning that leads to template formation is broadly *selective*: Only objects, or songs, that satisfy certain general characteristics will be accepted as models. These properties are not universal, however. It is likely that the song learning of doves does not involve any kind of stored perceptual model: Their simple song seems to be just a *motor program*, like walking or flying, that needs little practice and no specific external stimulation to develop normally.

These characteristics fit conveniently into the three learning stages illustrated in Figure 13.3. In the song learning of white-crowns, the first stage is the laying down of the template; we know very little about the details of this process, because no overt behavior accompanies it. This stage is inferred from the results of deafening and deprivation experiments. The learning may be laborious or quite rapid. The mechanism involved has the property of selectivity I have just described. I am assuming that imprinting also involves the formation of a template, but in this case overt behavior (following, learning responses that bring the imprinted object closer, etc.) is tied to the template as soon as it is formed. But in neither case is the organism's own behavior directly involved: Template formation is induced entirely by outside stimulation.

The second stage of song learning involves the circular interaction between the bird's own singing and the stored template. It is here that our ignorance of the nature of the template becomes most obvious. All we observe is a circular process that continues for several weeks and eventually stabilizes in adult ("crystallized") song. But whether the learning involves progressive approximation to a stored model (as the template idea implies), or whether the process converges by some completely different means, we do not know. We can be sure only that the final form of the song is determined by early acoustic experience. In imprinting, there is nothing corresponding to the second or third stages.

The third stage is the final, "automatic" song, that is almost unaffected by disruption of the feedback loop that was essential to its original development.

The song learning of chaffinches and white-crowned sparrows involves all three of these stages, but the development of song-sparrow song skips the first stage. Evidently the template here is innate and requires no specific environmental input. Deafening experiments show that the second stage is necessary, however.

The development of song in nonpasserines, such as chickens and doves, skips all but the last stage. In these species, the motor program itself is innate and requires no external input – either from the environment or the bird itself – to develop normally.

These three types of behavioral organization are based on experiments on imprinting and song learning in birds. Nevertheless, I believe that they are to be seen in other species learning quite different things in quite different situations. Indeed, I will argue that most learning of habits and skills follows just these three stages.

I turn now to a consideration of the many types of learning that involve some kind of "reinforcement": the occurrence of a hedonic event in systematic relation to other aspects of the environment or to the animal's own behavior.

REINFORCED LEARNING

Reinforced learning has been a major concern of psychologists since the time of Thorndike and Pavlov. Unfortunately, until recently, research tended to converge on ever more standardized experimental paradigms in which a simple, preferably unidimensional, response could be studied under highly controlled conditions. The field has become increasingly divided into two topics – classical (Pavlovian) conditioning, and operant conditioning. This division conforms to no characteristic of the processes involved; the distinction between classical and instrumental conditioning is much more a matter of procedure (open-loop vs. closed-loop, in the terminology of Chapter 5) than of process (see Chapter 4).

The division goes along with different theoretical and practical preconceptions. These different interests have led the operant conditioners almost to abandon work on the *process* of learning (i.e., of behavior change) and concentrate on steady-state adjustments. The same tendency has overtaken many classical conditioners, who, while affirming a deep interest in the learning process, present data only on steady states. The implicit notion that the learning process is something that should be describable by smooth curves has led others to ever more refined situations and ever larger groups of disposable subjects.

Learning means change, and change has given psychologists the same kind of trouble that motion gave Zeno: How can one study motion, when a body must be in one place or another? How can something move at all? Learning involves the same kind of moving target: At any time the organism "knows" some things and not others; we can clearly identify the beginning and ending stages (ignorance → knowledge), but what happens in between is often obscure.

The Procrustean solution is to act as if there is a single learning process and then design our experiments so as to preclude any other. Some of the more tightly controlled conditioning procedures have this flavor. A safer tack is to look at a range of learning phenomena in different species and see what useful generalizations emerge. In this section I shall look at the learning of bees, taste-aversion learning, and several experiments on classical conditioning. A fair conclusion is that although there are some general learning *principles*, there is no single learning *process*. Learning is almost certainly *discrete* rather than continuous, and seems to involve the assembly of many elements, both preformed and remembered, to form new programs for action.

The learning of insects is most easily analyzed. Careful experiments have often been able to break down apparently complex learning into simple elements. For example, honey bees (*Apis mellifera*) learn the color, shape, location, odor, and payoff schedule (many flowers secrete nectar only at certain times of day) of their food sources.[11] In their searches for new food sources and their exploitation of old ones they do not seem to be significantly inferior to many "higher" species even though their learning is built up out of very simple elements.

For example, the color of a flower is clearly visible to the bee before landing, while standing on the flower and while circling the source after takeoff. Yet the color of the flower is learned only during the final 3 sec before the bee lands. A naive bee carried to an artificial feeder from the hive will circle after taking on a load of sugar water, but will not be able to pick out the correct feeder color on a second trip a few minutes later. Yet if the bee is interrupted during its first feeding, so that it must take off and land again, it remembers the feeder color perfectly. In the same way, bees learn landmarks only when taking off – the circling flight that reliably follows takeoff presumably serves to scan the environment for this purpose. A bee feeding at a new location and removed before finishing has no recollection of the location, even though she had plenty of opportunity to scan landmarks when first arriving.

These peculiarities are necessary adaptations to the bee's limited mental capacity: Like a small computer, the bee must do one thing at a time, and cannot afford to store useless information. It makes functional sense for the animal to note landmarks only after leaving, because she then knows whether the place is valuable enough to be worth recording (bees don't circle when leaving a potential food site that provided no food). Presumably color is learned when landing because that could be done equally well before or after landing, but landmark-learning belongs better to the takeoff phase and preempts the "processor" at that time. While on the flower, the bee is looking for and gathering nectar; evidently this activity is sufficiently demanding that no processing resources are left over for color learning. Since the bee must obviously look at the flower in order to land on it, color learning can occur most efficiently at that time.

The learning of a bird or mammal is not so perfectly sequential as the learning of bees. Nevertheless, when a pigeon homes on its loft or a swallow returns to its nest after a foraging flight, it must process the same kind of information as the bee, even if not in such a rigid order. Moreover, the phenomenon of *attention* discussed in earlier chapters represents exactly the same kind of constraint as the compartmentalized learning of the bees. Mammals, birds, – and man – have limited processing resources, and if the task at hand is difficult, dealing with it effectively limits the amount of attention that can be paid to other tasks. We may all be able to walk and chew gum at the same time, but few can do mental arithmetic at the same time as they answer questions or read a book. The larger brains of mammals and birds give them this advantage over the bees: They can *shift* attention in relatively flexible ways, rather than being preprogrammed to attend to particular things at particular times.

The learning of bees provides two further lessons. First, the learning of an apparently straightforward task, such as the location and identity of a new food source, involves the encoding and storage of a number of elements, the solution of a number of subsidiary problems. Second, the actual storage of information – remembering a color or the properties of landmarks, for example – does not require a behavioral explanation. Remembering the color of a flower seems to be a relatively simple process for a bee. It takes place rather rapidly, has no behavioral accompaniment, and does not seem susceptible of further behavioral analysis. Such events might be termed *atomic acts*. Further analysis of such acts must be at the level of neurophysiology. For complete understanding we need to know how the visual system encodes physical properties of the visual world, and what changes take place in the animal's nervous system when it sees, and remembers, a particular colored object. These are fascinating problems for neuroscience, but take us to a nonbehavioral level of analysis.[12] The task for behavioral analysis is to discover the atomic acts, the program, involved in the learning of particular tasks.

Learning as program assembly

Analysis into elements solves the Zeno problem, since it reduces the process of learning to the building of a program, in which rules and elements are combined to produce a system that can solve a particular problem, such as getting to a particular food source. In bees the process is often relatively simple, since steps are added one at a time, as in a small digital computer. Moreover, their learning seems largely to involve defining *variables*, to be operated upon by built-in programs, rather than building the programs themselves. Thus, the worker honey bee is innately provided with a set of routines for reacting to landmarks and colored food sources; its learning consists largely in applying those routines to particular landmarks and flowers. The bee is like a hand-calculator that knows innately how to multiply and relies on the environment only for the input numbers.

In higher animals, the process is much harder to get at, because in addition to defining variables (the process termed *stimulus control*), program elements – rules – are combined in new ways during learning. Many things seem to be happening at once, or at least in rapid succession, often with no obvious behavioral accompaniment, which further complicates the task of analysis. These difficulties force us to deal with learning in birds and mammals at a more abstract level, in terms of poorly defined concepts such as "template," "coordinate system," "internal representation," and "response strength" even though our ultimate objective is to reduce this learning also to the level of atomic acts.

Reinforcement and learning

The bee learning provides some insight into an old issue in learning theory, the role of reinforcement in learning. The question was: Does reinforcement affect *learning* or just *performance*? We know that reinforcement affects performance, that is, action, because we can train animals to do things by rewarding them for it – and often they will not do certain things *unless* they are rewarded for it. But is this *all* that is affected by reinforcement?

This question was difficult to answer because learning is a vague concept, and reinforcement seems to be retroactive: What is *learning*? If we can't define it precisely, we can't expect to answer detailed questions about it. How is *performance* different from learning? Obviously we cannot assess learning other than by watching the animal's behavior, so how can we ever detect learning that produces no change in performance? Is the question even a meaningful one? The second problem is the indirect relation between reinforcement and its effects: A reward comes *after* the behavior being rewarded, so the immediate effect of reward must be on the animal's *internal state* (in the sense defined in Chapter 4), rather than directly on the behavior being rewarded. Presumably the animal's state is changed in such a way that the rewarded behavior becomes more likely. So perhaps the question should be rephrased as follows: Are changes in the animal's state (of the sort we might like to term *learning*) produced only by reward or punishment, or can they take place independently of reward and punishment?

The bee learning provides an answer: Bees store the color of a flower *before* they have landed on it, hence before they have any opportunity to be rewarded. Clearly color, an essentially neutral stimulus, can cause a change in the bee's state independently of reward. On the other hand, the bees store information about landmarks only after, and if, they have been rewarded. And they only demonstrate by their behavior that they have learned if they have eaten: It was necessary to reward the bees so that failure to choose correctly could be attributed to something other than lack of reward. Moreover, it is quite likely that information about flower color is *erased* from the bee's modest memory if the flower fails to yield nectar, although this might be hard to demonstrate behaviorally. Consequently it might well be impossible to demonstrate what was once termed *latent learning* in bees, even though other experiments have proved that the animals can remember things without being rewarded for doing so.[13]

The question as originally posed is an example of one of those undecidable propositions about black-box systems discussed in Chapter 4. Since we must infer learning from performance, and we know that reward is often necessary for performance, we simply cannot give a general answer to the question of whether reward is necessary for learning. As the bee example illustrates, it is quite easy to come up with learning processes in which reward is essential to the demonstration of learning, yet deeper analysis reveals memory effects that are independent of reward.

On the other hand, based upon both logic and the bee experiments, we can say with certainty that reward and punishment are not the only events that can cause a change in an animal's internal state. And we can decide experimentally whether a given thing is learned before or after the animal is rewarded. The bees encoded flower color before being rewarded by the flower, but it is conceivable that they could have learned it afterward, as they learn landmarks. It is logically possible that an animal notice nothing until it has obtained a reward, at which time it scans its environment and learns something about it: All learning could be like bee landmark learning. The converse is also a theoretical possibility: Animals could learn everything in advance of reward, using reward just as a guide to action, as in bee color learning. Neither of these extreme arrangements seems uniquely efficient. Real learning is surely a balance of both.

The Pavlovian, or classical-conditioning "paradigm," provides a convenient method for the study of problems such as the order in which things are learned, what is learned, innate biases in learning, and the ways in which causal relations are represented by animals.

Classical conditioning

The evolutionary function of "reinforced learning" is to detect regularities in the environment related to things of value to the animal. For example, the function of the bee learning just described is to identify and locate sources of food. Learning of this sort is a process of *inference*, in the sense that certain environmental features, such as landmarks (a particular location) and colors are taken as signals of food.

Any such inference can be mistaken: if the environment changes in between

Table 13.1. *Probability of food for feeders with different color-pattern combinations*

		Pattern		
		X	Y	
Color	A	1	0	.5
	B	1	0	.5
		1	0	

observations, or if too few observations are made, for example. If the experimenter changes the color of the feeder after the bee's first visit, the animal will be unable to locate the food source: Its initial inference that color A predicts food will have been proved wrong. If the feeder is only full some of the time, a single observation will give unreliable information.

Some cues are more predictive than others. One can imagine an experiment with artificial feeders having two properties, a color and a pattern. The experimenter can arrange a predictive relation between food and either color or pattern or both. Let there be two patterns, X and Y, and two colors, A and B. Suppose that pattern X predicts food (all X-type feeders contain sugar water), but pattern Y does not (all Y-type feeders are empty). Color is made irrelevant (A- and B-color feeders both contain food half the time). These contingencies (see Chapter 5) are illustrated in Table 13.1, which shows the four possible feeder types (AX, AY, BX, BY) and the probability of finding food at each. If the bee can learn about only one aspect at a time (as seems likely), and if it learns on first exposure, then its choice of whether to attend to color or pattern is critical. If it attends first to color, it will always be wrong. Obviously this is a case where the animal must make *several observations* before it can detect the real invariance in the relation between visual appearance and food potential.

This need for more observations is almost independent of the animal's computational capacity. Even if the bee could attend to *both* color and pattern on each trial, it still would not know which one was more important until it had sampled several flowers, that is, until it could fill in from its own experience the probabilities in Table 13.1.

This example shows why (in a functional sense) animals learn some things rapidly and others more slowly. Speed of learning is determined both by the processing limitations of the beast, and by the necessity to sample the environment a sufficient number of times to obtain a reliable estimate of invariant features.

The problem of precisely *how much* sampling an animal should do is an exceedingly difficult one theoretically, even in a simple environment with known properties.[14] Nevertheless, the main factors are straightforward: (a) How much does the animal already know? That is, what are the *prior probabilities*? to use the Bayesian term,[15] and how many of them are there? (b) How large is its computational capacity? That is, how many factors can it assess on each trial? (c) How important are the consequences of the learning? This will be determined by the relative costs and benefits of "hits," "misses," (failing to respond to a valid signal) and "false alarms" (responding to an invalid signal), together with the costs of sampling.[16]

Innateness of priors. Factor (a), the prior probabilities, is likely to be related to the variability of the environment: Properties that are variable should be learned more slowly than properties that are fixed because the animal can be less certain of the meaning of a given conjunction: Should the child bitten by a dog be afraid of all dogs, just German shepherds, or just *this* German shepherd? Obviously more data are needed: Providing the risk is not too high, he should sample a few more dogs – unless the genetically coded experience of his ancestors can bias him in one direction or another. The appearance of flowers, even of the same species, is quite variable: Blooms vary in size and tint even from day to day. On the other hand, the *odor* of a given flower doesn't vary at all. Odor is a much more *valid cue* to the identity of a flower than is color. Moreover, the number of possible flower colors is very large, whereas the number of possible odors is much smaller. Correspondingly, bees should take some time to learn the color of a flower, because of the need to experience a representative sample, whereas odor learning should be rapid. Bees show precisely this difference: Odor is learned on one visit, whereas color takes several and is never learned perfectly. Time of day is learned more slowly still, presumably reflecting typical variation in the time at which flowers secrete nectar.

In simple animals the priors are innate. Bees do not learn colors more rapidly if raised in an environment in which colors never vary and are always reliable predictors of food. Nor can they learn to be less impulsive in learning odors.[17] Even in mammals and birds, examples of innate bias of this sort are common. For example, a rat made sick by a harmless injection of lithium chloride half an hour after tasting a new food in a new cage will develop a strong aversion to the novel taste, but not to the new cage. *Taste-aversion learning*, of which this is an example, follows the classical-conditioning paradigm (an unconditioned stimulus [US], sickness, follows a conditioned stimulus [CS], the taste of food) but has several unusual properties.[18] It is relatively insensitive to long delays (conventional Pavlovian conditioning would be almost impossible with CS–US delay of 30 min or more), it is highly selective (a taste CS is much more effective than a CS of some other kind), and it occurs on a single trial.

The difficult experiment of raising rats in an environment in which something other than taste is correlated with gastrointestinal effects has not been done, so we do not know if the priority of taste as a CS for sickness is completely immune to alteration by experience. However, taste-aversion experiments have been done with very young animals, and here also taste is dominant over other stimuli. The dominance of taste is evidently innate.

Previous experience does play a key role in taste-aversion learning, however. If a rat is made sick some time after having tasted both a familiar and a novel food, an aversion develops only to the novel food. Here is a perfect example of Bayesian inference. The prior probability that the familiar food is poisonous is obviously low, since it has been eaten safely on previous occasions. If the animal must make a decision on the basis of a single exposure, then obviously the novel food is the preferred candidate.

The other characteristics of taste-aversion learning are also highly adaptive. The primacy of taste in general follows from the invariable relation (in the history of any species not subject to X-rays or inventive experimenters) between food and sickness: Animals may feel sick for reasons other than poison, such as disease, but nonfood stimuli never cause sickness. The rapid learning follows from the severe risk associated with eating poisonous substances – factor (c), above: "false-alarms" can be very costly. The cost of omitting a safe item from the diet will generally be small; the cost of eating poison may be very high. This skewed payoff matrix obviously biases selection in favor of conservatism: rapid taste-aversion learning and *neophobia*, avoidance of novel objects, particularly foods. And indeed, rats (especially feral rats) are very cautious when confronted with new foods; they eat only a little at first and wait some time, usually without eating any other food, before incorporating the new food into their diet.

The long delay sustainable by taste-aversion learning is probably not (or at least, not entirely) a specially programmed feature. Because taste is the effective stimulus, and tastes and other stimuli are not easily confused, a taste experienced 30 min before sickness will not be confused with all the nontaste stimuli experienced in the interim. The main reasons that long-delay Pavlovian conditioning with nontaste stimuli fails are: (a) Because there is nothing to distinguish the CS from concurrent nontaste stimuli and stimuli that occurred before and after the CS, the animal has no way of knowing which stimulus to associate with the US; and (b) Even with repeated trials in which the only invariant relation is between the CS and the (delayed) US, these other, irrelevant stimuli impair recall for the CS and thus prevent the animal making any connection between it and the US. If the animal cannot remember a particular stimulus that occurred 30 min ago, it is in a poor position to show a conditioned reaction to that stimulus as a consequence of a US that just occurred. (See the discussion in the previous chapter of the sources of proactive and retroactive interference in situations of this sort.)

Two kinds of experiment are necessary to back up these conclusions. One is to show that when interfering stimuli are eliminated, conventional Pavlovian CSs, like lights and tones, can become conditioned even with long delays. The second is to show that taste-aversion learning can be impaired if interfering tastes are allowed to occur before the initial taste or between it and the US.

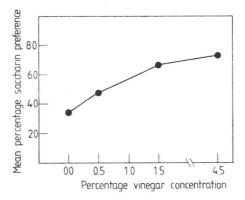

Figure 13.4. Preference of rats for drinking saccharin solution vs. a novel coffee solution, as a function of the concentration of vinegar drunk in between original experience with saccharin and lithium-chloride-induced sickness. The groups that tasted the stronger vinegar solutions showed greater preferences for (i.e., less aversion to) saccharin. (After Revusky, 1971.)

Both types of experiment have been done. For example, the experiments by Lett, discussed in the previous chapter, have shown that conditioning (her experiments were operant rather than classical, but this is not a critical point) with conventional visual signal stimuli can be successful with long delays if the stimuli are highly salient and if interfering stimuli are minimized.

The effects of interference on taste-aversion learning have been shown in several studies by Revusky (1971). In one study, four groups of rats were allowed to drink 2 ml of saccharin solution; then, 15 min later, different groups were allowed to drink water containing differing concentrations of vinegar (a novel taste). One hour after the saccharin drink, all the rats were made sick by an injection of lithium chloride. Three days later, each group received a series of preference tests in which the animals were offered a choice between saccharin water and dilute coffee. Figure 13.4 summarizes the results: The weaker the intervening vinegar solution, the less saccharin water drunk (i.e., the stronger the aversion to saccharin – here relatively weak because the alternative was coffee, not pure water). In a second experiment, Revusky showed that the vinegar interferes with the development of saccharin aversion even when it precedes, rather than follows, the saccharin. He also showed that the interfering effect of the vinegar is much reduced if it is a familiar rather than a novel taste.

There are two possible reasons for the interfering effect of the interpolated vinegar taste in Revusky's first experiment: (a) The vinegar coming after the saccharin interfered with the animal's recall of the saccharin (retroactive interference); consequently the animal could not associate the saccharin with the sickness 1 hr later. (b) The vinegar was closer in time to sickness than the saccharin, so that the animal associated the vinegar rather than the saccharin with sickness.

This would be an example of an *innate prior*, since recent events are much more likely to be the cause of something than more remote events. This dominance of one potential CS over another is termed *overshadowing*. In either case, we might expect the rats to form an aversion for vinegar, and this presumably occurred, although Revusky did not test for it.

The proximity explanation is ruled out by Revusky's second experiment, in which the vinegar interfered even when given before the saccharin. We have already seen a comparable result, however, in the discussion of temporal control in Chapter 12. Figure 12.1, for example, shows that temporal control by a brief, neutral stimulus is impaired by a preceding food delivery. I argued that this reflects the greater importance of food, a hedonic stimulus, that preempts the animal's attention after a few seconds and abolishes temporal control by the neutral stimulus. The animals in Revusky's experiment had only one opportunity to learn, so that the proximity of the saccharin to the sickness (*postremity* is the technical term) was outweighed by the higher prior probability that vinegar, a salient, novel taste, was the cause of later sickness. The animal had only two things to go on in the first experiment: Saccharin, though familiar, was closer to sickness than the novel vinegar, but the vinegar was both salient and novel. In the event, given one trial, the vinegar won out.

A couple of thought experiments are instructive here. We can ask: What would have happened had both vinegar and saccharin been familiar? And, what would happen if the experience, vinegar → saccharin → sickness, were repeated several times?

Common sense, and data from other experiments, can answer the first question at once: When the two events are equal in every other respect, the rats will avoid the taste closer in time to the sickness. The innate properties of memory, of which retroactive interference is one, reflect facts about the causal properties of the world: in particular, the fact that a more recent event is more likely than a more remote event to be the cause of a present event (sickness).[19] The temporal resolution of memory *should* reflect the probable usefulness of the information remembered. As in an office filing system, the older an event, and the more similar events that have occurred subsequently, the less useful information about it is likely to be – and the less need to retain a record of it.

The effect of repeated experience is also easy to predict. Even if the vinegar is novel at first, with repeated experience it necessarily becomes less so, and then the advantage of postremity is likely to favor the more recent stimulus, saccharin; eventually (and if it has no other basis for preference) the animal is likely to avoid saccharin, the stimulus closest to sickness.

The discussion of the temporal resolution of memory in the previous chapter suggests one final thought experiment. In Revusky's second experiment, the vinegar occurred 15 min before the saccharin, and the animal was made sick 60 min after that. Suppose the rat had been made sick only 10 min after the saccharin? There is little doubt that *some* time, shorter than 60 min, but probably more than a few seconds, could be found such that the familiar saccharin would be avoided, not the preceding, novel, vinegar.

The general conclusion is that the animal's "choice" of which stimulus to avoid is the outcome of a delicate balance in which all the potential factors that could determine cause – stimulus type, novelty (i.e., past experience with each stimulus) and postremity – are combined in an equation whose weighting factors are determined by heredity.

The order of learning. Learning that occurs on one trial must require at least some acquisition of information ("change of state" in the sense discussed earlier) before the occurrence of the reinforcer. The animal must record the novel taste if he is to acquire an aversion to it based upon delayed sickness (this is like color learning in the bee example). On the other hand, this information need not dominate, and additional processing may go on after a reward has occurred. For example, in one experiment (Keith-Lucas & Guttman, 1975) rats were given a single electric shock while taking a pellet from a small hole in the center of a novel stimulus display (the conditioned stimulus, CS). After a brief delay (different for different groups) a spiky toy rubber hedgehog was suddenly introduced into the chamber. The next day the rats were given choice tests in which the hedgehog was compared with the CS and the location in which the shock had been delivered. By a number of measures, the animals tended to avoid the hedgehog most, even though it had occurred *after* the shock; moreover, the degree of aversion was inversely related to the delay between shock and appearance of the hedgehog – indicating that proximity to the shock was a key factor. This is an example of *backward conditioning*, something which is not supposed to occur (and usually doesn't) and makes little causal sense – if B occurs after A it cannot be the cause of A.

From another point of view, of course, the animals were acting perfectly sensibly. All conditioning is a process of inference. It is impossible to be certain of the validity of a causal inference on the basis of a single instance. Animals make mistakes, and predators are often concealed before they attack. The prior probability (from the rat's vantage point) that a striped pattern (the CS in this experiment) will be dangerous is small; the prior probability that a lifelike dummy will be dangerous is much higher. Repeated experiences (denied the animal in this experiment) would serve to correct mistaken inferences drawn on the basis of a single experience. From this point of view, the animal's "guess" that the spiky hedgehog was the source of his pain must seem a very reasonable one indeed.

This result also shows that considerable processing goes on after the US has occurred. Here the rats evidently acquired some fear of the hedgehog; it is likely that a subsequent trial in which presentation of the hedgehog was soon followed by shock would rapidly confirm this fear and produce long-lasting avoidance. Thus, the first conditioning trial not only produces some "conditioning," it also creates a set of CS *candidates* to which the animal can attend on later trials. I return to this point in later discussion of theories of classical conditioning (see also note 22).

Other experiments have demonstrated the occurrence of post-US processing in conditioning. For example, Wagner[20] and his associates studied the effects of post-US events on the conditioning of the rabbit nictitating membrane. In the first phase, the animals were trained with a standard CS-US combination until they developed a reliable conditioned membrane response. In the second phase, the animals were divided into two groups. Both groups continued to receive the standard CS-US trials, but after each trial they also received one of two types of interference trials. Both types involved a second CS: either a previously established, different CS+ (i.e., a stimulus that had been paired with shock), or a previously established CS− (i.e., a previously established safety signal). *No* US occurred on these interference trials. Thus, for the added-CS+ group they involved a violation of expectations; for the added-CS − group they did not. The animals that always received the incongruous posttrial event (CS + → no shock) learned (i.e., developed a membrane response to the standard CS) more slowly than the animals that received the unsurprising sequence (CS− → no shock).

Evidently the post-US "surprising" event interfered with essential retrospective processing (termed "rehearsal" by Wagner) of the preceding trial.

What is learned? The effect of any event on an animal usually depends upon his current state of knowledge. For example, some years ago, a colleague was puzzled by the observation that a flock of feeding pigeons would sometimes be startled into general flight by the flight of one individual, whereas at other times, the takeoff of one bird had no effect. Perhaps pigeons have a special "alarm flight" that alerts others to the presence of danger and causes them to take flight, we thought. Reasonable as this hypothesis sounds, painstaking experimental work failed to find any distinguishing characteristic of "alarming" versus "nonalarming" takeoffs. The answer to the puzzle turned out to be not the property of the takeoff, as such – there was nothing to differentiate alarming from nonalarming flights – but in the "intention movements" birds make *before* they take off. A takeoff preceded by the proper intention movement (head raising, small wing movements) had no effect; but an *un*signaled takeoff caused alarm and the flight of other birds (Davis, 1975).

Birds in a flock closely monitor one another's behavior. Clearly the intention movements of a bird about to take off change the state of his fellows in such a way that his flight causes no alarm. In colloquial terms, the intention movements lead the other birds to *expect* a takeoff and only unexpected takeoffs cause alarm.

This same principle, that the effect of something depends upon the animal's state of knowledge, should obviously extend to learning. Thus an answer to the question, What are the necessary and sufficient conditions for learning? also requires an answer to the question, What is learned? Applying the rule suggests the general principle: *Animals learn only when something violates their expecta-*

tions, or when they have no expectations (as in a novel situation). All the examples of reinforced learning I have discussed fit this generalization: When two tastes have an identical temporal relation to poisoning, conditioning develops only to the novel taste. When a *compound stimulus*, such as a tone–light combination, is used in one of the standard classical-conditioning arrangements (see Chapter 5) the amount of conditioning that takes place to each element depends upon what the animal knows about each. For example, if, in a conditioned suppression (CER) paradigm, the tone had previously been paired with electric shock and the light was novel, no conditioning will take place to the light: The shock is already predicted by the tone, so no expectation is violated, and no new learning takes place. This is termed *blocking*. The Wagner et al. experiment is an explicit test of the role of surprise in conditioning, and I will discuss some others in a moment.[21]

In general, the greater the violation of expectation, the larger the learned change in behavior. For example, if two stimuli are presented in alternation with shocks occurring only in the presence of one of them (CS +), then the other (CS −) becomes a "safety stimulus," that is, a stimulus that signals the absence of shock. If such a CS − is now presented with a novel stimulus, *X*, as a compound and paired with shock, then *X* acquires a larger increment in its power to suppress lever pressing than when presented alone (so-called superconditioning). The expectation here was presumably for a reduction in the probability of shock, so that the occurrence of shock represents a larger violation of expectation than if the CS − stimulus had no prior significance. It is interesting that the increment in conditioning goes to *X* rather than to the safety signal, but this presumably reflects the Bayesian process we have already encountered, since CS − has already been established as a safety signal.

This same process accounts for the differential conditioning to taste, when the US is sickness, or to audiovisual stimuli, when the US is electric shock. The (innate) priors (in rats) favor taste as the predictor of sickness, and an audiovisual stimulus as a predictor of pain.

The first occurrence of a surprising event is critical. In a blocking experiment by Mackintosh and his collaborators (Mackintosh, Bygrave, & Picton, 1977), five groups of rats were pretrained with a light–shock combination in a standard CER arrangement. In the second phase of the experiment, a tone was added to the light, and different groups received either one or two additional trials, and a surprising additional shock just after the tone–light compound on the first or second of the additional trials. The six possibilities, five of which were tried, are summarized in Table 13.2. The entries in the table give the rank ordering of suppressive effect of the tone alone, as shown in a subsequent test. The group that received two tone–light trials with an extra shock on the first trial showed most suppression by the tone (rank 1). Next came the two groups that received two additional trials, but no additional shock on the first tone–light trial – evidently the added shock on the second tone–light trial had no immediate effect. Last were the two groups that received only one tone–light trial.

Table 13.2. *Rank-order of conditioning for five groups in an experiment by Mackintosh, Bygrave, and Picton (1977)*

		Neither	1	2
Number of tone-light trials	1	3	3	—
	2	2	1	2

Trial when added shock occurred (column header spanning Neither, 1, 2)

The added shock on the second tone–light trial evidently had *no direct effect*: The suppression after one trial was the same whether a shock was added or not (groups 1-neither and 1-1 in Table 13.2 showed equal suppression of rank order 3); and suppression after two trials was the same whether a shock was added or not, providing it was added only to the second trial (groups 2-neither and 2-2 were equal, rank 2). But the added shock did serve to create a candidate, the tone, that could be confirmed on a subsequent trial. Thus, the group that received *two* tone–light trials with the added shock occurring on the *first* trial (group 2-1, rank 1) showed the greatest suppression by the tone.

There are some puzzles left by this rather elaborate experiment, however. Does effectiveness of the added shock on the first trial depend upon a general increase in attentiveness that carries over to the second trial and causes the animal to attend to the presence of the tone on that trial, or is it really a retrospective or "backward-scanning" effect that causes the animal to recall the presence of the tone on the first trial (as suggested by the hedgehog experiment)? In the latter case, the second trial only serves to confirm a prior "hypothesis." A sixth group, in which the tone is omitted on the first added-shock trial but not the second, is necessary to answer this question.

Learning is initiated by violation of expectation – *surprise*; but the change that subsequently occurs depends on Bayesian factors. For example, blocking can be abolished by either increasing or decreasing the strength of the US. Recall, blocking refers to the outcome of a two-part experiment: In the first phase, a conditioned stimulus, A, is paired with an unconditioned stimulus (usually electric shock, in the CER paradigm) until it has acquired maximal power to suppress lever pressing. In the second phase, a neutral stimulus, X, is presented along with A, and the compound, AX, is paired with the US as before. X acquires little power to suppress responding, presumably because the occurrence of shock is perfectly predicted by A. If the strength of the US (the size of the shock or number of shocks) is either increased or decreased on the first AX trial, however, then X acquires some suppressive power. Presumably either of these changes violates the animals' expectations, based upon previous A → US pairings, so that some learning can now occur.

Consider the case where the US is incremented. The new learning could take any of three forms: an increment in the suppressive power of A; an increment in the suppressive power of X; an increment in both. There are comparable alternatives for US decrement. Yet in both cases, all the change (which is always an increment in suppression) takes place in the *novel* stimulus, X. This is perfectly Bayesian: The significance of A has already been established; a US change coincident with the introduction of a novel stimulus, X, is most plausibly attributed to X. Since shock continues to occur, and X does not suppress at first, the change is always an increase (Dickinson, Hall, & Mackintosh, 1976).

This separation between the surprise that initiates learning and the Bayesian processes that determine its direction can have unexpected implications. In one experiment (Hall & Pearce, 1979; see also Pearce & Hall, 1980) three groups of rats received standard CER training. One group $(T \rightarrow s)$ received pairings of a tone with weak electric shock; a second group $(L \rightarrow s)$ received pairings of a light and weak shock. The third group $(T \rightarrow 0)$ just heard the tone. In the second phase of the experiment, all three groups received pairings of the tone with a strong shock. The question is: Which group will learn most rapidly to suppress lever pressing in the presence of the tone?

The second phase is surprising for all groups, because of the increase in shock intensity, so all will learn something. But the tone is novel only for the light group $(L \rightarrow s)$; hence, these animals should learn fastest to suppress to the tone. The tone has some prior significance for both the tone-alone $(T \rightarrow 0)$ and tone-weak shock $(T \rightarrow s)$ groups. Hence, one might expect less rapid conditioning to them. Moreover, the tone-alone group has had an opportunity to learn that the tone signifies nothing, whereas the tone–weak-shock group has at least learned that the tone predicts something about shock. Hence the expected order of conditioning is $L \rightarrow s > T \rightarrow s > T \rightarrow 0$, which is what occurred. This result is puzzling at first, because the group with prior experience of tone–weak-shock pairings learned about the contingency between the tone and the strong shock more slowly than a group with no prior tone–shock experience. Yet it follows directly from Bayesian inference: A change in the US is more plausibly attributed to a novel stimulus (about which nothing is known) than to one with a prior history, even a prior history of association with (weak) shock. In the next chapter I describe a way of representing prior probabilities that makes these predictions in a more cut-and-dried way than this purely verbal argument.

Our techniques for answering subtle questions about when particular things are learned in the course of classical-conditioning experiments obviously leave something to be desired. Even quite simple questions, such as whether the effect of a surprising event (e.g., added shock in the Mackintosh et al. study) is immediate or delayed, lead at once to large experiments with many groups of animals. Yet as large as they are, there never seem to be enough groups to answer all the questions that arise. Moreover, it is not always clear that individual animals all behave as the group does. Nonetheless, the role of surprise, the importance of the first trial of any change, and the occurrence of processing both before and after the presentation of a US, are amply demonstrated.[22]

Expectation and classification

Learning evidently depends on surprise; and surprise implies a discrepancy between experience and the animal's representation of the world. To understand learning, therefore, we need to understand something about how animals represent things.

I argued in Chapter 10 that animals learn about objects and situations, not disembodied dimensions or features. One way to look at an object is as a stable set of properties or attributes. For example, dog Fido has a certain size, sex, color, temperament, physical location, number of legs, and so on. Some of these attributes are more variable than others (location and size are more variable than number of legs and sex, for example), and some are more *differential* than others: All dogs have four legs, but only Fido has his particular coat color and pattern of likes and dislikes.

But which of these features *really* defines Fido? This problem has exercised philosophers ever since Plato raised the problem of universals (What makes a table a table? Is a chair being used as a table really a table?). Yet I believe there is a pretty good answer to it, at least within the limited confines of animal behavior and artificial intelligence. It is captured by the epistemological aphorism: "If it walks like a duck, quacks like a duck, and lays eggs like a duck, it must be a duck." In other words, an object is defined by its attribute set. In a moment I will show one way to represent a set of attributes so that objects are easily differentiated from one another.

If an animal can distinguish just ten grades for any property, and it can assess ten different properties, then it is potentially capable of uniquely identifying 10^{10} different objects. This is a very large number; to store information about so many things would require much memory and be highly inefficient. Most of the potential entries would be blanks (i.e., the individual will never have encountered objects with just that set of properties), and there is no need to discriminate even among all nonblank entries because many items will be functionally equivalent.

The memory load can be substantially reduced if the animal resorts to *classification*: Fido may be unique, but there is no need to distinguish very carefully among the myriad other dogs with which one comes into contact. Certainly little purpose would be served by remembering every one of them individually. Moreover, as we have already seen, not all attributes are equally valid identifiers: Fido's location varies from minute to minute; it's almost useless as an identifier. His temperament varies from day to day over a narrower range; it's a bit better. Fido always has four legs, but so do many other dogs. On the other hand, an object with only two legs is very unlikely to be Fido. Fido answers to the name "Fido"; rather few other dogs do so. And Fido knows and loves his master, a quality shared by few other organisms, perhaps. These two attributes are obviously good identifiers.

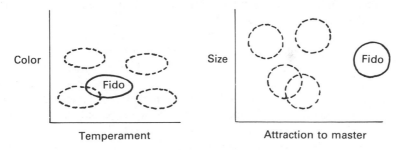

Figure 13.5. Two-dimensional slices through the regions of attribute (semantic) space containing the representation of dog Fido. Left panel: a slice parallel to the color-temperament plane. Right panel: a slice parallel to the size-"attraction to master" plane.

All these properties can be captured very simply in a multidimensional Euclidean representation of the sort discussed in Chapter 10. We can get the basic idea just by taking two-dimensional "slices" like those in Figure 13.5. The left panel shows a slice along the two dimensions of color and temperament (of course, color is itself multidimensional, but this is just an illustration). The cigar-shaped *region* corresponds to the representation of "Fido"; the dashed regions correspond to other dogs. The fact that the cigars are narrower in the vertical direction indicates that color is less variable (within individual dogs) than temperament. The fact that dashed regions are equally scattered through the space surrounding Fido indicates that neither Fido's color nor his temperament is unique.

The right panel in Figure 13.5 shows a slice through the space along the dimensions of size and "attraction to master." Here Fido is closer to being unique: Although there are many other dogs of similar size, none rivals him in affection for Fido's master. But notice that this method of representation can differentiate even among objects that are not unique on any single dimension.

The fact that objects are represented as regions (rather than points) takes account both of the variability of attributes and of the animal's limited storage capacity. Alternatively, one can imagine the two-dimensional slices in Figure 13.5 as being actually divided into discrete cells (like the memory cells of a computer), so that objects falling within the same cell are classified as equivalent. Or one can imagine the animal viewing the space with a metaphorical "mind's eye" of limited acuity. The point is that no two objects can be closer than some minimum distance without being treated as equivalent.

Learning and classification. As I pointed out in Chapter 10, the dimensions of semantic spaces such as those in Figure 13.5 do not correspond precisely to physical dimensions, although there is often a loose relation. One dimension, often the most salient, is completely nonphysical: good–bad. When the objects to be represented have any hedonic significance at all (animals or odors, for example), human similarity-scaling experiments always yield a space in which *value* is one of the major dimensions.

Value is something that must usually be learned.[23] Classical conditioning can be regarded as a procedure that leads an animal to assign a positive or negative value to an initially neutral stimulus. Unconditioned stimuli such as food, water, and electric shock, of course, differ in more than one way (food has a positive value for a hungry animal, as water does for a thirsty one, but their effects are not identical; see Chapter 7), and this difference can be accommodated within a multidimensional semantic space. Whether represented on one dimension or several, value is obviously an important, perhaps the most important, attribute of any object; and it is an attribute generally (but not invariably) acquired through experience.

A spatial representation has several advantages. It provides us with a potentially precise meaning for the term *expectation*, and it suggests that stimulus dimensions should be linked during learning. Bee learning again provides a simple example.

Learning about a particular food source corresponds to locating a *vector* of attributes $A(c, s, p, t, l \ldots)$, where c, s, p, t, and l represent the color, scent, pattern, time, and location of the object, along the value (v) dimension. In other words (this model suggests), each object is represented by the bee as a point, $A(c, s, p, t, l, \ldots, v)$ in N-dimensional space, where v is its value. Each learning trial with a given, constant, food source gives the bee additional confidence that an object with these coordinates exists. One can think of the first learning trial as creating a single point corresponding to the object $A(c, s, p, t, l, \ldots, v)$ in the appropriate region of the animal's memory. Each succeeding trial adds additional points until finally it reaches the level at which the animal is certain that a food source, with such and such properties, has a permanent existence.

A concrete model is as follows: Old information is less valuable than new information. This fading with time was represented by the perspective metaphor, and trace decay, in the previous chapter. But since I want something easy to visualize, let's just say that the point representing each valued event is like a spot of ink, the more valuable the original event, the denser the original ink spot. The ink diffuses at different rates along the different dimensions. The density of the ink at any time then represents the net value of that object (region of semantic space) at a given time. The spread along the odor axis will be slow, since odor is learned fast; spread along the time-of-day axis will be more rapid, as time of day is learned more slowly. Thus, after passage of time, the animal will still be able to recall the odor of the object, but its color, time of day and other variable attributes will be less certain.

However it is expressed, the rule that the *rate at which attributes are forgotten is inversely related to their importance* to the animal is a general one. Attributes of high a priori, that is, innate, validity are learned rapidly and forgotten slowly.

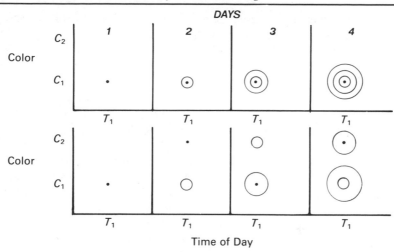

Figure 13.6. Diffusion model of object learning. Each encounter with object (T_1,C_1) or (T_1,C_2) adds a diffusing spot at that point in attribute space. Top panel: lines of equal concentration after 1–4 days of training with both attributes constant. Bottom panel: lines of equal concentration after 1–4 days of training with rewarded color alternating each day.

After several learning trials, an equilibrium between the rates of diffusion and the rate of replenishment (conditioning) will be reached, and the animal will be said to have learned about the food source.

In a system of this sort, the rate at which learning proceeds will depend upon the variability of the object; learning about one attribute will depend upon the variability of all attributes. For example, if the time of day at which the bees are fed remains constant, but the color varies from day to day, then the learning of *both* color *and* time of day is retarded. This is because a change in *any* dimension moves the point representing the learning to a new region in the space, that is, changes the object. Unless the new region is adjacent to the old, there will be no cumulative effect.

This dimensional interaction is illustrated in Figure 13.6. The top row of panels shows the appearance of the space after successive trials in which both color and time of day are constant. Each concentric ring represents an increment of strength. Thus on day 1, a point is created in the space at coordinate $(T_1C_1,)$. After the second conditioning day, it has diffused to the ring shown in panel 2, and a second dot is added. By the fourth day, there is a dot and three diffusion rings at that point. The bottom panels show the comparable changes when colors alternate from day to day, but time of day remains constant. Clearly, after four days, the concentration at the point (T_1,C_1) after consistent reinforcement (top panels) is much greater than the concentration at either point after inconsistent reinforcement (bottom panels). Granted that the bee's behavior is determined by the regions of highest concentration (it seeks the most promising sites) the accuracy with which it chooses either the color *or the time of day* in the variable case must be less than its accuracy in the constant case.

Bees show precisely this kind of linkage between attributes. For example, in one experiment (Bogdany, 1978) a colony of honey bees was trained to feed on sugar water at an artificial feeder some 400 m from the hive. At the feeder site, 16 color cards, each with a feeder in the center, were arranged in a square. Only one feeder contained sugar water. Two conditions were compared: both color and time of day constant; and time constant, color variable.

In both conditions, the bees were trained for five days. In the both-constant condition, the bees were rewarded for picking out the color blue (the other 15 cards were different shades of gray) between 11:00 and 12:00 hr each day. On day 6, the feeding station was set up and observed until 18 hr, but all 16 feeders were empty. Approaches to the blue card only were counted as correct.

The results are shown in the top panel of Figure 13.7. The bees were very accurate: Ninety-six percent of approaches (white squares) were correct; only 4% (gray squares) were not to the blue square. The animals' timing was also very good: Almost all the approaches were within 15 min of the 1-hr training "window" (11:00–12:00 hr).

In the time-constant, color-variable condition, the full feeder was on a different color card each of the five training days. The bottom panel of Figure 13.7 shows the test results. Color accuracy must obviously be poor here, and is not shown. Time accuracy is also much impaired, however. As the figure shows, the majority of approaches were outside the training window.

Other conditions of this experiment showed similar linkages between time and odor and between all three attributes, time, color, and odor.

I am not aware of any experiments of precisely this sort with birds or mammals. A few years ago, there was much interest in a related hypothesis: that attention of pigeons or rats to one stimulus dimension (measured as the slope of the generalization gradient; see Chapter 10) might be *inversely* related to attention to another. The results of this work were inconclusive; some people found the expected relation, whereas others found the opposite (see Mackintosh, 1977, for a review). The present argument suggests that one might well expect a positive relation if the two dimensions could be integrated as a single object (a color and a pattern located on the same response key, for example); whereas the expected negative relation might be found if the two dimensions are not integrable in this way (a tone and a spatially separated light, for example).

The effect of pairing a valuable outcome with a complex stimulus obviously depends critically on how the stimulus is represented by the animal. For example, suppose we pair a loud tone and dim light with shock and then later test the tone and light separately for their suppressive effects. You know the result: The loud tone will overshadow the dim light – the light will have little or no effect. But that's because tones and lights do not form integral combinations: "Tone–light" isn't in any sense an object for a rat. It is otherwise with tastes and odors: Garcia and his students (1981) have shown that if rats are made sick after experiencing a

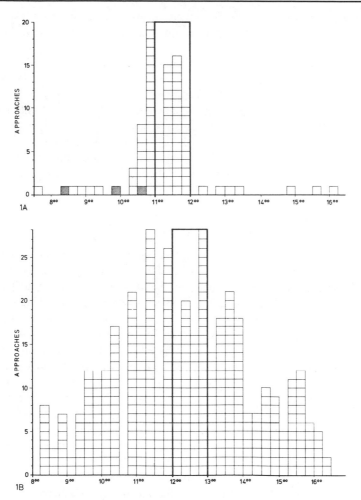

Figure 13.7. Top panel: number of bee approaches to an artificial feeder at different times of day after training with constant color and fixed time (11:00–12:00 hr). Bottom panel: Number of bee approaches to the feeder at different times of day after training at a fixed time, but with variable feeder color. (From Bogdany, 1978.)

weak odor and a strong taste, both odor and taste are later avoided. In another experiment, Garcia's group showed that hawks can't learn to avoid mice of a particular color when poisoned after eating them. But if black mice (say) taste different from white, black mice are violently rejected. In terms of my ink model, a given taste and odor, or taste and mouse color, seem to represent a single region in semantic space, whereas a tone and a light represent two separate regions.

Different species seem to differ in the dimensions that combine integrally or separably; indeed, as I argued earlier, much of what we term "species differences" seems to boil down to differences in the way animals represent their worlds – and these differences, in turn, no doubt reflect their different niches.

Recognition and expectations. I am arguing that the sensible attributes of an object, its color, location, odor, and so on, identify a region in the animal's semantic space. If that region has a known value, the animal can be said to *recognize* the object as a familiar one. This value constitutes the animal's *expectation* about the object. If, on a particular occasion, the appropriate positive or negative consequences fail to occur, there will be a discrepancy between the stored location of the object on the value axis, and the location indicated by the new experience. It is this discrepancy – violation of expectation – that will tend to change the animal's representation of the object, that is, produce new learning.

SUMMARY

Obviously a great deal remains to be understood about the acquisition of knowledge through conditioning. My purpose in elaborating this speculative scheme is to clarify the kinds of questions that must be answered before we arrive at full understanding. Terms like "memory," "expectancy," and "concept" have gained currency partly in reaction against the mindless stimulus-response psychology that dominated the study of animal learning until recently. However, these terms do not go far enough – and may mislead by encouraging a reliance on empathy with the animal. The terms must be given an exact, mechanical meaning if they are to be of much scientific use. We need to know both the form in which the world is represented by the animal, and the rules by which experience changes its representation. We must also recognize that these two things may not be separate. In simple animals, the rules of action may be the only form of representation the animal has. But in mammals and birds – and bees – there does seem to be a point in separating the two.

Classification based on some kind of semantic space looks like a good working hypothesis in answer to the representation question. The facts of song learning and imprinting fit into this kind of scheme, as does much of what we know about conditioning. In the song learning of white crowns, for example, a region of semantic space appears to be reserved for conspecific song. Anything within the boundaries of that region is accepted and is stored as the object or template by which later song will be guided. Songs lying outside the region are rejected. The imprinting object is similarly constrained to a broad region within semantic space.

The rules for changing the representation are Bayesian, with priors initially determined genetically. However, since a discrepancy between the current representation and experience is what drives any change, these priors also change with experience.

In this chapter I have dealt largely with learning that takes place on one or a few trials, and I have devoted little space to the action guided by that learning.

The discussions of bee learning do not conflict with the emphasis on classical, rather than instrumental, conditioning, because bee foraging seems to be rigidly tied to what they have learned of the value of food sources. For a bee, to know a good flower is to seek it out: The action does not have to be learned, only its object. We will see a few other such rigid links between knowledge and action in the next chapter, as well as cases where behavior is not forced by the nature of valued objects. I shall also there consider in more detail the problem of sampling and the nature of multitrial learning.

NOTES

1. Life span is obviously not the only factor that determines a species' ability to learn; its *niche* is also critical. Galapagos tortoises, for all their charm, are not great intellectual performers. Presumably their combination of large size, longevity, and stupidity is well suited to a stable and relatively undemanding environment. Size is also not critical. All ethologists are familiar with the "law" that no matter how clever the trick learned by any given mammalian species, behavior of comparable difficulty can be demonstrated in bees. The ways in which bees cope with their limited computational capacities will be discussed later. Nevertheless, other things being equal, longevity and large body size conduce to the evolution of learning ability.

2. The term *precocial* refers to the relative maturity of the birds on hatching, and distinguishes newly hatched chickens, ducks, and the like from newly hatched songbirds, pigeons, and doves. The young of *altricial* species like these are utterly helpless at birth. As one might expect, these differences relate to the species' niche: Mobile species with primitive nests tend to have precocial young; species that provide a more protected early environment have altricial young. There are comparable differences in mammals: The young of grazing animals, such as deer, guinea pigs, and antelope, "hit the ground running," whereas young rats, dogs, and cats are blind and helpless at birth.

3. Much of the work on imprinting has been done by German animal behaviorists, although the very earliest report seems to be by Spalding (1873). The original German work is that of Heinroth (1911) and Lorenz (1935; reprinted and translated in Lorenz, 1970). Some of this work is not yet translated into English: for example, the studies on shrews by Zippelius (1972), and the work on ducks by Schutz (1971; see note 5). The studies of auditory imprinting in ducklings are those of Gottlieb (e.g., 1971), and the work on maternal attachment in macaque monkeys is by Harlow (e.g., Harlow & Harlow, 1965). The studies of imprinting in goats are by Klopfer and Gamble (1966). Hess (e.g., 1973) has contributed much of the recent knowledge about imprinting in ducklings. A good general review of imprinting is Bateson (1974); Immelman (1972) has discussed the relation between filial and sexual imprinting, arguing that the two are largely independent.

4. This may not always be true, in the sense that for some species there may be imprinting stimuli more effective than the natural one. Such *supernormal stimuli* (Tinbergen, 1948) are not uncommon even when no learning at all is required. For example, male silver-washed fritillary butterflies (*Argynnis paphia*) are stimulated to court females by the flickering movement of their wings, and this response requires no learning. The natural frequency of flicker is about 8–10 Hz, yet in laboratory experiments Magnus (1958) found that male butterflies would court more vigorously to stimuli flickering at up to 140 Hz – frequencies much higher than anything encountered in the animals' natural environment. There are many other examples: birds preferentially retrieving extra large eggs, gull chicks preferring to peck at artificial bills that are longer, thinner and more brightly colored than their parents', and so on.

Many if not all of these instances seem to be explicable in terms of *asymmetrical selection pressure* (see Staddon, 1975b). Thus, the wings of predatory birds flicker at a rate lower than the flicker rate of the female fritillary, so that an error in the downward direction may cost a male butterfly dearly. Higher frequencies are never encountered, however, so that an error in this direction exerts no selection pressure. There is an obvious resemblance, therefore, between the conditions that produce peak shift in an operant-conditioning experiment (see Chapter 11) and those that produce supernormality in nature.

There is every reason to expect that the selection pressures that favor certain kinds of imprinting object are of exactly the same sort as those that produce innate reactions to stimuli. Hence we may expect to see instances of supernormality in both cases.

5. The mallard experiment was by Schutz (1971), described in Fantino and Logan (1979, p. 366). It is interesting that the selection by females of their proper mates was apparently *independent* of rearing condition: The females always chose males of their own species. There are two kinds of explanation that can be offered for this difference. One relates to the sexual dimorphism of ducks, and the bright plumage of the males: Perhaps the females simply have an easier job because the males of different species look so different. Since the task is easier, perhaps it makes more sense for it to be "prewired." I'm not so sure, given the earlier arguments (Chapter 1) against "simplicity" as a criterion for innateness: Some pretty complicated behavior patterns are innate, and some pretty simple ones depend heavily on learning.

A second possibility was foreshadowed in Chapter 1. The argument rests on two premises: (a) That the selectivity of all imprinting implies that species recognition, even in these cases, was at one time innate and independent of particular experience, and (b) that the reproductive cost to a male of making a species identification "error" (in its choice of target for copulation) is much lower than the cost for a female. The greater cost to the female is a consequence of the greater reproductive cost of egg-production as compared to sperm production, and the lesser parental investment of males. If hybrids have lowered fitness, then a female duck inseminated by a male of another species is committed to producing a lower-fitness offspring and suffers a delay (an *opportunity cost* in the language of economics) in her ability to produce additional high-fitness (nonhybrid) offspring. The greater fitness cost of error for females suggests that their mechanism for mate choice should leave less to chance. If innateness of response is the ancestral condition, strong selection for female selectivity would tend to preserve the innate process in females, but might permit the degeneration represented by imprinting in the case of the males of species where the male plays little or no parental role.

This premium on selectivity by females may also account for the distinctive plumage of the males, since it pays a male to identify itself as strongly as possible as *not* belonging to a species different from the female with which it is attempting to copulate. Perhaps the variety and brilliance of plumage and other accoutrements in males of many bird species is a consequence of the selection pressure on them to look as different as possible. The need for camouflage, *crypsis*, tends to produce convergence in the drab appearance of females of different species sharing the same habitat.

Several other mechanisms have been suggested in recent years to show that differential reproductive investment provides a basis for Darwin's (1871) suggestion that *sexual selection* is responsible for most cases of sexual dimorphism (see reviews in Krebs & Davies, 1978).

Not all birds are sexually dimorphic: Male and female geese, doves and pigeons, for example, are essentially identical, and unspectacular, in appearance. Cross-fostering experiments have shown that in doves females as well as males learn their species identity through early experience, but the altricial nature of these birds and the long rearing period make this learning less prone to error than the imprinting of precocial ducks and chickens. The symmetry here goes along with the symmetry of parental investment: Doves form permanent pair bonds and both parents are equally involved in the incubation and feeding of young.

6. Perception of flicker and perception of movement are closely related at the neural level, hence the behavioral equivalence of flicker and movement in imprinting studies.

7. Mulligan (1966). Kroodsma (1977) has discussed more recent evidence on the development of song-sparrow song.

8. The now-classical work by Marler and his associates on song development in white-crowns, and the earlier, related work by Thorpe on chaffinches, are widely reviewed. Good summaries appear in Hinde (1969), Thielcke (1976), Kroodsma (1978), and Catchpole (1979).

9. The term *passerine* refers to the *Passeriformes* or "perching birds," the order that contains most of the songbirds. Finches, swallows, tits, sparrows, and thrushes are all passerines.

10. A number of evolutionary hypotheses have been offered for the enormous variety of birdsong. Several identify song variety as a weapon in territorial defense. The most colorful of these is the "Beau Geste" hypothesis, which likens the defending bird to Beau Geste, of French Foreign Legion fame. Legionnaire Geste singlehandedly defended a fort, in which all but he had perished, by propping up the corpses in the battlements, and running from one to another firing each man's weapon himself. By singing different songs, from different places, a defending bird may convey the impression that there are many residents in a given area, thus deterring potential intruders more effectively than he could if he appeared as one. The evolutionary reasons for the form and diversity of birdsong are still a matter for active debate. See reviews in Catchpole (1979), and Krebs and Davies (1978).

11. The literature on bee behavior is large and much of it is in German. For a clear, readable summary of what is known of their mental life see Gould and Gould (1981), which also contains a number of other references. Gould's ethology text (1982) also has an excellent account of bee behavior.

12. In many early works on learning, it was customary to express wonder at the faculty of *memory*, and at the fact that animals as lowly as insects could possess it. The idea that learning is a continuous process, to be measured by smooth curves, may derive from this impression that the act of storage itself is a demanding one. Now that electronic gadgets such as tape recorders and computers are commonplace, the relative triviality of the processes involved in reversible storage of information should be apparent. The task for behavioral analysis is not to explain memory as such, but to understand the organizational and computational properties of organisms. It is not learning (in the sense of a mere effect of past experience) that is the problem (see Revusky, 1977), but rather the elementary rules and codes by which animals master complex tasks. I return to this theme later in connection with theories of classical conditioning.

13. *Latent learning*. The controversy about latent learning arose from the assertions of the early learning theorist Edward Tolman about maze learning. Tolman believed, quite reasonably, that rats learn about mazes as an automatic consequence of exploring them, which they will do without explicit reward. This learning was termed *latent*, but could be demonstrated by comparing groups of animals given different kinds of maze experience.

For example, in one famous study (Tolman & Honzik, 1930) three groups of rats were trained daily in a complicated maze with a single goal box. For one group, there was always food in the goal box; these were animals rewarded for learning the maze. For a second group, there was never any food in the goal box. For the third group, there was no food for the first 10 days, but food was introduced on the 11th day and thereafter. Obviously the no-food group would show no progressive improvement: With no food in the goal-box, there was nothing to differentiate it from the terminations of other alleys, and no reason the animals should enter it sooner or with fewer "errors" (blind-alley entries) than any other place in the maze. The critical comparison was between the food group and the group fed only after day 10: Would the group fed after the 10th day learn

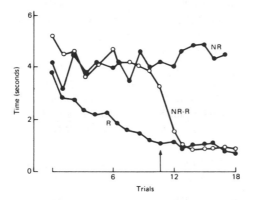

Figure 13.8. An early demonstration of latent learning in a maze. The curves show the average time to reach the goal box each day (trial) for three groups of rats. Control group NR was never fed in the maze; control group R always had food in the goal box. The experimental group, NR-R, had food in the goal box only after day 10. (From Tolman & Honzik, 1930.)

at the same rate as the food group (measured from day 1), or more rapidly, indicating some (latent) learning on the preceding days without food?

The results are shown in Figure 13.8. As expected, the no-reward group (NR) shows no improvement over days in time to reach the goal, and the rewarded group (R) shows a progressive improvement. The group rewarded only after day 10 (NR-R) shows an improvement after food is introduced that is much more rapid than the original improvement of the food group, which seemed to confirm Tolman's latent learning hypothesis.

This result might seem to settle the matter, but it did not. As often happens when terms are poorly defined, counterexplanations are offered, numerous experiments are carried out, and finally the issue either dies of boredom, or becomes changed into a question that can be answered experimentally. Opponents of Tolman's position pointed out that rats take time to *habituate* to a new maze. Initially they may be scared and the behavior associated with this fear may interfere with maze learning. Although (these critics continued) the NR-R group was learning nothing about the maze before food was introduced, it was getting over these initial, interfering response tendencies. Once reward was introduced the animals appeared to learn rapidly because their learning was *unimpeded*, not because they had already learned anything. These response tendencies could not be accurately measured, nor could their supposed interfering effect be objectively assessed. Nevertheless, all agreed that rats do behave differently on first exposure to a new maze, so this objection could not easily be ruled out.

Other critics, whose errors were of a more subtle sort, conceded the learning of the NR-R group, but pointed out that food could not be considered the only reward in the maze. Perhaps exploration has its own rewards ("seeing a new alley" may be intrinsically rewarding), so that the apparently unrewarded latent learning was not unrewarded at all. Thus free to invoke new rewards at will, Tolman's opponents rested their case, secure in a theoretical position incapable of disproof.

Later experiments, with much simpler mazes, showed that indeed rats do learn something about the configuration of a maze in advance of food reward. For example, in an experiment very similar in design to the bee-learning study, Seward (1949) allowed a hungry rat to explore a T-maze with two distinctive empty goal boxes. Then he placed the rat in one goal box with some food and allowed it to eat. When returned to the beginning of the T, rats so trained almost invariably selected the arm leading to the rewarded goal box.

For reviews of the literature on latent learning see Osgood (1953) and Mackintosh (1974).

14. This problem has been tackled in connection with the so-called "two-arm bandit" problem by Krebs, Kacelnik, and Taylor (1978). In their experiment great tits (*Parus major*) chose between two variable-ratio schedules with different mean values. The problem for the birds was to decide which of the two schedules paid off with the higher probability. This setup is formally similar to two Las Vegas-style one-armed bandit machines, hence the label. The birds sampled both choices for a while and then settled almost exclusively for one choice. Moreover, the longer the experimental session, the longer the birds sampled before settling down, just as theory suggests they should: the longer the session, the more time available to exploit the best choice, hence the greater the return on time invested in sampling. Even in this simple situation, theoretical analysis is quite tricky.

It is not yet clear exactly *how* the birds were able to devise their optimal strategy, but it seems likely that interactions between experimental sessions, according to the memory principles discussed in Chapter 12, may play a large role.

The optimal-sampling problem also turns up in connection with the problem of resistance to *extinction*: How long should an animal persist in a learned response when reward is no longer forthcoming? The so-called *partial reinforcement effect* – greater persistence following intermittent reward – obviously reflects the optimal policy of persisting longer when reward is uncertain (see Chapter 14). McNamara and Houston (1980) have tackled this difficult theoretical problem.

15. *Bayes' rule*. The term *Bayesian* refers to the Reverend Bayes, an eighteenth-century English clergyman and inventor of Bayes' rule, which allows the estimation of a *posterior probability* for any hypothesis, given its *prior probability* and some new data. The rule can be derived simply by rewriting basic probability identities, as follows: Let H be the hypothesis that some event is the cause of an observed outcome E. Let the probability that H is true be $p(H)$, and the probability that event E occurs under some defined set of circumstances be $P(E)$. Then if H and E are independent the probability of the joint event EH (i.e., both H is true and E occurs) can be written in two ways, in terms either of $P(E)$ and the conditional probability of E given H, $p(E|H)$, or in terms of $P(H)$ and the conditional probability of H given E, $p(H|E)$:

$$p(EH) = p(E)p(H|E),$$

or

$$P(EH) = p(H)P(E|H).$$

Eliminating $p(EH)$ and rearranging yields

$$p(H|E) = p(H)p(E|H)/p(E),$$

which is Bayes' rule. So much is tautology; things get tricky when we try to use Bayes' rule to infer the true value of $p(H|E)$, the probability H is true given E, from some estimate of $p(H)$ in the absence of E, an estimate of $p(E)$, and knowledge of the probability of E given that H is true, $p(E|H)$. The problem is that we cannot really estimate $p(H)$, the *prior probability* of H; nor do we really know how to compute $p(E)$ in the abstract. The rule works perfectly when we are dealing with a well-defined procedure (see, for example, Staddon, Hinson, & Kram, 1981, for a Bayesian analysis of some choice procedures). It is essentially unusable as a way of checking out real hypotheses in open-ended situations. I believe that the rule may be more valid as applied to animals' judgments, however, because the experience of an animal's ancestors potentially provides it with estimates for $p(H)$, $p(E|H)$, and $p(E)$.

Much of probability theory is mathematically trivial, but conceptually very difficult, however, so it is hard to say anything very conclusive on these points as yet.

The term *Bayesian inference* is broadly applied to any method of inference that takes account of prior information in its evaluation of new information. It is possible to show that Bayesian inference will usually lead to better decisions than any method that completely ignores prior information. As I show in this chapter and the next, animals certainly seem to use prior information, their own, and what we can infer of the genetically coded experience of their ancestors, in this way.

16. There will always be a tradeoff between misses and false alarms for any given level of accuracy (see the discussion of ROC analysis in Chapter 9). For any given level of accuracy (d'), the animal can pick an optimal criterion, which will yield a certain net benefit. However, the terms of the tradeoff can obviously be improved by additional sampling: the larger the number of trials, the more accurately the probabilities in Table 13.1 can be estimated. Natural selection will tend to produce an equality between the marginal costs of additional sampling and the marginal benefit of the increase in accuracy so obtained.

17. I make these assertions with a wary confidence based on general knowledge of insect behavior, but, in fact, the necessary experiments seem not to have been done.

18. The original experiments on long-delay taste aversion were done by John Garcia (e.g., Garcia, Clarke, & Hankins, 1973; Garcia, Kimeldorf, & Koelling, 1955). As with many novel observations, the phenomenon was discovered by accident, in this case as a by-product of applied research on the effects of gamma radiation (which causes nausea, among other disagreeable effects) on rats. The phenomenon had been previously known to rat catchers in the form of *bait-shyness* – the avoidance of novel baits and the consequent ineffectiveness of poisons associated with them. For a recent review of taste-aversion learning see Domjan (1980).

19. *Delay-of-reward gradient.* The causal status of events is reflected in the *delay-of-reward gradient*, that is, the function relating probability of conditioning to the delay between CS and US (see Staddon, 1973). For the standard classical-conditioning procedures this function peaks at about .5 sec, and declines thereafter (see Landauer, 1969), depending on the CR and other procedural features. Presumably the peak reflects the fact that causes take some time to produce their effects, so that the optimal delay is greater than zero. This need for delay is particularly clear in the case of poison-avoidance, since there will always be a substantial delay between ingestion of poison and the onset of its effects. Correspondingly the allowable CS–US interval is longer than usual. One would also expect that there should be a peak in the delay-of-reward function at a time considerably longer than .5 sec, but as far as I am aware this has not been firmly established.

20. Wagner, Rudy, and Whitlow (1973); see also Kamin (1969). Rabbits, cats, and many other mammals have a nictitating membrane behind their eyelids. The US for movement of the membrane is a localized electric shock applied at the periphery of the eye. The response of this membrane is readily conditioned to auditory, visual, and vibrotactual stimuli. The preparation is widely used in the study of classical conditioning.

21. Note that *some* learning will certainly take place on the first blocking trial, but it need not take the form of an association between the US and the novel stimulus. For example, it may be possible to show experimentally that after the first such trial, the animal is *less* likely to associate the new stimulus with either shock or its absence than if the stimulus had not been presented. The effect of the new stimulus will probably be different when presented as part of a compound than when presented alone. In fact Mackintosh, Bygrave, and Picton (1977) have shown that the novel element does acquire some suppressive power on the first blocking trial, but no further increments are added on subsequent trials.

In this section I am treating very briefly a large and highly technical literature on classical conditioning, in which the phenomena of blocking and overshadowing play a large part. The concepts of *surprise* (violation of expectation) and *expectancy* are also central to theoretical accounts of this work. For good reviews of current knowledge see Dickinson (1980), Dickinson and Mackintosh (1978), Rescorla (1975), and Mackintosh (1974). I return to these topics in the section on mathematical models for conditioning in note 22.

22. *Mathematical models for conditioning.* A number of mathematical models attempt to capture in some formalism the joint effects of *surprise* and *Bayesian inference* on conditioning. The effects are joint in the sense that the animal's expectation must be violated before any learning can occur.

The simplest way to represent the joint dependency is to assume that learning (i.e., a *change* in the power of a stimulus to affect behavior – usually termed *associative value*) is proportional to the *product* of two functions, one representing the surprise aspect, the other the inference aspect:

$$\Delta V = f(\text{surprise}) \cdot g(\text{inference}),$$

where ΔV is the change in associative value (V is almost invariably measured in terms of the *suppression ratio* in the CER experiment; see Chapter 5). If f is a monotonic function passing through the origin, then when surprise $= 0$, ΔV must be zero, which conforms to the axiom that surprise is necessary for learning.

Perhaps the simplest assumption we can make about surprise is that it corresponds to a discrepancy between the asymptotic associative value, λ, a CS would have after consistent pairing with a US and its actual value, V:

$$f(\text{surprise}) = \lambda - V. \tag{N13.1}$$

If the CS is a compound one, parsimony suggests that V is just the sum of the associative values of the elements: $V = \Sigma V_i$. For two elements, this yields

$$f(\text{surprise}) = \lambda - V_1 - V_2. \tag{N13.2}$$

The simplest inference assumption is that all stimuli are equally likely candidates for conditioning, except that some are more *salient* (more intense, more noticeable) than others, and so condition faster. This implies that the inference function is a constant which is different for different stimuli; whence

$$g(\text{inference}) = a_i, \tag{N13.3}$$

for the *i*th CS.

Combining Equations N13.2 and N13.3 yields

$$\Delta V_i = a_i(\lambda - V_1 - V_2), \tag{N13.4}$$

where ΔV_i is now the increment in associative strength of *one* of the elements of the compound stimulus. Equation N13.4 is the well-known model proposed by Rescorla and Wagner (1972). The effects of extinction (omission of the US) and US intensity are handled by setting λ equal to zero or different positive values; variations in stimulus properties are handled by the subscripted parameter a; and the different rates of learning associated with different levels of reinforcement are handled by means of a multiplicative parameter b, so that the final equation is

$$\Delta V_i = a_i b_j (\lambda_j - \Sigma V_i), \tag{N13.5}$$

where ΔV_i is the sum of the associative strengths of all stimulus elements.

Equation N13.5 is a direct descendant of the linear-operator learning model of Estes, discussed briefly in Chapter 8. The essential difference is the assumption that learning (ΔV) about each stimulus element is proportional to the difference between the asymptotic level (λ) and the *sum* of associative values of the stimulus elements. A less essential difference is the change from probability to associative value as the dependent measure; this change allows V to take on negative values (corresponding to inhibitory strength) and values greater than 1.

This model accommodates both blocking and overshadowing very well. In a compound made up of elements with high and low salience, when learning is complete ($\Delta V = 0$) $V_1 + V_2$ will equal λ, but most of the associative strength will have gone to the more salient stimulus: In fact, it is easy to show that at equilibrium, $V_1/(V_1 + V_2) = a_1/(a_1 + a_2)$.

Blocking is accommodated by the fact that after the first phase, in which S_1 alone is paired with the US, V_1 must approximately equal λ. Since V_2 is initially zero, $\lambda - V_1 - V_2$ is close to zero when S_1 and S_2 are presented together at the beginning of the second phase. Consequently, little or no further change can take place in either V_1 or V_2 in the second phase. Thus, the conditioning of S_2 is blocked by the prior conditioning of S_1.

The model also accounts for the results of Rescorla's "truly random control" procedure, discussed in Chapter 5. Single-stimulus conditioning can be considered as compound conditioning in which *contextual cues* (the physical environment of the test apparatus) constitute the other elements. In "true discrimination" where the US is paired only with the CS (stimulus 1), the CS and the contextual cues (stimulus 2) receive the same number of increments in associative value during conditioning, but the context also receives a substantial number of decrements (see Figure 5.8). For example, if the CS is present half the time, and a brief shock occurs on the average once a minute, then in an hour there will be 30 increments in the strength of both CS and background stimuli. However, there will be 30 missed opportunities for shock in the absence of the CS but in the presence of the context, which will cause reductions in the strength of the context. It is easy to show that this process will result in eventual exclusive control by the CS.

In the truly random control condition (sometimes called *pseudodiscrimination*), however, where the US is equally likely in the presence or absence of the CS, the CS continues to receive 30 increments in every hour, but the context now receives twice as many. Moreover, pairings of the US and context in the absence of the CS lead to larger increments in context strength than pairings of the US with the CS-context compound (because the surprise term is $\lambda - V_1 - V_2$ in the compound case, but $\lambda - V_2$ in the context-only case). This leads eventually to predominant control by the context and no conditioning (zero associative value) to the CS.

The Rescorla–Wagner equations, therefore, constitute a mechanism by which animals might detect correlations between conditioned and unconditioned stimuli. The process is computationally simple because the animal need keep track only of the values of V_i; it needs no memory for the specific conditioning history associated with each stimulus element.

Some otherwise-surprising experimental results are predicted by this model. For example, if S_2 has been established as a safety signal, its value must be negative so as to counteract the positive value of contextual cues. At the beginning of compound conditioning, therefore, the net value of the term $\lambda - V_1 - V_2$ is positive, rather than close to zero, as in the usual blocking arrangement. Consequently, positive increments will at first be added to both V_1 (which is already positive) and V_2. Since V_1 is already close to λ in value, this procedure leads at first to greater-than-maximal conditioning of S_1 – the *superconditioning* discussed in the text. This outcome depends on the assumption that "surprise" is given by the discrepancy between asymptotic associative value, λ, and the sum of associative values of all stimulus elements – the last term in Equation N13.5.

The assumption that surprise depends on the difference between λ and the sum of associative values leads to other surprising predictions. For example, consider an experiment with two compound CSs sharing a common element: S_{13} is a compound of elements 1 and 3; S_{23} is a compound of elements 2 and 3. Suppose that S_{13} and S_{23} are now presented equally often and S_{13} is consistently reinforced but S_{23} is not. This situation can be compared with a pseudodiscrimination in which S_{13} and S_{23} are both reinforced half the time. The reinforcement schedule for 3, the common element, is identical in both cases, yet the model correctly predicts that it will acquire much less strength in the true discrimination than the pseudodiscrimination.

In a related experiment, Rescorla (1976) used training with two overlapping compounds as a way to enhance the strength of one compound above its asymptotic level. In phase 1, compound S_{13} was conditioned in the normal way. The model predicts that at asymptote, the two elements, 1 and 3, will have strengths related to their saliences. In the second phase, element 3 is combined with a new element 2 and S_{23} is conditioned in the usual way. Since the associative value of 2 is initially zero, ΣV $(= V_2 + V_3)$ will be less than λ at the beginning of the second phase, allowing element 3 to gain strength. The model correctly predicts that in a subsequent test of the 1-3 compound, S_{13}, the added strength of element 3 should cause the total strength of S_{13} to rise above its normal asymptotic level.

The surprising feature of this experiment is that Rescorla did not use *explicit* compounds, like a tone and a light, but an *implicit* compound: He assumed that a high- and a low-frequency tone can be considered as compounds differing in frequency, but sharing other properties. Hence he was able to explain why asymptotic suppression to a high tone could be enhanced by subsequent training with a low tone. (A similar prediction follows from the discussion of integral stimuli at the end of the chapter.)

The Rescorla–Wagner model does a great deal with a very simple formalism. Nevertheless, it fails to account for some basic facts about conditioning. For example, blocking can be eliminated by a *decrease*, as well as an increase in the expected level of shock; and the new CS element gains positive associative value in both cases (Dickinson, Hall, & Mackintosh, 1976). The model predicts that the new CS element should gain positive value only if the change is an increase. The model predicts that a safety signal will have negative associative strength. Consequently, continued presentation of a safety signal in the absence of any US (a form of extinction) should lead eventually to a loss in its potency as its associative strength increases to zero, the asymptote for extinction. Yet safety signals retain their potency when presented without the US.

Most damaging is the finding that simple preexposure (presenting a stimulus without any consequence) to a CS can retard conditioning (*latent inhibition*). Since preexposure cannot change the associative value from zero (according to the model) there is no way to accommodate this result within the model.

The experiment by Hall and Pearce (1979), discussed in the text, in which pretraining with weak shock *retarded* conditioning to a strong shock, is also impossible to explain by the Rescorla–Wagner model.

These effects are comprehensible in Bayesian terms. For example, prior experience with a stimulus in the absence of shock should make it less likely that the stimulus has anything to do with shock. Hence subsequent conditioning should be retarded (latent inhibition). A safety signal presented without shock violates no expectation; hence no learning (no change in its properties) should occur.

These limitations have prompted a number of suggestions for improving the original model to allow it to deal with these exceptions without losing its ability to handle blocking and related effects. Since the "surprise" term in Equation N13.5 is essential for the prediction of blocking, most modifications concern the "inference" term.

An obvious possibility is to allow the stimulus saliences, a_i, to vary as a function of experience; this can accommodate the unique lability of novel stimuli (e.g., the special ease with which they become associated with changed conditions of reinforcement, and the retarded conditioning caused by simple preexposure: latent inhibition). For example, Mackintosh (1975) and Wagner (1978) have elaborated on a suggestion made by Rescorla and Wagner (1972) that stimulus salience may decrease with continued exposure to a stimulus. Wagner proposes that continued presentation of a CS in given context leads to associations between CS and context, and that the associability of the CS with a US is inversely related to the strength of these contextual associations: the more predictable the CS, the less the value of a.

Mackintosh proposed that the value of a_i decreases as stimulus i becomes a better predictor of the US. He also includes an explicit competition assumption (which is implicit in the surprise term of the Rescorla–Wagner model), proposing that increments in a occur only if stimulus i is the least predictive stimulus in the situation. This explicit competition assumption permits Mackintosh to give up the Rescorla–Wagner surprise term in favor of the original Estes version: $\Delta V_i = a_i(\lambda - V_i)$.

Most recently, Pearce and Hall (1980) have proposed that the predictive value of each stimulus is assessed independently (rather than all together, as in the Rescorla–Wagner model). Nothing further is learned about stimuli that predict their consequences accurately. Their model retains the surprise × inference form of Equation N13.1 yielding

$$\Delta V_i(n) = S_i\lambda(n)|\lambda(n-1) - V_i(n-1)|, \qquad (N13.6)$$

where $V_i(n)$ is the associative value of stimulus i on trial n, $\lambda(n)$ is the intensity of reinforcement on trial n and S_i is intensity of CS-element i. In words, this model says that learning is proportional to the absolute value (the term in $||$) of the discrepancy between the conditioning asymptote on the preceding trial and the associative value of the stimulus element on that trial (the surprise term), multiplied by the product of stimulus and US intensity (the inference term). The surprise term embodies the presumption that animals encode each individual CS in terms of the consequences that it predicts. The Rescorla–Wagner model, of course, only encodes the US. The Pearce–Hall model asks: Are the consequences of this CS accurately predicted? whereas the Rescorla–Wagner model asks: Was this US accurately predicted?

These modifications of the Rescorla–Wagner model are able to handle most of the discrepant facts I have described. The most effective so far seems to be the Pearce–Hall approach. For example, latent inhibition is explained as follows: When a novel stimulus is presented without a reinforcer, λ will be 0; since the associative value of the stimulus is also 0, the term in $||$ is zero and the future associability of the stimulus will soon be very low. When the stimulus is first paired with a US, no learning will occur on the first conditioning trial. Hence learning will be retarded relative to conditioning where the novel stimulus appears for the first time on the first conditioning trial. The same argument accounts for the retardation of conditioning to a strong shock by pretraining with a weak shock: on the first strong-shock trial, the term in $||$ will be zero, so that no learning will occur on that trial. In addition, the model makes the counterintuitive, but correct, prediction that some interpolated *extinction* after weak-shock training will facilitate subsequent training with the strong shock – because the extinction causes a discrepancy between $\lambda(n-1)$ and $V(n-1)$ in Equation N13.6, which allows learning to occur on the first strong-shock trial. (In colloquial terms, occasional extinctions teach the animal that it cannot trust to its earlier assessment of the significance of the CS.)

None of these modifications of the Rescorla–Wagner equations has the mathematical simplicity of the original – both inequalities (Mackintosh) and absolute-value expressions (Pearce–Hall) prevent simple solution of the finite-difference equations. And none deals with special relations between CS and US, as in taste-aversion learning, nor do they deal adequately with trial-by-trial changes or the different effects to be expected when stimuli are integral or separable. All assume a smooth learning process for which the evidence (in individual animals) is almost entirely negative. They account well for the effects of transitions between conditions, and for asymptotic levels of conditioning. They do not provide an accurate description of the real dynamics of learning.

In all the experiments dealt with by these models, the temporal relation between CS and US is standardized. For example, in the CER paradigm, shocks are paired (i.e., concurrent with) CS presentations. Yet we have already seen that some "conditioning" can take place even when the CS occurs after the US (Keith-Lucas & Guttman, 1975). If the CS and US are properly chosen, moreover, conditioning can occur over long delays, and some CSs can become preferentially conditioned over others that are (in terms of relative frequency of pairing) better predictors of the US. These CER-based models deal only with *relative frequency* of pairing between CS and US and *salience* (of CS and US) as the procedural features that determine conditioning, even though other experimental arrangements show that there are special relations between USs and CSs (as in taste-aversion learning). It seems clear that animals weigh several factors in deciding which of a number of candidate stimuli is the likely cause of a hedonic event such as food or electric shock: the type of stimulus in relation to the type of US, other properties of the CS such as its intensity, the time relation between CS and US, and the animal's prior history with both CS and US. Models of the Rescorla–Wagner type provide elegant summaries of the effects of frequency of pairing. But we need to know much more about how the physical events in conditioning experiments are represented by animals before all these other factors can be accurately incorporated.

23. It is generally assumed that unconditioned stimuli such as food, water and electric shock have innately determined values, but this is not always true. For example, chicks need experience to learn that seeds are food (have positive value). Hogan (1973) exposed young chicks to piles of sand and seeds. They pecked equally often at both. Then one group was offered just sand. Given a choice an hour later, they still showed no preference for seeds. However, another group, offered just seeds after the initial experience with both, when given the choice after an hour strongly preferred the seeds. The birds evidently could not learn that sand is *not* food, but given appropriate experience could learn that seeds *are* food. A further experiment in which the chicks were intubated with a high-calorie diet after eating sand showed that the preference for seeds in the first experiment was due to the nourishment derived from them, not to any special stimulus properties.

Since the development of this preference depends on the subsequent beneficial effect, a delay between the seeds-only exposure and the subsequent test is essential – presumably to allow time for some digestion and absorption to occur.

It makes great adaptive sense that chicks must learn the value of seeds, a motionless, inanimate stimulus similar to many nonnutritive objects in the environment. As one might expect, they require no learning to take live mealworms (although at first they show some conflicting fear responses to large and active worms). They almost immediately show a preference for mealworms over either seeds or sand.

14

LEARNING, II: THE GUIDANCE OF ACTION

The last two chapters showed how behavior in a range of situations, from fixed-interval schedules, through the radial maze to conditioning experiments with rats and bees, appears to be guided by the animal's representation of its world, together with principles of Bayesian inference. Studies of classical conditioning in the preceding chapter showed how these inference principles, in turn, allow the animal to incorporate into its representation something about the causal structure of the environment. In this chapter, I say more about the kinds of inference that seem to underlie both operant and classical conditioning and show how inference leads to action. The middle part of the chapter presents a view of learning as a circular process in which surprise and novelty cause an animal to update its representation of the situation, which in turn leads to new activity, hence to a new situation, perhaps more surprises, and so on in a spiral that usually converges on an adaptive pattern. The chapter ends with a discussion of several standard phenomena: conditioned reinforcement, conditioned suppression, avoidance and escape, set, and extinction. I begin with a brief account of the historical background to the contemporary study of operant learning. This account introduces some important empirical results and provides a context for the rather different approach to operant learning I propose later.

HISTORICAL BACKGROUND: OPERANT AND RESPONDENT BEHAVIOR

Several years ago a very simple experiment proved quite a shock to workers in animal learning. Brown and Jenkins (1968) presented hungry pigeons in a Skinner box with an 8-sec stimulus (a colored light) on the single response key at intervals of a minute or so. At the end of the 8 sec, the food hopper came up for 4 sec. Food delivery was independent of the animal's behavior; the pigeon could neither prevent nor aid operation of the feeder. The procedure is obviously Pavlovian – a CS (the key light) is reliably paired with a US (food) – even though the bird and the box are more usually associated with operant conditioning. The surprising feature of the experiment was not the procedure but the results. After as few as 10 light–food pairings, the pigeons all began to peck the lighted response key and continued to do so indefinitely thereafter.

Brown and Jenkins called this effect *autoshaping*; they used it instead of training by successive approximations, that is, *handshaping* (more on this later) to teach experimentally naive pigeons to peck a response key. I described autoshaping earlier, in Chapter 11, to illustrate how stimuli that predict something of value gain control over behavior.

Autoshaping is interesting and convenient but, one might think, hardly anything to ruffle the conceptual feathers of old-tyme behavior theorists. At first, most people assumed that autoshaping could be explained along the following lines: A hungry pigeon is in a boring little box with an occasional colored light flashing on and off and intermittent brief access to food – why not peck at the light? There isn't much else to do, and maybe the light has something to do with the food? Pecks, once they begin to occur, are reliably followed by food, so perhaps they are in some way "superstitiously" reinforced by the food.[1] This account doesn't really explain the very first peck – but Skinner's notion that operant behavior is "emitted" seemed to take care of that problem (but see note 9). This explanation seemed to resolve what had appeared to be a paradox: a classical-conditioning situation producing pecking, the most widely studied operant response. Before this plausible account had time to exert its calming effect, however, a second experiment, by Williams and Williams (1969) effectively demolished it.

Williams and Williams made a small change in the procedure, turning it from a Pavlovian into an operant one: On trials when the pigeon did peck the key, food did *not* occur and the key light was turned off. If the pigeon failed to peck on a trial, the key light stayed on and food arrived as usual after 8 sec. This is termed *omission training*, and it is a traditional control procedure in classical-conditioning experiments. Clearly it pays the pigeon *not* to peck under these conditions, since pecking prevents food delivery. If, as a multitude of previous experiments seemed to have shown, pecking really is an operant response (i.e., one whose rate is determined by its consequences), then key pecking should cease. It did not. The pigeons pecked less often than before, but then they were also getting fewer light–food pairings (because food was omitted on trials when the animal pecked). In any event, the purely "operant" nature of key pecking was seriously called into question.[2]

Autoshaping is not unique to pigeons and key pecking. Subsequent experiments under appropriate conditions have shown autoshaping with rats, monkeys, and even people, and responses such as lever pressing, nosing, and others previously used in standard operant-conditioning arrangements.[3]

To understand the impact of these experiments it helps to know something about the historical context. For many years it was thought that classical- and operant-conditioning arrangements exerted their effects on different types of behavior. The Pavlovian procedure was thought to be uniquely effective on behavior controlled by the autonomic nervous system, such as salivation, changes in skin resistance and heart rate, pupillary dilation, and so on. Skinner termed this *respondent* conditioning. He and many others distinguished it from *operant*

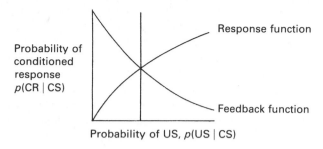

Figure 14.1. Theoretical analysis of partial reinforcement of a pure respondent under an omission schedule.

conditioning, which is the result of a closed-loop arrangement in which reinforcement depends upon the occurrence of a response – and which is effective only on skeletal muscle movements. The operant response is not elicited but *emitted*, said Skinner, and stimuli were said to "set the occasion for" responses made in their presence.

Respondents such as salivation were thought to be insensitive to purely operant contingencies. For example, if food is *omitted* on each trial when the animal salivates in the presence of the CS, he salivates anyway, the amount determined by the actual frequency with which food occurs in the presence of the CS. The theoretical picture is shown in Figure 14.1: The declining function is the feedback function for the *omission procedure*, in which the occurrence of the conditioned response (salivation) prevents occurrence of the US. The function is declining because the omission procedure is an avoidance (food-postponement) schedule (see Chapter 5), since the probability of reinforcement (food) is inversely related to the frequency of the conditioned response. The increasing, negatively accelerated function is the response function for salivation, which is presumed to increase as the probability of CS-US pairing increases: the higher the probability of food in the presence of the CS, the higher the probability of the conditioned response. The intersection of these two curves gives the equilibrium frequency of salivation.

This outcome is very different from what we would expect of an operant response (see Chapter 7). Granted that the preferred frequency of food is rather high, the usual operant response should cease to occur under a food-avoidance schedule, so that the rate of food delivery can be as high as possible.

The analysis in Figure 14.1 can be tested using the yoked-control procedure described in Chapter 5. The experimental animal should salivate about as often as a control animal that receives the same CS-US pairings but is not subject to the negative feedback of the omission procedure. (The feedback function for the yoked animal is indicated by the vertical line through the intersection point in the figure.) This is generally true in practice for salivation; the yoked animal salivates no more than the omission animal. Salivation shows only the *restriction* effect of the omission contingency, there is no *contingent* (reinforcing) effect (see Chapter 7). Autoshaped pecking seems to be an intermediate case: The yoked animal responds slightly faster than the omission animal (this difference is the contingent effect), but the omission animal continues to lose some food by responding to the key light. Evidently pecking has both operant and respondent properties – a most unsatisfactory conclusion, since the usefulness of the distinction depends on common activities falling entirely into one category or the other.

Obviously some skeletal responses are sensitive to Pavlovian contingencies. There is less evidence for crossovers in the other direction: operant conditioning of autonomic activities. The proposition that heartrate can be modified directly by operant means is difficult to test because changes in skeletal responses often have correlated autonomic effects. For example, changes in breathing pattern can have effects on heartrate; a person can change his heartrate and skin resistance by imagining (an operant, voluntary activity) a frightening scene, and so on.[4]

One way to test the direct reinforcibility of autonomic effects is to eliminate skeletal muscle activity by injecting a derivative of the paralytic drug curare (since the curarized animal is conscious, this would not eliminate "imagining," but rats were not thought capable of such subtleties). For example, an experiment by Miller and Banuazizi (1968) used curarized rats with electrical brain stimulation as the operant reinforcer. Heartrate changes or intestinal contraction or relaxation were the reinforced, autonomic responses for different groups of animals. Heartrate increased when the brain stimulation depended upon increases, and decreased when stimulation depended on decreases; intestinal motility showed similar effects. Unfortunately, the drastic curare treatment is subject to side effects that make positive results hard to interpret. Several of the dramatic early experiments of Miller and his associates have turned out to be seriously flawed. Workers in Miller's own laboratory have not been able to repeat some of these early results.[5]

Thus, autonomic responses do not seem to be directly affected by operant procedures – nor should we expect them to be, given their evolutionary function. Heartrate, for example, under normal circumstances never has any *direct* effect on the external environment; nor do skin resistance, intestinal motility, pupillary dilation, or any of the other autonomic responses used in Pavlovian experiments. There can have been no selection pressure favoring direct sensitivity of response systems such as these to external contingencies. Autonomic activities are concerned with regulating the internal, not external economy of the organism.

The internal economy is, of course, indirectly involved in operant behavior. "Fight or flight" reactions to external threat will be more efficient if the body is suitably prepared in advance: The animal will fight more efficiently if its heart and breathing rates are high, so the blood is well charged with oxygen, if its digestive and other vegetative functions are suppressed, releasing metabolic resources for the struggle, if its pupils are dilated for better vision, and if its body hair is erect ready to shed the heat from impending exertions. It makes perfect sense, therefore, that these functions should all be anticipatory and guided by the predictiveness of situational cues, as they almost invariably are (see also note 6). Comparable changes take place in anticipation of food, water, and other hedonic events.

The conclusion is that under the proper conditions operant responses such as key pecking and lever pressing are sensitive to purely Pavlovian contingencies, but autonomic activities are probably insensitive to operant contingencies. I argue in a moment that operant contingencies act partly through the Pavlovian contingencies that are inseparable from them; sensitivity to Pavlovian contingencies is an essential part of the reinforcibility of operant behavior.

Evolutionary function is a better guide to operant reinforcibility than the autonomic–skeletal distinction. For example, grooming (a skeletal activity) in hamsters is almost unaffected by food reinforcement (Shettleworth, 1978; see also Chapter 7). No doubt tail wagging in dogs would be hard to condition on a shock schedule, as would smiling in humans.

If the operant–respondent distinction doesn't hold water, we are left still with the problem of understanding the relations between classical- and operant-conditioning procedures. In Chapter 4 I proposed that classical conditioning sets the stage for operant conditioning by allowing the animal to identify the kind of situation he is in, to classify it. In the last chapter, I argued that classical conditioning is an inference process: Situations and objects are assigned values depending on the animal's prior knowledge and the temporal relations prescribed by the conditioning procedure. In the next section, I argue that the same principles – knowledge representation plus Bayesian inference – apply also to operant conditioning.

BEHAVIORAL VARIATION. THE ORIGINS OF OPERANT BEHAVIOR

A standing criticism of purely cognitive views of behavior is that they imply no action: Tolman's cognitive-map idea was criticized because it left the rat sitting in the maze "lost in thought" – the rat knew where everything was, but had no reason to go anyplace. I argue in this section that the same inference processes that allow the animal to identify potential conditioned stimuli can also permit him to identify potential operant responses. The rat knows not only what is, but also what should be done about it.

Representation is the key concept in the previous chapter. The function of Pavlovian learning seems to be the development by the animal of an accurate model of its world. Further learning is then driven by a *discrepancy* between the model and the actuality, that is, by surprise or violation of expectation. What is the role of *action* in all this?

Action forms part of the animal's representation. As we saw at the end of the last chapter, for a bee, to know a good flower is to seek it out. Certain kinds of situation automatically entail certain kinds of action. All Pavlovian conditioning carries implications for action, which were often concealed from Pavlov and his successors by the restrictions they imposed on their animals. Pavlov's dogs were restrained in a special harness so that their salivation could be measured easily. Yet salivation is probably the least important aspect of what they learned. A visitor to Pavlov's laboratory reports the reaction when a dog previously trained to salivate at the sound of a metronome was released: The animal at once approached the metronome, wagged its tail at it, barked, and behaved in every way as it might toward an individual who could feed it. Clearly the major effect of conditioning was to place the metronome in a certain category for the dog; the category then determined appropriate action. Whether the dog also salivated is not reported.[6]

Autoshaping is thus only a special case of the built-in tendency for all animals to approach objects of positive value and withdraw from objects of negative value (in Chapter 7 I argued that approach and withdrawal, in turn, just reflect the preferred level for different things: high for things approached, low for things avoided). The inference mechanisms of Pavlovian conditioning allow the animal to detect which objects have value. This value is sufficient by itself to produce approach and withdrawal as well as other, specialized patterns such as autoshaped pecking or lever pressing, flight or fight, "freezing," and other situation- and species-specific reactions. I provide some more specific examples in a moment.

What determines which action shall occur in a given situation? The rules are essentially the same as those discussed in the previous chapter. I show here how to represent the inference problem in the contingency space first discussed in Chapter 5, and then extend the same principles to operant conditioning.

Inference and classical conditioning

The occurrence of an hedonic event (i.e., a US), such as food or shock, poses a Bayesian problem for the animal: Which environmental feature, of those present now and in the recent past, is the most likely cause of the US? The animal has two kinds of information available: prior knowledge, and temporal relations between the valued event (US) and its potential cause (CS – or operant response). Let's look at how the animal might use both kinds of information:

Prior knowledge is of two kinds: innate priors, and previously formed representations. The animal's evolutionary history provides it with predispositions to connect certain kinds of event with certain kinds of outcome: Pain is more likely to be associated with an organism than with an inanimate stimulus – recall the hedgehog experiment. A related finding is *shock-induced fighting*: A pair of rats placed in a cage and briefly shocked through their feet will usually attack one another. Pain, in the presence of a probable cause, elicits the behavior that is preprogrammed for that situation, namely, attack directed at the cause (Azrin, Hutchinson, & Hake, 1967; see Morse & Kelleher, 1977, for a review). Both the quality and intensity of the US will determine the likely cause.

If the situation is familiar, or is similar (in terms of the animal's representation – see Chapter 10 for a discussion of similarity) to a familiar one, the probable cause may be partly determinable. For example, a rat shocked in the presence of

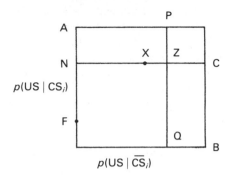

Figure 14.2. Representation of Bayesian inference in a contingency space. Points F, N, X, and Z represent different prior conditions discussed in the text.

two others, one familiar and the other not, is more likely to attack the novel animal. We have seen a similar effect in taste-aversion learning: Given two potential taste CSs, an aversion develops just to the novel one. Blocking is the converse effect: Given a CS-element already associated with the US, negligible conditioning takes place to the novel element.

The way in which prior knowledge guides future conditioning can be represented in the contingency space encountered first in Chapter 5, and shown again as Figure 14.2. Given the US, the animal has for each potential CS *prior estimates* of the probability of the US given that CS ($p(US|CS_i)$) and the probability of the US in the absence of the CS ($p(US|\overline{CS}_i)$). Complete conditioning (the consequence of a history of perfect prediction of the US by the CS) is indicated by point A in the figure, where $p(US|\overline{CS}_i) = 0$ and $p(US|CS_i) = 1$. Any other situation represents something less than complete positive conditioning. Complete conditioning of a *safety signal*, a CS−, is indicated by point B in the figure.

The ease with which a particular stimulus (or response, as we will see in a moment) can become conditioned as a CS+ should obviously depend on its distance from point A in the space. However, the conditionability of one stimulus is not independent of the conditionability of others. As we saw in the last chapter, cues compete with one another. This is implicit in most models of conditioning, and explicit in the Rescorla–Wagner model, one of the most influential (see Chapter 13, note 22). The function of conditioning is to determine action, which means the animal must generally decide which of two things is better, or worse – a conclusion that both are equal is not helpful as a guide to choice. Thus, ease of conditioning should depend on the distance of a cue from point A in the space *relative to the distances from A to other cues.*

Let's look at some of the cases described in the last chapter from this point of view:

Blocking. The already-conditioned element lies at point A, whereas the novel, unconditioned stimulus, X, lies at point X on line NC, which represents the prior probability (innately determined if this is an entirely novel stimulus) that this stimulus is associated with the US. The horizontal coordinate of X is determined by the animal's *preexposure* to the US, since it is the probability of the US in the absence of the novel CS. Since the animal has already had several conditioning trials, $p(US|\bar{X})$ is quite high. The vertical coordinate of X is set by prior experience, and innate factors that may link some classes of stimuli (e.g., tastes) to some classes of USs (e.g., sickness). Since stimulus A in Figure 14.2 is already perfectly conditioned and X is a long way from the point of perfect conditioning, the conditionability of X is obviously low.

Conditioning to novel tastes. If a familiar and a novel taste both precede poisoning, aversion develops to the novel taste. The familiar taste, F, can be represented by point F in Figure 14.2. $p(US|F)$ is low, and $p(US|\bar{F}) = 0$ if this is the animal's first exposure to sickness. The novel taste, N, can be represented by point N since its (innate) prior association with sickness is presumed high – or at least higher than for F. Clearly, point N is closer to A, complete conditioning, than point F, so that aversion develops to the novel taste, N.

Latent inhibition. Preexposure of the CS in the absence of the US retards conditioning. If a novel CS is located at N in the contingency space, then preexposure without the US must displace its location to some lower point on the vertical axis, say F, since experience of the CS in the absence of the US should reduce the animal's estimate of $p(US|CS)$. Consequently, CS preexposure should retard conditioning.

Preexposure to the US. This is a case not previously considered that also fits into this scheme. Given prior exposure to the US in the absence of the CS, the situation at the outset of conditioning can be represented by the intersection of lines *NC*, representing $p(US|CS)$ for the novel CS, and *PQ*, representing the above-zero probability of the US in the absence of the CS. The intersection of these lines at point Z is obviously more distant from point A than point N, the comparable intersection in the absence of US preexposure. Hence we might expect US preexposure to retard conditioning, as it usually does.

    Thus the contingency space provides a useful summary for the effects of novelty and preexposure on CS conditionability.

    Temporal relations between a potential CS and the US are the second source of information available to the animal in making its causal inference. As we saw in the last two chapters, and in Chapter 5 (Figure 5.13), relative proximity to the US is important, with better conditioning if the CS closely precedes the US than if it closely follows the US. The general form of the function relating CS-US delay (for a given US-CS time, that is, intertrial interval) and the inferential weighting given to the CS, the *delay-of-reinforcement gradient*,[7] is shown sche-

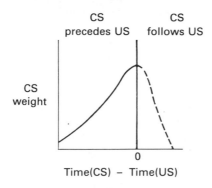

Figure 14.3. Schematic representation of the delay-of-reinforcement gradient.

matically in Figure 14.3. The figure summarizes two facts and a conjecture about the effect of CS-US delay: The optimal CS-US interval is slightly greater than zero; but otherwise, longer CS-US delays are less effective in producing conditioning. The part of the gradient to the right of zero is dashed, because I am uncertain of its meaning: Both forward and backward conditioning are possible, suggesting that there should be a gradient to the right of zero. On the other hand, it is not certain that backward conditioning ever persists beyond the first trial – this part of the gradient may not have a stable existence. Whether it does or not, it is clear that greater CS-US delays are tolerated in the forward direction, hence the steeper slope of the dashed portion.

As already mentioned, cues compete, so that other things being equal, of two stimuli both equally often paired with the US, the stimulus relatively higher up on the delay-of-reinforcement gradient is favored at the expense of the other.

As I pointed out earlier, the part of the delay-of-reinforcement gradient to the left of time zero (forward conditioning) probably reflects limitations on memory. Consequently, events such as food or shock (i.e., potential USs) that are easy to remember (see Chapter 12) should have more extended gradients than neutral events and be easier to condition over delays. And so it is: Trace conditioning, in which the CS occurs briefly and is followed by the US only after a delay (see Figures 5.6 and 5.13), fails at CS-US delays over a few seconds. But temporal conditioning, the periodic presentation of the US (the US also acts as a CS and the US-US interval corresponds to the CS-US interval in trace conditioning), is successful over delays of minutes or even hours.

Anything that makes an event easier to remember should facilitate learning. Increases in stimulus intensity or duration, elimination of sources of proactive and retroactive interference, should all help. We will see in a moment that *conditioned reinforcement* acts in this way to aid operant conditioning.

In summary, animals use prior information, either innate or based on previous experience, together with information about temporal relations between CS and US, to guide conditioning. *Candidates* for conditioning (i.e., potential causes) are identified either before or after US occurrence, as I described in the last chapter.[8] Identification of candidates is an active process – Pavlov termed the behavior associated with noticing and processing novel stimuli the *orienting reflex*. The orienting reflex and indeed all new learning seems to be motivated by surprise, a perceived discrepancy between reality and the animal's representation of it.

The details of the process by which candidates are selected, and the precise composition of the candidate set are not known. (It is even a simplification to talk of a candidate set as if it contained a number of independent elements, when what seems to be changing during conditioning is some form of representation in which stimulus elements are unlikely to be completely independent of one another). All we know for sure is that Pavlovian conditioning does involve selection from among a set of possibilities, that the selection is biased by innate factors and previous experience, and that the temporal relations between a putative CS and the US are important, although not necessarily decisive, determinants of the outcome.

Inference and operant conditioning. This scheme applies with little change to the learning of activities. The problem for the animal is still to decide which of several possibilities caused the US; the only difference is that now the possibilities are aspects of its own behavior. Figure 14.2 remains unchanged, except that the axes are now $p(US|\bar{R}_i)$ and $p(US|R_i)$, where R_i is the ith activity. The delay between response and reinforcer (US), and prior information about the response, both guide conditioning.

A complete, formal representation of operant conditioning therefore requires a four-dimensional contingency space, with two dimensions for the conditional probability of the US given the presence and absence of stimuli, and two more dimensions giving the conditional probability of the US given the presence and absence of responses.

The major difference between operant and classical conditioning is in the selection of candidates for conditioning. In the Pavlovian case, the candidates (potential CSs) in some sense already exist and merely have to be attended to by the animal. There is probably less to this difference than meets the eye, in the sense that external stimuli must be encoded by the animal before they become candidates: Stimuli and responses are both names for internal processes. The only real difference is that operant behavior has a feedback function: by doing something the animal may change his environment, but merely encoding something, without action, has no effect. Moreover, if there is a real dependency between reinforcer (i.e., the US) and behavior, the US will not occur unless the appropriate behavior occurs. This *priority of action* is the functional basis for spontaneous activity, which, as we saw in Chapter 4, underlies all operant behavior.

As in Pavlovian conditioning, there are priors associated with any candidate. For example, if food is given to a pigeon just after it has pecked and preened (in either order), it is much more likely to associate the food with the pecking than the preening, even though the preening is closer to the food. Priors contribute to the selection of one response over another as the probable cause of the US.

Priors can also be used to *originate* a response: This is the message of autoshaping. The CS in an autoshaping experiment is a signal for food; the pigeon comes equipped by its evolutionary and personal history with a repertoire of activities that have proven useful in food situations. These activities involve both the internal economy of the animal, and its relations with the external world. Activities of the first kind are the traditional province of Pavlovian conditioning. Autoshaping, and other examples I will describe in a moment, are examples of the second kind. Skeletal activities, such as pecking and a few others such as pacing and bill swiping, comprise a modest repertoire that has proven useful to ancestral pigeons in food situations. Consequently, pigeons have been designed by their evolution so that these activities are the first to occur in any situation where (by the animal's estimate) food is probable.

Behavior of this sort might be termed *situation-induced*. It forms the matrix out of which operants, as functional units with stimulus and response components, will emerge. Every situation generates its own range of activities, and from this set a single activity or a small number of activities will eventually emerge, over repeated trials, as *the* learned response. Induced behavior, in this sense, corresponds to Skinner's *emitted* behavior; the difference is that induced behavior is largely determined by the situation that produced it (not largely random, as the word "emission" implies), and the inducing factors continue to act and guide the process of operant conditioning. There may be no such thing as pure emitted behavior, occurring at random and free once it has occurred to become conditioned to any stimulus by any reinforcer.

Inducing situations are of two main types: the outcome of Pavlovian training, where the food-significance (or significance in terms of some other reinforcer) of a stimulus has been established by a prior predictive history, or simply the delivery of food. Because food often occurs in patches or "runs," finding one piece of food is often a signal for others (see the discussion of area-restricted search in Chapter 9). Many animals innately treat food as a signal for more food, therefore, so that food delivery (especially in a novel context) elicits the full range of food-related activities – a hungry chicken when given a little grain will at once become active, start scratching at the floor and pecking at small objects, for example. Or a rat receiving periodic food will spend much more time in food-related behavior following an occasional extra-large food portion – even if it gets the extra food every day for many days (Reid & Staddon, 1982). The unexpected occurrence of other hedonic stimuli, such as shock, water, or a conspecific, similarly arouses a set of specific activities.

It is hardly conceivable that animals could be constructed so as *not* to make use of prior information in this way, no matter how congenial such a tabula-rasa view might be to traditional stimulus-response theory.[9] Life in a changing environment (and the animal changes its environment merely by moving through it) constantly poses fresh problems. Finding food or a mate, and the avoidance of predators, are ever-present challenges. All problems can be thought of as a *search* through a set of potential solutions. Sometimes the set is very well defined, as in finding a source of food: The animal moves at such and such a speed in a uniform environment, can scan such and such an area per unit time, and the density of food sources is such and such. At other times, if the environment is not uniform, or if the type of response required to obtain food is not specified, the set of solutions is much harder to define. Animals have limited computational resources and limited time. Without some means of *limiting the search*, most such problems would take too long to solve. Animals can only reduce their problems to a reasonable level of difficulty by making use of prior knowledge – both individual and racial, ontogenetic and phylogenetic.

There are now numerous documented examples of the spontaneous emergence of food-related behaviors in food situations and fear-related behaviors in frightening situations. Some of the most dramatic come from attempts to train animals such as pigs and raccoons to do tricks. Marion and Keller Breland wrote a paper, ruefully entitled "The misbehavior of organisms,"[10] in which they described a number of failures to condition (by operant means) several "arbitrary" activities. For example, a raccoon was being trained to pick up a wooden egg and drop it down a chute. At first the animal easily released the egg and went at once to the feeder for its bite of food. But the second time the animal was reluctant to let go of the egg: "He kept it in his hand, pulled it back out, looked at it, fondled it, put it back in the chute, pulled it back out again, and so on; this went on for several seconds" (Breland & Breland, 1966, p. 67). Animals of other species – otters, pigs, and squirrel monkeys – are equally reluctant to let go of an object that has become a reliable predictor of food.

In another experiment, a chicken had been trained to make a chain of responses, leading at last to standing on a platform for 15 sec, when food was delivered. After training, the chicken showed vigorous ground scratching (an apparently innate food-seeking activity) while waiting on the platform.

The activity observed depends on all aspects of the situation. For example, in another experiment the Brelands trained a chicken to operate a firing mechanism which projected a ball at a target; a hit then produced food. All went well at first, but when the animal had learned the temporal association between ball and food, it could not refrain from pecking directly at the ball. Thus, when a chicken must stand and wait for food, it ground scratches, but when it has a foodlike object that signals food, it pecks at it. In mammals and birds information from several sources contributes to the selection of a response.

In situations associated with electric shock, animals show so-called species-specific defense reactions, that is, built-in reactions to potentially dangerous situations (Bolles, 1970). There are three main reactions to danger: flight, fight, or "freeze" (remain immobile). Which one is chosen depends in sensible ways on the magnitude and certainty of the threat, and the escape routes and cover available. For example, baby chicks, poor fighters at best, are likely to freeze in any threatening situation, unless cover is available, in which case they are likely to run to it. Rats will flee from a superior opponent, fight an inferior one, and freeze if the source of danger is uncertain.

The reactions are quite specific both to the type of hedonic stimulus and the other features of the situation. For example, even in the apparently simple pigeon-autoshaping situation, the type of US makes a difference. Careful observations show the form of the pigeon's peck differs depending on whether it expects food or water: Pigeons attack grain with an open beak, whereas they suck water with a closed beak. The autoshaped peck shows the appropriate correlations: The birds peck a food CS with an open beak, a water CS with a closed one. Other experiments have shown that types of food that require different handling also elicit matching autoshaped responses.

Behavioral variation and sampling

If a situation can be identified with precision and there is a unique solution, then evolution will tend to build it in to the organism – especially if the risks associated with rigidity are negligible and/or the costs of error are high.[11] Thus, species-identification is generally either innate or depends in a rigid way on early experience. Protective reflexes, as we saw Chapter 2, are almost impossible to modify.

What holds for phylogeny also goes for ontogeny. A situation in which the outcome is relatively certain and of high value (positive or negative), is not a situation that encourages variability in behavior. If the outcome is positive, then variation will incur a substantial loss in the form of positive outcomes foregone; if the outcome is negative (as on a shock-avoidance schedule) variations from the effective response will also incur a heavy loss in the form of negative outcomes encountered.

The rule that high-valued situations tend to produce stereotypy accounts for the two phases in the Brelands' experiments. For example, when the raccoon first learned to drop his egg down the chute, nothing much depended on it. But after a few tries, with food following every egg, the egg became a rich food predictor. The raccoon's genes tell him what to do with something that looks like food, is strongly associated with food, and can be handled like food: hang on to it! So he did; failed then to get more food; became less reluctant to let go of the egg; dropped a couple; got more food; the egg became valuable again; and so on. The rigidifying effect of high hedonic value also accounts for the feature-negative effect, which occurs when $S+$ is a rich predictor of food, not when it is a poorer predictor, and for omission autoshaping, which also occurs only when the stimulus is brief, hence a strong food predictor.

The same rule applies on reinforcement schedules. A hungry pigeon pecking a key for food on a relatively rich schedule should spend little time doing anything else: The food given up is certain, the prospect of a better alternative tenuous. Similarly, a well-trained squirrel monkey pressing a lever to avoid shock also has little incentive to explore: Time out from pressing will soon be punished, and long training has taught the animal the futility of looking for a better game in this situation.

This argument is really the same one we have already encountered in the form of the marginal-value theorem in Chapters 7, 8, and 9, except that the animal's choice here is not between staying or leaving a patch, but between allocating its time to one activity (the reinforced response – although this term needs some further definition) versus others – exploitation versus sampling. The general rule is: the "hotter" the situation, and the more experience the animal has with it, the less variable the behavior. Consequently, animals soon come to behave in rigid ways in situations of high hedonic value: on schedules involving electric shock or high food deprivation and frequent food, for example. We will see in a moment that this rigidity has some surprising and sometimes maladaptive consequences.

THE GUIDANCE OF ACTION

We are now in a position to pull together the threads of an account of operant conditioning. The account has three ingredients: (a) The *initial state* of the animal – this corresponds to the set of *candidates*, both external stimuli and aspects of the animal's own behavior, for predictors of the hedonic stimulus (the US or reinforcer); (b) the *selection rules* that determine how the set of candidates is changed; and (c) the *feedback function* that determines how the animal's behavior changes the external world, that is, the stimuli, both hedonic and otherwise that act on the animal. I will say something about each of these three ingredients, and then describe how they interact to produce both operant and classical conditioning. In the final section, I discuss some common learning phenomena in light of these principles.

Initial state

Situations vary in familiarity. If the situation is novel, then animals seem to have available a variety of built-in activities that serve the function of testing the situation in various ways. Small animals subject to predators will naturally be cautious at first. A rat, especially a feral rat, placed in a laboratory environment will at first hide, or at least remain immobile for some time. If nothing bad happens, the animal will probably explore, not entirely randomly, but according to rules that depend on the nature of the place. In a maze, as we have already seen (Chapter 12), rats will tend to avoid recently visited locations. They also have a fairly stereotyped routine elicited by novelty that includes rearing, sniffing, and nosing into holes (Kello, 1973). This activity is almost entirely endogenous and innate. If the environment contains another animal, then a repertoire of social "probes" is called into play that serves to establish the sex and status of the stranger.

These activities serve the function of exposing the animal to stimuli that allow it to build up an internal representation of the environment. As we have already seen, the experimental evidence for a maplike spatial representation is now quite

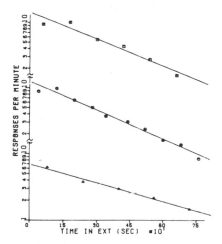

Figure 14.4. General activity as a function of time after a single, isolated feeding. Each curve is the average of several pigeons in similar experiments. The fitted straight lines correspond to the function $y = A \exp(-Bt)$, where A and B are constants. The y-axis is logarithmic. (From Killeen, Hanson, & Osborne, 1978.)

strong (Chapter 12). The behavior is also, of course, *guided* by this developing representation. Other things being equal, the animal is likely to seek out parts of the environment that are relatively less familiar, either because they have not been visited in some time or because they have never been visited.

At some point, perhaps as a consequence of action by the animal, perhaps for some other reason, something of value to the animal may occur – delivery of food or electric shock, or the appearance of a conspecific. The situation is now both more familiar (the animal has experience of other food situations, for example) and one where it pays the animal to invest additional efforts, both of thought (computation) and activity: "There is something good (or bad) here; I had better find out about it!"

This additional effort is termed *arousal*, and it can be measured both physiologically (via electrical measures of brain activity) and behaviorally. For example, in an experiment by Killeen and his associates (Killeen, Hanson, & Osborne, 1978; see also Killeen, 1982) a hungry pigeon was placed in a Skinner box instrumented so as to measure the amount of moving about by the bird. The floor was divided into six hinged plates, so that as the bird walked from plate to plate, switches were depressed and released. After about 30 min to settle down, the pigeon was given a single feeding and allowed to remain in the apparatus for a further 30 min. This procedure was repeated for several days. The records of switch operations for these days were then averaged to show how the amount of movement declined with postfood time. The results are strikingly regular. Food delivery produces an immediate increase in activity that declines exponentially with postfood time, as shown in Figure 14.4.

The regularity of these curves depends upon considerable averaging; the effect of a single food delivery on the behavior of an individual pigeon is by no means this predictable. This variability is not surprising. The functional argument tells us that the animal should "try harder" after food, but this effort involves both computation and activity. Hence activity is only an indirect measure of the total effect of the single feeding. There are also good reasons why both computation and activity should involve random elements. As we saw in earlier chapters, all adaptation involves selection from a variable set of possibilities. The arousal produced by food suggested that the animal is generating "candidate causes" for the food at a high rate that decreases as the food delivery recedes into the past.

Hedonic stimuli are not the only things that arouse animals. The food in Killeen's experiments was not delivered the instant the pigeon entered the apparatus because the first encounter each day with the Skinner box is itself arousing. Any situation reentered after a delay is to some degree novel and generates the exploratory thought and action appropriate to a novel situation. The animals were left in the box for an unpredictable time (so they could not anticipate when food would occur) long enough to allow the arousal elicited by the bird's initial encounter with the box each day to habituate.

The general rule is that arousal is elicited by situations of potential importance (in the evolutionary sense) to the animal. Such situations seem to be of two kinds: *novel* or *surprising* situations, and situations involving *hedonic stimuli* such as food, shock or a conspecific. Both surprise and novelty motivate learning (i.e.. changes in the animal's internal representation) which reduces surprise and novelty. Hence, their effects are self-limiting and the arousal they produce habituates with time. The distinctive thing about hedonic stimuli (reinforcers) is that their arousing effects do *not* habituate – although the *variety* of behavior decreases with experience.

Selection rules

If a reinforcer is delivered to an animal for the first time in a new situation, the animal becomes aroused: More behavior occurs and (we infer) the animal attends more to external stimuli. But these effects are selective, guided by the priors appropriate to the reinforcer: Certain kinds of stimuli will command more attention than others, certain kinds of action are more likely than others, and some actions are more likely than others to be considered as potential causes of the reinforcer.

As additional reinforcers occur, temporal relations – the delay-of-reinforcement gradient – also have an effect in reducing the set of candidates: candidates that are especially memorable, or closely precede the reinforcer being preferentially retained, according to the principles discussed in the previous chapter. Thus, at each instant the candidate set is determined by the priors and temporal relations associated with that situation. As further reinforcers occur, the candidate ·set is changed as both these factors change. If the reinforcers are delivered according to a simple rule, then the set of candidates will eventually be reduced so that stimuli and responses consistent with the rule are retained and others are lost.

The selection of response candidates is not independent of the selection of stimulus candidates, because conditioned stimuli redefine the situation and provide targets for responses. Key pecking does not occur in the autoshaping situation until the animal has learned about the predictive relation between the CS and food, for example. The CS now defines a food situation and provides a peckable target.

Response-reinforcer contiguity. We know little about the details of selection rules. For example, the temporal relation between response and reinforcer was once thought to be the only factor in conditioning, yet now we are less certain how it acts. Animals are sensitive to immediate contiguity between their actions and stimulus changes, but this is clearly not the only way they assess cause–effect relations.

When they have ample opportunity to sample, and factors like autoshaping don't create a strong bias in favor of a particular response, animals' ability to detect response-dependent stimulus changes is excellent. For example, in another experiment by Killeen (1978), pigeons pecked the center key of three and one in twenty pecks were unpredictably followed by darkening of the center key and illumination of the two outside keys. As the pigeon pecked the center key, the controlling computer was generating "pseudo pecks" at the same rate. Thus, half the time, the stimulus change depended upon the bird's peck (according to a variable-ratio 20 schedule). On the remaining occasions, the stimulus change occurred independently of the animal's behavior (i.e., a variable-*time* schedule). If the stimulus change on the center key was response-contingent, a peck on the right key gave immediate food, but a peck on the left key caused a brief timeout, thus delaying access to future food opportunities. Conversely, if the stimulus change was independent of responding, a peck on the left key produced food, a response on the right, timeout. The contingencies in this experiment are shown in Table 14.1. Pigeons had little difficulty in responding accurately, even though the only cue available to them was the very brief time between a peck and stimulus change, and even though the peck-stimulus times for some response-*in*dependent changes must have been less than the time between pecks and response-dependent changes.

Because these two peck–stimulus-change distributions were not completely separate,[12] perfect performance was impossible. Nevertheless, the pigeons' ability to detect the small time differences between response-dependent stimulus change and response-independent change was very good – and when Killeen varied the relative payoff for *hits* (correct identification of response-dependency) versus *correct negatives* (correct identification of response-independent stimulus change) the pigeons adjusted their bias appropriately. Figure 14.5 shows ROC curves (i.e., plots of the probability of a hit vs. the probability of a false alarm, see Chapter 9) traced out as relative payoff was varied, and it is clear that the pigeons behaved very much like the theoretical "ideal detector."

Table 14.1

Stimulus change was	Choice	
	Left	Right
Response dependent	Timeout	Food
Response independent	Food	Timeout

Thus, the time between a response and a positive reinforcer that follows it must play a role in response selection.[13] Response-reinforcer contiguity can play only an indirect role in *avoidance* schedules, however. In shock-postponement procedures, for example, brief shocks occur periodically (every t_1 sec) unless the animal makes a response; each response postpones the next shock for t_2 sec (see Chapter 5). Although rats find such schedules difficult (and pigeons find them almost impossible), most of the "standard" experimental animals can eventually learn to make the required response. Response-shock contiguity can play a role only by selectively eliminating from the candidate set those responses that are ineffective in putting off shock. Of course, adaptation to shock-postponement schedule is much easier if the effective response is a natural defense reaction, but then temporal relations have almost nothing to do with response selection.

So response selection by positive reinforcement is partly guided by the temporal relations between responses and reinforcers – and these relations may have an indirect effect on response selection by negative reinforcement. Nevertheless, once a particular response has been selected, response-reinforcer delays may have little effect (as we will see in a moment). Response *strength*, defined in the various ways discussed in Chapter 11, is *not* affected by response-reinforcer contiguity. The role of contiguity seems to be more one of guiding responding on a moment-by-moment basis rather than in some way connecting a particular response to a particular stimulus, as was at one time supposed.[14]

Shaping. The moment-by-moment effect of contiguity is demonstrated dramatically in the well-known animal-training technique of *shaping by successive approximations*. Shaping, a standard classroom demonstration as well as a traditional method of circus trainers, proceeds in two steps. First a hungry pigeon or rat is allowed to become familiar with its environment, the presence of the trainer, and so on. Then the trainer settles on some trick he wishes the animal to perform. The trick will either be something that animals occasionally do spontaneously

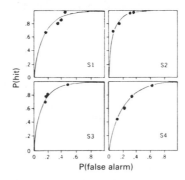

Figure 14.5. Receiver operating characteristic (ROC curves) for four pigeons in an experiment to measure pigeons' ability to detect response-dependency. P(hit) is the proportion of times that a pigeon indicated "yes" (by pecking the right key) when a stimulus change was caused by the response; P(false alarm) is the proportion of times that a pigeon indicated "yes" when a stimulus change was caused by the computer. (From Killeen, 1978.)

(most circus tricks are of this sort – seals balancing balls, dolphins leaping, etc.); more complex acts are built up out of elements, as I describe in a moment. If the animal is inactive, it may be given some food, which will usually produce activity via the arousal mechanism described earlier. If the target act then occurs it is immediately reinforced by feeding the animal as soon as it behaves in the desired fashion. Consistent reinforcement for the act, interspersed with judicious "free" reinforcers if activity wanes, soon suffices to produce an animal behaving in the appropriate fashion. Little really new has been created, however. The training has merely brought an existing behavior under the stimulus control of the trainer – the hungry seal now balances the ball when the trainer feels like it, not at the whim of the seal.

But this is not the limit of shaping. All too often, the desired act does not occur spontaneously in its entirety. This is where the "successive approximations" come in. As before, the trainer decides on a target behavior, but instead of waiting for perfection, he reinforces behavior that in his judgment approximates, or is a component of, the desired behavior. For example, if we want a pigeon to walk in a figure-eight pattern, we begin by reinforcing turning in one direction. Once that is well established, we wait until the animal shows an incipient turn in the opposite direction after completing the first turn, then reinforce it. And so on, until the whole pattern is established. Before autoshaping was discovered, key pecking was shaped in this way. First the bird was fed after any movement in the general direction of the key. Then, once it had learned to position itself in front of the key, any head movement was reinforced, then movement directed toward the key, and finally pecking the key.

The possibility of shaping indicates that the set of stimulus and response candidates is not closed: The animal does not at once show everything of which he is capable in a given situation.

Shaping takes a certain amount of intuition on the part of the trainer, because he must guess which movements will be precursors of the desired pattern. The trainer must, in fact, have a crude idea of the way the animal's movement system is represented. Sometimes the task is trivial – in training the animal to approach a certain place, for example. Here physical and psychological distance are equivalent. It may be less clear how to proceed if the desired behavior is some complex movement.

Most of human education is of course a process of shaping: The problem is not so much "strengthening" the behavior as getting it to occur for the first time. Even the term *behavior* is misleading, since the objective of other than skill learning is the attainment of that special change in internal state known as "understanding." The point about understanding something is not any particular behavior, but rather the creation of a whole new *repertoire* of behavior. The difficulty in getting children to understand something like long division is finding the necessary precursor behaviors – the proper sequence of subtasks that will lead to mastery. If the subject is well structured (like mathematics, or shaping of location in the pigeon), there is less difficulty than if the subject has no obvious structure. In teaching mathematics, addition and subtraction is an essential prerequisite to mastery of division and multiplication, for example. Even so, one can imagine several ways to get children to understand addition itself – by manipulation of counting sticks, by logical explanation, by induction from examples, by rote memorization of tables, and so on: There is an order at a higher level, but nothing to guide teaching of the elements. There is no unique sequence at any level to guide the teaching of history, or writing, or even psychology. Different students may have different initial expectations (priors) about such fields, so that different sequences might be appropriate for different people. The problem of the proper sequence is at least as important to the success of shaping as the timely delivery of reinforcers. In human learning, it is much more important.[15]

In summary, we know that both priors and temporal relations contribute to the selection of stimuli and responses but temporal relations do not always dominate. We also know that the more valuable the reinforcer, the smaller the final set of candidates, that is, the more stereotyped the behavior.

Feedback function

Candidate activities interact with the environment through a feedback function. In the natural environment, feedback functions are generally positive and the interaction is almost invariably productive. For a raccoon, manipulation of something associated with food, like a nut or a corncob, will usually help him to get at the eatable portion. For a pigeon, finding grain under leaves will lead to pecking at leaves, which is likely to reveal more food. In these cases, the response induced by food-predicting stimuli will often produce more food – and will never *prevent* food. Such behavior looks food-reinforced: Because the behavior is effective, its induced nature is concealed. Only the omission contingency reveals that the dominant factor is not the temporal relation between response and food, the response contingency, but the prior associated with the predictive relation between stimulus and food, the stimulus contingency.

Every reinforcement schedule involves stimulus as well as response contingencies. Even the simple fixed-interval schedule, in which the first response T sec after reinforcement produces the next reinforcement, provides two valid predictors of the reinforcer: the response, and postreinforcement time. Consequently, every operant conditioning procedure has the potential for conflict between inducing factors and the temporal relations (response-reinforcer contiguity) that select for one response over another.

Inducing factors arc of two kinds. The first we have just encountered: the priors (innate and otherwise) that prescribe the occurrence of food-related activities in a situation where food is likely, of agonistic activities in a situation where conflict is likely, and so on. The second inducing factor derives from activities that compete with the dominant, reinforcer-related activity: A hungry rat needs to eat, and in the presence of a signal for food it will show food-related activities such as gnawing and pawing at the feeder (see Staddon & Ayres, 1975; Locurto, Terrace, & Gibbon, 1981). However, it also needs to spend time in other activities such as grooming, drinking, running, and so on (see Chapters 7 and 11). These activities compete with food-related activities and tend to displace them in the present of stimuli that signal a low probability of food (see the discussion of *behavioral competition* in Chapter 11). These two factors – induction of US-related activities in US-predictive stimuli, and US-unrelated activities in stimuli predicting US absence – are different sides of the same coin; both are aspects of mechanisms that allow animals to allocate their time and energy efficiently.

Schedule-induced behavior. On fixed-interval and fixed-time schedules, these processes lead to the partitioning of the fixed period between food deliveries into two main periods: the time just before food, which is devoted to food-related activities (the *terminal response*), and the time just after food, which is devoted to other activities such as drinking and running (*interim activities*).[16] The resulting distribution of activities for a rat on a 30-sec periodic food schedule is shown in Figure 14.6 (see also Figure 11.4). The left panel shows the apparatus in which the food was delivered; the right panel shows the average distribution of activities in between food deliveries.

The terminal response and some of the interim activities occur at a higher rate than they would in the absence of the periodic food schedule. The terminal response is always reinforcer-related, but the interim activities differ from species to species. If water is available, for example, rats and monkeys on a periodic-food schedule will drink to excess during the interim period (*schedule-induced polydipsia*). Pigeons, on the other hand, show no excess drinking, but will attack a conspecific or a stuffed model during this time. The excessive nature of many

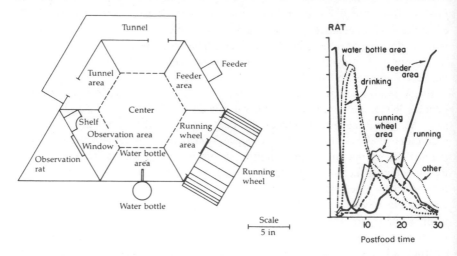

Figure 14.6. Left panel: a hexagonal apparatus providing a rat with opportunities to run, drink and eat, look at another rat, and hide in a tunnel. Right panel: average behavior within each interfood interval of a single rat after extended training on a schedule in which a single food pellet was delivered every 30 sec in this apparatus. (From Staddon & Ayres, 1975.)

interim activities indicates that they cannot be explained solely by efficient time allocation: It makes sense that the rat should drink just after, rather than just before, food on a fixed-time schedule. But the allocation argument does not require the animal to drink to excess. Yet rats on fixed-time schedules in the 1–4-min range can drink three to four times their normal daily amount within a 1-hr session.[17]

There is still controversy on the causes of schedule-induced interim activities. It is easy to understand the effect of periodic food on the terminal response. From a functional point of view, it pays the animal to expend more effort exploiting a richer resource; hence the positive relation between rate of the terminal response and rate of food delivery is understandable. (Chapter 7 is devoted to this general argument.) The generally increasing interval-schedule response function may be related to the arousal effect of single reinforcers discussed earlier. For example, Killeen (1982) has shown that cumulating arousal alone is sufficient to account for the negatively accelerated relation between food rate and rate of the terminal response on interval schedules: If each food delivery adds to the total level of arousal according to the exponential function shown in Figure 14.4, then as food deliveries come more frequently, there is less time for the arousal elicited by each food delivery to decay before the next food delivery and the average arousal level increases. Given a ceiling on response rate, the predicted relation between food rate and (terminal) response is an increasing, negatively accelerated one which conforms quite well to the observed response function (see Equation 11.8, the hyperbolic relation proposed by Herrnstein).

We saw in Chapter 7 that the interval-schedule response function, though increasing over much of the range, does turn down at very high food rates. The comparable relation between rate of induced drinking (the best-studied interim activity) and food peaks at much lower food rates than the response function for the terminal response. This difference can be understood as follows. Both the terminal response and schedule-induced drinking take up a larger and larger percentage of the interfood interval as the interfood interval decreases (food rate increases). At some point, activities other than these two are completely excluded. Further decreases in the interfood interval mean that the time taken up by the terminal response increases at the expense of drinking, leading to a downturn in the drinking function.

The best evidence for this interpretation is that activities such as running in a wheel, that occur on schedules but are not facilitated (induced) by them (these have been termed *facultative activities*), are progressively suppressed at high food rates. The effect is illustrated in Figure 14.7, which shows the rates of drinking and wheel running in the apparatus shown in Figure 14.6 on different periodic food schedules, ranging from fixed-time 3 min to fixed-time 5 sec. Drinking increases as food rate increases, but running decreases.

It is still not clear why interim drinking tends to increase along with the terminal response on periodic food schedules. There is obviously some relation between the increase in drinking and the cumulative increase in arousal produced by the periodic food, but why drinking should be specially favored in rats and attack in pigeons, and why either should be facilitated rather than suppressed by increases in the rate of periodic food, is not known.

Conclusion

Reinforced learning proceeds through four repeating stages, three in the animal, one in the environment: surprise (or novelty) → inference → action → new situation → surprise.... A new situation elicits a program of first protective and then exploratory activities that serve to build up a representation, and perhaps change the value of the situation – by causing the delivery of food or electric shock, for example. Causes for the change, in the environment and the animal's own behavior, are identified by the inference principles described in this and the previous chapter: Candidates are weighted according to their salience, prior relation to valued events, and relative temporal proximity to the US. Identification of causes allows hill-climbing processes to act: If the change produced is an improvement, and stimulus A is identified as the probable cause, A can be approached. If further instances confirm A's predictive value, perhaps some response may be made to A (autoshaping). If response B is identified as the probable cause, B can be repeated. B then acts to change the environment according to the prevailing feedback function. If the outcome is surprising (e.g., no further improvement) the inference rules are applied again, the predictive relations change and some other stimulus may be attended to, and some other activity may occur.

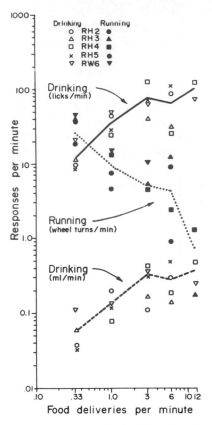

Figure 14.7. Drinking rate (licks/min and ml/min) and running rate (turns of a 27-cm diameter running wheel) as a function of food rate on five fixed-time schedules for five individual rats run in the apparatus shown in Figure 14.6. Points show data for individuals; lines are means. Each rat was exposed to four of the five food rates; hence each line is the average of four animals. (From Staddon, 1977a.)

The whole process stabilizes when the action induced by these principles maintains the conditions that gave rise to it: The animal is no longer surprised by what happens to him, and the pattern of action generates a history that sustains that action.

The largest unknown in all this is obviously the animal's representation. The arguments in this chapter and the last suggest quite strongly that both action and learning are guided by the animal's knowledge of the world: Learning happens when knowledge fails to match reality; action happens when knowledge (innate and learned) indicates something to approach or avoid, or do.

EXPERIMENTAL APPLICATIONS

This last section of the chapter discusses a number of "standard" phenomena from the point of the principles derived in this and earlier chapters.

Conditioned reinforcement

Schedule-induced behavior occurs when periods of food and no-food are signaled by time. What happens when periods with different payoff conditions are signaled by nontemporal stimuli, like lights, sounds, or spatial locations? The simplest case is when stimulus changes are independent of responding; these are the *multiple schedules* discussed at length in Chapter 11. The allocation of behavior here parallels the fixed-time schedule: Interim (facultative) activities predominate in the components associated with a low rate of reinforcement; terminal (instrumental) responses predominate in components associated with a high reinforcement rate.

More complex effects come into play when the stimulus changes depend on the instrumental response. There are two main effects to consider: stimulus change as an aid to *memory*, and stimuli as guides to behavioral *allocation*. Sometimes these two effects work together, and sometimes they conflict.

The memory effects are seen most clearly in the acquisition of a response. For example, suppose we attempt to train a pigeon to peck a white key for food reward, but delay the food for 5 sec after each peck. Even if the animal does eventually peck the key, the effect of each reward is likely to be small, the animal will obtain little food, and training may be unsuccessful. The likely reason is that it is difficult for the animal to pick out the peck, a brief event preceded and followed by other activities, as the best predictor of food, unless peck and food occur close together in time.

We can make the bird's task much easier in the following way: We train him in two phases. In phase 1, the white response key occasionally turns green for 5 sec, after which food is delivered. No response is required. This is of course an autoshaping procedure, and the pigeon will soon be pecking the green key, but this is not in itself important (the results will be the same even if we prevent the animal from pecking the key during this phase, either by restraining him or by covering the key with a clear window). In the second phase, pecks on the white key are immediately followed by the green stimulus plus food after 5 sec. The time relations between pecking and food are exactly the same in the second phase as in the first – yet the pigeons will rapidly learn to peck the white key if given 5 sec of green as a signal. It is not even necessary to begin with the autoshaping procedure: Pigeons dumped directly into phase 2 will also learn. Why is this added-stimulus procedure more effective than the simple delay procedure?

The two-stimulus procedure in phase 2 is termed a *chained schedule*: peck → stimulus → food. The green stimulus is used as a reinforcer and seems to act like one. On the other hand, it gains its power not innately (or early in development), but by virtue of its pairing with food. Hence it is termed a *conditioned* or *secondary* reinforcer. The green-key conditioned reinforcer aids conditioning for two reasons: First, it bridges the temporal gap between the peck (the real cause of the food) and its consequence (the food). Rather than having to remember a brief event occurring 5 sec before its consequence, the animal has only to remember that pecking leads to stimulus change – since the peck → stimulus-change delay is negligible, this presents no difficulty. Second, the rate of food delivery in the presence of the green stimulus is much higher than the average rate of food delivery in the apparatus. Hence, the green stimulus is a predictor of food and has value of its own, although the value is relatively transient and depends upon reliable delivery of the food. If food ceases to occur, the green stimulus will soon lose value, as choice tests have repeatedly demonstrated. The ubiquitous hill-climbing tendency will lead the pigeon to peck at the white key so as to produce the higher-valued green key.

Looking at the green key as a memory aid leads to different predictions than looking at it as a surrogate for food, a "reinforcer" that can "strengthen" behavior. Let's look at some of the problems with the reinforcer model:

If a one-link chained schedule is effective in maintaining key pecking, why not two, three, or N links? Two links are generally effective. For example, suppose we train a pigeon on the following sequence, where the notation S_i:peck → S_j indicates that a peck in the presence of stimulus S_i produces stimulus j:

$$S_1\text{:peck} \rightarrow S_2\text{:peck} \rightarrow S_3 \rightarrow \text{food} \rightarrow S_1,$$

which corresponds to a two-link chain. Providing the durations, t_i, of each stimulus are appropriate (t_2 and t_3 should not be much longer than t_1; the ideal arrangement is for t_1 to be much longer than t_2 and t_3) pigeons will learn to peck S_1 to produce S_2, and peck S_2 to produce S_3. But the process cannot be continued indefinitely. Imagine a chained schedule of the following form:

$$S_1\text{:}t > T, \text{ peck} \rightarrow S_2\text{:}t > T, \text{ peck} \rightarrow \ldots S_N\text{:}t > \text{food} \rightarrow S_1\text{:}\ldots$$

The term S_i:$t > T$, peck → S_j denotes a fixed-interval T-sec schedule in the presence of S_i (a peck more than T sec after the onset of S_i is reinforced by the appearance of S_j). The whole sequence denotes a chained schedule totaling N fixed-interval links. How many such fixed-interval links can be strung together and still maintain responding in S_1, the stimulus most remote from food?

The answer is, not more than five or six. The top record in Figure 14.8 shows a cumulative record from a pigeon trained with six fixed-interval 15-sec links (the recorder pen reset after food). Long pauses occur after food, and the times between food delivery are always much longer than the 90-sec minimum prescribed by the schedule. An additional link would have caused the pigeon to stop responding altogether. With longer fixed intervals, five links are the upper limit.

The middle record in Figure 14.8 shows a typical performance when the stimulus change from one fixed-interval to the next is eliminated, but all else remains the same (this is termed a *tandem schedule*). There are two notable differences between performance on the tandem and chained schedules: The tandem schedule is not subject to the long postfood pauses on the chained schedule; and the terminal response rate (i.e., the slope of the cumulative record

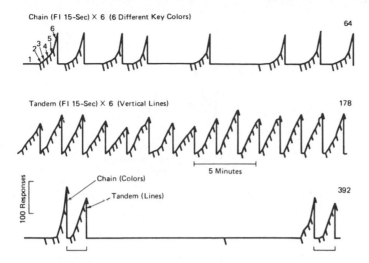

Figure 14.8. Top: Cumulative record from a pigeon well trained on a 5-link (6-stimulus) chained fixed-interval 15-sec schedule (the recorder pen reset after food, and blips indicate stimulus changes). Center: record of a pigeon trained on the tandem schedule equivalent to the 6 stimulus chain (a single stimulus, vertical lines, was on the key throughout, but the six successive fixed-interval schedules were in effect between food deliveries). Bottom: record of a pigeon exposed to both the chained and tandem procedures in alternation. (From Catania, 1979, Figure 8–12.)

in the last fixed interval before food) is much higher in the chained schedule. These differences appear even more clearly in the bottom record, which shows both types of schedule successively in the same pigeon.

Obviously one problem with the simple conditioned-reinforcement idea is that it cannot explain results like those in Figure 14.7: Why should five or six links be the limit? Why should the pigeons respond *less* in the first link than if there were no stimulus change? These problems are usually handled by adding another process to conditioned reinforcement. Thus, two things are usually thought to be responsible for the differences between chained and tandem schedules illustrated in Figure 14.8: (a) the relative proximity to reinforcement (food) of each stimulus in the series; and (b) conditioned reinforcement, that is, the contiguity between pecking and the transition to a stimulus closer to food.

Proximity to food clearly exerts the major effect. It is illustrated in Figure 14.9, which plots the serial order of each stimulus against its temporal proximity to food. The *C*'s (for "chain") in the figure denote stimuli in the chained schedule, where serial order is perfectly correlated with proximity to food. The *"T"* denotes the serial order and average proximity to food of the single tandem stimulus. Clearly, response rate in the presence of each stimulus is directly related to its relative proximity to food: Rate in the chained procedure is higher, for

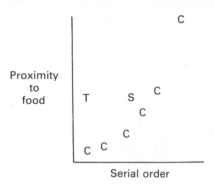

Figure 14.9. Stimuli arranged according to their serial order and proximity to food in three procedures: chained schedule (C's), tandem schedule (T), and scrambled chain schedule (S).

the terminal link, and lower, in the initial link, than the average rate in the tandem schedule. This conclusion is confirmed by the *S* in the figure, which shows the effect of scrambling the order of the chained stimuli, so that each one appears equally often in each serial position. Pigeons so trained respond at essentially the same rate in the presence of each stimulus, as suggested by the single value for both serial order and proximity to food.

Time by itself has a small effect on the tandem schedule, as indicated by the gradual increase in response rate with postfood time.

The concept of conditioned reinforcement (that is, the response contingency between pecking and stimulus change) adds nothing to our understanding of chained schedules. Conditioned reinforcement is a redundant notion. Providing the response contingency for food in the terminal link is maintained, it can be omitted in earlier links with little effect on key pecking, as long as stimulus changes continue to take place as before (i.e., the fixed-interval contingency is replaced with a fixed-*time* contingency – this is termed a fixed-interval *clock*, because the successive stimulus changes signal the lapse of postfood time). Behavior on chained schedules is determined by temporal proximity to food in the same way as behavior on multiple schedules.

The pattern of terminal responding on chained schedules is complementary to the pattern of interim activities, which occur preferentially in components whose reinforcement rate is below the average. The increasing pattern of terminal responding, and the tendency to long pauses in the first link, can be derived by the same argument used to explain behavior contrast in Chapter 11. For example, the immediate postfood period (i.e., the first link) predicts the absence of food, so that interim activities will tend to occur preferentially at that time. Increases in interim activities in the first link reduce the probability of a terminal response. Since a response is necessary for the transition to the second link, this tends to

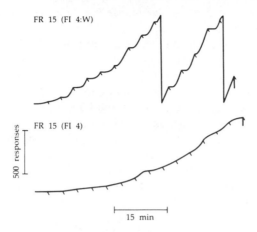

Figure 14.10. Top record: cumulative record of the performance of a single pigeon on a second-order FI 60-min (FI 4-min) schedule. Every 4 min, a peck produced a brief .7-sec stimulus on the response key (diagonal blips on the record; this is the FI 4 component); at the end of the 15th such presentation, food was also delivered (the record resets; FI 60). Thus, the brief stimulus was paired with food once an hour – and also signaled a food opportunity after 4 min. Bottom record: performance on the same schedule, but with no brief-stimulus presentations (tandem schedule); blips show FI-4 components. (After Kelleher, 1966.)

lengthen the first link. The longer the first link, the lower its proximity to reinforcement relative to others, which further reduces the probability of a terminal response, further lengthening that link, and so on. Moreover, the more time devoted to interim activities in the first link, the less likely they are to occur in later links (because of the satiation–deprivation processes described in Chapter 11), allowing more time for the terminal response. The whole process, then, favors the kind of temporal gradient that is actually observed, in which terminal response rate increases through the chain, winding up highest in the last link.[18]

Conditioned reinforcement in two-link chained schedules facilitates the acquisition and maintenance of behavior because it aids memory. A related procedure, *second-order schedules*, has its effects by impairing memory. The basic phenomenon is illustrated in Figure 14.10. The top cumulative record shows the relatively high response rate and scalloped pattern produced in a well-trained pigeon by splitting up a fixed-interval 60-min schedule into fixed-interval 4-min components, each terminated by a response-contingent brief (.5-sec) stimulus. Food follows (i.e., is paired with) the last such brief stimulus in the 15-component cycle. The bottom record shows the low, unpatterned responding generated by the comparable tandem schedule. The tandem performance shows that without the brief stimuli, response rate is very low; and in the absence of time markers, there can be no 4-min scallops.

The brief stimuli seem to act by interfering with the animal's recall for the most recent food delivery. In Chapter 12, I showed how the scalloped pattern on fixed-interval schedules depends upon the animal's ability to recall the most recent time marker. The best temporal predictor of food in the second-order procedure shown in Figure 14.10 is food – but it is temporally remote (the interval is 60 min long) and the time is filled with periodic brief stimuli, the last-but-one of which is also a temporal predictor of food: Food is always preceded by a brief stimulus 4 min earlier. The greater validity of food ($p(F|t > 60) = 1$) seems to be outweighed by the greater frequency and closer proximity to food of the less valid ($p(F|t > 4) = 1/16$) brief stimulus – so that the brief stimulus seems to overshadow food as the effective time marker. Since the animal does not know where it is in the interval (in confirmation of this, the temporal gradient after the first postfood brief stimulus in second-order schedules is much less than in comparable fixed-interval schedules), and anticipates food every 4 min, response rate is naturally higher in the second-order than the tandem schedule. As this argument leads one to expect, brief stimuli have their largest effects with long interfood intervals.

The effects of brief stimuli were, at one time, attributed to the pairing of the final stimulus with food. Subsequent work has devalued the importance of pairing, which seems now to be important only for the initial acquisition of the pattern (see Squires, Norborg, & Fantino, 1975; Stubbs, 1971; and Staddon, 1974).

As with *primary* reinforcers, such as food or electric shock, conditioned reinforcers seem to exert their major effect through inducing factors – the Pavlovian relations between particular synchronous or trace stimuli and the frequency of primary reinforcement. The response contingency plays a role in two ways. It may aid acquisition of the correct response and correct any tendency for the response to drift away from the effective form; and through the feedback function it maintains the stimulus–reinforcer relations that sustain the response.

Conditioned emotional response

Many, perhaps most, contemporary experiments on classical conditioning use a mixed procedure in which the effects of a CS are assessed indirectly. Rather than measuring the direct effect of a CS on an autonomic response such as salivation or pupillary dilation, it has turned out to be more convenient to study its indirect effects on food-reinforced lever pressing. Rats are first trained to press a lever for food, delivered on a variable-interval (VI) schedule. As we have seen (Chapter 5), after a little training this schedule maintains a moderate, steady rate of response – an ideal baseline for the study of other variables. Stimuli of perhaps 60-sec duration are occasionally superimposed on this VI baseline. Rats soon habituate to the added stimulus, and response rate is the same in the presence of the superimposed stimulus as in the background stimulus before and after. In the final phase, electric shock is delivered in the presence of the superimposed stimulus (CS). Within a few stimulus-shock pairings, the onset of the CS causes a reduction in the rate of lever pressing, which has been termed *conditioned suppression*, or the *conditioned emotional response* (CER). The amount of suppression in the CS, relative to the just-preceding baseline period, then provides a measure of the amount or strength of conditioning.

On the face of it, this suppression makes little adaptive sense. The rat has ample opportunity to learn that it cannot avoid or escape from the shock. By reducing its lever-press rate, the animal loses needed food reinforcers. What causes this apparently maladaptive suppression?

By now, the answer should be clear. The superimposed stimulus is a CS for shock. Once the animal has learned this relation, therefore, we may expect to see candidate defense reactions induced by the CS, according to the Pavlovian inference mechanisms I have discussed in this and the preceding chapter. These reactions will generally interfere with lever pressing, hence show up as a depression in lever pressing – the CER. These reactions persist, even though they have no effect on shock, for the same reason that key pecking persists under an omission contingency. The priors associated with a stimulus that predicts imminent shock simply outweigh the opposing effects of the food contingency for lever pressing.

This argument, and data of the Brelands discussed earlier, suggests that similar suppression effects might be produced even by a food stimulus on a food-reinforced baseline. For example, a feeder light that comes on 5 sec before the feeder operates will soon cause a hungry rat to approach the feeder as soon as the light comes on. If such stimulus–food pairings are superimposed on a VI 60-sec baseline, no one would be startled to observe a reduction in lever pressing during the stimulus, and this is indeed what occurs. Conversely, if the light were on the response lever rather than the feeder, lever pressing might well *increase* during the CS – because the rat now approaches the lever (and presses it) rather than approaching something incompatible with lever pressing.

Both these effects of a food CS have been widely observed.[19] As these examples suggest, the magnitude and direction of the effect depends on the duration of the CS (both absolute, and relative to the time between CS presentations), the type (e.g., tone, light, localizable vs. unlocalizable) and location of the CS relative to the feeder and the lever, and the magnitude and frequency of food in the presence of the CS. The general rule is that if the food rate in the presence of the stimulus is significantly higher than the rate in its absence (i.e., on the baseline VI schedule), the stimulus will tend to induce food-related behavior. The effect on lever pressing then depends on the nature of the induced behavior: If it is physically compatible with lever pressing, the effect of the CS is to facilitate lever pressing; if not, it is to suppress lever pressing.

Avoidance and escape

Electric shock has a number of paradoxical effects, which all derive from its strong inducing effects. Shock, and stimuli that signal shock, produce very stereotyped reactions from most animals, and immobility ("freezing") is often dominant, especially if the shock is severe. For example, if a rat is presented with a train of brief electric shocks that can be turned off by some predesignated response, he may only learn the necessary escape response if it is part of the induced defense reactions. If the required response is to press a lever, it will be learned much more easily if lever *holding* is sufficient than if active *depression* is required. Holding is compatible with the induced reaction of "freezing" on the lever, whereas lever pressing requires that the animal periodically release the lever so that it can be depressed again. The inducing effects of shock make shock-motivated operant conditioning difficult for rats and pigeons for reasons not directly related to the supposedly indirect nature of avoidance schedules.

In all shock experiments, the animal's highest priority is to escape from the apparatus entirely. This is precisely what one might expect not only from common sense, but from the principles of positive reinforcement. In a food situation, animals detect the stimulus most predictive of food and approach it. In an aversive situation, they detect the stimulus or situation most predictive of the *absence* of shock, and approach that – this rule takes them at once out of the experimental apparatus, which is the only place they normally experience shock. Since escape is always prevented, anything they do in the experimental situation is, in a sense, second best, and not an ideal measure of the effect of the shock schedule. Studies of positive reinforcement do not suffer from this problem: A hungry animal is happy to be in a box where he gets food.

So-called one-way shuttlebox avoidance shows the importance of withdrawal. Animals shocked on one side of a long box will immediately run to the other side when placed in the box. In contrast, "two-way" avoidance, in which the animal must run back and forth from one side of the shuttlebox to the other whenever a signal sounds, is harder for animals to learn. There is no safe place (within the shuttlebox), the animal's dominant response (escape) is blocked, and so learning is difficult.

Free-operant, nonlocomotor, avoidance procedures are of two main types: *discriminated avoidance*, and *shock-postponement* (Sidman-avoidance) schedules. In discriminated avoidance, a stimulus is occasionally presented; shock occurs after a few seconds, and then shock and stimulus both terminate (Figure 14.11, top). Granted the difficulty in ensuring that lever pressing is part of the animal's candidate set, in other respects the explanation for discriminated avoidance is the same as the explanation for conditioned reinforcement: The animal is presented with two situations, CS and CS-absence ($\overline{CS}$), one higher valued than the other. A response in the CS (the lower-valued situation) produces $\overline{CS}$, the higher-valued one. The response-contingent transition from CS to $\overline{CS}$, ensures that the animal can remember the effective response, and the higher value of (lower shock rate in) $\overline{CS}$ ensures that once the response occurs, it will be maintained. Experiments with shock-postponement schedules have shown that the pairing of the CS with shock is much less important than its effect in making

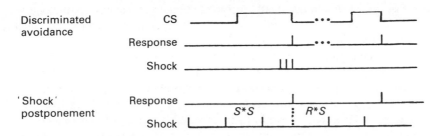

Figure 14.11. Shock-avoidance procedures. Top: discriminated avoidance – a response during the CS turns it off (avoidance); shock occurs at the end of the CS until a response occurs (escape). Bottom: shock postponement. Shock occurs at fixed intervals (the shock–shock – $S*S$ – interval) unless a response occurs; each response postpones the next shock for a fixed interval (the response–shock – $R*S$ – interval).

the avoidance response easy to remember. Thus the improvement in shock rate (detected we know not how) is the factor that maintains avoidance responding.

There are two main shock-postponement procedures. One is illustrated in Figure 14.11 (bottom): Brief shocks occur at fixed time intervals of say 10 sec (this is termed the shock–shock or $S*S$ interval); if the animal responds, usually by pressing a lever, the next shock (only the next) is postponed for a fixed time, say 15 sec (the response–shock or $R*S$ interval), which may be the same as or different from the $S*S$ interval. This procedure is termed *Sidman avoidance*, after its inventor. A more descriptive label might be *fixed-interval shock-postponement*. No matter what the value of the $R*S$ interval relative to the $S*S$ interval, it pays the rat to respond at least once every t seconds, where t is less than the $R*S$ interval. By so doing, it can avoid all shocks. The second procedure, variable- (or random-) interval shock-postponement, is very similar. If the rat does not respond, it receives random, brief shocks at a certain rate, say two per minute, defining an average intershock interval t_S. If it does respond, the *next* shock occurs at an average time t_R after the response, where t_R is greater than t_S.

Rats learn to respond on both procedures, albeit with some difficulty. Judged by the proportion of animals that completely fail to learn, these procedures are more difficult even than the two-way shuttlebox, and much more difficult than one-way shuttle avoidance, or avoidance where the response is running in a wheel. The source of the difficulty seems to be that lever pressing is often not one of the candidate responses made by animals in these situations. Weaker shock often aids learning, presumably by diminishing the tendency for strong shock to induce rigidly stereotyped behavior. Both shock-postponement procedures provide an opportunity for response selection by relative contiguity: Of all activities, the effective response will, on the average, be the one most remote from shock. If proximity to shock tends to exclude an activity from the candidate set, then the animal should eventually arrive at the correct response as the one *least contiguous* with shock – and this response will generally be effective in reducing overall shock rate.

Relative contiguity is, of course, the process by which *punishment* (response-contingent negative reinforcement) selectively eliminates the punished activity.

Rats will sometimes respond on a shock-postponement schedule that has no effect on shock rate, providing each response produces a brief shock-free period. For example, Hineline[20] trained rats in an extensive series of experiments based on a 60-sec cycle. A response lever was present at the beginning of each cycle. If no lever press occurred, the lever retracted after 10 sec, and a single brief shock occurred at the end of 11 sec. The lever reappeared after 60 sec and the cycle resumed. If a lever press occurred, the lever at once retracted, the 11-sec shock was omitted, but a shock occurred at sec 39. Thus, a lever press had no effect on the overall shock rate, which was always 1 every 60 sec, but always produced a shock-free period of $39-t$ sec after lever retraction (vs. 1 sec if no response occurred), where t is the time in the cycle when a response occurred. In later experiments, Hineline was able to train some rats to respond even if responding actually *increased* shock rate (several shocks, rather than just one, occurred after sec 39). Control conditions, in which shocks occurred independently of responding, confirmed that the shock-free postresponse period was critical to maintenance of lever pressing in these animals.

It is hard to know how to interpret these studies. Not every rat shows the effects, and to explain response selection by the production of a safety period depends upon the rat's representing the situation in a certain way. If the animal times everything from lever retraction (a very salient stimulus, because of the loud sound of the solenoid mechanism), then a lever press does indeed delay shock. On the other hand, if the animal assesses intershock time in some way, responding has no effect. Although shock never occurs when the lever is present, it does occur within 1 sec of lever retraction, so that the lever does predict shock if the animal fails to respond. A 1-sec delay after a 10-sec stimulus is very short and perhaps small enough for the animal to treat the lever as a trace CS, so that the experiment is really a case of discriminated avoidance, rather than one of shock-postponement. A modified procedure in which the lever is present throughout each cycle, but all else remains the same, would almost certainly fail to sustain lever pressing.

What may be required as an adequate test of the delay-reduction idea is a modification of the variable-interval shock-postponement procedure in which a response produces a transient shock-rate reduction after variable amounts of delay. Instead of selecting the *first* shock after a response from a distribution with a lower mean rate, one might select the second, third, or some later shock in this way. This technique encounters all the familiar problems of delay-of-reinforcement studies and might be difficult to design cleanly.[21] In this way one might perhaps trace out the negative delay-of-reinforcement gradient.

The conclusion of this discussion of avoidance is rather unsatisfactory. In a

general sense, this behavior seems to follow the same rules as positive reinforcement: Shock is an hedonic stimulus and arouses the animal – although
passive behavior ("freezing") often dominates unless special steps are taken to
prevent it. The commonest active behaviors are attempts to withdraw from the
source of shock or escape from the experimental situation. Since these tendencies
are invariably blocked, the set of remaining candidates may be small. From this
set, effective members are selected in ways not fully understood. The usual result
is reduction of shock rate, but this cannot be used as a mechanistic explanation
since we know little about how shock rate is assessed by the animal: The temporal relations between response and shock play some role, but it is clearly a rather
weak one under many conditions. And some rats, under some conditions, can be
trained to respond even if overall shock rate is thereby increased, so that shock-
rate reduction is not acceptable as a general optimality account for avoidance
behavior.

The small set of candidate activities induced by shock implies stereotyped
behavior; as we will see shortly, this stereotypy can often be highly maladaptive.[22]

Set, response-produced shock, and "learned helplessness"

Two necessary features of learning can produce maladaptive behavior under
some conditions: (a) Learning depends upon *surprise*; and (b) Long training in
situations with strong reinforcers (a very hungry animal and large or frequent
food portions, strong electric shock) greatly reduces the candidate set. The
stereotyped and inflexible behavior so produced has traditionally been known as
behavioral *set*. These two characteristics can allow changes in the feedback
function to go undetected, resulting in unnecessary, or even counterproductive,
behavior.

Effective avoidance performance on a shock-postponement schedule means
that the animal gets very few shocks. The few shocks that are received are
important, however. Even well-trained rats show a reliable "warm-up" effect in
each session, responding slowly or not at all at first. The few shocks received at
this time induce avoidance responding so that later in the session almost all
shocks are avoided. Suppose we take such an animal and turn off the shock
generator, except for a few response-independent shocks at the beginning of each
session. The animal now has almost no information to tell him that the world has
changed and he need no longer respond: responding fails to occur at first, so he
cannot learn it is ineffective; and never fails to occur later, so he cannot learn it is
unnecessary. There is nothing to produce surprise, the first requirement for
learning. The animal's only protection against this bind is to *sample* his environment either by responding early, or by occasionally *not* responding late, in a
session. But sampling is just the overt manifestation of a large candidate set, and
we have already established that schedules of severe shock, and protracted training, greatly reduce the candidate set. The animal does not sample, and if for
some extraneous reason he fails to respond for a while later in the session, the

absence of shock goes unnoticed. Avoidance responding persists almost indefinitely under these conditions.

Behavior maintained by the *production* of electric shock provides the most striking example of the effects of shock schedules in minimizing behavioral variation. An experiment by McKearney (1969; reviewed by Morse & Kelleher, 1977) is typical. Squirrel monkeys restrained in a chair were first trained to lever press on a schedule of shock postponement ($S*S = 10$ sec, $R*S = 30$ sec). When the typical steady rate had developed, a 10-min fixed-interval (FI) schedule of shock *production* was superimposed. Eventually, the shock-postponement schedule was phased out. The monkeys, nevertheless, continued to respond on the fixed-interval shock-production schedule and behavior soon became organized with respect to it: Lever pressing followed the typical scalloped pattern, increasing up until the moment of shock. As with food reinforcement, the shorter the fixed-interval duration, the higher the response rate. When shock was omitted entirely, the behavior soon ceased.

Rats in the typical Skinner box are less prone to the kind of rigidity shown by monkeys and cats in these experiments. But rats can be trained to produce electric shocks in the shuttlebox apparatus. Similar behavior has been established in humans and even goldfish. So-called self-punitive behavior is not an isolated phenomenon.

In McKearney's experiment, long training, physical restraint, and the highly aversive electric shock, all act to reduce behavioral variation. In addition, the transition from simple shock postponement to shock postponement plus FI shock is barely perceptible. The monkeys occasionally failed to avoid shocks on the postponement schedule, so an additional shock every 10 min made little difference. Moreover, even if the shock were detected, the animal could not cease responding without receiving many more shocks through the shock postponement contingency. All thus conspires to maintain responding in the face of occasional response-contingent shocks.

When the shock-postponement schedule was eliminated, leaving only response-produced FI shock, the sustaining factor for action, occasional shock, continued to occur. Moreover, given the relative effectiveness of avoidance behavior in well-trained animals, the omission of the shock-postponement schedule must have been imperceptible: Many, perhaps most, of the shocks received were from the FI contingency before, and all were after, the change. Given a sufficient reservoir of behavior variation, the monkey might have detected that almost any new behavior would lead to less shock than lever pressing. The monkeys' failure to do so suggests that the previous training had so reduced the candidate set that the necessary variation was lacking, so that responding continued indefinitely.

In some response-produced shock experiments, the process of systematically eliminating all candidates but the desired response is made quite explicit by a "saver" provision: For example, if the animal fails to respond at the end of a fixed interval, after a brief delay, several response-independent shocks may be delivered. This feature may later be phased out, but while it acts the animal is confronted with an avoidance schedule where responding makes perfect sense.

Behavior maintained by response-produced shock is not masochistic; the shock does not become pleasurable or positively reinforcing. Squirrel monkeys are transported to the experimental apparatus in restraining chairs partly because they would not otherwise choose to stay. The behavior is a misfiring of mechanisms that normally lead to escape and avoidance.

Shock in these studies seems to play two roles: (a) It provides a discriminative stimulus. The major discriminative property is to define the situation (see Chapters 10 and 13). Its most important characteristic for this purpose is its aversiveness, as we see in a moment. When shock is periodic (as in McKearney's experiment) it also provides a time marker. (b) Shock also motivates the behavior. Even highly effective avoidance behavior eventually extinguishes when shock is entirely omitted, but behavior maintained by response-produced shock persists indefinitely.

I have been arguing that all learning is selection from among a set of stimulus and response candidates (but see note 15), and we have seen that surprising results can come from procedures that seem to reduce severely the size of the candidate set. What if the set of response candidates is reduced to zero? The requirements for such a reduction should now be obvious: very severe, response-independent shock. Severity ensures that the candidate set will be small; response-independence that it will be empty, or at least contain only the "behavior of last resort" (usually freezing). These two characteristics define the phenomenon known as *learned helplessness*. In the original experiment (Seligman & Maier, 1967; for reviews see Maier & Jackson, 1979; Alloy & Seligman, 1979; and Glazer & Weiss, 1976) dogs were first restrained in a harness and given a series of severe, inescapable shocks. The next day, they were placed in a simple discriminated avoidance situation: in each trial, when a CS came on, shock followed after 10 sec unless the dogs jumped over a low barrier. If they failed to jump, the CS remained on and shocks occurred for 50 sec. Thus the animals had an opportunity either to avoid or escape from the shock by jumping the barrier.

Normal dogs have no difficulty learning first to escape from shock, and then to avoid it by jumping as soon as they hear the CS. But the dogs pretrained with inescapable shock almost invariably failed to jump at all. Similar effects have been shown with a variety of species and aversive stimuli in appropriately engineered situations. The effects often generalize from one highly aversive stimulus, such as water immersion (very unpleasant for rats), to another, such as shock – emphasizing that the aversive property of the situation is the defining one for most animals.

Learned helplessness is a dramatic example of the impairment of learning by *US preexposure*, discussed earlier: The dogs that received shock in the absence of the CS subsequently found it more difficult to detect the predictive properties of the CS. The magnitude of the impairment is surprising, but the severity of the shock can perhaps account for it.

The result also follows from the candidate-selection idea: Pretraining with inescapable shock reduces the candidate set; subsequent training in a situation that permits avoidance or escape from shock is perceived as being essentially the same (aversiveness defines the situation), so inaction persists. Shock continues to occur because the animal fails to avoid, so the process is stable – a self-fulfilling prophecy.

Some have argued that learned helplessness goes beyond the effect of CS preexposure. For example, a number of experiments have shown that the damaging effects of inescapable shock can be mitigated if animals have some prior experience with escapable shock. But the inferential explanation for US preexposure and related effects given earlier implies that the animal's candidate set should be determined by *all* its exposures to shock, not just the most recent, so this result does not justify placing learned helplessness in a special category.[23]

The effects of prior experience on learned helplessness are just what one would expect from the principles of memory: Animals exposed only to response-independent shock, looking back will see no effective response; animals with some experience of escaping shock see some possibilities for active escape that should transfer to the new shock situation. The importance of memory is also emphasized by the time limitations on the transfer from the response-independent shock to the test situation: If the two are separated by more than 24 hr, the dogs succeed in learning the avoidance response. As the memory of response-independent shock fades, the situation becomes more novel, the candidate set is in consequence larger, and the animal learns.

Learned helplessness and response-produced-shock behavior can both be abolished by increasing behavioral variation. For example, helplessness can be overcome by physically helping the dog over the barrier – forcing it to sample. We have already seen that prior training with escapable shock leads via transfer to a reservoir of behavior that prevents helplessness. Other experiments have shown that changes in shock intensity and other features of the situation may mitigate helplessness – presumably by reducing transfer from the inescapable-shock situation. All generalization decrement (of which this is an example) represents increase in behavioral variation.

As this analysis leads one to expect, these maladaptive behaviors are metastable; that is, if behavior is disrupted in some way, it may not return to its previous form. For example, if responding that produces electric shock is somehow prevented, when the block is removed, the behavior is likely not to recover.

Learning is a way for animals to use their limited resources for action and computation as efficiently as possible. Particular activities are more likely to occur in situations where they will be of use – food-related activities when food is likely, defense reactions where there is threat, and so on. In this way, energy is conserved and risk minimized. *Sampling*, variation in behavior so as to discover the properties of a situation, and *exploitation*, making use of what is learned, are necessarily in opposition. Without sampling, nothing can be learned; but without exploitation, there is no benefit to learning. Shock-produced behavior and learned helplessness are simply illustrations of one of the two kinds of mistake an animal can make during the process of learning: It can sample too much, and waste resources, or it can sample too little, and make mistakes. Lacking omniscience, there is no way for an animal, including man, to be certain of avoiding these two errors. The very best guess a creature can make is that when the cost of errors is severe, and he has developed some way of coping, he should stick with it: Don't mess with a winning (or at least, not-losing) combination. In human social life, the more serious the decision, the more it is surrounded with form and ritual: from cockpit drills to marriage – when the outcome is momentous, a rigid pattern of coping tends to emerge.

This proneness to fixity in critical situations reflects not stupidity, but intelligence. Simple animals don't show these rigidities because their representation of the world is so primitive they cannot afford to rely on it. In place of rigid patterns, tied to well-specified situations, they must waste resources in random searching – which protects them from traps, but limits their ability to allocate their resources efficiently (see Chapter 3). No one has shown learned helplessness in a pigeon, and rats only show it in situations where even unshocked animals learn slowly. It is easy to show in people, and terrifying shock is not needed. The phenomenon is well known to sociologists under the name of *self-fulfilling prophecy*.

Recent work in human decision making has amply documented how people limit their sampling in ways that can lead to poor decisions. Consider, for example, how competitive research grants are awarded, how students are admitted to a selective college, or how personnel are selected. In every case, the thing on which a decision is based (the *selection measure*) is necessarily different from the final behavior that is the goal of the selection (the *performance measure*): research performance for the research grant, success in college and in later life for college admission, or on-the-job performance for personnel selection. Since performance cannot be measured at the time of selection, some indirect measure must be used, such as evaluation of a research proposal by a peer group (for a research award) or some combination of test scores and scholastic record (for college admission). Invariably a selection measure is chosen that will allow applicants to be ranked. This procedure assumes that the measure is positively related to the criterion; that is, people with high measure scores are likely also to score highly in terms of the performance measure: Highly ranked students should do well in college; highly ranked research proposals should lead to high-quality research.

The decision situation is illustrated in Figure 14.12. The horizontal axis shows the selection measure (test score, for example); the vertical axis shows the performance measure (grade-point average, say). The performance- and selection-criterion values are shown by the crossed lines. Each individual can be represented by a point in this space whose coordinates are his scores on the test and in reality (performance score). The oval area shows the region where most individuals will lie, assuming that the test has some validity, but is not perfect (all points would lie on a line for a perfect measure).

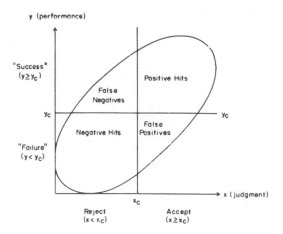

Figure 14.12. The relation between selection and performance measures, and criteria for selection and performance, given an imperfect selection measure. (After Einhorn & Hogarth, 1978.)

The decision maker obviously has two tasks here: to find the best rule for picking people, and to pick the best people. These two tasks are in opposition; they are just the dichotomy between sampling and exploitation we have already discussed: To pick the best people, the decision maker need only set his selection criterion as high (as far to the right) as possible. In this way his proportion of *positive hits* (people above the *performance* criterion) will be as high as possible. But to estimate the validity of his selection measure, he also needs to sample individuals to the left of the criterion in Figure 14.12; that is, he needs an estimate of the proportion of false negatives – which means admitting some students with poor scores, or awarding research grants for poorly rated proposals. The necessary sampling of people to the left of the selection criterion almost never occurs,[24] partly because of the cost of false positives to the decision maker (bad things happen to the personnel manager who hires a loser, but a winner missed costs him little), but partly also because many decision makers are unaware of the need to keep up to date on the validity of their decision rules.

People are just as dumb as animals in their unawareness of how poorly they are sampling. For example, in one experiment, subjects were presented with four cards lying on a table. A single letter or number (a, b, 2, or 3) was visible on each card. They were then told to check the truth of the statement "All cards with a vowel on one side have an even number on the other." To check positive instances, the card with "a" should be turned over; all subjects did this. But to check negative instances, the appropriate choice is the card with "3" visible; none turned over the "3." Many chose the "2" card, which of course adds nothing to the choice of the "a" card.

Both animals and people show this hill-climbing bias in favor of positive hits in experiments designed to test the so-called information hypothesis for *observing behavior*. In an observing-behavior experiment, subjects are offered the opportunity to produce a stimulus that tells them whether a reinforcer is likely or not, but has no effect on the actual availability of the reinforcer. For example, suppose that food for hungry pigeons is scheduled on a variable interval 60-sec (VI 60) schedule for pecking on the left key, which is normally white. Pecks on the right key have no effect on food delivery, but turn the left key green if the VI is due to make food available within 30 sec. Under favorable conditions, pigeons will soon learn to peck the right key because the rate of food delivery in the presence of green is higher than its rate in the situation as a whole, the standard conditioned-reinforcement result. And they peck for green on the right even though the observing response has no effect on the overall rate of reinforcement. A procedure that can give the animals essentially the same information is one where the left key is normally white, as before, but a peck on the right, "observing" key turns it green if food is *not* to become available in the next 30 sec. Pigeons will not peck the "observing" key under these conditions because it is associated with a rate of reinforcement lower than the overall average. This preference for "good news" is the same as the preference for positive hits in the human experiments. It reflects the universality of the hill-climbing heuristic (follow the direction of improving outcomes) discussed in connection with choice in Chapter 9.

Extinction

Extinction is both the abolition of an existing reinforcement schedule (procedural definition), and the decrease in the previously reinforced response that usually follows (behavioral definition). It involves the same learning mechanisms as the original acquisition of the response. The major difference is in the repertoire with which the animal confronts the changed situation, and the feedback function. In acquisition, the initial repertoire is variable and exploratory, appropriate to a novel situation. At the beginning of extinction, a single response class typically dominates. In acquisition, the animal's task is to detect a positive contingency between behavior and reinforcer; in extinction to detect the absence of any effective response. The task in extinction is more difficult, in the sense that in acquisition there is a "right answer," whereas in extinction there is none. Some say that "extinction is slower than conditioning," but there is no real evidence for this: The tasks are different, not the processes.

Extinction is rarely immediate. For example, given a pigeon well-trained to peck on a variable-interval schedule, the cumulative record of responding on the first day of extinction will follow a negatively accelerated pattern of slow decline, with a complete cessation of responding only after several hours. When the animal is returned to the apparatus the next day, however, responding will resume at almost its old pace: This is termed *spontaneous recovery*. Spontaneous recovery makes great sense from a functional point of view – after all, things may have changed after a delay, and perhaps the mass of previous successful experience before the single extinction day is, in some sense, worth more than that one negative experience. The resemblance to habituation, and recovery from habituation after lapse of time, is also striking.

In terms of mechanism, spontaneous recovery seems best explained as a reflection of memory processes. It is a consequence of Jost's law, the gain in influence of old experiences at the expense of more recent ones with the passage of time. Thus, at the end of the first extinction session, the recent experience of (procedural) extinction is decisive, and the animal ceases to respond. But at the beginning of the next session, the experience of reinforcement in numerous past sessions reasserts its influence, and responding resumes. Soon, the growing period without reinforcement again exerts its effect and responding ceases. The same process is repeated, with diminished effect, the next day. As a backlog of days without reinforcement is accumulated, these experiences exert a growing effect at the beginning of each session, and behavior is finally abolished.

The memory property described by Jost's law is also involved in sequential effects: As we will see in a moment, an intermittently reinforced response usually takes longer to extinguish than a continuously (fixed-ratio one) reinforced response. But if an animal is trained first with intermittent reinforcement, then with continuous, and then reinforcement is withdrawn, the earlier experience has an effect: The response extinguishes slowly, at a rate appropriate to intermittent reinforcement.

If, during initial training, reinforcement is predictable – delivered for every response, or at regular intervals, or only in the presence of a specific signal – then its omission will be easy to detect and extinction should be rapid. Conversely, if reinforcement occurs at unpredictable times, its omission will be hard to detect and extinction should be slow. For example, consider two groups of rats, one trained to run down an alley for food delivered on every trial (*continuous* reinforcement), the other trained to run to food delivered only on half the trials, randomly determined (*partial* reinforcement). How long will each group continue to run if food is omitted? It is hard to put oneself in a frame of mind where the actual result, the *partial* group runs longer, is surprising. Yet in the dawn days of learning theory, this unsurprising outcome was known as "Humphreys' paradox"; it is now known as the *partial-reinforcement effect* (PRE). The paradox came from the early idea that reinforcement was a "strengthener," which seemed to imply that the more frequently reinforced *continuous* group should have a stronger habit than the *partial* group, so should run longer. As we have seen, animals adapt to reinforcement contingencies in much more subtle ways than the old strength model implies.

The effect of training procedure on resistance to extinction (persistence) is only one of the questions one might ask about extinction. Other questions concern the effects of training pattern on behavior in extinction and the effects of extinction on relearning of the same and related tasks; the first question gets at the processes involved in original learning ("what is learned," this topic has already been discussed in Chapters 10, 11, 12, and 13), the second can tell us something about the way the organism is changed by extinction.

Persistence is perhaps the most tractable problem; most work has been done on it, and the questions about it can be posed most clearly. Consider again the continuous- versus partial-reinforcement alley study. How much more effective is partial than continuous reinforcement in building persistence? The first question that comes up concerns the number of initial training trials: Should we equate the two groups in terms of total number of trials—in which case the *continuous* group will get twice as many reinforcers—or in terms of number of reinforcers— in which case the *partial* group will get twice as many trials? The first choice is the usual one, but there is obviously no right answer to this question. A better question may be: What is the effect of number of reinforcements on resistance to extinction? If we can't control for something, then just take the bull by the horns and measure it directly. This was done in several classic studies,[25] with mixed results. The earliest studies seemed to show a monotonic effect: The more rewards received, the greater subsequent resistance to extinction. But later work sometimes suggested a nonmonotonic effect: Resistance to extinction rises at first, but then declines in very well-trained animals. And, of course, in these studies, number of *trials* is perfectly confounded with number of reinforcements; to attempt a deconfounding would just bring us back to the partial versus continuous reinforcement experiment. Evidently there is no solution to the problem here.

An alternative tack is to look directly at the effect of the training procedure (number of trials, frequency of reinforcement) on the level of *behavior* – speed of running, rate of lever pressing, probability of a conditioned response. This is a familiar problem: In free-operant experiments, the relation between frequency of reinforcement and rate of response is the *response function*, discussed extensively in earlier chapters. Similar functions can be obtained for any response and reinforcement measures. If we restrict ourselves to *asymptotic* behavior, that is, behavior after sufficient training that it shows no systematic change, response functions are stable and well defined. For example, for variable-interval reinforcement, over most of the range the response function is negatively accelerated: As obtained rate of reinforcement increases, response rate increases, linearly at first, and then at a decelerating rate up to a maximum. For ratio schedules, over a similar range, the function has a negative slope; for spaced-responding (DRL) schedules it is linear, and so on (see Chapter 7).

We can get at the resistance-to-extinction problem by considering the effective *stimuli* controlling this behavior. There are two kinds of discriminative stimuli: the reinforcer itself (i.e., food, shock), and everything else. We have already seen considerable evidence for the discriminative function of reinforcers, and there is additional evidence. For example, if a food-reinforced activity is thoroughly extinguished, and then free food is given to the animal, responding usually resumes at once. If the temporal pattern of reinforcement is maintained in extinction, but on a response-independent basis (e.g., a variable-time schedule for animals trained on variable-interval), extinction is greatly retarded. This persistence is evidence both that the role of response contingency is minimal once the effective response has been acquired and that periodic food delivery plays a major role in sustaining the instrumental response. Even stronger proof of both these points is provided by a number of studies comparing persistence of a group of animals trained with an omission contingency (a food-avoidance schedule similar to shock postponement) in extinction with yoked animals that receive the same number and distribution of food deliveries, but independently of responding (Rescorla & Skucy, 1969; Uhl, 1974). The general result of these experiments is that the rate of extinction is the same in both groups; in other words, it is the absence of food as a stimulus (rather than absence of the response-contingency) that is responsible for most, perhaps all, the reduction of responding in extinction. I described earlier similar results from experiments on autoshaped pecking. The presence or absence of reinforcers, quite apart from any response contingency, is the major variable that determines persistence.

Suppose we assume, in line with the argument in Chapter 11, that the two sources of stimulus control, reinforcement and everything else, are additive. These assumptions can be summarized thus:

$$x = aR(x) + S_x \qquad (14.1)$$

where x is the rate of response X, $R(x)$ is the reinforcement rate for X, S_x is the (nonreinforcement) discriminative stimulus for X, and a is a constant proportional to reinforcement magnitude representing the relative contribution of reinforcement and stimuli other than reinforcement to discriminative control of X: This formulation assumes that the contribution of reinforcement to total stimulus control of x is directly proportional to reinforcement rate. x is related to $R(x)$ by the response function f: $x = f(R(x))$. Hence the reduction in x associated with omission of reinforcement $(R(x) = 0)$ is given by

$$\Delta x = f(R(x)) - aR(x), \qquad (14.2)$$

which is a reasonable first-approximation estimate for resistance to extinction: directly proportional to base rate of behavior, and inversely related to base rate of reinforcement – I give a slightly more rigorous derivation of equation 14.2 in the notes.[26]

These two assumptions are illustrated graphically in Figure 14.13; the four data points are from a rabbit nictitating-membrane classical-conditioning experiment by Gibbs, Latham, and Gormezano (1978), but the general form of the response function is a familiar one. The straight line through the origin is just the function $x = a R(x)$, which is the hypothesized contribution of reinforcement to total response rate according to my descriptive model. The difference between these two functions (e.g., line segment AB for the 50%-reinforcement condition) represents the net response strength at the outset of conditioning, that is, our estimate of persistence. Note the ranking of these distances for each reinforcement-percentage condition: 25 > 50 > 15 > 100. For the most part, resistance to extinc-

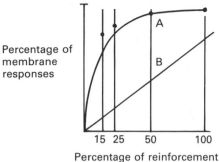

Percentage of membrane responses

Percentage of reinforcement

Figure 14.13. The negatively accelerated curve is the estimated response function relating percentage of conditioned responses to percentage of reinforcement from a rabbit-nictitating-membrane conditioning experiment by Gibbs, Latham, and Gormezano (1978) – the four group-data points are shown. The ray through the origin represents the hypothetical contribution of reinforcement to total response strength, according to the descriptive model discussed in the text.

tion is predicted to be greater the less frequent reinforcement during training; there is a partial reversal for the 15% condition, reflecting the relatively low base response probability during training. This apart, it is clear that this approach does predict the partial-reinforcement effect.

Figure 14.14 shows the rest of the data from the Gibbs et al. experiment. The average maintenance levels in phase II (maintenance) are shown across trials in the center panels, and as overall averages in the previous figure. The left panel in Figure 14.14 shows the changes during acquisition of the membrane response; the right panel shows the changes during extinction. There is a PRE, in the sense that the 100% group extinguishes most rapidly, the 25% group least rapidly. The rank ordering in the right panel of Figure 14.14 is the same as the ordering just derived from Figure 14.13.

This descriptive approach can handle a number of variables that have been shown to affect persistence, such as species, type of response, and amount of reinforcement. For example, for a given reward frequency, resistance to extinction is reduced by increasing reward size. The constant, a, in Equation 14.2 represents the relative contribution to response strength of reinforcement factors, the slope of the straight line in Figure 14.13. Increasing reward size increases a, and thus term $aR(x)$, in direct proportion. The term $f(R(x))$ is also increased, but because f is a negatively accelerated function, this increase will always be less than proportional. Hence, the net effect will be to reduce Δx, our estimate of resistance to extinction, which is the usual empirical result. This result is intuitively plausible, in the sense that the change in extinction from some reward to none should be more easily discriminated if the original reward is large.

The model assumes that reward amount has two effects: to increase a, and to increase the rate of approach of the response function to its fixed maximum. The

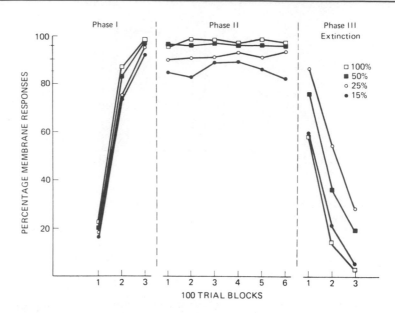

Figure 14.14. Three phases of a classical-conditioning experiment with the rabbit-nictitating-membrane response. Phase I (left): acquisition with 100% reinforcement. Phase II (center): maintenance with 100, 50, 25, or 15% reinforcement. Phase III (right): extinction. (After Gibbs, Latham, & Gormezano, 1978.)

first assumption I make here for the first time; but the second assumption follows from optimality principles discussed earlier. The effect of reward amount on resistance to extinction is shown in Figure 14.15. The dashed straight line is the contribution to response strength of reinforcement rate, as before. The figure shows that at low amounts of reinforcement a PRE should be much more difficult to obtain than at high amounts. The reason is that the PRE depends upon the curvature of the response function, almost independently of the slope of the straight line representing the contribution of reinforcement rate or probability to response strength. Since reducing reinforcement amount reduces the curvature of the function (see note 27), the PRE should be correspondingly reduced. This effect of large rewards in promoting the PRE is well known.

It is likely that this descriptive approach can also account for species and response-type differences. For example, response types differ in their sensitivity to reinforcement. Pecking and lever pressing are very sensitive and occur at high rates, even at low levels, of food reinforcement. On most schedules, their response functions rise steeply at first and then flatten out; they show considerable curvature, and, therefore, the PRE is easy to obtain. Other responses, such as classically conditioned salivation, or treadle pressing for food by pigeons, show a more proportional relation to reinforcement probability; that is, their response functions are closer to being linear. Hence, a PRE is difficult or impossible to

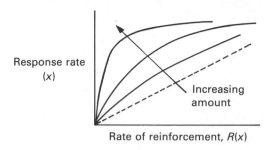

Figure 14.15. Effect of amount of reinforcement on a typical negatively accelerated response function.

demonstrate. Swimming by goldfish is perhaps more like treadle pressing than key pecking, hence it not surprising that a PRE has proven difficult to find, except when reward magnitude is large.

Under appropriate conditions, response functions of almost any form can be found. For example, consider the concurrent variable-interval situation discussed extensively in Chapter 8: The animal responds to two keys, each providing food on independent VI schedules. If the total reinforcement rate is held constant ($R(x)$ + $R(y) = K$), then response rate on one key should be directly proportional to reinforcement rate on that key.[27] We would predict, therefore, that there should be no partial reinforcement effect here. If food is discontinued for responding on one alternative, responding should cease sooner the lower the reinforcement rate in training. Unfortunately, precisely this experiment has not been done, but the prediction certainly accords with one's intuition that when another choice is available, a less-frequently reinforced choice should be more readily given up.

This approach seems to apply equally well to positive and negative reinforcement. We have already seen that avoidance behavior is highly persistent in well-trained animals. Since shock occurs rarely under training conditions, it can make little contribution to response strength; hence its absence in extinction results in little reduction in the animal's tendency to respond, and behavior persists. The greatest persistence has been shown in experiments with variable-interval shock postponement. In this procedure, in the absence of responding, brief shocks occur at an average interval t_S; if a response occurs, the next shock occurs at an average time t_R where $t_R > t_S$. If $t_S = t_R$, there is no benefit to responding and the situation is analogous to the delivery of free food in extinction. Rats well trained to avoid take a very long time to cease responding under these conditions, attesting to the importance of shock as part of the stimulus complex sustaining avoidance responding.

Resistance-to-extinction experiments are expensive to carry out because a separate group of animals is required for each condition. A theory that depends upon parametric information about response functions is, therefore, difficult to test, and this one has not been adequately tested. Nevertheless, it summarizes

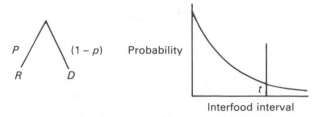

Figure 14.16. Left: Decision tree for extinction. Right: distribution of interfood times on a random-interval schedule.

much of a very large and confusing literature and is worth presenting on that account. I hope that more adequate tests will be forthcoming.

Extinction provides an animal with a detection task. Consequently, optimality theory is a natural way to look at resistance to extinction. Consider an animal on a variable-interval schedule. After much training, the feeder mechanism is turned off: When should he quit? The diagram on the left in Figure 14.16 shows a simple way of representing the animal's problem. He has two hypotheses to consider: (a) That the VI schedule is still in effect, but he has encountered an extra-long interfood interval; or (b) that the VI schedule is no longer in effect. If (a) is true, than he can expect to get food with a frequency R, *where R* is 1/VI value. If (b) is true, then he can expect to get food at some residual rate, D, depending on his priors and past experience.

The mutually exclusive probabilities p, and $1 - p$ associated with these two hypotheses depend upon the distribution of interfood intervals the animal has experienced in the past in the apparatus. Since we are assuming an experienced animal, these are given by the actual distribution of interfood intervals prescribed by the VI schedule. This distribution, for a random-interval schedule, is shown on the right. The area to the right of the vertical line at time t, labeled $p(> t|A)$, represents the proportion of intervals longer than t. If we time extinction from the last food delivery, then the animal must estimate p, the probability of hypothesis A, given a time t since the last food delivery. This can be done theoretically using Bayes' rule, which gives $p(A|t)$ as a function of the animal's prior estimate of A (initially close to one), $p(> t|A)$, and $p(> t|B)$ (which equals one, since no food is delivered if B is true). The problem is tricky, because the analysis must take into account that $p(A|t)$ is constantly updated as t increases.[28]

Despite the formal complexities, it is easy to see intuitively that the animal's estimate of $p(A|t)$ must slowly decrease as t increases, because it becomes less and less likely that an interval this long without food could occur under the prior VI schedule. Thus, the animal should increasingly spend its time doing things other than making the previously reinforced response, leading to the typical negatively accelerated extinction curve. Analyses of this sort can be carried out for any instrumental conditioning arrangement.

The virtue of this type of analysis, pursued rigorously, is that it can provide precise predictions not only about the form of the extinction curve, but also about the resistance to extinction to be expected from different types of schedules, that is, different distributions of interfood intervals, response ratios or trials. Violations of these predictions point to constraints that can provide useful information about learning mechanisms. I expect to see more theoretical and experimental studies along these lines.

SUMMARY

The first step in the learning of birds and mammals seems to be their representation of the situation. If it is incomplete – things about the situation are unexpected – animals set about updating their knowledge. This updating is not desultory, aimless, or unmotivated; checking out things you don't know about is serious business: A starving rat will return to the place where it has just been shocked, or avoid a place where it has just been fed, in order to build up a complete picture of a new environment. Bolles (1969) did an experiment in which two groups of rats were trained to avoid shock in a running wheel. For one group, the avoidance response was running; for the other, it was standing up. Both these responses occur naturally in this situation, yet only the running group learned. The animals that were required to stand up in order to avoid shock not only failed to learn, but the frequency of the response actually decreased with training. The explanation seems to be that the standing response is part of the animal's exploratory repertoire: The rats were looking for an escape route. Once this task is completed – and whether an escape is found or not – the response decreases in frequency, even though it is effective in avoiding shock. Intelligent animals will not act until they know what they are doing.

Once a representation has been formed, it guides action according to the inference principles I have discussed: Stimuli that predict good things will be approached, pecked, or handled, depending on the kind of stimulus, the kind of animal, and the animal's previous experience. This initial behavior may change the situation (via a feedback function), produce good or bad things, which leads the animal to update its representation, which leads to further action, and so on, in a circular process that eventually converges – usually on a behavior that makes things better or prevents them from getting worse.

The smarter the animal, the more comprehensive its representation – and the more it relies upon it. Having a richer mental world allows intelligent animals to respond selectively: Each action can be perfectly suited to circumstances and little effort and time need be wasted. The final stage of learning is a kind of automatization: in song learning, the song becomes almost independent of feedback from its template – the deafened bird continues to sing almost perfectly; in operant conditioning, the well-learned response is reduced to a stereotyped form guided by minimal stimuli. The more intelligent the animal, the more automatic an action seems to become. Less intelligent animals can be less sure about what is appropriate in any given situation (situations associated with instinctive actions are, of course, exceptions to this: Evolution has ensured that even the simplest animals have fail-safe methods for identifying mates, essential foods, and so on) and are accordingly more variable in their behavior, constantly sampling to make sure that an opportunity has not been missed.

But intelligence carries a penalty: The very precision with which an intelligent animal can identify situations means that it will sometimes fail to try actions that could reveal a change in feedback function. The helpless dog acts as if it has perfect confidence that the escapable-shock situation today is the same as the inescapable-shock situation yesterday and does not try again escape responses it has already learned are ineffective. Rats, quite properly less confident of their less sophisticated representation of the world, don't give up sampling so readily, and so, in a comparable situation, do discover that the rules have changed and do learn to escape the shock. People, most intelligent of all, show these *set* effects most dramatically, and do not need the severe variation-reducing manipulations, such as subtetanizing shock, necessary to impair the exploratory behavior of animals. Simple tasks involving negative information, or the application of an untested selection rule, find people at least as reluctant to explore relevant alternatives as dogs.

This opposition between the exploratory and exploitative functions of action makes the analysis of operant learning difficult. For example, bill swiping is part of the exploratory (sampling) repertoire of pigeons, but it can also be learned and used to exploit a food resource. At one time it is guided by the exploratory function; at a later time, it may be guided by the availability of food. Sometimes, as in Bolles' experiment just described, the two functions can be dissociated, but animals are rarely this inflexible, even in shock situations. When we measure just a single response, therefore, as almost all learning studies do, what we are seeing is the effect of a mixture of causes and consequently an apparently chaotic pattern of waxing and waning in the level of the response – until it suddenly appears at essentially full strength (see Figures 5.1 and 14.17).

The ethologist Tinbergen (1963) described four questions that can sensibly be asked of behavior (he was thinking primarily of instinctive behavior): (a) its selective value or function; (b) its causation (controlling stimuli and motivational factors); (c) its development (ontogeny); and (d) its evolutionary history (phylogeny). These questions can be asked of any species-typical behavior, whether learned or innate. I have discussed behavior from all four points of view in this book, although the emphasis is on the first three.

The study of learning mechanisms combines two of Tinbergen's factors: The study of causation, of controlling stimuli, is the study of "What is learned," in psychological terminology. But the study of the *process* of learning is essentially developmental. Operant learning is not (as was once widely believed) a gradual accretion of "strength" by a stimulus-response system akin to the reflex, but rather the building up of a program for action, with inputs both from the present environment and a representation of the past. The study of the learning process is, therefore, more a question of identifying necessary stages through which this process of representation building and program assembly must pass, than the tracing out of some smooth curve. As in development, environmental feedback is involved at each stage. For example, the development of operant pecking involves first learning of predictive stimulus-reinforcer relations. This is usually termed *classical conditioning*, but it has little in common with reflex strengthening: The animal is learning what leads to what, not building a mental muscle. Under uncontrolled conditions, this learning is inextricably mixed up with the occurrence of exploratory responding and the stimulus changes it produces. The separate role of the process of representation building only became clear when pecking was studied in open-loop situations where it was permitted no effect. Then the two-stage process was revealed: First the bird learns what predicts food, only then does it begin to peck. We know rather little about the further stages, except that selection mechanisms, involving both the temporal (predictive) relations between activities and reinforcers and priors derived from heredity and past experience, act to winnow out ineffective variants from the pool provided by classical conditioning.

Future work proceeds on two fronts – which roughly define the divisions in the field – to understand more about the properties of animals' representations, how they are formed, and what they are: What do animals know, and how do they learn it? And to understand more about the selection of responses: What determines the pool of variants, and how are some activities selected over others?

NOTES

1. *Contiguity vs. contingency*. The term *superstition* here was first used by Skinner (1948) to describe the behavior he observed in a most ingenious experiment. Hungry pigeons were placed in a box and given brief access to food at fixed periods – 15 sec for some animals, longer periods for others. This is *temporal conditioning*, which is a Pavlovian procedure because the animal's behavior has no effect on food delivery (other than the necessity to approach the feeder and eat when food appears). Despite the absence of an operant contingency, all the animals developed striking stereotyped, "superstitious" activities in between feeder operations. On its face, this experiment might seem a considerable embarrassment to Skinner's theory, which places great emphasis on the role of the response contingency.

It is a mark of his intellectual resourcefulness that Skinner was not only unworried by this result, but was able to present it as support for his views ("Every crisis is an opportunity" said Winston Churchill). Superstitious behavior was not devastating to Skinner because reinforcement theory existed in two forms, contingency theory and contiguity theory. The first emphasized the importance of *contingency*, i.e., the dependence of reinforcement on the occurrence of the response: Reinforcement strengthens the behavior on which it is dependent (see Chapter 5). In all practical matters, this is the view that prevails. In clinical behavior modification, for example, an undesirable behavior is abolished by omitting reinforcers normally contingent on it, or by delivering contingent punishment; a desirable behavior is strengthened by making reward contingent on it.

Response contingency is a procedural feature, not a behavioral process – contingencies must act through some proximal mechanism. The second view of reinforcement emphasized the mechanism, which was widely thought to be just response-reinforcer *contiguity* (i.e., temporal proximity). Response contingencies work, this view implies, because they ensure that when the reinforcer is delivered, the contingent response is always contiguous with it. As we saw in the last chapter, a similar view once prevailed about the learning of CS-US relations in classical conditioning. At the time of Skinner's original superstition experiment, the operant equivalent of Rescorla's (1967) "truly-random control" experiment had not been run, so the empirical inadequacy of simple contiguity theory was not apparent.

Skinner explained the vigorous, stereotyped behavior of his pigeons in between periodic food deliveries by means of *adventitious reinforcement*, that is, accidental contiguity between food and a behavior that originally occurs for "other reasons." The argument ran as follows: The pigeon is not a passive creature, especially when hungry and in a situation where it receives occasional food. Suppose it happens to be doing something toward the end of an interfood interval and food is delivered. The behavior will be contiguous with the food and so (according to the contiguity theory of reinforcement) will be more likely to occur again. If the next food delivery comes quite soon, this same behavior might still be occurring, and so receive another accidental pairing with food, be further strengthened, occur again in the next interval, and so on. By means of this positive feedback process, some behavior might be raised to a very high probability.

Since there is no real causal relation between behavior and reinforcer, Skinner called the behavior "superstitious," by analogy with human superstitions – which he believed to arise in a similar way.

Skinner's plausible account was not based on direct observation of the process he described. No one had actually recorded these accidental response-reinforcer contiguities or the progressive increase in response strength that was supposed to follow them.

The contiguity view was often defended in a way that rendered it almost immune to disproof. For example, even in 1948 there was some understanding that the delivery of "free" reinforcers on a response-contingent schedule might tend to weaken the instrumental (i.e., reinforced) response (this was demonstrated experimentally several years later). This effect was explained by the hypothesis that if, for some unexamined reason, a behavior other than the instrumental response should happen to occur, then by chance it would sometimes be contiguous with the occasional free reinforcers. This would tend to strengthen "other" behavior and thus, by competition, weaken the reinforced response. No careful observations were done to substantiate this hypothesis (Did such behaviors occur? How often were they contiguous with free food? What is the form of behavioral competition? Did the frequency of the instrumental response drop before or after such accidental pairings?), and since no quantitative details were given (How many accidental pairings are needed to produce how much increment in strength? How often should unreinforced "other" behaviors be expected to occur?) the view was difficult to refute. Subsequently, a few substantiating observations have appeared (see Henton & Iversen, 1978) but the weight of evidence is against Skinner's idea.

The contiguity account of reinforcement in its original form poses a difficult methodological problem. It is a causal hypothesis where the cause, response-reinforcer contiguity, *is not under experimental control*: The experimenter can control the contiguity aspect, but the *animal* determines when he will make the response. Inability to control the occurrence of the response makes it impossible to be sure of the effect of response-reinforcer contiguity. For example, suppose the response occurs and we at once deliver a reinforcer, thus ensuring response-reinforcer contiguity. Suppose that additional responses then occur: Have we demonstrated a strengthening effect of contiguity? Not at all; perhaps this response just happens to occur in runs, so that one response is usually followed by others, quite apart from any reinforcing effect. This is not uncommon – a pigeon will rarely make just one peck, for example. Suppose we reinforce a second response, just to be sure. After a few repeats, no doubt the pigeon is pecking away at a good rate: Have we then demonstrated an effect? Again, not really. By repeating the pairings of response and reinforcer we have now established a real *dependency* between response and reinforcer; perhaps the increase in pecking is just the result of some *other* process that allows the pigeon to detect such *molar contingencies*, as they are called. (We have also established a stimulus contingency between situational stimuli and food, and of course this will also tend to facilitate pecking.)

These difficulties would not be serious if careful observation revealed the kind of process implied by Skinner's argument. If each response-reinforcer pairing produced a predictable and measurable increase in the rate of the reinforced response, his account would be assured. There have been too few careful observational studies; but all those familiar to me have completely failed to find the systematic effects implied by the contiguity view. Instead the reinforced response emerges in an apparently chaotic manner from an unpredictable mix of variable activities; and moreover, the pattern of its emergence shows almost no uniformity from individual to individual.

Staddon and Simmelhag (1971; some aspects of this study were described in Chapter 11) repeated Skinner's superstition experiment, and carefully observed the pigeons' behavior in each interfood interval from the very beginning of training. They found three things that differ from Skinner's account: (a) The activities that developed are of two kinds: *interim activities*, that occur in the first two-thirds or so of the interfood interval, and a single *terminal response*, that occurs during the last third (see Figures 11.4 and 14.6). (b) The terminal response is

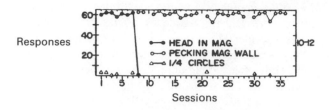

Figure 14.17. Behavior of an individual pigeon during the last 2 sec of a 12-sec interfood interval throughout an experiment in which 3-sec access to response-independent food occurred every 12 sec. (After Staddon & Simmelhag, 1971.)

either pecking or stereotyped pacing activity obviously related to it; the terminal response does not differ from animal to animal in the irregular way implied by Skinner's account. (c) Terminal pecking does not develop in the accidental way implied by the adventitious reinforcement hypothesis.

Figure 14.17 shows experimental data for one pigeon. The figure shows the behavior in the last 2 sec of a 12-sec interfood interval over the whole course of the experiment. For the first seven sessions, when food was delivered the pigeon had its head in the food magazine. We know from other studies that this is a perfectly reinforcible response – if pigeons are required to get their food by putting their heads into the feeder opening, they have no difficulty in learning to do so. For the first seven sessions here, this behavior occurs reliably within 2 sec or less of food delivery. Yet on day 8 it ceases to occur and is completely replaced by the pecking response. Obviously contiguity with food is both ineffective in maintaining head-in-feeder and makes essentially no contribution to the strengthening of pecking. The pecking here is in fact an instance of *autoshaping*, with a temporal stimulus (postfood time) rather than a visual one.

Contiguity also makes no contribution to the genesis of the interim activities, since these are almost never contiguous with food.

Data like these, together with the manifest incompleteness of the contiguity account of reinforcement, have led to the abandonment of "adventitious reinforcement" as an explanation.

I argue in the text that animals detect contingencies with the aid of a mixture of mechanisms, one of which involves the temporal relation between response and reinforcer. The contiguity vs. contingency debate is now widely seen as an ill-formed contest, for two reasons. First, the two explanations are of different types: Contiguity is a mechanism, albeit a primitive and incomplete one, for the detection of contingent relations, a procedural feature. To show the dominance of contingency over contiguity is not to provide an alternative mechanism for conditioning, but just to show the inadequacy of contiguity as a mechanism. Second, the adventitious-reinforcement account is incomplete as an explanation, because the original cause of the adventitiously reinforced behavior is undefined, and because the quantitative properties of the contiguity strengthening process are completely unspecified; as originally stated, adventitious reinforcement can account for anything. The whole issue is now of largely historical interest.

2. There is now an extensive experimental and theoretical literature on autoshaping. Good reviews are by Hearst and Jenkins (1974) and Schwartz and Gamzu (1977). A book edited by Locurto, Terrace, and Gibbon (1981) provides a comprehensive summary of current work.

3. *Optimizing mechanisms and optimal results.* Pigeons that continue to peck in the omission procedure are obviously not maximizing reinforcement rate. Does this result, and a number of others briefly mentioned at the end of Chapter 7, invalidate the optimality approach? Of course not: Even if pigeons had managed to solve the omission-training problem, some other situation could certainly have been found to defeat them. Optimal behavior is always subject to *constraints* and in the limit these constraints define the *mechanism* that animals use to achieve something like reinforcement maximization under many, but never all, conditions. As we saw in earlier chapters, rats and pigeons seem to behave as hill climbers, going for the best option. Autoshaping is one aspect of this rule: The pigeon's evolution tells him that a stimulus as predictive of food as an autoshaping CS (or the positive feature in the feature-negative effect of Jenkins and Sainsbury, discussed in Chapter 11) is worth pecking. Under most conditions, this rule is a good one and does maximize payoff. It happens to fail under the very specialized conditions of the Williams' experiment.

Optimality theory remains useful because when it fails it points to limits on animals' ability to optimize, and thus to constraints and mechanisms. When it succeeds it is useful because it makes sense of behavior in many situations: In Chapter 7, for example, I showed how the broad differences among reinforcement schedules are all explicable in optimality terms. Mechanisms alone cannot integrate a range of observations in this way: For example, we may at some point be able to explain every detail of the development of a bat's wing in terms of its genotype and the biochemical interactions during ontogeny, and similarly for the wings of pigeons, hawks, and hummingbirds. But until we recognize the *function* of a wing as an aerodynamic structure, the similarities among wings must remain incomprehensible.

It should be unnecessary to make this very obvious point (which is just one aspect of the larger question of *reductionism*), but the persistent confusion surrounding it, and the failure of a few behaviorists of literal bent to see the virtues of functional as well as mechanistic approaches to biological problems, makes it necessary to reiterate the basic argument on an intermittent schedule.

4. *Voluntary and involuntary behavior.* The concept of *voluntary* vs. *involuntary* activity bears a close but untidy relationship to the distinction between operants and respondents. At some commonsense level, things like lever pressing, typing, speaking, playing the piano, and so on – conventional operant habits – are also preeminently voluntary. Conversely, we feel that we have little direct control over autonomic activities such as heartrate, the secretion of tears, skin resistance (the main component of lie-detector tests), and so forth. These are normally considered involuntary, even though actors and savvy criminals are able to gain some indirect control over some of them.

The voluntary–involuntary distinction is based on introspection; it cannot be objectively verified. Some professedly voluntary activities, such as the creation of mental images, cannot even be objectively measured: If the act itself is private, any discussion of its voluntariness or involuntariness is obviously futile. Moreover, if we are dealing in subjective impressions, then most would agree that the end-state of learning is usually an essentially *in*voluntary or automatic habit; unless our attention is specifically drawn, few are aware of the routine activities they carry out each day: brushing teeth, flushing the toilet, opening and closing doors, tying a necktie, and so on. Obviously we must distinguish between voluntary and involuntary as *potential* properties of some activity and the voluntariness of *particular instances* of any activity.

Despite these philosophical difficulties, the evidence from human experience is that activities easily modified by operant contingencies are (potentially) voluntary in the instrospective sense; and conversely, involuntary activities are hard to modify in this way. See Kimble and Perlmutter (1970) and Hearst (1975) for more extended discussion of these issues.

5. See Miller and Dworkin (1974) and other papers in the volume edited by Obrist, Black, Brener, and DiCara for a review of some of these problems.

6. *Relation between classical and instrumental conditioning.* Probably the dog did not salivate until it reached the food bowl. Several elegant experiments have established that autonomic responses, such as salivation, tend to occur at the last predictable moment before food delivery. For example, Ellison and Konorski (1964) trained dogs to press a panel nine times to produce a signal; after 8 sec of signal, food was delivered. The dogs pressed but did not salivate before the

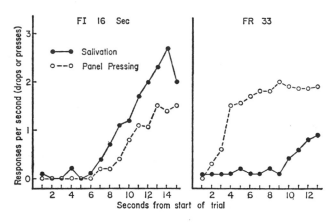

Figure 14.18. The time course of salivation and an operant response during two operant-conditioning procedures. Left panel: during the postfood interval after fixed-interval training. Right panel: during the postfood interval after fixed-ratio training. (After Williams, 1965.)

signal, then salivated, but did not press, afterward. In a similar experiment, Williams (1965) trained dogs on a fixed-interval 16-sec schedule or a fixed-ratio 33-sec schedule and measured both lever pressing and salivation. Time is the best predictor of food in the FI case, and pressing and salivation tended to increase together (Figure 14.18, left panel); but lever pressing is the best predictor of food in the ratio case, and correspondingly, salivation occurred well after the onset of lever pressing (Figure 14.18, right panel).

These and many other studies make the point that autonomic conditioned responses are fundamentally anticipatory.

7. The discussion of event memory in Chapter 12 showed that the effect of a past stimulus depends on both its recency (i.e., delay) and the time between it and potentially interfering events before and after. Hence there is probably nothing corresponding to *the* delay-of-reward gradient, but rather a set of such gradients, one for each interference condition.

8. Of course, backward conditioning is only possible over a small number of trials (see the Keith-Lucas & Guttman experiment discussed in the previous chapter). Other things being equal, over many trials, a CS that reliably occurs after the US will be treated as a CS−, i.e., a predictor of US absence. A backward CS with high prior probability may show some conditioning after one or two trials, but this will usually disappear as the animal gains more experience. For more information on backward conditioning see Mahoney and Ayres (1976) and Wagner and Terry (1975).

9. *Radical behaviorism.* The traditional radical behaviorist position, now of largely historical interest, was that operant conditioning is the strengthening of an essentially arbitrary response in the presence of an equally arbitrary stimulus. The hereditary slate was if not clean at least cluttered only with sets of elementary sensations and muscle movements. These arbitrary responses were "emitted" by the organism, much as alpha particles are emitted by radium. Perhaps all responses were not emitted equally often, but no explicit account was taken of built-in situational biases, of the type just discussed.

Effects of past experience were recognized only in the form of the empirically defined concepts of stimulus and response generalization: A response might be expected to occur in a new situation similar (in some sense) to a familiar one "controlling" that particular response. But "similarity" is not an objective, physical property of events: Like beauty, similarity is in the eye (and brain) of the beholder (Chapter 10). The relation between the purely empirical results of generalization experiments and the internal representation of which they are a manifestation was specifically excluded from behaviorist theory. All was unstructured. Any response might be controlled by any stimulus and strengthened by any reinforcer.

Skinner in 1966 retreated somewhat from this impractical environmentalism, but provided no constructive alternative to it. The original scheme was simple and comprehensible, and its difficulties were not obvious to its adherents – partly because they constructed a language within which questions of representation

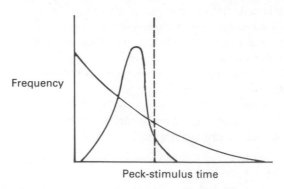

Frequency

Peck-stimulus time

Figure 14.19. Hypothetical distributions illustrating the problem of discriminating between response-dependent and response-independent stimulus changes in the experiment of Killeen (1975). The narrow distribution on the left depicts response-dependent delays; the broad distribution on the right depicts response-independent delays.

and computation could not be expressed. In consequence many honest toilers were won over by the disarming simplicities of stimulus-response behaviorism.

10. Breland and Breland (1961). Keller Breland was an early associate of Skinner's; the title of the Brelands' paper is, of course, a parody of Skinner's great work *The behavior of organisms* (1938). The Brelands also wrote a book, published in 1966, that summarizes their views on animal behavior.

11. This statement is a bit of a simplification. There will be selection for canalized development of a behavior pattern (i.e., for innateness) to the extent that the expected fitness benefit of engaging in the activity is discriminably greater than the expected benefit of not engaging in it. Over a population of individuals, there will be some probability distribution of outcomes associated with each of these two mutually exclusive and exhaustive courses of action. For any pair of such hypothetical distributions it is possible to arrive at a measure of difference, analogous to the d' of signal-detection theory, which will be related to both the means and variances of the two distributions. The larger this difference, the greater the selection pressure for innate determination of the activity.

12. The problem posed by Killeen's procedure is illustrated in Figure 14.19. The time between a peck and a response-dependent stimulus change is shown by the distribution on the left, which has two main properties: low variance (the delay is almost constant, since it is determined by the scheduling apparatus) and small mean; the delay is short and constant. The distribution of times between a peck and a response-independent stimulus change is shown on the right, which has high variance and mean: The delay is variable and long. The bird's task is to set a criterion (the vertical line, for example) that minimizes the two kinds of error that are inevitable with two such overlapping distributions. The pigeons came very close to the ideal choice.

13. *Response selection.* The rules of response selection are a major unsolved problem in learning. It is clear by now that stimulus contingencies alone can induce behavior, according to the priors appropriate to the motivational system involved. It is also clear that these priors can be overridden. For example, pigeons prefer to peck for food, and pecking will often occur to a food signal ("often" must be replaced by "invariably" if the signal is relatively short and the pigeon very hungry). Even so, if they are not too hungry, and the food rate not too high, pigeons can learn to press a treadle or turn a pirouette for food. This is presumably an effect of response-reinforcer contiguity, which overrides their initial tendency to peck – although, as I argue in the text, contiguity can be distinguished from some other method for detecting *molar dependency* only if the number of trials is small. As the bird is trained, only treadle pressing is contiguous with food; other activities are temporally further removed. This is evidently sufficient to strengthen treadle pressing at the expense of pecking and other predisposed activities.

Nevertheless, the treadle-press response is fragile: Sometimes the birds come to peck at the treadle, even if this interferes with food delivery. For example, on a spaced-responding schedule, treadle pecking tends to occur at a rate high enough to miss many potential reinforcements; when the birds treadle press with their feet, the rate is lower, and they consequently get more food. Nevertheless, some animals switch from treadle pressing to treadle pecking – and lose food thereby.

Contiguity acts selectively in two ways: on the *form* of the response, and on the *place* that the response occurs or the *stimulus* to which it is directed. The great majority of conditioning experiments deal exclusively with the second type of selection. Most operant-conditioning studies study the pecking of pigeons; the reinforcement contingencies are used not to vary the topography of the response, but the stimuli to which it is made. Most rat studies involve either lever pressing or locomotion. In these cases, the response contingency is not in conflict with the stimulus contingency: The response is one that naturally tends to occur in food situations, and providing the contingency is positive (i.e., the response produces, not prevents, food) the stimulus to which the animal is attending (the lever, the key) will also form part of a stimulus contingency. Training only breaks down when the contingency is negative (as in omission training) or some other stimulus is a better predictor of food than the one to which the effective response must be directed.

For example, imagine an experiment in which the pigeon faces two response keys. On the left, an autoshaping procedure is in effect: The key is dark for random times averaging 1 min, then an 8-sec green stimulus comes on at the end of which food is delivered. The right key is always white, but food can be obtained by a peck during the 8-sec illumination of the left key. This experiment has not been done because the result is a foregone conclusion. There is little doubt that even if the pigeon were pretrained to peck the right key, exposure to this procedure would soon lead to exclusive pecking of the left key during the green stimulus – even though the animal thereby loses the additional food delivery it could obtain by pecking the right key.

There are situations (we don't really know how common) where temporal relations can override predisposition. For example, pigeons can be trained to behave adaptively in the omission-training procedure by turning off the CS before the bird has a chance to peck at it: A criterion distance between bird and key is established and the CS is turned off if the bird gets closer than the criterion. By progressively reducing the criterion distance pigeons can be brought to the point where they refrain from pecking, although they usually continue to approach the CS. In many cases (and this sometimes happens in omission training even without explicit training) the birds adapt by pecking to one side of the key – a selection for *place* that allows the predisposed response to occur without preventing food (see Barrera, 1974; Lucas, 1975; Peden, Browne, & Hearst, 1977; Wessells, 1974).

We do not know how priors (i.e., the inducing effect of a CS) and contiguity combine to guide behavior. An economical possibility is that the priors act partly through the delay-of-reinforcement gradient. For example, suppose for a given activity that the breadth of the gradient in Figure 14.3 depends on the motivational properties of the situation. For a pigeon in food situations, the gradient might be quite broad for pecking, but quite steep for some other response, like treadle pressing. Given that pecking is both more likely to occur in such a situation, and more tolerant of both forward and backward time differences, pecking might well predominate even if food delivery is sometimes prevented. Conversely, in some other situation, say one involving electric shock, the gradient for pecking might be steep, for wing striking (an aggressive response) quite broad, so that pecking may fail to occur even if it is the effective avoidance response. This speculation has some appeal but is difficult to test, because the contiguity variable cannot act until the response has occurred – inducing factors must always have priority and cannot be entirely eliminated experimentally. Consequently, it is almost impossible to test the effect of contiguity independent of inducing factors.

14. The immediate effect of response–reinforcer contiguity can sometimes be quite dramatic. Who has not had the experience of making some small action – turning on a light, say – that happens to coincide with some large but unconnected event – a power failure, for example: The impression that one's action caused the contiguous event is very strong.

15. The problem of proper sequencing is related to the notion of priors. The objective of the initial reinforcements must be to create a series of situations where at each step the set of induced behaviors (i.e., the response candidates) will contain members increasingly close to the desired form. Or, more generally, it is to find a sequence of states such that the *final* state will contain the desired response. We assume that this final state is best approached via intermediates that are increasingly similar (in our human judgment) to the desired form, but of course this need not always be true. Learning, especially in subhuman animals', usually proceeds by gradual steps, but so-called *insight* learning shows that it does not always do so. *Incubation* in human creative endeavor shows that problems are sometimes solved after delays of days or weeks without explicit training. Solutions so arrived at often involve combinations of disparate elements. The optimal training sequence may sometimes take the animal away from the target response.

These problems arise from the nature of the candidates that are selected by any conditioning process. For convenience, I speak of "response" and "stimulus" candidates, but my own speculation is that what is selected are *elements* analogous to the instructions and "macros" of a computer program. Sometimes these elements correspond pretty closely to "input" (stimulus) or "output" (response) properties; sometimes they correspond to stimulus-response *links*, so that the presence of a given stimulus is sufficient to elicit a given response; and at other times they may serve intermediate (computational) functions and have no overt accompaniment. The final skill or concept (program) involves elements at all levels, so that efficient shaping must sometimes bear no obvious relation to the target behavior.

The new field of cognitive science is much concerned with the representation of knowledge. Cognitive science could help solve these problems, but unfortunately its proponents are interested almost exclusively in people or machines. People represent the most difficult problem to solve, and machines a set of problems remote from biology. These emphases make cognitive science both less successful, and less useful to animal psychologists, than it might be if it were less anthropocentric. As a consequence, our current understanding of the process of learning is still primitive. It would be nice to see a more concerted attack on problems of representation and learning in animals, where they should be simpler to solve than in people. For an excellent, brief survey of cognitive science see Posner and Shulman (1979).

16. Food-related activities can also occur just *after* food (Penney & Schull, 1977), so that the interfood interval can be divided into three periods. The critical variable seems to be the magnitude of the stimuli associated with food delivery: If food delivery, and especially the *end* of food delivery, is clearly signaled, less food-related behavior occurs just after food (Reid & Staddon, 1982). For reasons given in the text, food is often a signal for more food; consequently unless food offset is signaled clearly there is some tendency for food-related activities to persist for a while after feeding.

17. Schedule-induced polydipsia was discovered by Falk (1961). For a general review of schedule induced behavior, see Staddon (1977a).

18. *Concurrent-chained schedules.* There has been considerable interest in chained schedules over the years because they seem to offer a way to measure reinforcing value independently of reinforcement schedule. Concurrent-chained schedules comprise two independent, two-link chained schedules in which the first links are usually the same (typically VI 60-sec schedules) and the second links are different. Reinforcement for pecking a first-link choice is the appearance of the second-link choice on that key; the other choice is disabled until the animal collects one or a few second-link food reinforcers, after which both first links reappear. Typically, first-link reinforcements are programmed by a single interval timer, with random, equal assignment to each side. The second-link food reinforcement is separately programmed for each key, by interval timers that run only while that component is in effect. The attraction of this procedure is the notion that the proportion of responses on each of the first-link choices can provide a measure of the conditioned-reinforcing effectiveness of the second links that is independent of the rates and patterns of responding maintained by the second-link schedules. The promise has not really been fulfilled, because the strength concept of reinforcement has not lived up to expectations, and because conditioned reinforcement, in particular, has not turned out to be a very useful concept.

The concurrent-chains procedure is complicated. The second-link contingencies favor exclusive choice: Since each timer runs only so long as the animal is in that component, there is no point entering the component with the longer average time to food. This expectation is borne out by experiments in which separate VIs (rather than a single timer) were used for the first link: If the first links are relatively short, pigeons tend to fixate on the key with the better second link. But the more typical, single-assignment first-link VI favors nonexclusive choice, since it ensures that fixation on one key will soon lead to extinction (see Chapter 8, and Staddon, Hinson, & Kram, 1981). It is not easy to anticipate the resultant of these two conflicting contingencies. The results of concurrent-chained-schedule experiments accordingly are often less reliable than the results of the simple concurrent studies discussed earlier.

In an early experiment, Herrnstein (1964) found that the relative frequency of pecking on the two first-link VIs matched the relative rates of reinforcement in the second-link VIs; for example, the birds would peck 2:1 on the left key in the first link if the two second-link VIs were VI 30 sec and VI 60 sec. This seemed an encouraging confirmation of the generality of the *matching law* (see Chapters 8 and 11) and supported the view that the procedure could provide a way of measuring the effectiveness of conditioned reinforcement.

Later work has shown that Herrnstein's result is not general. More commonly, if the first link is not too long relative to the second links, pigeons tend to *overmatch*, that is, disproportionately favor the better second-link VI: If the second-link VI reinforcement rates are in the ratio 2:1, choice proportions might be in the ratio 3:1 (e.g., MacEwen, 1972). In addition, first-link choice depends upon the absolute values of both first and second links. For example, Fantino (1969) showed that preference for the richer of the two terminal VIs diminishes as the length of the first-link VIs increases. This makes adaptive sense: If the first-link VIs are much longer than the second-link VIs, then the animal's objective should be to get into *either* second link, since both are so much closer to food than either first link. Hence, the longer the first links relative to the second links, the closer to indifference should the animal become.

An optimality analysis of concurrent-chain schedules with independent (i.e., dual-assignment, not single-assignment) VIs in the first link is, in fact, quite straightforward. There are only two feasible strategies: either respond exclusively to one side, or respond to both (i.e., alternate between the first links so as to pick up either second link when it becomes available). The analysis, an extension of the analysis of delay situations in Chapter 8, is as follows: The expected food rate, given exclusive choice and assuming that each interval schedule runs only when its stimulus is on is

$$1/(T_1 + t_1), \tag{N14.1}$$

where T_1 is the interval value in the first link, and t_1 the interval value in the second link. If the animal approximately alternates in the first link so that either second link is entered as soon as it becomes available, then the expected food rate is the sum of the average time in the first link, plus the expected time to food in the second links, weighted by the proportion of time the animal enters each of the second links. The average time in the first link is then the reciprocal of the sum of the rates, that is, $T_1 T_2/(T_1 + T_2)$. The average time in the second link is just $t_1 p + t_2(1 - p)$, where $p = T_2/(T_1 + T_2)$. If alternative 1 is the majority choice, then the switching condition for responding exclusively to 1, versus responding to both, is the sum of these two equations, versus Equation N14.1, which boils down to

$$1/(T_1 + t_1) \geq (T_1 + T_2)/(T_1 T_2 + T_2 t_1 + T_1 t_2), \qquad t_1 < t_2$$

which reduces to

$$t_2 \geq T_1 + t_1. \tag{N14.2}$$

Equation N14.2 has the surprising feature that the decision whether to respond to one or both alternatives depends on only one of the first-link interval values: the value associated with the shorter of the second links. The reason is that after a certain amount of time without choosing alternative 2, the expected time to enter its second link will be close to zero; hence it always pays to choose 2 at some time, so long as t_2, the expected time to food in its second link, is less than the expected time to food on side 1, which is the sum of the two expected delays on that side.

Equation N14.2 can be represented in the usual way as a switching line in t_1/t_2 space, as shown in Figure 14.20. The results of two kinds of experiment can be predicted at once from the figure. First, increasing T, the length of the initial links, moves the switching line up the t_2 axis, and thereby increases the range of t_1 and t_2 values for which the animal should choose both alternatives rather than fixating on one. As this would lead one to expect, pigeons are increasingly indifferent between the two alternatives as T increases. The second prediction is indicated by the dashed line in the figure, which represents the case where $t_1 < t_2$

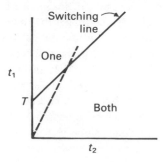

Figure 14.20. Switching-line analysis of concurrent-chained schedule. Solid line is Equation N14.2 in the text, with $T_1 = T_2 = T$ (i.e., equal first links). The dashed line shows the effect of increasing the absolute values of the second links (i.e., t_1 and t_2) while maintaining their ratio constant.

and t_1/t_2 is held constant while the absolute values of t_1 and t_2 are both increased. Clearly as the absolute values of t_1 and t_2 are both increased, preference should shift toward the shorter second link, and this has also been found (see MacEwen, 1972; review in Fantino, 1977).

Equation N14.2 can also predict what should happen when the first-link VIs are varied. For example, suppose both first-link VIs are reduced in value while their ratio remains the same: What will be the effect on preference? If T_1 becomes negligible, then Equation N14.2 reduces to $t_2 \geq t_1$, that is, exclusive choice of the shorter second-link VI. (It may surprise that the optimal solution does *not* converge on an approximation to matching as the first-link VIs are reduced in value, but of course the reason is that the second-link VIs do not run concurrently – hence there is no point entering the longer second-link VI.)

The simple switching-line analysis cannot account for variations in the degree of preference within the "both" region, although it is obviously possible in principle to extend it in several ways, none of them inviting. A few descriptive models of the matching-law variety have been proposed. The simplest is that of Fantino (e.g., 1981), as follows:

$$x_1/x_2 = (D - t_1)/(D - t_2), \quad t_1, t_2 < D, \tag{N14.3}$$

where D is the overall average time to food at the beginning of a cycle (onset of the first link), and x_1 and x_2 are rates of response in the first link. If either t_1 or $t_2 > D$, then exclusive choice is predicted. This model can both explain the effects of increasing T and of increasing the absolute values of t_1 and t_2 while leaving their ratio constant.

A weakness of Equation N14.3 is that it predicts indifference whenever $t_1 = t_2$, irrespective of the value of the first links. Squires and Fantino (1971) therefore proposed a modified version where the delay-reduction ratio (the right-hand side of Equation N14.3) is weighted by the overall food rate on each side:

$$x_1/x_2 = (D - t_1)R_1(D - t_2)R_2, t_1, \qquad t_2 > D, \qquad \text{(N14.4)}$$

which ensures that the equation converges on matching as the second-link VIs approach zero.

Killeen (1982) has extended his arousal model to a wide range of choice procedures, including concurrent chains. Fantino's model and any optimality analysis, is parameter free, but says nothing about the means animals use to adapt to these procedures. Killeen's model has two parameters, but promises to relate choice to processes of arousal. It is too early to say how this model will fare, but its initial predictions seem accurate and quite comprehensive.

A parenthetical note on the virtues of optimality analysis. It took several years after the first experiment with the concurrent-chains procedure before someone noticed that the value of the first-link VI schedules should make a difference to preference. It took a few more years before they noticed that the values of the first-link VIs need not be the same, and that this also might make a difference to choice. There is still no clear distinction drawn between the results to be expected from single-assignment (sometimes called *interdependent*) first-link VIs and dual-assignment (independent) VIs. All these factors come to immediate attention as soon as we ask: What is the optimal strategy? Hence, a major virtue of optimality analysis, quite apart from its theoretical merits, is that it draws attention to what are very likely to be the critical experimental variables.

19. Reviews of experimental work on positive conditioned suppression appear in Lolordo, McMillan, and Riley, 1974; Schwartz and Gamzu, 1977; and Staddon, 1972b. See also Buzsáki (1982) and Buzsáki, Grastyán, Winiczai, and Mód (1979).

20. This experiment and the lever-press studies discussed here are reviewed in Hineline (1977).

21. For example, in delay studies with positive reinforcement, it is always necessary to decide what to do about responses after the effective response (i.e., the response that starts the delay timer that produces reinforcement): Should they be allowed to occur, in which case the reinforcer may follow some responses after less than the prescribed delay? Or should later responses cancel or put off reinforcement, in which case the delay contingency may have a substantial effect on reinforcement *rate*? The compromise of retracting the lever or preventing later responses in some other way introduces an additional stimulus, with its attendant problems of behavioral reallocation. These same problems would confront any attempt to measure a delay gradient for negative reinforcement.

22. It is worth noting that the present treatment of reinforcement in terms of selection from among a stimulus and response candidate set avoids the problems posed for traditional reinforcement theory by avoidance procedures. The creation of the candidate set via surprise, arousal, and inference mechanisms (termed *mechanisms of behavioral variation* in an earlier account, see Staddon & Simmelhag, 1971) had priority over selection. Consequently, the absence in shock postponement schedules of contiguity between response and a stimulus signaling shock reduction just emphasizes the role of inducing mechanisms in the behavior. And as we have seen, these mechanisms are indeed of primary importance in avoidance: The type of response required is critical, as is the animal's past history and the intensity and frequency of shock. Contiguity is probably involved in the sense that the behavior least contiguous with shock tends to be favored, but its effect is indirect and subordinate to inducing factors.

23. *Cognition and learned helplessness.* There has been much controversy over whether the learned-helplessness effect represents a *cognitive* or *associative* deficit – or just an effect of conditioned passivity. There is perhaps less to this argument than meets the eye. Few now believe that all the effects of past experience on an animal are entirely manifest in its overt behavior. Hence all learning by mammals and birds is "cognitive," if by that term we mean nonmotivational internal changes more complex than links between simple stimuli and responses. The argument, therefore, boils down to a difference of degree: *How much* of the effect is explainable solely by motivational or activity-level changes? The answer is far from clear. Some authors are not convinced that *any* of the animal results require more than this; all agree that many of them can be explained in this way. A set of carefully controlled transfer experiments with rats seems to provide the best evidence for some kind of generalized "associative deficit" (Alloy & Seligman, 1979).

The vigor of the controversy perhaps derives not so much from unthinking behaviorists' disbelief in cognition, as from skepticism about the *particular kind* of rationalistic cognitive account most frequently offered. For example, dogs are said to have "acquired the expectation that shock termination would be independent of their responses." This profundity is termed "helplessness theory." The emptyness of the account is obvious once we replace the word "dog" with "computer." Given this account of a computer program, would we feel confident of understanding how it works? Such a statement simply restates the problem; the questions of how the animals learn, what they learn, and what general principles apply to the learning, remain. We saw in the last chapter that the term *expectation*, convenient though it is (and I have made use of it several times in this book), implies some form of representation; without some attempt to specify that representation, the word has no scientific meaning. Regrettably, almost none of the learned-helplessness research has begun to approach this question. Instead ever-more-ingenious experiments obsessively flog once again the dead horse of a stimulus-response account now believed by almost no one. The answer to stimulus-response theory is not yet another control group, but some positive evidence on the form of the animal's representation. Data and theory on this point are sadly lacking.

24. For reviews of the fascinating recent work on human decision making see Einhorn and Hogarth (1978); Kahneman and Tversky (1979); and the book edited by Wallsten (1980).

25. See the books by Mackintosh (1974), Osgood (1953), or Hulse, Deese, and Egeth (1975) for good reviews of these old studies, and of the whole topic of extinction and partial reinforcement.

26. The status of reinforcement rate as a discriminative stimulus is a little different from the status of the usual synchronous stimulus, present all the time. Part of the problem is that the notion of discriminative stimulus is not itself well understood. I am inclined to think of a discriminative stimulus as akin to a memory-retrieval "address" that reinstates a program assembled during previous learning to deal with the situation identified by the stimulus. This is a speculative view, however, and the direct action of most discriminative stimuli makes it easier to think of them almost as elicitors of the behavior under their control – although Skinnerian terminology avoids this by means of the elliptical usage that the discriminative stimulus "sets the occasion for" the operant response.

Since reinforcement occurs only episodically in most operant procedures, reinforcement rate cannot usually act directly to produce responding. Instead it may be that the synchronous stimuli act in some way to reinstate the action program, which also contains some representation for the expected frequency and distribution of reinforcement. Resistance to extinction is then inversely related to the *discrepancy* between the actuality which, in extinction, contains no reinforcement, but does contain the nonreinforcement discriminative stimuli, and the representation, which contains both: the smaller the discrepancy, the greater the resistance to extinction. This situation can be symbolized as follows:

$$S^* = aR(x) + S_x, \tag{N14.3}$$

and

$$S = S_r, \tag{N14.4}$$

where S^* is the representation built up during training, S the situation in extinction, S_x the contribution to S^* of nonreinforcement factors, and $aR(x)$ the contribution of reinforcement factors; a is a constant, so Equation N14.3 assumes that the contribution of reinforcement factors to S^* is proportional to the rate of reinforcement.

Representation S^* then generates a rate of responding, x, which follows the appropriate response function. When reinforcement is omitted in extinction, $R(x) = 0$, and S and S^* differ by the term $aR(x)$. Resistance to extinction presumably depends both on the discrepancy – the larger the discrepancy, the easier it is to detect the change, hence the faster the extinction – and on the base rate of responding – the higher the rate, the longer it should take for responding to disappear. Under training conditions, S^* is associated with a rate of responding x, which is related to $R(x)$ by the response function f: $S^* \to x = f(R(x))$. A simple relation that incorporates the effect of both discrepancy ($S^* - S = aR(x)$) and base rate ($x = f(R(x))$) on persistence (P) is, therefore,

$$P = f(R(x)) - aR(x), \tag{N14.5}$$

which is Equation 14.2 in the text.

27. See Herrnstein (1961). This result follows from the matching relations discussed in Chapter 10: $x = kR(x)/(R(x) + R(y)) = kR(x)/K$ when total reinforcement is constant.

The effect of reinforcement magnitude on the curvature of the variable-interval response function can be derived as follows: The descriptive equation for the VI response function is $x = kR(x)/(R(x) + R(z))$, where $R(z)$ is the reinforcement for "other" behavior. The term $R(z)$ also gives the value of $R(x)$ for which x is half the maximum value; the smaller the value of $R(x)$ at which this occurs, the greater the curvature of the function over a given range. Since the function deals only with relative measures, increasing the magnitude of reinforcement for x is equivalent to reducing the value of the reinforcement for z, hence to reducing term $R(z)$; a reduction in $R(z)$ corresponds to an increase in the curvature of the response function. Hence, increasing reinforcement magnitude increases the predicted curvature of the response function.

28. McNamara and Houston (1980) have tackled this difficult problem in a theoretical paper.

REFERENCES

Allison, J. Paired baseline performance as a behavioral ideal. *Journal of the Experimental Analysis of Behavior*, 1981, *35*, 355–366.

Allison, J., Miller, M., & Wozny, M. Conservation in behavior. *Journal of Experimental Psychology: General*, 1979, *108*, 4–34.

Alloy, L. B., & Seligman, M. E. P. On the cognitive component of learned helplessness and depression. In G. H. Bower (Ed.), *The psychology of learning and motivation* (Vol. 13). New York: Academic Press, 1979.

Anderson, M. C., & Shettleworth, S. J. Behavioral adaptation to fixed-time food delivery in golden hamsters. *Journal of the Experimental Analysis of Behavior*, 1977, *25*, 33–49.

Arend, L. E., Buehler, J. N., & Lockhead, G. R. Difference information in brightness perception. *Perception & Psychophysics*, 1971, *9*, 367–370.

Ashby, W. R. *An introduction to cybernetics*. London: Chapman & Hall, 1956.

Atkinson, J. W., & Birch, D. *The dynamics of action*. New York: Wiley, 1970.

Attneave, F. Dimensions of similarity. *American Journal of Psychology*, 1950, *63*, 516–556.

Azrin, N. H., Hutchinson, R. R., & Hake, D. F. Attack, avoidance, and escape reactions to aversive shock. *Journal of the Experimental Analysis of Behavior*, 1967, *10*, 131–138.

Bacotti, A. V. Matching under concurrent fixed-ratio variable-interval schedules of food presentation. *Journal of the Experimental Analysis of Behavior*, 1977, *25*, 171–182.

Barrera, F. J. Centrifugal selection of signal-directed pecking. *Journal of the Experimental Analysis of Behavior*, 1974, *22*, 341–355.

Bateson, P. P. G. The imprinting of birds. In S.A. Barnett (Ed.), *Ethology and development*. London: Heinemann, 1974.

Baumol, W. J. *Economic theory and operations analysis* (4th ed.). Englewood Cliffs, N.J.: Prentice-Hall, 1977.

Beatty, W. W., & Shavalia, D. A. Rat spatial memory: resistance to retroactive interference at long retention intervals. *Animal Learning and Behavior*, 1980, *8*, 550–552.

Békésy, G. von. *Sensory inhibition*. Princeton: Princeton University Press, 1967.

Bell, P. R. (Ed.) *Darwin's biological work*. Cambridge: Cambridge University Press, 1959.

Benson, W. W. Evidence for the evolution of unpalatability through kin selection in the Heliconiinae (Lepidoptera). *American Naturalist*, 1971, *105*, 213–226.

Bitterman, M. E. Phyletic differences in learning. *American Psychologist*, 1965, *20*, 396–410.

Bitterman, M. E. Habit-reversal and probability learning: rats, birds, and fish. In R. M. Gilbert & N. S. Sutherland (Eds.), *Animal discrimination learning*. New York: Academic Press, 1969.

Blough, D. S. Attention shifts in a maintained discrimination. *Science*, 1969, *166*, 125–128.

Blough, D. S. Steady-state data and a quantitative model of operant generalization and discrimination. *Journal of Experimental Psychology: Animal Behavior Processes*, 1975, *104*, 3–21.

Blough, D. S. Effects of the number and form of stimuli on visual search in the pigeon. *Journal of Experimental Psychology: Animal Behavior Processes*, 1979, *5*, 211–223.

Bogdany, F. J. Linkage of learning signals in honey bee orientation. *Behavioral Ecology and Sociobiology*, 1978, *3*, 323–336.

Bolles, R. C. Avoidance and escape learning: Simultaneous acquisition of different responses. *Journal of Comparative and Physiological Psychology*, 1969, *68*, 355–358.

Bolles, R. C. Species-specific defense reactions and avoidance learning. *Psychological Review*, 1970, 77, 32–48.

Boren, J. J. Resistance to extinction as a function of the fixed ratio. *Journal of Experimental Psychology*, 1961, *61*, 304–308.

Boring, E. G. *Sensation and perception in the history of experimental psychology.* New York: Appleton-Century-Crofts, 1942.

Boring, E. G. *A history of experimental psychology.* New York: Appleton-Century-Crofts, 1957.

Bouzas, A., & Baum, W. M. Behavioral contrast of time allocation. *Journal of the Experimental Analysis of Behavior*, 1976, *25*, 179–184.

Bower, G. H., & Hilgard, E. R. *Theories of learning.* Englewood Cliffs, N.J.: Prentice-Hall, 1975.

Brandt, H. Die lichtorientierung der mehlmotte, *Ephestia kuehniella. Zeitschrift für Vergleichende Physiologie*, 1934, *20*, 646–673.

Brazier, M. B. The historical development of neurophysiology. In J. Field, H. Magoun, & V. E. Hall (Eds.), *Handbook of physiology, Section I.* Washington, D.C.: American Physiological Society, 1959.

Breland, K., & Breland, M. The misbehavior of organisms. *American Psychologist*, 1961, *16*, 661–664.

Breland, K., & Breland, M. *Animal behavior.* New York: Macmillan, 1966.

Brobeck, J. R. Effects of variations in activity, food intake, and environmental temperature on weight gain in the albino rat. *American Journal of Physiology*, 1945, *143*, 1–5.

Brown, P. L., & Jenkins, H. M. Auto-shaping of the pigeon's key-peck. *Journal of the Experimental Analysis of Behavior*, 1968, *11*, 1–8.

Buck, S. L., Rothstein, B., & Williams, B. A. A re-examination of local contrast in multiple schedules. *Journal of the Experimental Analysis of Behavior*, 1975, *24*, 291–301.

Buddenbrock, W. von. Mechanismus der phototropen bewegungen. *Wissenschaft Meeresuntersuch. N. F. Abteilung Helgoland*, 1922, *15*, 1–19.

Bush, R. R., & Mosteller, F. *Stochastic models for learning.* New York: Wiley, 1955.

Buszâki, G. The "where is it?" reflex: autoshaping the orienting response. *Journal of the Experimental Analysis of Behavior*, 1982, *37*, 461–484.

Buzsáki, G., Grastyán, E., Winiczai, Z., & Mód, L. Maintenance of signal directed behavior in a response dependent paradigm: a systems approach. *Acta Neurobiologiae Experimentalis*, 1979, *39*, 201–217.

Campbell, D. T. On the conflicts between biological and social evolution and between psychology and moral tradition. *American Psychologist*, 1975, *30*, 1103–1126.

Caplan, A. I., & Ordahl, C. P. Irreversible gene repression model for the control of development. *Science*, 1978, *201*, 120–130.

Caraco, T., Martindale, S., & Whitham, T. S. An empirical demonstration of risk-sensitive foraging preferences. *Animal Behaviour*, 1980, *28*, 820–830.

Catania, A. C. Concurrent performances: reinforcement interaction and response independence. *Journal of the Experimental Analysis of Behavior*, 1963, *6*, 253–264.

Catania, A. C. Self-inhibiting effects of reinforcement. *Journal of the Experimental Analysis of Behavior*, 1973, *19*, 517–526.

Catania, A. C. *Learning*. Englewood Cliffs, N.J.: Prentice-Hall, 1979.

Catania, A. C. Preference for free choice over forced choice in pigeons. *Journal of the Experimental Analysis of Behavior*, 1980, *34*, 77–86. (a)

Catania, A. C. Freedom of choice: a behavioral analysis. In G. H. Bower (Ed.), *The psychology of learning and motivation* (Vol. 14). New York: Academic Press, 1980. (b)

Catania, A. C. Contingency, contiguity, correlation, and the concept of causation. In P. Harzem & M. D. Zeiler (Eds.), *Advances in the study of behaviour. Vol. 2: Predictability, correlation and contiguity*. Chichester, UK: John Wiley, 1981.

Catania, A. C., & Gill, C. A. Inhibition and behavioral contrast. *Psychonomic Science*, 1964, *1*, 257–258.

Catania, A. C., & Reynolds, G. S. A quantitative analysis of the behavior maintained by interval schedules of reinforcement. *Journal of the Experimental Analysis of Behavior*, 1968, *11*, 327–383.

Catania, A. C., Silverman, P. J., & Stubbs, D. Concurrent performances: stimulus-control gradients during schedules of signalled and unsignalled concurrent reinforcement. *Journal of the Experimental Analysis of Behavior*, 1974, *21*, 99–107.

Catchpole, C. K. *Vocal communication in birds*. Baltimore: University Park Press, 1979.

Cerella, J. Visual classes and natural categories in the pigeon. *Journal of Experimental Psychology: Human Perception and Performance*, 1979, *5*, 68–77.

Charnov, E. L. Optimal foraging: the marginal value theorem. *Theoretical Population Biology*, 1976, *9*, 129–136.

Cherry, C. *On human communication*. New York: Science Editions, 1961.

Chiang, A. C. *Fundamental methods of mathematical economics*. Tokyo: McGraw-Hill Kogakusha, 1974.

Chomsky, N. A review of B. F. Skinner's *Verbal behavior. Language*, 1959, *35*, 26–58.

Church, R. M. Systematic effect of random error in the yoked control design. *Psychological Bulletin*, 1964, *62*, 122–131.

Church, R. M. The internal clock. In S. H. Hulse, H. Fowler, & W. K. Honig (Eds.), *Cognitive processes in animal behavior*. Hillsdale, N. J.: Erlbaum, 1978.

Church, R. M. Short-term memory for time intervals. *Learning and Motivation*, 1980, *11*, 208–219.

Collier, G., Hirsch, E., & Hamlin, P. H. The ecological determinants of reinforcement in the rat. *Physiology and Behavior*, 1972, *9*, 705–716.

Collier, G., Hirsch, E., & Kanarek, R. The operant revisited. In W. K. Honig & J. E. R. Staddon (Eds.), *The handbook of operant behavior*. Englewood Cliffs, N. J.: Prentice-Hall, 1977.

Cook, R. M., & Cockrell, B. J. Predator ingestion rate and its bearing on feeding time and the theory of optimal diets. *Journal of Animal Ecology*, 1978, *47*, 529–547.

Cooper, L. A. Internal representation. In D. R. Griffin, (Ed.) *Animal mind – human mind*. Berlin/Heidelberg/New York: Springer-Verlag, 1982.

Corning, W., Dyal, J., & Willows, A. O. D. (Eds.) *Invertebrate learning* (Vols. 1–3). New York: Plenum, 1973.

Cowie, R. J. Optimal foraging in Great Tits (*Parus major*). *Nature*, 1977, *268*, 137–139.

Crespi, L. P. Quantitative variation in incentive and performance in the white rat. *American Journal of Psychology*, 1942, *5*, 467–517.

Crovitz, H. F. *Galton's Walk: methods for the analysis of thinking, intelligence, and creativity*. New York: Harper & Row, 1970.

Croze, H. *Searching image in carrion crows*. Berlin: Paul Parey, 1970.

Crozier, W. J., & Cole, W. H. The phototropic excitation of *Limax. Journal of General Physiology*, 1929, *12*, 669–674.

Dale, R. H. I., & Staddon, J. E. R. A temporal theory of spatial memory. Unpublished.

D'Amato, M. R. Delayed matching and short-term memory in monkeys. In G. H. Bower (Ed.), *The psychology of learning and motivation: advances in research and theory.* New York: Academic Press, 1973.

Darwin, C. *The descent of man, and selection in relation to sex.* London: John Murray, 1871.

Darwin, C. *The origin of species.* Oxford: Oxford University Press, 1951. (Reprinted from the 6th ed., 1872.)

Darwin, C. *The expression of the emotions in man and animals.* London: John Murray, 1872.

Darwin, C. *The variation of animals and plants under domestication.* London: John Murray, 1875. (a)

Darwin, C. *The movements and habits of climbing plants.* London: John Murray, 1875. (b)

Darwin, C. *The power of movement in plants.* London: John Murray, 1880.

Davies, N. B. Prey selection and the search strategy of the spotted flycatcher (*Muscicapa striata*), a field study on optimal foraging. *Animal Behaviour*, 1977, *25*, 1016–1033.

Davies, N. B., & Houston, A. I. Owners and satellites: the economics of territory defence in the Pied Wagtail, *Motacilla alba. Journal of Animal Ecology*, 1981, *50*, 157–180.

Davis, J. D., & Levine, M. W. A model for the control of ingestion. *Psychological Review*, 1977, *84*, 379–412.

Davis, M. Socially induced flight reactions in pigeons. *Animal Behaviour*, 1975, *23*, 597–601.

Dawkins, M. Perceptual changes in chicks, another look at the "search image" concept. *Animal Behaviour*, 1971, *19*, 566–574. (a)

Dawkins, M. Shifts of "attention" in chicks during feeding. *Animal Behaviour*, 1971, *19*, 575–582. (b)

Dennett, D. C. *Brainstorms.* Montgomery, Vt.: Bradford, 1978.

de Ruiter, L. Some experiments on the camouflage of stick caterpillars. *Behaviour*, 1952, *4*, 222–232.

de Villiers, P. A. The law of effect and avoidance: a quantitative relationship between response rate and shock-frequency reduction. *Journal of the Experimental Analysis of Behavior*, 1974, *21*, 223–235.

de Villiers, P. A. Choice in concurrent schedules and a quantitative formulation of the Law of Effect. In W. K. Honig & J. E. R. Staddon (Eds.), *Handbook of operant behavior.* Englewood Cliffs, N.J.: Prentice-Hall, 1977.

de Villiers, P. A., & Herrnstein, R. J. Toward a law of response strength. *Psychological Bulletin*, 1976, *83*, 1131–1153.

Dickinson, A. *Contemporary animal learning theory.* Cambridge University Press, 1980.

Dickinson, A., Hall, G., & Mackintosh, N. J. Surprise and the attenuation of blocking. *Journal of Experimental Psychology: Animal Behavior Processes*, 1976, *2*, 213–222.

Dickinson, A., & Mackintosh, N. J. Classical conditioning in animals. *Annual Review of Psychology*, 1978, *29*, 587–612.

Domjan, M. Ingestional aversion learning: unique and general processes. In J. S. Rosenblatt, R. A. Hinde, & M. C. Busnel (Eds.), *Advances in the study of behavior* (Vol. 11). New York: Academic Press, 1980.

Dunham, P. J. Punishment: method and theory. *Psychological Review*, 1971, *78*, 58–70.

Dunham, P. J. The nature of reinforcing stimuli. In W. K. Honig & J. E. R. Staddon (Eds.), *Handbook of operant behavior.* Englewood Cliffs, N.J.: Prentice-Hall, 1977.

Edmon, E. L. Multiple schedule component duration: a reanalysis of Shimp and Wheatley (1971) and Todorov (1972). *Journal of the Experimental Analysis of Behavior*, 1978, *30*, 239–241.

Einhorn, H. J., & Hogarth, R. M. Confidence in judgment: persistence of the illusion of validity. *Psychological Review*, 1978, *85*, 395–416.

Ekman, G. Dimensions of color vision. *Journal of Psychology*, 1954, *38*, 467–474.

Ellison, G. D., & Konorski, J. Separation of the salivary and motor responses in instrumental conditioning. *Science*, 1964, *146*, 1071–1072.

Estes, W. K. Effects of competing reactions on the conditioning curve for bar-pressing. *Journal of Experimental Psychology*, 1950, *40*, 200–205.

Estes, W. K. The problem of inference from curves based on group data. *Psychological Bulletin*, 1956, *53*, 134–140.

Estes, W. K. The statistical approach to learning theory. In S. Koch (Ed.), *Psychology: a study of a science* (Vol. 2). New York: McGraw-Hill, 1959.

Estes, W. K., & Skinner, B. F. Some quantitative properties of anxiety. *Journal of Experimental Psychology*, 1941, *29*, 390–400.

Estes, W. K., Koch, S., MacCorquodale, K., Meehl, P. E., Mueller, C. G., Schoenfeld, W. N., & Verplanck, W. S. *Modern learning theory*. New York: Appleton-Century-Crofts, 1954.

Ettinger, R. H., & Staddon, J. E. R. The operant regulation of feeding: a static analysis. *Behavioral Neuroscience*, in press.

Falk, J. L. Production of polydipsia in normal rats by an intermittent food schedule. *Science*, 1961, *133*, 195–196.

Fantino, E. Choice and rate of reinforcement. *Journal of the Experimental Analysis of Behavior*, 1969, *12*, 723–730.

Fantino, E. Conditioned reinforcement. In W. K. Honig & J. E. R. Staddon (Eds.), *Handbook of operant behavior*. Englewood Cliffs, N.J.: Prentice-Hall, 1977.

Fantino, E. Contiguity, response strength, and the delay-reduction hypothesis. In P. Harzem & M. D. Zeiler (Eds.), *Advances in the study of behaviour*. Vol. 2: *Predictability, correlation, and contiguity*. Chichester: Wiley, 1981.

Fantino, E., & Logan, C. A. *The experimental analysis of behavior: a biological perspective*. San Francisco: Freeman, 1979.

Farley, J. Automaintenance, contrast, and contingencies: effects of local vs. overall and prior vs. impending reinforcement context. *Learning and Motivation*, 1980, *11*, 19–48.

Feller, W. *An introduction to probability theory and its applications* (2nd ed.). New York: John Wiley, 1957.

Ferster, C. B., & Skinner, B. F. *Schedules of reinforcement*. New York: Appleton-Century-Crofts, 1957.

Fletcher, H. J. The delayed-response problem. In A. M. Schrier, H. F. Harlow, & F. Stollnitz (Eds.), *Behavior of nonhuman primates*. New York: Academic Press, 1965.

Ford, E. B. Polymorphism. *Biological Review*, 1945, *20*, 73–88.

Fraenkel, G. S. Die mechanik der orientierung der tiere im raum. *Biological Review*, 1931, *6*, 36–87.

Fraenkel, G. S., & Gunn, D. L. *The orientation of animals*. Fair Lawn, N.J.: Oxford University Press, 1940. (Dover edition: 1961.)

Frank, J., & Staddon, J. E. R. The effects of restraint on temporal discrimination behavior. *Psychological Record*, 1974, *23*, 123–130.

Friedman, M., & Savage, J. The utility analysis of choices involving risk. *Journal of Political Economy*, 1948, *56*, 279–304.

Friedman, M. I., & Stricker, E. M. The physiological psychology of hunger: a physiological perspective. *Psychological Review*, 1976, *83*, 409–431.

Gallistel, C. R. *The organization of action: a new synthesis*. Hillsdale, N.J.: Erlbaum, 1980.

Garcia, J. Tilting at the paper mills of academe. *American Psychologist*, 1981, *36*, 149–158.

Garcia, J., Clarke, J., & Hankins, W. G. Natural responses to scheduled rewards. In P. P. G. Bateson & P. Klopfer (Eds.), *Perspectives in ethology*. New York: Plenum, 1973.

Garcia, J., Kimeldorf, D. J., & Koelling, R. A. A conditioned aversion towards saccharin resulting from exposure to gamma radiation. *Science*, 1955, *122*, 157–159.

Gendron, R. P. Foraging for cryptic prey. Doctoral dissertation, Duke University, 1982.

Gendron, R. P., & Staddon, J. E. R. Searching for cryptic prey: the effect of search speed. *American Naturalist*, 1983, *121*, 172–186.

Gibbon, J. Scalar expectancy and Weber's law in animal timing. *Psychological Review*, 1977, *84*, 279–325.

Gibbon, J., Berryman, R., & Thompson, R. L. Contingency spaces and measures in classical conditioning. *Journal of the Experimental Analysis of Behavior*, 1974, *21*, 585–605.

Gibbon, J., Locurto, C., & Terrace, H. Signal food contingency and signal food frequency in a continuous-trials autoshaping paradigm. *Animal Learning and Behavior*, 1975, *3*, 313–324.

Gibbs, C. M., Latham, S. B., & Gormezano, I. Classical conditioning of the rabbit nictitating membrane response: effects of reinforcement schedule on response maintenance and resistance to extinction. *Animal Learning and Behavior*, 1978, *6*, 209–215.

Gill, F. B., & Wolf, L. L. Economics of feeding territoriality in the golden-winged sunbird. *Ecology*, 1978, *56*, 333–345.

Glaister, S. *Mathematical methods for economists*. London: Gray-Mills, 1972.

Glazer, H. I., & Weiss, J. M. Long-term interference effect: an alternative to "learned helplessness." *Journal of Experimental Psychology: Animal Behavior Processes*, 1976, *2*, 201–213.

Gottlieb, G. *Development of species identification in birds*. Chicago: University of Chicago Press, 1971.

Gottlieb, G. On the acoustic basis of species identification in wood ducklings (*Aix sponsa*). *Journal of Comparative and Physiological Psychology*, 1974, *87*, 1038–1048.

Gould, J. L. *Ethology: the mechanisms and evolution of behavior*. New York: Norton, 1982.

Gould, J. L., & Gould, C. G. The insect mind: physics or metaphysics? In D. R. Griffin (Ed.), *Animal mind – human mind*. Berlin/Heidelberg/New York: Springer-Verlag, 1982.

Gould, S. J. *Ontogeny and phylogeny*. Cambridge: Harvard University Press, 1977.

Granit, R. *Charles Scott Sherrington: an appraisal*. London: Nelson, 1966.

Grant, D. S. Delayed alternation in the rat: effect of contextual stimuli on proactive interference. *Learning and Motivation*, 1980, *11*, 339–354.

Green, D. M., & Swets, J. A. *Signal detection theory and psychophysics*. New York: Wiley, 1966.

Green, L., & Snyderman, M. Choice between rewards differing in amount and delay: toward a choice model of self control. *Journal of the Experimental Analysis of Behavior*, 1980, *34*, 135–147.

Greenwood, M. R. C., Quartermain, D., Johnson, R. R., Cruce, J. A. F., & Hirsch, J. Food motivated behavior in genetically obese and hypothalamic-hyperphagic rats and mice. *Physiology and Behavior*, 1974, *13*, 687–692.

Griffin, D. R. *The question of animal awareness*. New York: Rockefeller University Press, 1976. (2nd edition, 1981.)

Grindley. G. C. The formation of a simple habit in guinea pigs. *British Journal of Psychology*, 1932, *23*, 127–147.

Grodins, F. S. *Control theory and biological systems*. New York: Columbia University Press, 1963.

Grossberg, S. Psychophysiological substrates of schedule interactions and behavioral contrast. *SIAM-AMS Proceedings*, 1981, *13*, 157–186.

Grossberg, S. *Studies of mind and brain*. Boston: D. Reidel, 1982.

Guttman, N. Effects of discrimination formation on generalization. In D. I. Mostofsky (Ed.), *Stimulus generalization*. Stanford, Calif.: Stanford University Press, 1965.

Guttman, N., & Kalish, H. I. Discriminability and stimulus generalization. *Journal of Experimental Psychology*, 1956, *51*, 79–88.

Hall, G., & Pearce, J. M. Latent inhibition of a CS during CS-US pairings. *Journal of Experimental Psychology: Animal Behavior Processes*, 1979, *5*, 31–42.

Hamilton, B. E., & Silberberg, A. Contrast and autoshaping in multiple schedules varying reinforcer rate and duration. *Journal of the Experimental Analysis of Behavior*, 1978, *30*, 107–122.

Hanson, H. M. The effects of discrimination training on stimulus generalization. *Journal of Experimental Psychology*, 1959, *58*, 321–334.

Harlow, H. F. The formation of learning sets. *Psychological Review*, 1949, *56*, 51–65.

Harlow, H. F., & Harlow, M. K. The affectional systems. In A. M. Schrier, H. F. Harlow, & F. Stollnitz (Eds.), *Behavior of nonhuman primates* (Vol. 2). New York: Academic Press, 1965.

Hassell, M. P. *The dynamics of competition and predation*. London: Edward Arnold, 1976.

Hassell, M. P. *The dynamics of arthropod predator–prey systems*. Princeton, N.J.: Princeton University Press, 1978.

Hatten, J. L., & Shull, R. L. Pausing on fixed-interval schedules: effects of the prior feeder duration. *Behaviour Analysis Letters*, 1983, in press.

Hearst, E. The classical–instrumental distinction: reflexes, voluntary behavior, and categories of associative learning. In W. K. Estes (Ed.), *Handbook of learning and cognitive processes* (Vol. 2). Hillsdale, N.J.: Erlbaum, 1975.

Hearst, E. Stimulus relationships and feature selection in learning and behavior. In S. H. Hulse, H. Fowler, & W. K. Honig (Eds.), *Cognitive processes in animal behavior*. Hillsdale, N.J.: Erlbaum, 1978.

Hearst, E. *The first century of experimental psychology*. Hillsdale, N.J.: Erlbaum, 1979.

Hearst, E., & Jenkins, H. M. Sign-tracking: the stimulus-reinforcer relation and directed action. *Psychonomic Society Monograph*, 1974.

Hearst, E., Koresko, M. B., & Poppen, R. Stimulus generalization and the response-reinforcement contingency. *Journal of the Experimental Analysis of Behavior*, 1964, *7*, 369–380.

Heinroth, O. Beiträge zur Biologie, namentlich Ethologie und Psychologie der Anatiden. Proceedings of the 5th International Ornithological Congress, 1911, 589–702.

Henderson, J. M., & Quandt, R. E. *Microeconomic theory: a mathematical approach*. New York: McGraw-Hill, 1971.

Henton, W. W., & Iversen, I. H. *Classical conditioning and operant conditioning*. New York: Springer-Verlag, 1978.

Herrnstein, R. J. Relative and absolute strength of response as a function of frequency of reinforcement. *Journal of the Experimental Analysis of Behavior*, 1961, *4*, 267–272.

Herrnstein, R. J. Secondary reinforcement and rate of primary reinforcement. *Journal of the Experimental Analysis of Behavior*, 1964, *7*, 27–36.

Herrnstein, R. J. On the law of effect. *Journal of the Experimental Analysis of Behavior*, 1970, *13*, 243–266.

Herrnstein, R. J. Self control as response strength. In C. M. Bradshaw, C. F. Lowe, & E. Szabadi (Eds.), *Recent Developments in the quantification of steady-state operant behavior*. Amsterdam: Elsevier/North-Holland, 1981.

Herrnstein, R. J., & Boring, E. G. *A source book in the history of psychology*. Cambridge: Harvard University Press, 1965.

Herrnstein, R. J., & Heyman, G. M. Is matching compatible with reinforcement maximization on concurrent variable interval, variable ratio? *Journal of the Experimental Analysis of Behavior*, 1979, *31*, 209–223.

Herrnstein, R. J., & Loveland, D. H. Complex visual concept in the pigeon. *Science*, 1964, *146*, 549–551.

Herrnstein, R. J., & Loveland, D. H. Hunger and contrast in a multiple schedule. *Journal of the Experimental Analysis of Behavior*, 1974, *21*, 511–517.

Herrnstein, R. J., & Loveland, D. H. Maximizing and matching on concurrent ratio schedules. *Journal of the Experimental Analysis of Behavior*, 1975, *24*, 107–116.

Herrnstein, R. J., Loveland, D. H., & Cable, C. Natural concepts in pigeons. *Journal of Experimental Psychology: Animal Behavior Processes*, 1976, *2*, 285–302.

Hess, E. H. Space perception in the chick. *Scientific American*, 1956, *195*, 71–80.

Hess, E. H. *Imprinting*. New York: Van Nostrand/Reinhold, 1973.

Hetherington, A. W., & Ranson, S. W. Hypothalamic lesions and adiposity in the rat. *Anatomical Record*, 1940, *78*, 149–172.

Hetherington, A. W., & Ranson, S. W. The spontaneous activity and food intake of rats with hypothalamic lesions. *American Journal of Physiology*, 1942, *136*, 609–617.

Heyman, G. M. Matching and maximizing in concurrent schedules. *Psychological Review*, 1979, *86*, 496–500.

Hicks, J. R. *A revision of demand theory*. Oxford: Clarendon Press, 1956.

Hilgard, E. R., & Marquis, D. G. *Conditioning and learning*. New York: Appleton-Century-Crofts, 1940.

Hinde, R. A. *Bird vocalizations*. Cambridge: Cambridge University Press, 1969.

Hinde, R. A. *Animal behaviour: a synthesis of ethology and comparative psychology* (2nd ed.). New York: McGraw-Hill, 1970.

Hinde, R. A., & Hinde, J. (Eds.), *Constraints on learning: limitations and predispositions*. New York: Academic Press, 1973.

Hineline, P. N. Negative reinforcement and avoidance. In W. K. Honig & J. E. R. Staddon (Eds.), *Handbook of operant behavior*. Englewood Cliffs, N.J.: Prentice-Hall, 1977.

Hinson, J. M., & Staddon, J. E. R. Behavioral competition: a mechanism for schedule interactions. *Science*, 1978, *202*, 432–434.

Hinson, J. M., & Staddon, J. E. R. Maximizing on interval schedules. In C. M. Bradshaw, C. F. Lowe, & E. Szabadi (Eds.), *Recent developments in the quantification of steady-state operant behavior*. Amsterdam: Elsevier/North-Holland, 1981.

Hinson, J. M., & Staddon, J. E. R. Hill-climbing by pigeons. *Journal of the Experimental Analysis of Behavior*, 1983, *39*, 25–47.

Hirsch, E., & Collier, G. The ecological determinants of reinforcement in the guinea pig. *Physiology and Behavior*, 1974, *12*, 239–249.

Hodos, W. & Trumbule, G. H. Strategies of schedule preference in chimpanzees. *Journal of the Experimental Analysis of Behavior*, 1967, *10*, 503-514.

Hoebel, B. G., & Teitelbaum, P. Weight regulation in normal and hypothalamic hyperphagic rats. *Journal of Comparative and Physiological Psychology*, 1966, *61*, 189–193.

Hogan, J. A. How young chicks learn to recognize food. In R. A. Hinde & J. Stevenson-Hinde (Eds.), *Constraints on learning*. New York: Academic Press, 1973.

Hogan, J. A., Kleist, S., & Hutchings, C. S. L. Display and food as reinforcers in the Siamese fighting fish (*Betta splendens*). *Journal of Comparative and Physiological Psychology*, 1970, *70*, 351–357.

Hogan, J. A., & Roper, T. J. A comparison of the properties of different reinforcers. *Advances in the Study of Behavior*, 1978, *8*, 156–255.

Hollings, C. S. The components of predation, as revealed by a study of small mammal predation of the European pine sawfly. *Canadian Entomologist*, 1959, *91*, 293–332.

Hollings, C. S. The functional response of predators to prey density and'its role in mimicry and population regulation. *Memoirs of the Entomological Society of Canada*, 1965, *45*, 5–60.

Holman, G. L. Intragastric reinforcement effect. *Journal of Comparative and Physiological Psychology*, 1969, *69*, 432–441.

Honig, W. K. Studies of working memory in the pigeon. In S. H. Hulse, H. Fowler, & W. K. Honig (Eds.), *Cognitive processes in animal behavior*. Hillsdale, N.J.: Erlbaum, 1978.

Honig, W. K., Boneau, C. A., Burstein, K. R., & Pennypacker, H. S. Positive and negative generalization gradients obtained after equivalent training conditions. *Journal of Comparative and Physiological Psychology*, 1963, *56*, 111–116.

Honig, W. K., & Urcuioli, P. J. The legacy of Guttman and Kalish (1956): 25 years of research on stimulus generalization. *Journal of the Experimental Analysis of Behavior*, 1981, *36*, 405–455.

Houston, A. I., & McNamara, J. How to maximize reward rate on two variable-interval paradigms. *Journal of the Experimental Analysis of Behavior*, 1981, *35*, 367–396.

Hovland, C. I. Human learning and retention. In S. S. Stevens (Ed.), *Handbook of experimental psychology*. New York: Wiley, 1951.

Hull, C. L. *Principles of behavior*. New York: Appleton-Century, 1943.

Hull, C. L. *A behavior system*. New Haven: Yale University Press, 1952.

Hull, C. L., Hovland, C. I., Ross, R. T., Hall, M., Perkins, D. T., & Fitch, F. B. *Mathematico-deductive theory of rote learning*. New Haven: Yale University Press, 1940.

Hulse, S. H., Deese, J., & Egeth, H. *The psychology of learning*. New York: McGraw-Hill, 1975.

Hutchinson, J. W., & Lockhead, G. R. Similarity as distance: a structural principle for semantic memory. *Journal of Experimental Psychology: Human Learning and Memory*, 1977, *3*, 660–678.

Huxley, J. *Problems of relative growth*. London: Methuen, 1932. (Dover edition, 1972.)

Immelman, K. Sexual imprinting in birds. *Advances in the Study of Behavior*, 1972, *4*, 147–174.

Ingle, D. Focal attention in the frog: behavioral and physiological correlates. *Science*, 1975, *188*, 1033–1035.

Intriligator, M. D. *Mathematical optimization and economic theory*. Englewood Cliffs, N.J.: Prentice-Hall, 1971.

Jenkins, H. M. Generalization gradients and the concept of inhibition. In D. I. Mostofsky (Ed.), *Stimulus generalization*. Stanford, Calif.: Stanford University Press, 1965.

Jenkins, H. M., & Sainsbury, R. S. Discrimination learning with the distinctive feature on positive or negative trials. In D. I. Mostofsky (Ed.), *Attention: contemporary theory and analysis*. New York: Appleton-Century-Crofts, 1970.

Jennings, H. S. *Behavior of the lower organisms*. Bloomington: Indiana University Press, 1976. (Reprint of the 1906 edition.)

Kahneman, D., & Tversky, A. Prospect theory: an analysis of decision-making under risk. *Econometrica*, 1979, *47*, 263–291.

Kamil, A. C. Systematic foraging by a nectar-feeding bird, the Amakihi (*Loxops virens*). *Journal of Comparative and Physiological Psychology*, 1978, *92*, 388–396.

Kamil, A. C., & Sargent, T. D. (Eds.), *Foraging behavior: ecological, ethological, and psychological approaches*. New York: Garland Press, 1981.

Kamin, L. J. Selective attention and conditioning. In N. J. Mackintosh & W. K. Honig (Eds.), *Fundamental issues in associative learning*. Halifax: Dalhousie University Press, 1969.

Keesey, R. E., & Powley, T. L. Hypothalamic regulation of body weight. *American Scientist*, 1975, *63*, 558–565.

Keith-Lucas, T., & Guttman, N. Robust-single-trial delayed backward conditioning. *Journal of Comparative and Physiological Psychology*, 1975, *88*, 468–476.

Kelleher, R. T. Chaining and conditioned reinforcement. In W. K. Honig (Ed.), *Operant behavior: areas of research and application*. New York: Appleton-Century-Crofts, 1966.

Kello, J. Observation of the behavior of rats running to reward and nonreward in an alleyway. Doctoral dissertation, Duke University, 1973.

Kello, J. E., Innis, N. K., & Staddon, J. E. R. Eccentric stimuli on multiple fixed-interval schedules. *Journal of the Experimental Analysis of Behavior*, 1975, 23, 233–240.

Kelsey, J. E., & Allison, J. Fixed ratio lever pressing by VMH rats: work vs. accessibility of sucrose reward. *Physiology and Behavior*, 1976, 17, 749-754.

Kennedy, G. C. The hypothalamic control of food intake in rats. *Proceedings of the Royal Society of London, Series B*, 1950, *137*, 535–549.

Kennedy, J. S. Coordination of successive activities in an aphid: reciprocal effects of settling on flight. *Journal of Experimental Biology*, 1965, *43*, 489–509.

Kennedy, J. S. Behaviour as physiology. In L. B. Browne (Ed.), *Insects and physiology*. New York: Springer-Verlag, 1967.

Killeen, P. R. On the temporal control of behavior. *Psychological Review*, 1975, *82*, 89–115.

Killeen, P. R. Superstition: a matter of bias, not detectability. *Science*, 1978, *199*, 88–90.

Killeen, P. R. Incentive theory. In D. Bernstein (Ed.), *Nebraska Symposium on Motivation, 1981*. Lincoln: Nebraska University Press, 1982.

Killeen, P. R. Incentive theory: II. models for choice. *Journal of the Experimental Analysis of Behavior*, 1982, *38*, 217–232.

Killeen, P. R., Hanson, S. J., & Osborne, S. R. Arousal: its genesis and manifestation as response rate. *Psychological Review*, 1978, *85*, 571–581.

Kimble, G. A. *Hilgard and Marquis' conditioning and learning* (2nd ed). New York: Appleton-Century-Crofts, 1961.

Kimble, G. A., & Perlmutter, L. C. The problem of volition. *Psychological Review*, 1970, *77*, 361–384.

Klopfer, P. H., & Gamble, J. Maternal "imprinting" in goats: the role of chemical senses. *Zeitschrift für Tierpsychologie*, 1966, *25*, 588–592.

Koch, S. (Ed.) *Psychology: A study of a science* (Vol. 2). New York: McGraw-Hill, 1959.

Kolata, G. B. Obesity: a growing problem. *Science*, 1977, *198*, 905–906.

Konishi, M., & Nottebohm, F. Experimental studies in the ontogeny of avian vocalizations. In R. A. Hinde (Ed.), *Bird vocalizations*. Cambridge: Cambridge University Press, 1969.

Koshland, D. E. A response regulator model in a simple sensory system. *Science*, 1977, *196*, 1055–1063.

Koshland, D. E. *Bacterial chemotaxis as a model behavioral system*. New York: Raven Press, 1980.

Kraly, F. S., & Blass, E. M. Increased feeding in rats in a low ambient temperature. In D. Novin, W. Wyrwicka, & G. A. Bray (Eds.), *Hunger: basic mechanisms and clinical implications*. New York: Raven Press, 1976.

Krebs, J. R. Behavioural aspects of predation. In P. P. G. Bateson & P. H. Klopfer (Eds.), *Perspectives in ethology*. New York: Plenum, 1973.

Krebs, J. R. Optimal foraging: decision rules for predators. In J. R. Krebs & N. B. Davies (Eds.), *Behavioural ecology: an evolutionary approach*. Sunderland, Mass.: Sinauer, 1978.

Krebs, J. R., & Davies, N. B. (Eds.), *Behavioural ecology*. Sunderland, Mass.: Sinauer, 1978.

Krebs, J. R., & Davies, N. B. *An introduction to behavioural ecology*. Sunderland, Mass.: Sinauer, 1981.

Krebs, J. R., Kacelnik, A., & Taylor, P. Tests of optimal sampling by foraging great tits. *Nature*, 1978, *275*, 27–31.

Kreider, D. L., Kuller, R. G., Ostberg, D. R., & Perkins, F. W. *An introduction to linear analysis*. Reading, Mass.: Addison-Wesley, 1966.

Kroodsma, D. A reevaluation of song development in the song sparrow. *Animal Behaviour*, 1977, *25*, 390–399.

Kroodsma, D. A. Aspects of learning in the ontogeny of bird song: where, from whom, when, how much, which, and how accurately? In G. Burghardt & M. Bekoff (Eds.), *Development of behavior*. New York: Garland Press, 1978.

Kuo, Z-Y. *The dynamics of behavior development: an epigenetic view*. New York: Random House, 1967.

Kuo, Z-Y. The need for coordinated efforts in developmental studies. In L. R. Aronson, E. Tobach, D. S. Lehrman, & J. S. Rosenblatt (Eds.), *Development and evolution of behavior: essays in memory of T. C. Schneirla*. San Francisco: Freeman, 1970.

Lancaster, K. *Mathematical economics*. New York: Macmillan, 1968.

Land, E. H., & McCann, J. J. Lightness and retinex theory. *Journal of the Optical Society of America*, 1971, *61*, 1–11.

Landauer, T. K. Reinforcement as consolidation. *Psychological Review*, 1969, *76*, 82–96.

Lashley, K. S. The problem of serial order in behavior. In L. A. Jeffress (Ed.), *Cerebral mechanisms and behavior*. New York: Wiley, 1951.

Lea, S. E. G. Titration of schedule parameters by pigeons. *Journal of the Experimental Analysis of Behavior*, 1976, *25*, 43–54.

Lea, S. E. G. Concurrent fixed-ratio schedules for different reinforcers: a general theory. In C. M. Bradshaw, E. Szabadi, & C. F. Lowe (Eds.), *Quantification of steady-state operant behavior*. Amsterdam: Elsevier/North-Holland, 1981.

Lea, S. E. G., & Roper, T. J. Demand for food on fixed-ratio schedules as a function of the quality of concurrently available reinforcement. *Journal of the Experimental Analysis of Behavior*, 1977, *27*, 371–380.

Lehrman, D. S. Semantic and conceptual issues in the nature–nurture problem. In L. R. Aronson, E. Tobach, D. S. Lehrman, & J. S. Rosenblatt (Eds.), *Development and evolution of behavior: essays in memory of T. C. Schneirla*. San Francisco: Freeman, 1970.

Lett, B. T. Long-delay learning in the T-maze. *Learning and Motivation*, 1975, *6*, 80–90.

Lewin, R. Seeds of change in embryonic development. *Science*, 1981, *214*, 42–44.

Lockhead, G. R. Identification and the form of multidimensional discrimination space. *Journal of Experimental Psychology*, 1970, *85*, 1–10.

Lockhead, G. R. Processing dimensional stimuli: a note. *Psychological Review*, 1972, *79*, 410–419.

Locurto, C., Terrace, H. S., & Gibbon, J. (Eds.), *Autoshaping and conditioning theory*. New York: Academic Press, 1981.

Loeb, J. *Forced movements, tropisms, and animal conduct*. Philadelphia: Lippincott, 1918. (Dover edition, 1973.)

Lolordo, V. M., McMillan, J. C., & Riley, A. L. The effects upon food-reinforced pecking and treadle-pressing of auditory and visual signals for response-independent food. *Learning and Motivation*, 1974, *5*, 24–41.

Lorenz, K. Companions as factors in the bird's environment. In R. Martin (trans.), *Studies in animal and human behaviour*, Vol. I, Cambridge: Harvard University Press, 1970.

Lowe, C. F., Davey, G. C., & Harzem, P. Effects of reinforcement magnitude on interval and ratio schedules. *Journal of the Experimental Analysis of Behavior*, 1974, *22*, 553–560.

Lucas, C. A. The control of keypecks during automaintenance by pre-keypeck omission training. *Animal Learning and Behavior*, 1975, *3*, 33–36.

Luce, R. D. *Individual choice behavior: a theoretical analysis*. New York: Wiley, 1959.

Luce, R. D. Detection and recognition. In R. D. Luce, R. R. Bush, & E. Galanter, *Handbook of mathematical psychology* (Vol. 1). New York: Wiley, 1963.

Luce, R. D. The choice axiom after twenty years. *Journal of Mathematical Psychology*, 1977, *15*, 215–233.

Luce, R. D., Bush, R. R., & Galanter, E. *Handbook of mathematical psychology* (3 vols.). New York: Wiley, 1963.

Luce, R. D., & Raiffa, H. *Games and decisions*. New York: Wiley, 1957.

Ludlow, A. R. The behaviour of a model animal. *Behaviour*, 1976, *108*, 131–172.

Ludlow, A. R. The evolution and simulation of a decision maker. In F. M. Toates & T. R. Halliday (Eds.), *The analysis of motivational processes*. London: Academic Press, 1980.

MacColl, L. A. *Fundamental theory of servomechanisms*. New York: Van Nostrand, 1945.

MacEwen, D. The effects of terminal-link fixed-interval and variable-interval schedules on responding under concurrent chained schedules. *Journal of the Experimental Analysis of Behavior*, 1972, *18*, 253–261.

Mackintosh, N. J. Comparative psychology of serial reversal and probability learning: rats, birds, and fish. In R. Gilbert & N. S. Sutherland (Eds.), *Animal discrimination learning*. London: Academic Press, 1969.

Mackintosh, N. J. *The psychology of animal learning*. New York: Academic Press, 1974.

Mackintosh, N. J. A theory of animal attention: variations in the associability of stimuli with reinforcement. *Psychological Review*, 1975, *82*, 276–298.

Mackintosh, N. J. Stimulus control: attentional factors. In W. K. Honig & J. E. R. Staddon (Eds.), *Handbook of operant behavior*. Englewood Cliffs, N.J.: Prentice-Hall, 1977.

Mackintosh, N. J., Bygrave, D. J., & Picton, B. M. B. Locus of the effects of a surprising reinforcer in the attenuation of blocking. *Quarterly Journal of Psychology*, 1977, *29*, 327–336.

Macnab, R. M., & Koshland, D. E. The gradient-sensing mechanism in bacterial chemotaxis. *Proceedings of the National Academy of Sciences USA*, 1972, *69*, 2509–2512.

Magnus, D. Experimentelle untersuchungen zur bionomie und ethologie des kaisermantels *Argynnis paphia* L. (Lep. Nymph.) *Zeitschrift für Tierpsychologie*, 1958, *15*, 397–426.

Mahoney, W. J., & Ayres, J. J. B. One-trial simultaneous and backward fear conditioning as reflected in conditioned suppression of licking in rats. *Animal Learning and Behavior*, 1976, *4*, 357–362.

Maier, S. F., & Jackson, R. L. Learned helplessness: all of us were right (and wrong): learned helplessness has multiple effects. In G. H. Bower (Ed.), *The psychology of learning and motivation* (Vol. 13). New York: Academic Press, 1979.

Maller, O. The effect of hypothalamic and dietary obesity on taste preference in rats. *Life Science*, 1964, *3*, 1281–1291.

Malone, J. C., & Staddon, J. E. R. Contrast effects in maintained generalization gradients. *Journal of the Experimental Analysis of Behavior*, 1973, *19*, 167–179.

Marler, P. R., & Hamilton, W. J. *Mechanisms of animal behavior*. New York: Wiley, 1966.

Marler, P., Kreith, M., & Tamura, M. Song development in hand-raised Oregon juncos. *Auk*, 1962, *79*, 12–30.

Marler, P., & Tamura, M. Culturally transmitted patterns of vocal behavior in sparrows. *Science*, 1964, *146*, 1483–1486.

Marshall, A. *Principles of economics* (8th ed.). London: Macmillan, 1925.

Marwine, A., & Collier, G. The rat at the waterhole. *Journal of Comparative and Physiological Psychology*, 1979, *93*, 391–402.

Mast, S. O. *Light and the behavior of organisms.* New York: Wiley, 1911.

Maxwell, J. C. On governors. *Proceedings of the Royal Society of London*, 1868, *16*, 270–283.

May, R. M. Simple mathematical models with very complicated dynamics. *Nature*, 1976, *261*, 459–467.

Maynard Smith, J. The theory of games and the evolution of animal conflicts. *Journal of Theoretical Biology*, 1974, *47*, 209–221.

Maynard Smith, J. Evolution and the theory of games. *Scientific American*, 1976, *64*, 41–45.

Maynard Smith, J. Theory in evolution. *Annual Review of Ecology and Systematics*, 1978, *9*, 31–56.

McCleery, R. On satiation curves. *Animal Behaviour*, 1977, *25*, 1005–1015.

McCleery, R. Optimal behaviour sequences and decision making. In J. R. Krebs & N. B. Davies (Eds.), *Behavioural ecology.* Sunderland, Mass.: Sinauer, 1978.

McFarland, D. J. *Feedback mechanisms in animal behaviour.* New York: Academic Press, 1971.

McFarland, D. J. (Ed.). *Motivational control systems analysis.* London: Academic Press, 1974.

McFarland, D. J., & Houston, A. *Quantitative ethology: the state space approach.* London: Pitman, 1981.

McFarland, D. J., & Sibly, R. The behavioural final common path. *Philosophical Transactions of the Royal Society of London, Series B*, 1975, *270*, 265–293.

McGill, W. J. Stochastic latency mechanisms. In R. D. Luce, R. R. Bush, & E. Galanter (Eds.), *Handbook of mathematical psychology* (Vol. I). New York: Wiley, 1963.

McGinty, D., Epstein, A. N., & Teitelbaum, P. The contribution of oropharyngeal sensations to hypothalamic hyperphagia. *Animal Behaviour*, 1965, *13*, 413–418.

McKearney, J. W. Fixed-interval schedules of electric shock presentation: extinction and recovery of performance under different shock intensities and fixed-interval durations. *Journal of the Experimental Analysis of Behavior*, 1969, *12*, 301–313.

McNamara, J., & Houston, A. I. The application of statistical decision theory to animal behaviour. *Journal of Theoretical Biology*, 1980, *85*, 673–690.

Medin, D. L. Form perception and pattern reproduction by monkeys. *Journal of Comparative and Physiological Psychology*, 1969, *68*, 412–419.

Meschowski, H. *Ways of thought of great mathematicians.* San Francisco: Holden-Day, 1964.

Miles, R. C. Discrimination-learning sets. In A. M. Schrier, H. F. Harlow, & F. Stollnitz (Eds.), *Behavior of nonhuman primates* (Vol. 1). New York: Academic Press, 1965.

Milich, R. S. A critical analysis of Schachter's externality theory of obesity. *Journal of Abnormal Psychology*, 1975, *84*, 586–588.

Miller, G. A. The magical number seven, plus or minus two: some limits on our capacity for processing information. *Psychological Review*, 1956, *63*, 81–97.

Miller, N. E., & Banuazizi, A. Instrumental learning by curarized rats of a specific visceral response. *Journal of Comparative and Physiological Psychology*, 1968, *65*, 1–7.

Miller, N. E. & Dworkin, B. R. Visceral learning: recent difficulties with curarized rats and significant problems for human research. In P. A. Obrist, A. H. Black, J. Brener, & L. V. DiCara (Eds.), *Cardiovascular physiology: current issues in response mechanisms, biofeedback, and methodology.* Chicago: Aldine, 1974.

Milsum, J. H. *Biological control systems analysis*. New York: McGraw-Hill, 1966.

Minsky, M. *Computation: finite and infinite machines*. Englewood Cliffs, N.J.: Prentice-Hall, 1967.

Mittelstaedt, H. Basic solutions to a problem of angular orientation. In R. F. Reiss (Ed.), *Neural theory and modeling*. Palo Alto: Stanford University Press, 1964.

Mittelstaedt, H. Kybernetik der schwereorientierung. *Verhandlungen der Deutschen Zoologischen Gesellschaft, 185–200*.

Moore, E. F. Gedanken-experiments on sequential machines. In C. E. Shannon & J. McCarthy (Eds.), *Automata studies*. Princeton, N.J., Princeton Annals of Mathematics Studies, 1956.

Moore, G. E. *Principia ethica*. Cambridge: Cambridge University Press, 1903.

Moorhouse, J. E., Fosbrooke, I. H. M., & Kennedy, J. S. "Paradoxical driving" of walking activity in locusts. *Journal of Experimental Biology*, 1978, *72*, 1–16.

Morse, W., & Kelleher, R. Determinants of reinforcement and punishment. In W. K. Honig & J. E. R. Staddon (Eds.), *Handbook of operant behavior*. Englewood Cliffs, N.J.: Prentice-Hall, 1977.

Morse, W. H., & Skinner, B. F. A second type of superstition in the pigeon. *American Journal of Psychology*, 1957, *70*, 308–311.

Mrosovsky, N., & Powley, T. L. Set points for body weight and fat. *Behavioral Biology*, 1977, *20*, 205–223.

Müller, A. Über lichtreaktionen von landasseln. *Zeitschrift für Vergleichende Physiologie*, 1925, *3*, 113–144.

Mulligan, J. A. Singing behavior and its development in the song sparrow, *Melospiza melodia*. *University of California Publications in Zoology*, 1966, *81*, 1–76.

Murdoch, W. W. Switching in general predators: experiments on predator specificity and stability of prey populations. *Ecological Monographs*, 1969, *39*, 335–354.

Murdoch, W. W., & Oaten, A. Predation and population stability. In A. Macfadyen (Ed.), *Advances in ecological research* (Vol. 9). New York: Academic Press, 1975.

Myerson, J., & Miezin, F. M. The kinetics of choice: an operant systems analysis. *Psychological Review*, 1980, *87*, 160–174.

Navarick, D. J., & Fantino, E. Self-control and general models of choice. *Journal of Experimental Psychology: Animal Behavior Processes*, 1976, *2*, 75–87.

Neimark, E. D., & Estes, W. K. *Stimulus sampling theory*. San Francisco: Holden-Day, 1967.

Neuringer, A. J., & Chung, S.-H. Quasi-reinforcement: control of responding by a percentage reinforcement schedule. *Journal of the Experimental Analysis of Behavior*, 1967, *10*, 45–54.

Nevin, J. A. Response strength in multiple schedules. *Journal of the Experimental Analysis of Behavior*, 1974, *21*, 389–408.

Nevin, J. A. Overall matching versus momentary maximizing: Nevin (1969) revisited. *Journal of Experimental Psychology: Animal Behavior Processes*, 1979, *5*, 300–306.

Nevin, J. A., & Shettleworth, S. J. An analysis of contrast effects in multiple schedules. *Journal of the Experimental Analysis of Behavior*, 1966, *9*, 305–315.

Nicolaïdis, S., & Rowland, N. Intravenous self-feeding: long-term regulation of energy balance in rats. *Science*, 1977, *195*, 589–591.

Nisbett, R. E. Hunger, obesity, and the ventromedial hypothalamus. *Psychological Review*, 1972, *79*, 433–453.

Novin, D., Wyrwicka, W., & Bray, G. A. (Eds.). *Hunger: basic mechanisms and clinical implications*. New York: Raven Press, 1976.

Olton, D. S. Characteristics of spatial memory. In S. H. Hulse, H. Fowler, & W. K. Honig (Eds.), *Cognitive processes in animal behavior*. Hillsdale, N.J.: Erlbaum, 1978.

Olton, D. S., & Samuelson, R. J. Remembrance of places passed: spatial memory in rats. *Journal of Experimental Psychology: Animal Behavior Processes*, 1976, *2*, 97–116.

Osgood, C. E. *Method and theory in experimental psychology*. New York: Oxford University Press, 1953.

Oster, J. F., & Wilson, E. O. *Caste and ecology in the social insects*. Princeton: Princeton University Press, 1978.

Panksepp, J. Hypothalamic regulation of energy balance and feeding behavior. *Federation Proceedings*, 1974, *33*, 1150–1165.

Parker, G. A., & Stuart, R. A. Animal behavior as a strategy optimizer: evolution of resource assessment strategies and optimal emigration thresholds. *American Naturalist*, 1976, *110*, 1055–1076.

Pavlov, I. P. *Conditioned reflexes* (G. V. Anrep, trans.). London: Oxford University Press, 1927.

Pearce, J. M., & Hall, G. A model for Pavlovian learning: variations in the effectiveness of conditioned but not of unconditioned stimuli. *Psychological Review*, 1980, *87*, 532–552.

Peck, J. W. Situational determinants of the body weights defended by normal rats and rats with hypothalamic lesions. In D. Novin, W. Wyrwicka, & G. A. Bray (Eds.), *Hunger: basic mechanisms and clinical implications*. New York: Raven Press, 1976.

Peden, B. F., Browne, M. P., & Hearst, E. Persistent approaches to a signal for food despite food omission for approaching. *Journal of Experimental Psychology: Animal Behavior Processes*, 1977, *3*, 377–399.

Penney, J., & Schull, J. Functional differentiation of adjunctive drinking and wheel running in rats. *Animal Learning and Behavior*, 1977, *5*, 272–280.

Pietrewicz, A. T., & Kamil, A. C. Search image formation in the blue jay (*Cyanocitta cristata*). *Science*, 1979, *204*, 1332–1333.

Pittendrigh, C. S. Adaptation, natural selection, and behavior. In A. Roe & G. G. Simpson (Eds.), *Behavior and evolution*. New Haven: Yale University Press, 1958.

Popper, K. The philosophy of Karl Popper. In P. Schilpp (Ed.), *The library of living philosophers* (Vol. 14). LaSalle, Ill.: Open Court, 1974.

Popper, K. R., & Eccles, J. C. *The self and its brain*. Berlin, W. Germany: Springer, 1977.

Posner, M. I. *Chronometric explorations of mind*. Hillsdale, N.J.: Erlbaum, 1978.

Posner, M. I., & Shulman, G. L. Cognitive science. In E. Hearst (Ed.), *The first century of experimental psychology*. Hillsdale, N.J.: Erlbaum, 1979.

Postman, L. The history and present status of the law of effect. *Psychological Bulletin*, 1947, *44*, 489–563.

Poulton, E. B. Proof of the protective value of dimorphism in larvae. *Transactions of the Entomological Society*, London, 1888, *9*, 50–56.

Poulton, E. C. The new psychophysics: six models for magnitude estimation. *Psychological Bulletin*, 1968, *69*, 1–19.

Powers, W. T. Quantitative analysis of purposive systems: some spadework on the foundations of scientific psychology. *Psychological Review*, 1978, *85*, 417–435.

Powley, T. L., & Opsahl, C. A. Autonomic components of the hypothalamic feeding syndromes. In D. Novin, W. Wyrwicka, & G. A. Bray (Eds.), *Hunger: basic mechanisms and clinical implications*. New York: Raven Press, 1976.

Premack, D. Reinforcement theory. In D. Levine (Ed.), *Nebraska Symposium on Motivation* (Vol. 13). Lincoln: University of Nebraska Press, 1965.

Premack, D., & Premack, A. J. Increased eating in rats deprived of running. *Journal of the Experimental Analysis of Behavior*, 1963, *6*, 209–212.

Pulliam, H. R. Diet optimization with nutrient constraints. *American Naturalist*, 1975, *109*, 765–768.

Pyke, G. H., Pulliam, H. R., & Charnov, E. L. Optimal foraging: a selective review of theory and tests. *The Quarterly Review of Biology*, 1977, *52*, 137–154.

Rabbitt, P. Sorting, categorization, and visual search. In E. C. Carterette & M. P. Friedman (Eds.), *Handbook of perception*, Vol. XI. New York: Academic Press, 1978.

Rachlin, H. *Introduction to modern behaviorism*. San Francisco: Freeman, 1970.

Rachlin, H. Contrast and matching. *Psychological Review*, 1973, *80*, 217–234.

Rachlin, H. A molar theory of reinforcement schedules. *Journal of the Experimental Analysis of Behavior*, 1978, *30*, 345–360.

Rachlin, H., & Burkhard, B. The temporal triangle: response substitution in instrumental conditioning. *Psychological Review*, 1978, *85*, 22–48.

Rachlin, H., & Green, L. Commitment, choice and self-control. *Journal of the Experimental Analysis of Behavior*, 1972, *17*, 15–22.

Rachlin, H., Green, L., Kagel, J. H., & Battalio, R. C. Economic demand theory and psychological studies of choice. In G. Bower (Ed.), *The psychology of learning and motivation* (Vol. 10). New York: Academic Press, 1976.

Ranson, S. W., & Clark, S. L. *The anatomy of the nervous system*. Philadelphia: Saunders, 1953. (Reprinted, 1958.)

Ratliff, F. *Mach bands: quantitative studies on neural networks in the retina*. San Francisco: Holden-Day, 1965.

Ratliff, F. *Studies in excitation and inhibition in the retina: a collection of papers from the laboratories of H. K. Hartline*. New York: Rockefeller University Press, 1974.

Real, L. A. The kinetics of functional response. *American Naturalist*, 1977, *111*, 289–300.

Real, L. A. On uncertainty and the law of diminishing returns in evolution and behavior. In J. E. R. Staddon (Ed.), *Limits to action: the allocation of individual behavior*. New York: Academic Press, 1980.

Reid, A. K., & Dale, R. H. I. Dynamic effects of food magnitude on interim-terminal interactions. *Journal of the Experimental Analysis of Behavior*, 1983, *39*, 135–148.

Reid, A. K., & Staddon, J. E. R. Schedule-induced drinking: elicitation, anticipation, or behavioral interaction? *Journal of the Experimental Analysis of Behavior*, 1982, *38*, 1–18.

Rescorla, R. A. Pavlovian conditioning and its proper control procedures. *Psychological Review*, 1967, *74*, 71–80.

Rescorla, R. A. Probability of shock in the presence and absence of CS in fear conditioning. *Journal of Comparative and Physiological Psychology*, 1968, *66*, 1–5.

Rescorla, R. A. Pavlovian excitatory and inhibitory conditioning. In W. K. Estes (Ed.), *Handbook of learning and cognitive processes* (Vol. 2). Hillsdale, N.J.: Erlbaum, 1975.

Rescorla, R. A. Stimulus generalization: some predictions from a model of Pavlovian conditioning. *Journal of Experimental Psychology: Animal Behavior Processes*, 1976, *2*, 88–96.

Rescorla, R. A., & Skucy, J. C. Effect of response-independent reinforcers during extinction. *Journal of Comparative and Physiological Psychology*, 1969, *67*, 381–389.

Rescorla, R. A., & Wagner, A. R. A theory of Pavlovian conditioning: variations in the effectiveness of reinforcement and nonreinforcement. In A. Black & W. R. Prokasy (Eds.), *Classical conditioning II*. New York: Appleton-Century-Crofts, 1972.

Revusky, S. H. The role of interference in association over delay. In W. K. Honig & P. H. R. James (Eds.), *Animal memory*. New York: Academic Press, 1971.

Revusky, S. H. Learning as a general process with emphasis on data from feeding experiments. In N. W. Milgram, L. Krames, & T. M. Alloway (Eds.), *Food aversion learning*. New York: Plenum, 1977.

Reynolds, G. S. Attention in the pigeon. *Journal of the Experimental Analysis of Behavior*, 1961, *4*, 203–208. (a)

Reynolds, G. S. Behavioral contrast. *Journal of the Experimental Analysis of Behavior*, 1961, *4*, 57–71. (b)

Reynolds, G. S. On some determinants of choice in pigeons. *Journal of the Experimental Analysis of Behavior*, 1963, *6*, 53–59. (a)

Reynolds, G. S. Some limitations on behavioral contrast and induction during successive discrimination. *Journal of the Experimental Analysis of Behavior*, 1963, *6*, 131–139. (b)

Richards, J. J., & Blackman, D. E. Effect of time-out duration after food reinforcement in a fixed-ratio schedule with pigeons. *Behaviour Analysis Letters*, 1981, *1*, 81–87.

Richardson, M. W. Multidimensional psychophysics. *Psychological Bulletin*, 1938, *35*, 659–660.

Richelle, M., & Lejeune, H. *Time in animal behaviour*. Oxford: Pergamon, 1980.

Riley, D. A., & Roitblat, H. L. Selective attention and related processes in pigeons. In S. H. Hulse, H. Fowler, & W. K. Honig (Eds.), *Cognitive processes in animal behavior*. Hillsdale, N.J.: Erlbaum, 1978.

Rilling, M. Stimulus control and inhibitory processes. In W. K. Honig & J. E. R. Staddon (Eds.), *Handbook of operant behavior*. Englewood Cliffs, N.J.: Prentice-Hall, 1977.

Roberts, W. A. Spatial separation and visual differentiation of cues as factors influencing short-term memory in the rat. *Journal of Comparative and Physiological Psychology*, 1972, *78*, 281–291.

Roberts, W. A., & Dale, R. H. I. Remembrance of places lasts: proactive inhibition and patterns of choice in rat spatial memory. *Learning and Motivation*, 1981, *12*, 261–281.

Roberts, W. A., & Grant, D. S. Short-term memory in the pigeon with presentation time precisely controlled. *Learning and Motivation*, 1974, *5*, 393–408.

Rodgers, W., & Rozin, P. Novel food preferences in thiamine-deficient rats. *Journal of Comparative and Physiological Psychology*, 1966, *61*, 1–4.

Roitblat, H. L. The meaning of representation in animal memory. *Behavioral and Brain Sciences*, 1982, *5*, 353–406.

Roper, T. J. What is meant by the term "schedule-induced," and how general is schedule induction? *Animal Learning and Behavior*, 1981, *9*, 433–440.

Rothkopf, E. Z. A measure of stimulus similarity and errors in some paired-associate learning tasks. *Journal of Experimental Psychology*, 1957, *53*, 94–101.

Rosenblueth, A., Wiener, N., & Bigelow, J. Behavior, purpose and teleology. *Philosophy of Science*, 1943, *10*, 18–24.

Rozin, P., & Kalat, J. W. Specific hungers and poison avoidance as adaptive specializations of learning. *Psychological Review*, 1971, *78*, 459–486.

Russek, M. A conceptual equation of intake control. In D. Novin, W. Wyrwicka, & G. A. Bray (Eds.), *Hunger: basic mechanisms and clinical implications*. New York: Raven Press, 1976.

Russell, B. *History of western philosophy*. London: Allen & Unwin, 1946.

Russell, B. *The basic writings of Bertrand Russell*. In R. E. Egner & L. E. Denonn (Eds.), New York: Simon & Schuster, 1961.

Sainsbury, R. S. Effect of proximity of elements on the feature-positive effect. *Journal of the Experimental Analysis of Behavior*, 1971, *16*, 315–325.

Samuelson, P. A. *Foundations of economic analysis*. Cambridge, Mass.: Harvard University Press, 1965.

Schachter, S., & Rodin, J. *Obese humans and rats*. Potomac, Md.: Erlbaum, 1974.

Schneider, B. A. A two-state analysis of fixed-interval responding in the pigeon. *Journal of the Experimental Analysis of Behavior*, 1969, *12*, 677–687.

Schneider, W., & Shiffrin, R. M. Controlled and automatic human information processing: I. detection, search and attention. *Psychological Review*, 1977, *84*, 1–66.

Schoener, T. W. Theory of feeding strategies. *Annual Review of Ecology and Systematics*, 1971, *2*, 369–404.

Schutz, F. Prägung des sexual verhaltens von enten und gansen durch sozialeindrücke wärhend der jugendphase. *Journal of Neurovisceral Relations*, Supplementum, 1971, *10*, 399–457.

Schwartz, B., & Gamzu, E. Pavlovian control of operant behavior. In W. K. Honig & J. E. R. Staddon (Eds.), *Handbook of operant behavior*. Englewood Cliffs, N.J.: Prentice-Hall, 1977.

Sclafani, A., & Springer, D. Dietary obesity in adult rats: similarities to hypothalamic and human obesity syndromes. *Physiology and Behavior*, 1976, *17*, 461–471.

Sclafani, A., Springer, D., & Kluge, L. Effects of quinine adulterated diets on the food intake and body weight of obese and non-obese hypothalamic hyperphagic rats. *Physiology and Behavior*, 1976, *16*, 631–640.

Seligman, M. E. P., & Hager, J. L. (Eds.). *Biological boundaries of learning*. New York: Appleton-Century-Crofts, 1972.

Seligman, M. E. P., & Maier, S. F. Failure to escape traumatic shock. *Journal of Experimental Psychology*, 1967, *74*, 1–9.

Seward, J. P. An experimental analysis of latent learning. *Journal of Experimental Psychology*, 1949, *39*, 177–186.

Shannon, C., & Weaver, W. *The mathematical theory of communication*. Urbana: University of Illinois Press, 1949.

Shaw, G. B. *The black girl in search of God: and some lesser tales*. Harmondsworth, U.K.: Penguin, 1946. (Reprinted from *The adventures of the black girl in her search for God*, first published in 1932.)

Shepard, R. N. Analysis of proximities as a technique for the study of information processing in man. *Human Factors*, 1963, *5*, 33–48.

Shepard, R. N. Attention and the metric structure of the stimulus space. *Journal of Mathematical Psychology*, 1964, *1*, 54–87.

Shepard, R. N. Approximation to uniform gradients of generalization by monotone transformations of scale. In D. I. Mostofsky (Ed.), *Stimulus generalization*. Stanford: Stanford University Press, 1965.

Shepard, R. N. Form, formation, and transformation of internal representations. In R. Solso (Ed.), *Information processing and cognition: the Loyola symposium*. Hillsdale, N.J.: Erlbaum, 1975.

Shepard, R. N. Multidimensional scaling, tree-fitting, and clustering. *Science*, 1980, *210*, 390–398.

Shepard, R. N., & Cooper, L. A. *Mental images and their rotations*. Cambridge: MIT Press, 1982.

Shepard, R. N., & Metzler, B. Mental rotation of three-dimensional objects. *Science*, 1971, *171*, 701–703.

Sherrington, C. S. *The integrative action of the nervous system*. New Haven: Yale University Press, 1906. (Reprinted, 1947.)

Shettleworth, S. J. Constraints on learning. In D. S. Lehrman, R. A. Hinde, & E. Shaw (Eds.), *Advances in the study of behavior* (Vol. IV). New York: Academic Press, 1972.

Shettleworth, S. J. Reinforcement and the organization of behavior in golden hamsters: hunger, environment and food reinforcement. *Journal of Experimental Psychology*, 1975, *104*, 56–87.

Shettleworth, S. J. Reinforcement and the organization of behavior in golden hamsters: punishment of three action patterns. *Learning and Motivation*, 1978, *9*, 99–123.

Shiffrin, R. M., & Schneider, W. Controlled and automatic human information processing: II. Perceptual learning, automatic attending and a general theory. *Psychological Review*, 1977, *84*, 127–190.

Shimp, C. P. Probabilistically reinforced choice behavior in pigeons. *Journal of the Experimental Analysis of Behavior*, 1966, *9*, 443–455.

Shimp, C. P. Optimal behavior in free-operant experiments. *Psychological Review*, 1969, *76*, 97–112.

Shimp, C. P., & Wheatley, K. L. Matching to relative reinforcement frequency in multiple schedules with a short component duration. *Journal of the Experimental Analysis of Behavior*, 1971, *15*, 205–210.

Sibly, R., & McFarland, D. On the fitness of behavior sequences. *American Naturalist*, 1976, *110*, 601–617.

Sidman, M. *Tactics of scientific research: evaluating experimental data in psychology.* New York: Basic Books, 1960.

Silberberg, A., Hamilton, B., Ziriax, J. M., & Casey, J. The structure of choice. *Journal of Experimental Psychology: Animal Behavior Processes*, 1978, *4*, 368–398.

Simon, H. A. A note on Jost's law and exponential forgetting. *Psychometrika*, 1966, *31*, 505–506.

Simon, H. A. *Models of thought.* New Haven: Yale University Press, 1979.

Singh, D. Effects of preoperative training on food-motivated behavior of hypothalamic hyperphagic rats. *Journal of Comparative and Physiological Psychology*, 1973, *84*, 47–52.

Skinner, B. F. On the rate of formation of a conditioned reflex. *Journal of General Psychology*, 1932, *7*, 274–286.

Skinner, B. F. The generic nature of the concepts of stimulus and response. *Journal of General Psychology*, 1935, *12*, 40–65.

Skinner, B. F. *The behavior of organisms.* New York: Appleton-Century, 1938.

Skinner, B. F. "Superstition" in the pigeon. *Journal of Experimental Psychology*, 1948, *38*, 168–172.

Skinner, B. F. Are theories of learning necessary? *Psychological Review*, 1950, *57*, 193–216.

Skinner, B. F. The phylogeny and ontogeny of behavior. *Science*, 1966, *153*, 1205–1213.

Snowdon, C. T. Motivation, regulation, and the control of meal parameters with oral and intragastric feeding. *Journal of Comparative and Physiological Psychology*, 1969, *69*, 91–100.

Spalding, D. A. Instinct: with original observations on young animals. *Macmillan's Magazine*, 1873, *27*, 282–293.

Spence, K. The differential response in animals to stimuli varying in a single dimension. *Psychological Review*, 1937, *44*, 435–444.

Sperry, R. W. Mechanisms of neural maturation. In S. S. Stevens (Ed.), *Handbook of experimental psychology.* New York: Wiley, 1951.

Sperry, R. W. A modified concept of consciousness. *Psychological Review*, 1969, *76*, 532–536.

Squires, N., & Fantino, E. A model for choice in simple concurrent and concurrent-chains schedules. *Journal of the Experimental Analysis of Behavior*, 1971, *15*, 27–38.

Squires, N., Norborg, J., & Fantino, E. Second-order schedules: discrimination of components. *Journal of the Experimental Analysis of Behavior*, 1975, *24*, 157–171.

Staddon, J. E. R. Some properties of spaced responding in pigeons. *Journal of the Experimental Analysis of Behavior*, 1965, *8*, 19–27.

Staddon, J. E. R. Asymptotic behavior: the concept of the operant. *Psychological Review*, 1967, *74*, 377–391.

Staddon, J. E. R. Multiple fixed-interval schedules: transient contrast and temporal inhibition. *Journal of the Experimental Analysis of Behavior*, 1969, *12*, 583–590.

Staddon, J. E. R. Temporal effects of reinforcement: a negative "frustration" effect. *Learning and Motivation*, 1970, *1*, 227–247. (a)

Staddon, J. E. R. Effect of reinforcement duration on fixed-interval responding. *Journal of the Experimental Analysis of Behavior*, 1970, *13*, 9–11. (b)

Staddon, J. E. R. Reinforcement omission on temporal go–no-go schedules. *Journal of the Experimental Analysis of Behavior*, 1972, *18*, 223–229. (a)

Staddon, J. E. R. Temporal control and the theory of reinforcement schedules. In R. M. Gilbert & J. R. Millenson (Eds.), *Reinforcement: behavioral analyses.* New York: Academic Press, 1972. (b)

Staddon, J. E. R. On the notion of cause, with applications to behaviorism. *Behaviorism*, 1973, *1*, 25–63.

Staddon, J. E. R. Temporal control, attention and memory. *Psychological Review*, 1974, *81*, 375–391.

Staddon, J. E. R. Limitations on temporal control: generalization and the effects of context. *British Journal of Psychology*, 1975, *66*, 229–246. (a)

Staddon, J. E. R. A note on the evolutionary significance of supernormal stimuli. *American Naturalist*, 1975, *109*, 541–545. (b)

Staddon, J. E. R. Schedule-induced behavior. In W. K. Honig & J. E. R. Staddon (Eds.), *Handbook of operant behavior*. Englewood Cliffs, N.J.: Prentice-Hall, 1977. (a)

Staddon, J. E. R. Behavioral competition in conditioning situations: notes toward a theory of generalization and inhibition. In H. Davis & H. M. B. Hurwitz (Eds.), *Operant-Pavlovian interactions*. Hillsdale, N.J.: Erlbaum, 1977. (b)

Staddon, J. E. R. On Herrnstein's equation and related forms. *Journal of the Experimental Analysis of Behavior*, 1977, *28*, 163–170. (c)

Staddon, J. E. R. Theory of behavioral power functions. *Psychological Review*, 1978, *85*, 305–320.

Staddon, J. E. R. Operant behavior as adaptation to constraint. *Journal of Experimental Psychology: General*, 1979, *108*, 48–67. (a)

Staddon, J. E. R. Regulation and time allocation: a commentary on "conservation in behavior." *Journal of Experimental Psychology: General*, 1979, *108*, 35–40. (b)

Staddon, J. E. R. (Ed.). *Limits to action: the allocation of individual behavior.* New York: Academic Press, 1980. (a)

Staddon, J. E. R. Optimality analyses of operant behavior and their relation to optimal foraging. In J. E. R. Staddon (Ed.), *Limits to action: the allocation of individual behavior*. New York: Academic Press, 1980. (b)

Staddon, J. E. R. On a possible relation between cultural transmission and genetical evolution. In P. Klopfer & P. P. G. Bateson (Eds.), *Perspectives in ethology. Vol. 4: Advantages of diversity*. London: Plenum, 1981. (a)

Staddon, J. E. R. Cognition in animals: learning as program assembly. *Cognition*, 1981, *10*, 287–294. (b)

Staddon, J. E. R. Behavioral competition, contrast, and matching. In M. L. Commons, R. J. Herrnstein, & H. Rachlin (Eds.), *Quantitative analyses of operant behavior: matching and maximizing accounts. Quantitative analyses of behavior, Vol. II* (five-volume series). Cambridge, Mass.: Ballinger, 1982.

Staddon, J. E. R., & Ayres, S. Sequential and temporal properties of behavior induced by a schedule of periodic food delivery. *Behaviour*, 1975, *54*, 26–49.

Staddon, J. E. R., & Frank, J. Mechanisms of discrimination reversal. *Animal Behaviour*, 1974, *22*, 802–828.

Staddon, J. E. R., & Gendron, R. P. Optimal detection of cryptic prey may lead to predator switching. *American Naturalist*, 1983, in press.

Staddon, J. E. R., Hinson, J. M., & Kram, R. Optimal choice. *Journal of the Experimental Analysis of Behavior*, 1981, *35*, 397–412.

Staddon, J. E. R., & Innis, N. K. Reinforcement omission on fixed-interval schedules. *Journal of the Experimental Analysis of Behavior*, 1969, *12*, 689–700.

Staddon, J. E. R., & Motheral, S. On matching and maximizing in operant choice experiments. *Psychological Review*, 1978, *85*, 436–444.

Staddon, J. E. R., & Motheral, S. Response independence, matching, and maximizing: a reply to Heyman. *Psychological Review*, 1979, *86*, 501–505.

Staddon, J. E. R., & Simmelhag, V. L. The "superstition" experiment: a reexamination of its implications for the principles of adaptive behavior. *Psychological Review*, 1971, *78*, 3–43.

Starr, B., & Staddon, J. E. R. Temporal control on fixed-interval schedules: signal properties of reinforcement and blackout. *Journal of the Experimental Analysis of Behavior*, 1974, *22*, 535–545.

Starr, B. C., & Staddon, J. E. R. Sensory superstition on interval schedules. *Journal of the Experimental Analysis of Behavior*, 1982, *37*, 267–280.

Stevens, S. S. On the averaging of data. *Science*, 1955, *121*, 113–116.

Stigler, G. J. The development of utility theory. II. *The Journal of Political Economy*, 1950, *58*, 373–396.

Stubbs, D. A. Second-order schedules and the problem of conditioned reinforcement. *Journal of the Experimental Analysis of Behavior*, 1971, *16*, 289–313.

Stubbs, D. A. Response bias and the discrimination of stimulus duration. *Journal of the Experimental Analysis of Behavior*, 1976, *25*, 243–250.

Sutherland, N. S., & Mackintosh, N. J. *Mechanisms of animal discrimination learning*. New York: Academic Press, 1971.

Suzuki, S., Augerinos, G., & Black, A. H. Stimulus control of spatial behavior on the eight-arm maze in rats. *Learning and Motivation*, 1980, *6*, 77–81.

Swazey, J. P. *Reflexes and motor integration: Sherrington's concept of integrative action*. Cambridge, Mass.: Harvard University Press, 1969.

Symposia of the Society for Experimental Biology, No. XVIII: *Homeostasis and feedback mechanisms*. Cambridge University Press, 1964.

Teghtsoonian, R. On the exponents in Stevens' law and the constants in Ekman's law. *Psychological Review*, 1971, *78*, 71–80.

Teitelbaum, P., & Epstein, A. N. The lateral hypothalamic syndrome. *Psychological Review*, 1962, *69*, 74–90.

Terrace, H. S. Stimulus control. In W. K. Honig (Ed.), *Operant behavior: areas of research and application*. New York: Appleton-Century-Crofts, 1966.

Thielcke, G. A. *Bird sounds*. Ann Arbor: University of Michigan Press, 1976.

Thompson, D. W. *On growth and form*. Cambridge: Cambridge University Press, 1961. (Abridgment of the 1917 and 1942 editions, edited by J. T. Bonner.)

Thompson, R. F., & Spencer, W. A. Habituation: a model phenomenon for the study of neuronal substrates of behavior. *Psychological Review*, 1966, *73*, 16–43.

Thorndike, E. L. Animal intelligence: an experimental study of the associative processes in animals. *Psychological Monographs*, 1898, *2*, 109.

Thorndike, E. L. *Animal intelligence*. New York: Macmillan, 1911.

Timberlake, W. A molar equilibrium theory of learned performance. In G. H. Bower (Ed.), *The psychology of learning and motivation* (Vol. 14). New York: Academic Press, 1980.

Timberlake, W., & Allison, J. Response deprivation: an empirical approach to instrumental performance. *Psychological Review*, 1974, *8*, 146–164.

Tinbergen, L. The natural control of insects in pinewoods: I. factors affecting the intensity of predation by songbirds. *Archives Neerlandaises de Zoologie*, 1960, *13*, 265–343.

Tinbergen, N. Social releasers and the experimental method required for their study. *Wilson Bulletin*, 1948, *60*, 6–51.

Tinbergen, N. *The study of instinct*. New York: Oxford University Press, 1951.

Tinbergen, N. *Curious naturalists*. New York: Basic Books, 1958.

Tinbergen, N. On aims and methods of ethology. *Zeitschrift für Tierpsychologie*, 1963, *20*, 410–433.

Tinbergen, N., Impekoven, M., & Franck, D. An experiment on spacing out as a defense against predators. *Behaviour*, 1967, *28*, 307–321.

Toates, F. *Control theory in biology and experimental psychology*. London: Hutchinson Educational, 1975.

Toates, F. M., & Archer, J. A comparative review of motivational systems using classical control theory. *Animal Behaviour*, 1978, *26*, 368–380.

Toates, F. M., & Halliday, T. R. (Eds.), *The analysis of motivational processes*. London: Academic Press, 1980.

Todorov, J. C. Component duration and relative response rates in multiple schedules. *Journal of the Experimental Analysis of Behavior*, 1972, *17*, 45–49.

Tolman, E. C. *Purposive behavior in animals and men*. New York: Appleton-Century-Crofts, 1932. (Reprinted, University of California Press, 1949.)

Tolman, E. C., & Honzik, C. H. Introduction and removal of reward, and maze performance in rats. *University of California Publications in Psychology*, 1930, *4*, 257–275.

Torgerson, W. S. Multidimensional scaling: theory and method. *Psychometrika*, 1952, *17*, 401–419.

Toulmin, S. Neuroscience and human understanding. In G. Quarton, F. Schmitt, & E. Melnechuk (Eds.), *Neuroscience: a study program*. New York: Rockefeller University Press, 1967.

Uexküll, J. von. *Streifzuge durch die Umwelten von Tieren und Menschen*. Berlin: Springer, 1934.

Uhl, C. N. Response elimination in rats with schedules of omission training, including yoked and response-independent reinforcement comparisons. *Learning and Motivation*, 1974, *5*, 511–531.

Verplanck, W. S. Since learned behavior is innate, and vice versa, what now? *Psychological Review*, 1955, *52*, 139–144.

Von Neuman, J., & Morgenstern, O. *Theory of games and economic behavior*. Princeton: Princeton University Press, 1947.

Waddington, C. H. Genetic assimilation of the *Bithorax* phenotype. *Evolution*, 1956, *10*, 1–13.

Waddington, C. H. Evolutionary adaptation. In S. Tax (Ed.), *Evolution after Darwin. Vol. 1: The evolution of life*. Chicago: University of Chicago Press, 1960.

Waddington, C. H. *New patterns in genetics and development*. London: Allen & Unwin, 1962.

Wagner, A. R. Expectancies and the priming of STM. In S. H. Hulse, H. Fowler, & W. K. Honig (Eds.), *Cognitive processes in animal behavior*. Hillsdale, N.J.: Erlbaum, 1978.

Wagner, A. R., Rudy, J. W., & Whitlow, J. W. Rehearsal in animal conditioning. *Journal of Experimental Psychology*, 1973, *97*, 407–426.

Wagner, A. R., & Terry, W. S. Backward conditioning to a CS following an expected vs. a surprising UCS. *Animal Learning and Behavior*, 1975, *3*, 370–374.

Wallsten, T. S. (Ed.). *Cognitive processes in choice and decision behavior*. Hillsdale, N.J.: Erlbaum, 1980.

Walsh, V. C. *Introduction to contemporary microeconomics*. New York: McGraw-Hill, 1970.

Watson, J. B. *Psychology from the standpoint of a behaviorist*. Philadelphia: Lippincott, 1919.

Werner, E. E., & Hall, D. J. Optimal foraging and the size selection of prey by the Bluegill Sunfish (*Lepomis macrochirus*). *Ecology*, 1974, *55*, 1216–1232.

Wessells, M. G. The effects of reinforcement upon the pre-pecking behaviors of pigeons in the autoshaping experiment. *Journal of the Experimental Analysis of Behavior*, 1974, *21*, 125–144.

West, M. J., King, A. P., Eastzer, D. H., & Staddon, J. E. R. A bioassay of isolate cowbird song. *Journal of Comparative and Physiological Psychology*, 1979, *93*, 124–133.

Whipple, W. R., & Fantino, E. Key-peck durations under behavioral contrast and differential reinforcement. *Journal of the Experimental Analysis of Behavior*, 1980, *34*, 167–176.

White, K. G. Behavioral contrast as differential time allocation. *Journal of the Experimental Analysis of Behavior*, 1978, *29*, 151–160.

Whitham, T. G. Coevolution of foraging in Bombus and nectar dispensing in Cilopsis: a last dreg theory. *Science*, 1977, *197*, 593–596.

Wickens, D. D., & Wickens, C. D. Some factors related to pseudoconditioning. *Journal of Experimental Psychology*, 1942, *31*, 518–526.

Wilcoxon, H. C. Historical introduction to the problem of reinforcement. In J. C. Tapp (Ed.), *Reinforcement and behavior*. New York: Academic Press, 1969.

Williams, D. R. Classical conditioning and incentive motivation. In W. F. Prokasy (Ed.), *Classical conditioning: a symposium*. New York: Appleton-Century-Crofts, 1965.

Williams, D. R., & Williams, H. Auto-maintenance in the pigeon: sustained pecking despite contingent non-reinforcement. *Journal of the Experimental Analysis of Behavior*, 1969, *12*, 511–520.

Wilson, M. P. Periodic reinforcement interval and number of periodic reinforcements as parameters of response strength. *Journal of Comparative and Physiological Psychology*, 1954, *47*, 51–56.

Wirtshafter, D., & Davis, J. D. Set points, settling points, and the control of body weight. *Physiology and Behavior*, 1977, *19*, 75–78.

Yates, F. A. *The art of memory*. London: Routledge & Kegan Paul, 1966.

Zeiler, M. D., & Buchman, I. B. Response requirements as constraints on output. *Journal of the Experimental Analysis of Behavior*, 1979, *32*, 29–49.

Zeiler, M. D., & Harzem, P. (Eds.). *Reinforcement and the organization of behavior*. New York: Wiley, 1979.

Zener, K. The significance of behavior accompanying conditioned salivary secretion for theories of the conditioned response. *American Journal of Psychology*, 1937, *50*, 384–403.

Zippelius, H. Die karawanenbildung bei feld- und hausspitzmaus. *Zeitschrift für Tierpsychologie*, 1972, *30*, 305–320.

Ziriax, J. M., & Silberberg, A. Discrimination and emission of different key-peck durations in the pigeon. *Journal of Experimental Psychology: Animal Behavior Processes*, 1978, *4*, 1–21.

NAME INDEX

Allison, 1981, criticism of optimality: 229
Allison & Kelsey, 1976: 168
Allison, Miller, & Wozny, 1979: 168, 186
Allison & Timberlake, 1974: 168; regulation: 221
Alloy & Seligman, helplessness: 475, 504
Anderson & Shettleworth, 1977: 318
Archer & Toates, theoretical studies: 85
Arend, Buehler, & Lockhead, 1971: 81
Aristotle, and "the good": 121; and final cause: 158; four kinds of causes: 157
Ashby, law of requisite variety: 89
Atkinson & Birch, self-inhibition: 352
Attneave, 1950, stimulus representation: 310
Augerinos, Black, & Suzuki, 1980: 310
Ayres & Staddon, 1975, food-related activities: 318, 459; competition: 221
Azrin, Hutchinson, & Hake, 1967: 444

Bacotti, variable-interval schedule: 191–192
Baerends, digger wasp: 308
Barrera, 1974, centrifugal swing: 498
Bateson, imprinting, a review, 1974: 19, 425
Battalio, Rachlin, Green, & Kagel: 226
Baum & Bouzas, contrast, 1976: 344
Baumol, 1977, economic theory: 225–226
Beatty & Shavalia, recency and: 380
Békésy, G. von, 1967: 81
Bell, 1959: 48
Benson, 1971, aposematic prey: 268
Bernard, Claude, and milieu interne: 162
Berryman, Thompson, & Gibbon, 1974: 143, 159
Bigelow, Rosenblueth, & Wiener, 1942: 79
Birch & Atkinson, self-inhibition: 352
Bitterman, 1965, 1969: 252
Black, Brener, DiCara, & Obrist: 494
Black, Suzuki, & Augerinos, 1980: 310
Blough, experimental data, visual search: 278, 342
Bogdany, bee experiment: 422-423
Bolles, 1970: 228, 450; 1980: 183; shock avoidance: 487; exploratory behavior: 488

Boneau, Burstein, Pennypacker, & Honig: 330
Booth, 1980: 183
Booth, Toates, & Platt, 1980: 185
Boren, 1961: 211
Boring, 1942: 49; 1957: 17
Boring & Herrnstein, 1965: 17; behaviorism and the law of effect: 158
Bouzas, & Baum, contrast, 1976: 344
Bowe & Hilgard, 1981: 156, 251
Brandt, 1934: 57f
Bray, Novin, & Wyrwicka, 1976: 185
Brazier, 1959, and the reflex: 49
Breland & Breland, 1966: 228, 450–451, 469, 496
Brener, DiCara, Obrist, & Black: 494
Bridgman, P. W., founder, operationism: 352
Brobeck, 1945, temperature and feeding: 186
Brown & Jenkins, 1968: 308, 439
Browne, Hearst, & Peden, 1977: 498
Buchman & Zeiler, 1979: 228
Buck, Rothstein, & Williams, 1975: 353
Buddenbrock, 1922: 58f
Buckhard & Rachlin, economic theory: 226
Burstein, Pennypacker, Honig, & Boneau: 330
Bush, Galanter, & Luce, 1963: 251
Bush & Mosteller, 1955: 251
Buzsáki, 1982: 503
Buzsáki, Grastyán, Winiczai, & Mód: 503
Bygrave, blocking experiment, 1977: 415

Campbell, 1975, epistemology and: 50
Cannon, W., and homeostasis: 162
Caplan & Ordahl, development and: 9
Caraco, Martindale, & Whitham, 1980: 186
Casey, Silberberg, Hamilton, & Ziriax: 254
Catania, 1963, 1973: 251; 1981: 160; 1980a,b, pigeons, choice: 259; cumulative record, pigeon: 465
Catania et al., and linear constraint: 333
Catania & Gill, and discriminability: 352; and local contrast: 342
Catania & Reynolds, and VI schedules: 359; and postfood time: 391
Catania, Silverman, & Stubbs, 1974: 333

SUBJECT INDEX